Developments in Applied Phycology

Volume 13

Series Editor

Michael A. Borowitzka, Algae R&D Centre, School of Veterinary and Life Sciences, Murdoch University, Murdoch, WA, Australia

Aims and Scope

Applied Phycology, the practical use of algae, encompasses a diverse range of fields including algal culture and seaweed farming, the use of algae to produce commercial products such as hydrocolloids, carotenoids and pharmaceuticals, algae as biofertilizers and soil conditioners, the application of algae in wastewater treatment, renewable energy production, algae as environmental indicators, environmental bioremediation and the management of algal blooms. The commercial production of seaweeds and microalgae and products derived there from is a large and well established industry and new algal species, products and processes are being continuously developed.

The aim of this book series, Developments in Applied Phycology, is to present state-of-the-art syntheses of research and development in the field. Volumes of the series will consist of reference books, subject-specific monographs, peer reviewed contributions from conferences, comprehensive evaluations of large-scale projects, and other book-length contributions to the science and practice of applied phycology.

Prospective authors and/or editors should consult the Series Editor or Publishing Editor for more details. Series Editor: Michael A. Borowitzka - M.Borowitzka@murdoch.edu.au Publishing Editor: Éva Loerinczi - eva.loerinczi@springer.com

Alfredo de Jesús Martínez-Roldán
Editor

Biotechnological Processes for Green Energy, and High Value Bioproducts by Microalgae, and Cyanobacteria Cultures

Springer

Editor
Alfredo de Jesús Martínez-Roldán
Investigador Nacional Nivel I, Catedrático CONAHCYT - TecNM/ITD
Departamento de Ingenierías Química y Bioquímica
División de Estudios de Posgrado
Maestría en Sistemas Ambientales
Durango, Mexico

ISSN 2543-0599 ISSN 2543-0602 (electronic)
Developments in Applied Phycology
ISBN 978-3-031-43968-1 ISBN 978-3-031-43969-8 (eBook)
https://doi.org/10.1007/978-3-031-43969-8

This Springer imprint is published by the registered company Springer Nature Switzerland AG
The registered company address is: Gewerbestrasse 11, 6330 Cham, Switzerland

Paper in this product is recyclable.

To Sofía, María Paula, Arya, and Jorge

Acknowledgement

Thanks to all those who at some point arrived and remained, to those who arrived and left, and to all those who will arrive.

Gracias a todos aquellos que en algún momento llegaron y permanecen, a los que llegaron y se marcharon y a todos aquellos que llegarán

Contents

1 Introduction to Environment-Friendly Bioprocesses by Microalgae and Cyanobacteria

Alfredo de Jesús Martínez-Roldán

Abstract

Microalgae include diverse organisms with different cellular structures (prokaryotic and eukaryotic), the capability to grow in diverse ecosystems (sea, rivers, lakes, lagoons, soil, etc.), and the possibility of performing autotrophic, mixotrophic, and heterotrophic metabolism. This diversity is the reason for their ability to produce and accumulate different compounds, many of which have the potential to be used in industrial processes. These compounds include lipids, proteins, amino acids, carotenoids, biofuels, adsorbents (carbons), polyunsaturated fatty acids, and animal feeds. In addition, processes based on microalgae can capture carbon dioxide (CO_2), eliminate pollutants (such as nitrogen, phosphorous, and heavy metals), or even utilize biomass as feed for livestock. Nevertheless, the growth conditions, induction process, and extraction and purification strategies are specific to every strain. This book aims to include recent developments in environment-friendly processes derived from microalgae and cyanobacteria.

Keywords

Microalgae · Cyanobacteria · High-value bioproducts · Biotechnology

1.1 Introduction

Microalgae are photosynthetic microorganisms with many metabolic pathways very similar to superior plants; nevertheless, they have several advantages compared with terrestrial plants, such as the possibility to be cultivated and reaching massive cultures in photobioreactors and the fact that the development of fruits, seeds, or a specific tissue is not necessary, as in the case of vegetable crops, to obtain a high-value product (Barsanti and Gualtieri 2014). Historically, microalgae have been used in experimental studies to elucidate metabolic pathways, specifically for the description of oxygenic photosynthesis. The experiments performed to describe the route of carbon fixation in oxygenic photosynthesis, commonly known as the Calvin–Benson–Bassham cycle, were developed using *Chlorella* cultures exposed to light for small periods and subsequently inactivated by dropping in hot methanol. This experiment allowed us to determine all the molecules produced in the Calvin–Benson cycle and describe all the chemical reactions involved (Biel and Fomina 2015).

Since then, microalgae and cyanobacteria have been proposed to develop diverse bioprocesses with two main objectives: eliminating different pollutants from liquid and gaseous effluents and producing high-value products by taking advantage of specific metabolic pathways (Borowitzka 2013). Recently, genetic modifications of microalgae and cyanobacteria were carried out to either increase the amount of a specific metabolite or improve the performance of the culture under special operational conditions (Barati et al. 2021; Beacham et al. 2017).

Around the 1950s, the majority of the research related to microalgae and cyanobacteria focused on their role in the facultative lagoon and the tertiary treatment of wastewaters, as symbiosis was observed between aerobic bacteria and photosynthetic microorganisms and an increase in the efficiency of organic matter removal was observed in the presence of microalgae (Oswald et al. 1957; Oswald and Golueke 1960; Oswald and Gotaas 1957).

Later, the development of technological devices for microalgal culture started; their containers were called photobioreactors and today their variety is huge (Martínez-Roldán and Cañizares-Villanueva 2015). Some photobioreactors include configurations, such as fermenters,

A. de Jesús Martínez-Roldán (✉)
Investigador Nacional Nivel I, Catedrático CONAHCYT - TecNM/ITD, Departamento de Ingenierías Química y Bioquímica, División de Estudios de Posgrado, Maestría en Sistemas Ambientales, Durango, Mexico
e-mail: adjmartinezro@conacyt.mx

A. de Jesús Martínez-Roldán (ed.), *Biotechnological Processes for Green Energy, and High Value Bioproducts by Microalgae, and Cyanobacteria Cultures*, Developments in Applied Phycology 13, https://doi.org/10.1007/978-3-031-43969-8_1

tubular horizontal and/or vertical airlift mixing, columns, flat panels, and thin layers, all of which satisfy specific culturing requirements of the microorganisms, such as mixing, shear stress, and light supply. In all configurations, the main objective is to maximize biomass production, biomass productivity, or even the production of a specific metabolite or a high-value product (Acién et al. 2017; Chini Zittelli et al. 2013; Torzillo and Chini Zittelli 2015). Recently, microalgal biotechnology has focused on environmental applications or the production of high-value bioproducts, but always from a sustainability perspective.

Some environmental applications take advantage of the diverse qualities of microalgal cultures, e.g., their capability to fix carbon dioxide (CO_2), which is useful for the developing processes to capture CO_2 or to reduce the CO_2 concentration in fuel gases from diverse industrial processes. However, there are no technological developments at the commercial scale, because there are numerous obstacles related to the engendering of the process (Solovchenko and Khozin-Goldberg 2013; Wang et al. 2008; Zhou et al. 2017). Another characteristic used for environmental applications of microalgae is the fast consumption of nutrients from its culture media (nitrogen and phosphorus), which has been fully studied, allowing us to describe the role of microalgae and cyanobacteria in the stabilization of lagoons or even the use of microalgal cultures for the tertiary treatment of domestic wastewater. The inclusion of microalgae and cyanobacteria in wastewater treatment permits the elimination of nitrogen and phosphorus, which cannot be eliminated by aerobic and anaerobic processes for organic matter elimination (Martínez-Roldán and Ibarra-Berumen 2019; Olguin 2003).

With regard to the use of microalgae and cyanobacteria in wastewater treatment, the proposal is to eliminate specific contaminants from water, some of which are heavy metals and semimetals. Microalgal biomass has a huge potential for the removal of ions because it is possible to use both live and dead biomass as adsorbents. The use and process of microalgal biomass as ion adsorbents is very efficient, and recovery of the removed ions is quite simple (Cañizares-Villanueva 2000; Perales-Vela et al. 2006). Owing to their capability to remove pollutants from the culture medium, the microalgae are proposed to eliminate specific pollutants recently detected in urban wastewaters and denominated as emerging contaminants; the major problem with this type of compound is its wide variety because the sources are very diverse (Peña-Guzmán et al. 2019).

Some emerging pollutants have actually reached high concentrations and are further increasing, causing concern to the scientific community. Therefore, many studies have focused on the development of processes to eliminate them. The emerging contaminants include colorants, drugs, hormones, healthcare products, cosmetics, and antibiotics. Microalgae have proven to eliminate the contaminants by the process of adsorption/absorption or even biotransformation; however, in this case, it is possible to obtain subproducts with higher toxicity than the original ones (Geissen et al. 2015; Jain et al. 2022; Keen et al. 2014; Peña-Guzmán et al. 2019). Therefore, regardless of the potential of microalgae and cyanobacteria to eliminate these contaminants, there are no commercial-scale treatment processes, and there is an unknown economic cost and real efficiency.

The potential environmental application of microalgae is not only the elimination of pollutants from liquid and gaseous effluents. Since the biomass is a source of a large number of different molecules, several of them have high market value (Borowitzka 2013). Some of these bioproducts include pigments, antioxidants, fatty acids, oils, polyunsaturated fatty acids, and the lipid fraction of the biomass, which can be converted into liquid fuels, such as biodiesel or jet fuel (Cañizares-Villanueva et al. 2022).

The high-value bioproducts are very diverse, but some of them have higher potential for pigment production because there are strains with the capability to produce high amounts of different carotenoids or xanthophylls, such as beta-carotene, astaxanthin, lutein, violaxanthin, and antheraxanthin, as well as many other molecules with similar chemical properties. In addition, it is possible to obtain molecules with antioxidant properties, such as polyphenols, tocopherols, and ascorbic acid (Safafar et al. 2015). The lipid fraction of the biomass can be used to obtain a specific fatty acid (oleic, linoleic, arachidonic, etc.), or subjected to a chemical process to obtain a specific type of fuel, such as biodiesel or jet fuel (Rodolfi et al. 2009). Nevertheless, the number of processes at the production scale is small, and economic feasibility has not been proven.

There are many possible applications of microalgae and cyanobacteria, but it is necessary to develop processes from the perspective of sustainability. This has led to an increase in the proposal of processes based on the use of wastewater as a nutrient source and the complete exploitation of the biomass in the biorefinery concept (because of its similarity with oil refinery). The biorefinery processes propose to reduce the effect of the processes on the environment and reduce the generation of residues and reach a positive life-cycle assessment.

This book analyzes many examples of biotechnological applications of microalgae and cyanobacteria cultures, some of them with experimental data, and other chapters that include reviews with a general overview of innovative and promising applications.

References

Acién FG, Molina E, Reis A, Torzillo G, Zittelli GC, Sepúlveda C, Masojídek J (2017) Photobioreactors for the production of microalgae. In: Muñoz R, Gonzalez-Fernandez C (eds) Microalgae-based biofuels and bioproducts: from feedstock cultivation to end-products. Elsevier, pp 1–44

Barati B, Zeng K, Baeyens J, Wang S, Addy M, Gan SY, El-Fatah Abomohra A (2021) Recent progress in genetically modified microalgae for enhanced carbon dioxide sequestration. Biomass Bioenergy 145:105927

Barsanti L, Gualtieri P (2014) Algae: anatomy, biochemistry, and biotechnology, 2nd edn. CRC Press, p 362

Beacham TA, Sweet JB, Allen MJ (2017) Large scale cultivation of genetically modified microalgae: a new era for environmental risk assessment. Algal Res 25:90–100

Biel K, Fomina I (2015) Benson-Bassham-Calvin cycle contribution to the organic life on our planet. Photosynthetica 53:161–167

Borowitzka MA (2013) High-value products from microalgae-their development and commercialisation. J Appl Phycol 25:743–756

Cañizares-Villanueva RO (2000) Biosorption of heavy metals by microorganisms. Rev Latinoam Microbiol 42:131–143

Cañizares-Villanueva RO, Rojo-Gómez E, Arroyo-Sánchez BI, Martínez-Roldán AJ (2022) *Chlorella* as a source of antioxidant compounds. In: Marutholi M (ed) *Chlorella* and its health benefits. Nova Science Publishers

Chini Zittelli G, Rodolfi L, Bassi N, Biondi N, Tredici MR (2013) Photobioreactors for microalgal biofuel production. In: Borowitzka MA, Moheimani NR (eds) Algae for biofuels and energy. Springer, Dordrecht, pp 115–131

Geissen V, Mol H, Klumpp E, Umlauf G, Nadal M, van der Ploeg M, van de Zee SEATM, Ritsema CJ (2015) Emerging pollutants in the environment: a challenge for water resource management. Int Soil Water Conserv Res 3:57–65

Jain M, Khan SA, Sharma K, Jadhao PR, Pant KK, Ziora ZM, Blaskovich MAT (2022) Current perspective of innovative strategies for bioremediation of organic pollutants from wastewater. Bioresour Technol 344:126305

Keen OS, Bell KY, Cherchi C, Finnegan BJ, Mauter MS, Parker AM, Rosenblum JS, Stretz HA (2014) Emerging pollutants – Part II: Treatment. Water Environ Res 86:2036–2096

Martínez-Roldán AJ, Cañizares-Villanueva RO (2015) Photobioreactors: improving the biomass productivity. In: Torres-Bustillos L (ed) Microalgae and other phototrophic bacteria: culture, processing, recovery and new products. Nova Science Publishers, pp 145–170

Martínez-Roldán AJ, Ibarra-Berumen J (2019) Employment of wastewater to produce microalgal biomass as a biorefinery concept. In: Alam A, Wang Z (eds) Microalgae biotechnology for development of biofuel and wastewater treatment. Springer, Singapore, pp 487–504

Olguin E (2003) Phycoremediation: key issues for cost-effective nutrient removal processes. Biotechnol Adv 22:81–91

Oswald WJ, Golueke CG (1960) Biological transformation of solar energy. Adv Appl Microbiol 2:223–262

Oswald WJ, Gotaas HB (1957) Photosynthesis in sewage treatment. Trans Am Soc Civil Eng 122:73–105

Oswald WJ, Gotaas HB, Golueke CG, Kellen WR, Gloyna EF, Hermann ER (1957) Algae in waste treatment. Sewage Ind Wastes 29:437–457

Peña-Guzmán C, Ulloa-Sánchez S, Mora K, Helena-Bustos R, Lopez-Barrera E, Alvarez J, Rodriguez-Pinzón M (2019) Emerging pollutants in the urban water cycle in Latin America: a review of the current literature. J Environ Manag 237:408–423

Perales-Vela HV, Peña-Castro JM, Cañizares-Villanueva RO (2006) Heavy metal detoxification in eukaryotic microalgae. Chemosphere 64:1–10

Rodolfi L, Chini Zittelli G, Bassi N, Padovani G, Biondi N, Bonini G, Tredici MR (2009) Microalgae for oil: strain selection, induction of lipid synthesis and outdoor mass cultivation in a low-cost photobioreactor. Biotechnol Bioeng 102:100–112

Safafar H, Van Wagenen J, Møller P, Jacobsen C (2015) Carotenoids, phenolic compounds and tocopherols contribute to the antioxidative properties of some microalgae species grown on industrial wastewater. Mar Drugs 13:7339–7356

Solovchenko A, Khozin-Goldberg I (2013) High-CO_2 tolerance in microalgae: possible mechanisms and implications for biotechnology and bioremediation. Biotechnol Lett 35:1745–1752

Torzillo G, Chini Zittelli G (2015) Tubular photobioreactors. In: Prokop A, Bajpai RK, Zappi ME (eds) Algal biorefineries: Volume 2: Products and refinery design. Springer, Cham, pp 187–212

Wang B, Li Y, Wu N, Lan CQ (2008) CO_2 bio-mitigation using microalgae. Appl Microbiol Biotechnol 79:707–718

Zhou W, Wang J, Chen P, Ji C, Kang Q, Lu B, Li K, Liu J, Ruan R (2017) Bio-mitigation of carbon dioxide using microalgal systems: advances and perspectives. Renew Sust Energ Rev 76:1163–1175

CO_2 Bio-capture by Microalgae and Cyanobacteria Cultures

2

Cigdem Demirkaya and Hector De la Hoz Siegler

Abstract

Climate change is a global problem caused by the rise of carbon dioxide (CO_2) concentration in the atmosphere. To limit global warming to less than 2 °C, large-scale deployment of technologies to remove CO_2 from the air will be needed. As highly efficient, photosynthetic, single-cell factories, microalgae and cyanobacteria can play a critical role among carbon-negative technologies. Bio-capture of CO_2 using photosynthetic microbes is a viable method for recycling CO_2 into biomass, which can subsequently be utilized to produce bioenergy, fertilizers, biomaterials, and other high-value products. This chapter provides an overview of the different strategies for utilizing microalgae and cyanobacteria for CO_2 capture directly from the atmosphere or stationary point sources with minimal environmental impacts. Challenges, research needs, and opportunities for the integration of CO_2 bio-capture within a biorefinery perspective are discussed.

Keywords

Carbon capture · Microalgae · Cyanobacteria · Photosynthesis · Bioconversion · Biorefinery · Climate change

2.1 Introduction

Due to the role of carbon dioxide (CO_2) in driving global climate change, there is an increasing global pressure to limit CO_2 emissions, particularly at large-emission source points. In 2015, with the signing of the Paris Agreement, nations committed to reduce global emissions annually by 3% to avoid a global climate catastrophe. However, this has not been achieved and the path that is being followed, which includes mainly treatment of point sources, such as flue gas, is not enough to meet the target of limiting global average temperature below a 2 °C increase. Thus, there is a rising urgency for innovative methods to mitigate new emissions and to remove the CO_2 already in the atmosphere.

Photosynthetic microbes, including microalgae, cyanobacteria, and diatoms have a great potential for mitigating and abating CO_2 emissions, while producing valuable products (Moreira and Pires 2016; Vale et al. 2020) and fostering a more sustainable bio-economy. If part of the CO_2 captured in the biomass is used to make products with relatively long life (i.e., years), or if they are permanently stored, then the cultivation of microalgae and cyanobacteria can become a key carbon-negative technology to address the climate change crisis.

Biological carbon capture is an effective and simple approach with potentially much lower energy needs compared to physical and chemical carbon capture methods. For instance, the standard technology for carbon capture in post-combustion processes is amine scrubbing, using primary or secondary amines. This technology is the basis of several megaton-scale carbon-capture projects (Feron et al. 2020). The regeneration of the amine, however, is very energy-intensive, introducing a significant energy penalty and reducing the overall mitigation potential of this method (Alesi and Kitchin 2012; Stern et al. 2013).

Photosynthesis is nature's carbon capture solution. Photosynthetic organisms utilize the energy from light to drive the reaction of CO_2 and water and form biomolecules. In this way, carbon is removed from the atmosphere and stored in biomass. Gross primary production (GPP) refers to the amount of CO_2 removed from the atmosphere by photosynthesis. This is known to be one of the main fluxes controlling the carbon balance in the atmosphere and has a significant potential to offset anthropogenic carbon emissions (Beer et al. 2010). Terrestrial GPP is estimated at about 120 Pg of carbon per year (Beer et al. 2010), while marine phytoplankton are estimated to account for an additional 50 Pg of carbon per year (Yang et al. 2020). Global anthropogenic energy-

C. Demirkaya · H. De la Hoz Siegler (✉)
Department of Chemical and Petroleum Engineering, University of Calgary, Calgary, AB, Canada
e-mail: h.siegler@ucalgary.ca

A. de Jesús Martínez-Roldán (ed.), *Biotechnological Processes for Green Energy, and High Value Bioproducts by Microalgae, and Cyanobacteria Cultures*, Developments in Applied Phycology 13, https://doi.org/10.1007/978-3-031-43969-8_2

related CO_2 emissions in 2020 were estimated at 8.6 Pg of Carbon (IEA 2021), or roughly 5% of the carbon naturally fixed by photosynthesis. Thus, it is conceivable that technological solutions based on photosynthesis will be able to offset anthropogenic carbon emissions.

Microalgae and cyanobacteria are rapidly growing microorganisms able to fix CO_2 with efficiency 10 to 50 times higher than that of terrestrial plants (Cheah et al. 2015; Raheem et al. 2018; Zhang and Liu 2021); they have high areal productivity, and high lipid and/or carbohydrate content. They are able to grow in nonarable land, with minimal nutrient inputs, and in wastewater, saline, brines, or halo-alkaline waters (Moreira and Pires 2016). Thanks to their high areal productivity and ability to grow in hostile environments, photosynthetic microbes are more suitable for biological carbon capture technologies than terrestrial plants. The use of dedicated crops for industrial purposes has previously resulting in the diversion of arable lands away from traditional food crops, creating unintended impacts on food cost and supply, resulting in the well-known food versus fuel dilemma (Darnoko and Cheryan 2000; Issariyakul and Dalai 2012). This dilemma is avoided when using photosynthetic microbes.

In addition to their role of fixing CO_2 emissions, microalgae and cyanobacteria can be used to remove nitrogen and phosphorous from agricultural and industrial effluents, reducing eutrophication of receiving water bodies (Fal et al. 2021; Guo et al. 2018; W. Zhang et al. 2020). The microbial biomass produced can be used for several applications including the production of biofuels, bioplastics, food supplements, animal feed, cosmetic additives, pharmaceutical products, and building materials (Venkata Mohan et al. 2016; Singh and Dhar 2019; Daneshvar et al. 2022). Thus, biological carbon capture with microalgae and cyanobacteria offers a wide range of opportunities for building sustainable integrated processes to support a bioresource-based circular economy (Venkata Mohan et al. 2016; Hemalatha et al. 2019; Vale et al. 2020).

This chapter presents an overview of the factors affecting the performance of carbon capture using photosynthetic microorganisms and the different strategies for utilizing microalgae and cyanobacteria for CO_2 capture directly from the atmosphere or stationary point sources with minimal environmental impacts. The challenges, research needs, and opportunities for the integration of CO_2 bio-capture from a biorefinery perspective are discussed.

2.2 Photosynthesis: Natural Carbon Capture

Photosynthesis is a natural way of capturing CO_2 and is the process responsible for transforming Earth's atmosphere from CO_2-rich, more than 2 billion years ago when CO_2 atmospheric concentration was about 10 to 200 times the present level, to a relatively CO_2-depleted one (Kaufman and Xiao 2003). Photosynthesis is carried out in two phases, see Fig. 2.1. In the first phase, light-dependent reactions capture light energy and convert it into chemical energy that is ultimately stored within nicotinamide adenine dinucleotide phosphate (NADPH) and adenosine triphosphate (ATP). The light reactions occur in the photosynthetic unit (PSU), a light-harvesting complex and reaction center located within the thylakoid membrane. The NADPH and ATP energetic molecules are then consumed in the second phase, where light-independent reactions are used to convert CO_2 into sugars (Barsanti and Gualtieri 2005; Jensen et al. 2017; Sánchez-Baracaldo and Cardona 2020).

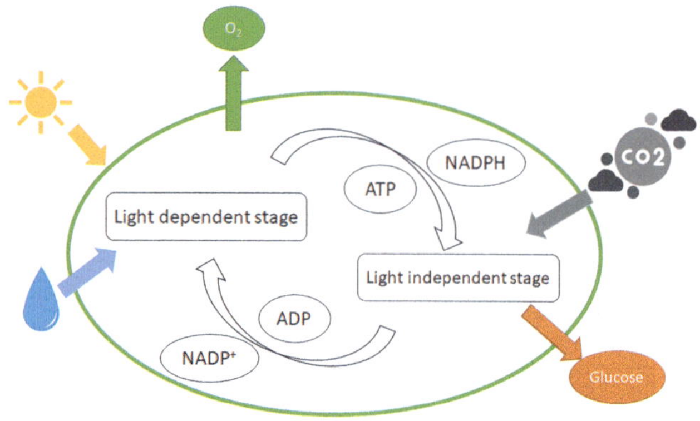

Fig. 2.1 Light-dependent and light-independent stage during photosynthesis (Adapted from Cheah et al. 2015)

$$2NADP + 2H_2O + 2ADP + 2P_i \longrightarrow 2NADPH_2 + 2ATP + O_2 \text{ Light} - \text{dependent reaction}$$

$$CO_2 + 4H^+ + 4e^- \longrightarrow CH_2O + H_2O \text{ Light} - \text{independent reaction}$$

Although photosynthesis originated in an environment with much higher CO_2 concentrations, microalgae and cyanobacteria cells have developed biological adaptations to survive under low CO_2 concentrations. The carbon concentrating mechanism (CCM) allows to increase the concentration of CO_2 within the cells relative to the normal CO_2 concentration in the air (300–400 ppm). The CCM improves photosynthetic efficiency by increasing the available CO_2 for ribulose bisphosphate carboxylase-oxygenase (RuBisCO). RuBisCO is an important enzyme that converts CO_2 into organic carbon (Gruber and Feiz 2018). Another important enzyme in the CCM is carbonic anhydrase (CA), which catalyzes the reversible conversion of CO_2 into HCO_3^- (DiMario et al. 2018).

Several different CCMs have been identified in microalgae and cyanobacteria. In cyanobacteria, there are two types of carboxysomes (α-type and β-type), which are specialized compartments for the accumulation of HCO_3^-. The accumulated HCO_3^- is then converted into CO_2 by the action of carboxysomal CA (Moroney and Ynalvez 2007). In the case of microalgae, *Chlamydomonas reinhardtii* has been studied as the model organism to understand the action of the CCM. In this microalga, the CCM can be divided into two phases. In the first phase, inorganic carbon is gathered from the environment in the form of CO_2 and HCO_3^- by transporter proteins. In the second phase, as the concentration of HCO_3^- increases in the chloroplast, HCO_3^- is converted into CO_2 by the action of CA (Wang et al. 2015).

Although microalgae and cyanobacteria have in general faster growth rate and higher light conversion efficiency than plants, the efficiency of large cultivation is hindered by the low CO_2 gas–liquid mass transfer rate and reduced light penetration and shading. In a fast-growing culture, the CO_2 transfer rate between the gas phase (i.e., air above culture or bubbles sparged) and the liquid medium phase media is too low to compensate for the CO_2 uptake by the cells (Zuccaro et al. 2020), resulting in carbon limitation and slower photosynthetic rate.

Light limitation also affects photosynthetic efficiency negatively (Brennan and Owende 2010). Although theoretical photosynthetic efficiency ranges between 8% and 12%, practical photosynthesis efficiency is rarely above 1.5–2%. This efficiency loss is primarily caused by light scattering and nonproductive absorption, which causes light to be exponentially attenuated as it travels along the optical path (Nwoba et al. 2019). During photosynthesis, the PSU can be in either a resting or nonactivated state or an activated state. A resting PSU is activated by the absorption of a photon. The absorption of excess photons converts functional PSUs into nonfunctional PSUs, resulting in photoinhibition (Camacho-Rubio et al. 2003). In a culture, the cells closer to the light source are more prone to experience photoinhibition as they are exposed to a higher light intensity, while the cells further down the optical path may not receive enough light. Thus, overall photosynthetic efficiency is affected by both light attenuation and scattering and photoinhibition.

2.3 CO_2 Bio-capture from Different Sources

Microalgae and cyanobacteria can capture CO_2 from stationary point emission sources, such as power plants or other carbon-intensive industrial processes, or directly from the atmosphere. The CO_2 concentration in the atmosphere is 0.03–0.06% (v/v), while for stationary point sources the CO_2 concentration can vary between 6 and 15% (v/v) (Rahaman et al. 2011). The cost and energy needed for capturing CO_2 is inversely proportional to concentration, the lower the concentration of CO_2 in a given source, the more expensive the capture process. Thus, capture from large-point sources is one of the best and more efficient options to abate CO_2 emissions, as the effluent streams from combustion and industrial processes have higher CO_2 concentrations.

Figure 2.2 shows how different CO_2 sources can be integrated with microalgae and cyanobacteria biomass cultivation for CO_2 bio-capture. The following sections present an overview of technologies for CO_2 bio-capture and a discussion of the efficiency of these bio-capture methods.

2.3.1 Bio-capture of CO_2 from Stationary Point Sources

Flue or stack gases released by various stationary point sources, such as industrial complexes and power plants, have relatively high CO_2 concentrations ranging from 6 to 15% (v/v) (Thomas et al. 2016). These flue gases can be used to boost the productivity of microalgal and cyanobacteria cultures. The high concentration of CO_2 in the flue gas allows for a faster mass transfer rate, higher photosynthetic efficiency, and support a higher final cell density in the cultures.

Because of the low CO_2 solubility, the flue gas needs to be injected or bubbled directly into the cultivation medium, adding to electricity demands. The energy spent in bubbling and mixing the CO_2 in the media represents up to 27% of the overall production cost; at the same time, typically between 55 and 90% of the CO_2 injected in the culture is lost to the atmosphere (Markou et al. 2014; Caia et al. 2018). Consequently, significant research efforts have been dedicated to

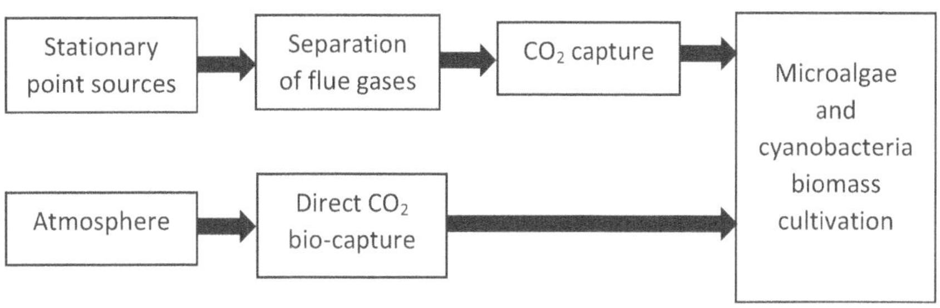

Fig. 2.2 CO_2 bio-capture from different sources and utilization in biomass

improving CO_2 diffusion rates (see Sect. 2.5) and increasing CO_2 utilization efficiency.

Microalgae and cyanobacteria strains with high CO_2 uptake rate and high biomass productivity are desirable to ensure an efficiency CO_2 capture process. Sepulveda et al. (2019) assessed the ability of 11 different microalgae and cyanobacteria strains to capture CO_2 and produce high biomass productivity. They reported that *Scenedesmus almeriensis* and *Neochloris oleoabundans* were the most productive strains when used in CO_2 capture processes compared to the cyanobacteria strains. Park et al. (2021) investigated CO_2 fixation at a CO_2 concentration ranging from 5 to 40% from biogas in five pure microalgal cultures and a mixed microalgal culture, including *Chlorella sp., Anabaena variabilis, Chlamydomonas iyengarii, Chlorella vulgaris,* and *Chlorella sorokiniana*. The highest CO_2 fixation rate was reported for *Chlorella sp.* at 1.785 g $L^{-1}d^{-1}$ at a CO_2 concentration of 15%. Additional studies on CO_2 capture and uptake by microalgae and cyanobacteria are summarized in Table 2.1.

Although the high CO_2 level in flue gas is beneficial for microalgae and cyanobacteria growth, these gases usually contain substances that can be inhibitory (Lam et al. 2012; Vale et al. 2020). In particular, unfiltered flue gas from coal combustion can have high concentration of SO_x and NO_x, microparticles, and heavy metals, such as mercury, which can present a challenge to biomass growth (Napan et al. 2015; Thomas et al. 2016). As the concentration of SO_x and NO_x increases, the acidity of the culture medium increases and this lowers the pH (Vale et al. 2020). Low pH values may inhibit microalgal growth or even result in cell death. Duarte et al. (2016) evaluated the tolerance of microalgae and cyanobacteria to the presence of NO_x and SO_x and found that strains were able to tolerate those gases at concentration of up to 400 ppm. Aslam et al. (2017) demonstrated the adaptation of mixed microalgal communities to growth in unfiltered flue gas from coal combustion. This microalgal community was dominated by *Desmodesmus* spp., which was the most resilient species. Radmann et al. (2011) evaluated the NO_x and SO_x tolerance of *C. vulgaris,*

Table 2.1 Application of microalgae in CO_2 capture from the atmosphere and CO_2 reach sources

CO_2 source	Microorganism	CO_2% (v/v)	CO_2 fixation rate (g L^{-1} d^{-1})	Culture conditions	Reference
Atmospheric CO_2	*Chlorella vulgaris* and *Pseudokirchneriella subcapitata*	Air	0.305	OECD medium, $T = 22$ °C, different dark/light cycles at 126 μmol photons m^{-2} s^{-1}	Pires et al. (2014)
	Dunaliella tertiolecta	0.04	0.07	Artificial sea water, $T = 26$ °C, continuous illumination at 350 μmol photons m^{-2} s^{-1}	Hulatt and Thomas (2011)
	C. vulgaris	0.09	3.45	Artificial sea water, $T = 25$ °C, continuous illumination at ~50 μmol photons m^{-2} s^{-1}	Fan et al. (2008)
	Anabaena sp.	0.03	1.45	Allen and Arnon medium, $T = 23$ °C, light/dark cycles with 900 μmol photons m^{-2} s^{-1}	Ramkrishnan et al. (2014)
Enriched CO_2 supply	*Spirulina* sp. DUT001	2	1.0	Zarrouk medium, $T = 25$ °C, photoperiod = 12:12, 188.7 μmol photons m^{-2} s^{-1}	Zhu et al. (2020)
	Chlorella vulgaris	15	1.0	BBM medium, $T = 28$ °C, membrane PBR, photoperiod = 12:12, 120 μmol photons m^{-2} s^{-1}	Senatore et al. (2021)
	Chlorella vulgaris, Synechocystis salina, Microcystis aeruginosa,	5	0.101	OECD test medium, $T = 24$ °C, continuous illumination at 120 μmol photons m^{-2} s^{-1}	Gonçalves et al. (2014)
	Scenedesmus obliquus	12	22.8	Soil extract medium, $T = 26$ °C, outdoor airlift PBR, 220–240 μmol photons m^{-2} s^{-1}	Li et al. (2011)
	Chlorella vulgaris	5–25	0.27–0.47	ESP-31 medium, $T = 28$ °C, continuous illumination at 50 μmol photons m^{-2} s^{-1}	Chou et al. (2019)
	Microalgae consortia	5.5	0.09–0.12	BBM medium, $T = 30$ °C, photoperiod = 12:12 at 1650.3 μmol photon m^{-2} s^{-1}	Aslam et al. (2018)
	Scenedesmus almeriensis, Neochloris oleoabundans	Flue gas	2.8–2.64	Natural water from the river Seine and the artificial Seine river water, $T = 25$ °C, continuous illumination at 390 μmol photons m^{-2} s^{-1}	Sepulveda et al. (2019)
	Spirulina sp.	2	0.81	Modified Zarrouk medium, $T = 20$ °C, pH 9, continuous illumination at 188.7 μmol photons m^{-2} s^{-1}	Zhu et al. (2020)
	Chlorella sp	15	1.785	BG-11 medium, pH 8.2–8.7, $T = 25$ °C, photoperiod = 12:12 at 171.91 μmol photon m^{-2} s^{-1}	Park et al. (2021)

Scenedesmus obliquus, and *Synechococcus nidulans* by using a simulated gas from coal combustion, containing 12% (v/v) CO_2, 100 ppm NO_x, and 60 ppm SO_x. They reported that the growth of *C. vulgaris* and *S. obliquus* was not inhibited, but this was not the case for *S. nidulans*.

In short, stationary point sources are excellent for supplying the required CO_2 concentration for carbon capture and biomass production in microalgae and cyanobacteria. However, direct use of flue gas is not, in general, possible without any separation or treatment. Identifying robust microalgae and cyanobacteria strains capable of high CO_2 bio-capture and adapted to the high concentration of other gases present in flue gas should be further explored to maximize the CO_2 bio-capture potential of microalgae and cyanobacteria culture. Furthermore, despite the large energy requirements for supplying CO_2 to the cultivation medium, a significant amount of the CO_2 provided is released into the atmosphere, decreasing net capture, and incurring inefficient energy use. Thus, additional research efforts must be directed at improving CO_2 diffusion rates and integrating different CO_2 capture techniques with microalgae and cyanobacteria cultures.

2.3.2 Biological Direct Air Capture with Microalgae and Cyanobacteria

Although capture from concentrated large-point sources of CO_2 is the most desirable and efficient option, about half of CO_2 emissions are from diffuse sources (Moreira and Pires 2016). Moreover, the CO_2 already accumulated in the atmosphere will continue to negatively contribute to climate change (Keith 2009). Thus, capture of atmospheric CO_2 using negative emissions technologies is needed to address emissions from diffuse sources and to restore the carbon balance in the atmosphere. Direct air capture (DAC) refers to technologies that directly remove CO_2 from the atmosphere. These technologies also offer the advantage of deployment in any location, independent of a specific source and without added costs for CO_2 transportation.

In the case of microalgal and cyanobacterial cultures, to capture 1 million ton of CO_2 per year, between 70 and 86 km^2 of the culture is needed, assuming an average productivity of 20 g m^{-2} day^{-1} of dry weight biomass and considering that about 1.6 to 2 grams of CO_2 are captured for every gram of biomass (Sayre 2010). Given the large land requirements for cultivation, it is more likely to find suitable land for deployment of large-scale cultures far away from industrial areas or population centers, where land may be scarce or expensive.

The low concentration of CO_2 in the atmosphere, however, is a major drawback as it limits CO_2 solubility and mass transfer rate into the cultivation media (Kumar et al. 2010). Carbon utilization has been shown to be more efficient when the supply rate of CO_2 matches closely with the demand of the growing biomass (Sobczuk et al. 2000; Vale et al. 2020). For DAC, active bubbling is not desirable as it requires a high energy input and will increase water evaporation. For a cost and energy-effective carbon capture process, the CO_2 supply to the cultivation needs to be improved by passive means.

To compensate for the low solubility of CO_2 in natural waters, several microalgae and cyanobacteria strains rely on the CCM to increase the intracellular concentration of bicarbonate ions and use CA to convert the HCO_3^- back to CO_2 to be used in photosynthesis. The ability of some microalgae and cyanobacteria to utilize HCO_3^- have prompted several researchers to explore the use of alkaline culture conditions to enhance CO_2 mass transfer rate and total inorganic carbon concentration in the culture media (Chi et al. 2013; Canon-Rubio et al. 2016). Alkalinity is defined as the sum of the concentration of hydroxyl ions, bicarbonate ions, and carbonates ions, times the corresponding ion charge. As alkalinity increases, so does the concentration of dissolved inorganic carbon. High alkalinity also improves CO_2 mass transfer rate from the gas phase to the cultivation medium, as there is an increased driving force (Vadlamani et al. 2019). In addition, it provides a higher buffering capacity enabling the uncoupling of CO_2 absorption from biomass growth (Chi et al. 2013; Santos et al. 2013), as illustrated in Fig. 2.3.

Because alkalinity can inhibit cell growth, it is necessary to operate at relatively low alkalinity or use alkali-tolerant or alkaliphilic microalgae or cyanobacteria strains. Extreme alkaline conditions together with alkaliphilic microalgae and cyanobacteria have been suggested for large-scale cultivation (Piiparinen et al. 2018; Song et al. 2019; Zhu et al. 2020). In soda lakes, at pH > 10, high concentrations of bicarbonate are present supporting a high growth rate of CO_2 fixation by photosynthetic microbes, while the consumed CO_2 is spontaneously replenished by passive diffusion from the air above the lakes (Sharp et al. 2017).

Vadlamani et al. (2017) demonstrated high biomass productivity by cultivating *C. sorokiniana* (>16 g m^{-2}·d^{-1}) in a 4.2 m^2 raceway pond using an alkaline cultivation medium and atmospheric CO_2 alone. In another study, Zhu et al. (2020) used extreme alkaline conditions with pH ranging between 10.0 and 12.5 for DAC using *Spirulina* sp. DUT001. Effective CO_2 bio-capture was reported with maximum biomass productivity about 1.00 g $L^{-1}d^{-1}$ and carbon-capture rate of 0.81 g L^{-1} d^{-1}.

2.4 Integrated Biorefinery for a Carbon-Neutral Circular Bioeconomy

To foster the development of an integrated, sustainable, and robust biological CO_2 capture process, circular economy principles must be applied to ensure the efficient processing

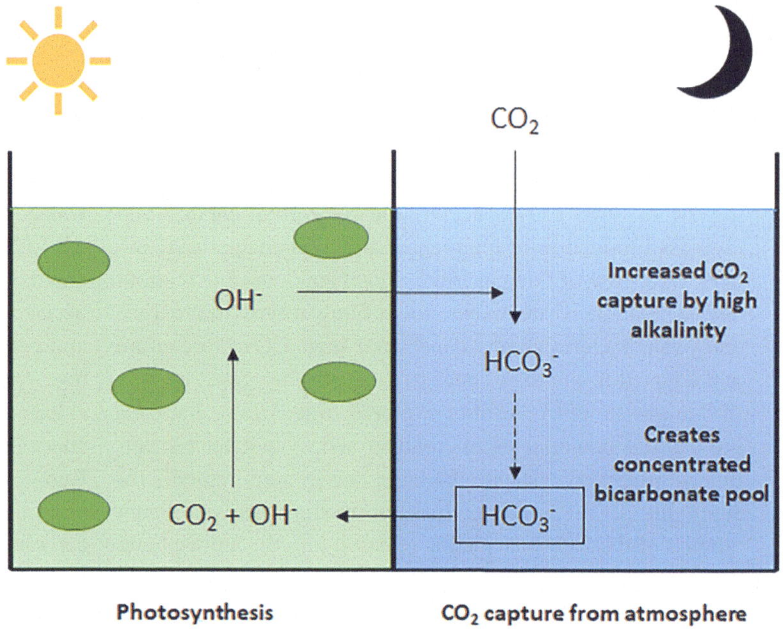

Fig. 2.3 Mechanism of the bicarbonate pool's role in the efficient capture of CO_2 from the air and rapid carbon supply for photosynthesis (Adapted from Zhu et al. 2020)

and conversion of the generated biomass, while designing out or minimizing waste, maximizing the reutilization of resources, and regenerating natural systems. The microalgal or cyanobacterial biomass produced from the CO_2 capture process consists of several biochemical compounds, including lipids, proteins, polysaccharides, and pigments. These compounds can be extracted and converted into biobased products which, in turn, displace alternative products obtained from non-sustainable sources or that have a high carbon footprint (Daneshvar et al. 2022).

An integrated biorefinery can be conceived where bio-capture of CO_2 occurs simultaneously with the production of valuable products, thus converting waste CO_2 emissions into carbon-neutral products. As microalgae or cyanobacteria require several nutrients for growth, the biorefinery concept starts with cultivation using a nutrient-rich waste stream, such as wastewater, as the primary source of nitrogen, phosphorus, sulfur, and trace metals; thus, allowing the recycling and reclamation of these materials and reducing nutrient supply costs (Razzak et al. 2013; Whitton et al. 2015; Yen et al. 2015; Singh et al. 2016).

Conventional downstream processing involves harvesting and separating biomass from the cultivation media, followed by biomass pretreatment by homogenization, beading, and chemical hydrolysis to extract products of interest, and finish with the upgrading to the final products (Khoo et al. 2020). This traditional downstream processing approach is wasteful and expensive. Thus, in the biorefinery approach, the goal is to utilize the biomass to generate multiple products within a single process.

The approach in microalgal biorefineries is the cascade system (Francavilla et al. 2015; Hemalatha et al. 2019), which allows for different biomass fractions to be extracted either simultaneously or separately by different methods (Monlau et al. 2021). Selecting the most suitable downstream processing strategy depends on the nature of the bioproducts, the required energy, and technology availability (Bastiaens et al. 2017). The use of mild separation technologies, which require low pressure, less energy, and less chemicals, is preferred for downstream to increase energy efficiency and avoid damaging the most sensitive products, which are often the most valuable. Figure 2.4 presents a possible pathway to produce multiple products from microalgal and cyanobacterial biomass.

Carbajal Tejada et al. (2020) studied five different biorefinery scenarios using *Scenedesmus dimorphus* biomass as feedstock to produce biodiesel, dihydroxyacetone, fishmeal, glycerol, and vegetable oil. Their results showed that integrated reactive distillation with the biological oxidation of glycerol to produce dihydroxyacetone was the most efficient biorefinery scenario. In another study, Moncada et al. (2014) simulated two different integrated biorefineries using *Chlorella* sp. grown with a CO_2-rich stream and sugar cane to determine the most promising scenario. The use of *Chlorella* sp. biomass to produce biodiesel, glycerol, ethanol, sugar, and electricity was found more environmentally and economically viable option than just using sugarcane alone.

Microalgal and cyanobacterial biomass is mainly composed of lipids (7–60%), proteins (6–71%), and carbohydrates (5–60%), depending upon the species and culture conditions (García-Garibay et al. 2003; Chen et al. 2013; Aziz et al. 2020). These macromolecules can be converted into several different products. The most effective use of the produced biomass will be the one that displaces

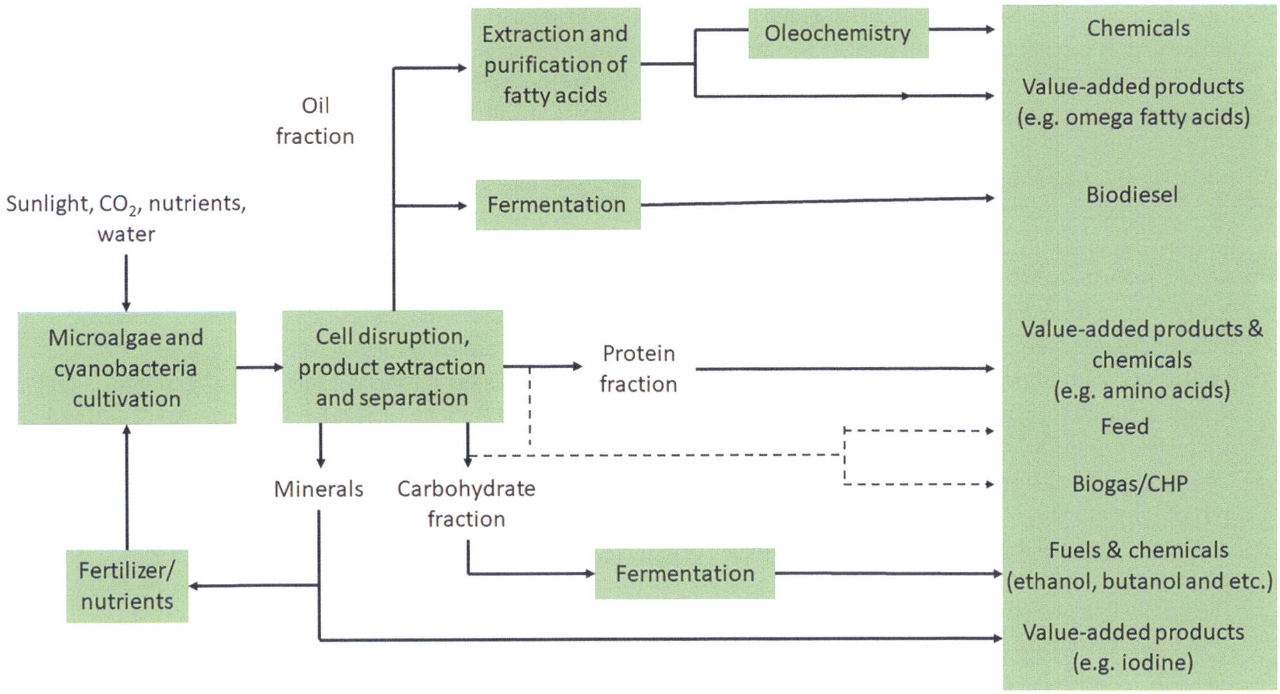

Fig. 2.4 Application of different downstream processing methods for integrated biorefinery (Adapted from Chew et al. 2017)

unsustainable feedstock or existing products with a high carbon footprint. In the following subsections, we discuss some of the products that can be more directly targeted as they either have a high carbon footprint or are high value, and therefore can help to improve the economics of the bio-capture process.

2.4.1 Biofuels

The high lipid content (15–60%) in many microalgae makes them very attractive for biodiesel production (Converti et al. 2009; Yeh and Chang 2011; Moazami et al. 2012; Polat and Altınbaş 2020). Lipid content can be further increased by manipulating several cultivation parameters (Huang and Su 2014); Polat and Altınbaş 2020). Although biodiesel production from microalgae is quite advantageous, it has not yet been commercially deployed due to the high energy-intensive downstream processing methods.

The carbohydrates in microalgal and cyanobacterial biomass are a suitable feedstock for hydrogen, bioethanol, and biogas production by fermentative pathways (Ho et al. 2013; Lakatos et al. 2019; Nagappan et al. 2019). Microalgal and cyanobacterial biomasses do not contain rigid cell wall components such as lignin, which makes them easier to process. Biodiesel production can be integrated with fermentation and anaerobic digestion to simultaneously produce a variety of energy products (Harun et al. 2011; González-González et al. 2018). After extraction of lipids for biodiesel production, the rest of the biomass, including carbohydrates, can be used as a feedstock for hydrothermal liquefaction, fermentation, or anaerobic digestion to produce biodiesel, bioethanol and/or biogas respectively. A recent technoeconomic analysis has shown that when microalgal biofuel production was integrated with other processes to obtain multiple valuable products (polyhydroxy butyrate and astaxanthin), the biofuel price was reduced to a competitive value of $0.54/L (Rafa et al. 2021).

2.4.2 Bioactive Compounds

The main bioactive compounds produced by microalgae include polyunsaturated fatty acids (PUFAs), carotenoids, chlorophylls, phycobiliproteins, polysaccharides, and proteins.

Several microalgae produce PUFAs with known bioactive properties, such as eicosapentaenoic acid (EPA, C20:5 $\omega-3$), docosahexaenoic acid (DHA, C22:6 $\omega-3$), arachidonic acid (ARA, 20:6 $\omega-6$), and γ-linolenic acid (GLA, 18:3 $\omega-6$) (López et al. 2019). Many of these fatty acids have been studied for their anti-inflammatory activity and have been shown to prevent many diseases such as asthma, diabetes, and cardiovascular diseases (Cheng et al. 2018; Hess et al. 2018).

Some microalgal polysaccharides are considered biologically active molecules with promising applications in food, cosmetic additives, and pharmaceutical products (De Jesus

Raposo et al. 2013; Barkia et al. 2019; Gouda et al. 2022). Sulfated polysaccharides are prominent for having antioxidant, anti-inflammatory, antitumoral, antiviral, antibacterial, and immunomodulatory activities (De Jesus Raposo et al. 2013). Among these, sulfated polysaccharides extracted from *Porphyridium* sp. and *Nannochloropsis oculata* have been shown to have antiviral, antitumoral, and immunostimulatory properties in pharmaceutical and therapeutic applications (Custódio et al. 2015; Casas-Arrojo et al. 2021). Apart from these benefits, polysaccharides obtained from microalgae are used as stabilizers, thickening agents, emulsifiers, and lubricants in foods, cosmetics, and textiles (Costa et al. 2021). Because some of these polysaccharides are released into the growth media during cultivation, their recovery and purification is much simpler than in the case of intracellular compounds.

Microalgae and cyanobacteria also produce essential amino acids, peptides, and proteins (Amorim et al. 2020). These valuable compounds have found multiple applications in the pharmaceutical, cosmetic, and food industries (Costa et al. 2021). *Spirulina platensis* and *C. vulgaris* are known for their high protein contents of 46–71% (Lupatini et al. 2017; Tokuşoglu and Ünal 2003). They have been used as food and feed supplements for decades, as they contain several essential amino acids, such as threonine, methionine, isoleucine, valine, leucine, lysine, and histidine (Wang et al. 2021). Montalvo et al. (2019) reported antioxidant, chelating, antimicrobial, anti-inflammatory, and anti-collagenase activity of three biopeptide fractions from *Arthrospira maxima* OF15 for potential applications in the pharmaceutical, cosmetic, and food industries.

Although all these metabolites have various applications, there are several limitations to be addressed. The main problem is the high cost of cultivation and extraction. Applying the biorefinery concept to obtain more than one product is essential to reach a cost-effective production system (Balasubramaniam et al. 2021).

2.4.3 Pigments

The three main pigment classes obtained from microalgal and cyanobacterial biomass include chlorophylls, carotenoids, and phycobilins (Koyande et al. 2019). These pigments are mainly utilized as food and feed supplements, food coloring, and as pharmaceutical and cosmetic additives because of their high antioxidant action (Begum et al. 2016; Morocho-Jácome et al. 2020). β-carotene produced by microalgae can be used as a food colorant, food and feed additive, or as precursor for vitamin A, and antioxidants (Wolf et al. 2021). Lutein and zeaxanthin are other carotenoids that are produced by microalgae and cyanobacteria and have been used as food additives due to their antioxidant activities (Granado-Lorencio et al. 2009). Phycocyanin is a blue-colored pigment-protein complex that is extracted from cyanobacteria species such as *Arthrospira platensis* (Zeng et al. 2012). It has antioxidant, anticancer, and anti-inflammatory properties and helps to improve immune function and inhibit cancer cell growth (Zeng et al. 2012). It has been used as a food ingredient and as an additional supplement to fight or prevent cancer. Furthermore, due to its naturally blue color, it has been used as a colorant for the textile and food industries to replace synthetic colorants (Rahman et al. 2017).

Pigments are one the most valuable products that can be obtained from microalgae. Although market size and product price are higher than other products, pigment extraction methods are expensive and involve the use of toxic materials. Economic feasibility and sustainability need to be improved by, e.g., investigating more environmentally friendly and non-toxic extraction methods (Rajesh et al. 2020).

2.4.4 Plastics

The global demand for plastics has increased exponentially since large-scale production of plastics started in the 1950s (Geyer et al. 2017). The carbon footprint of plastic production was estimated at 1.7 Gigaton (Gt) CO_2 equivalent in 2015 (Cabernard et al. 2021), representing 4.5% of global greenhouse gas (GHG) emissions. Because many plastics have long life spans, with some taking tens to hundreds of years to decompose, replacing fossil fuel–derived plastics with microalgal plastics is a feasible strategy for long-term carbon sequestration.

Some microalgae and cyanobacteria species produce metabolites that can be used directly for the fabrication of bioplastics, such as polyhydroxyalkanoates (PHAs) (Balaji et al. 2013), while other plastics can be obtained by chemical routes using the lipid, protein, and carbohydrate fraction of the microalgal and cyanobacterial biomass. Plastics obtained from microalgae and cyanobacteria can be designed to have properties comparable to those of fossil fuel–derived plastics (Rahman and Miller 2017).

Polyhydroxybutyrate (PHB), a type of PHA, is frequently found in cyanobacteria as an energy and carbon storage compound. Several cyanobacteria, such as *Synechocystis*, *Synochoccocus*, *Nostoc*, and *Spirulina*, are known PHB producers (Yashavanth et al. 2021), while *Synechocystis* PCC6803 has been used as a model organism to study the production of PHB (Singh et al. 2019; Koch et al. 2020). PHB is a biodegradable alternative to thermoplastics, such as polyethylene and polypropylene, and it is being commercially produced for applications in disposable food ware (McAdam et al. 2020). Recent studies have focused on the use of genetic engineering to increase PHB yield and integrate cultivation system with wastewater to reduce cultivation cost (Larkum et al. 2012; Katayama et al. 2018; López Rocha et al. 2020; Chong et al. 2021).

The triglycerides accumulated by many microalgae can be used as feedstock for the synthesis of different polyols, by chemically attaching hydroxyl groups to the unsaturated bonds in the fatty acid chains. These polyols can be converted into polyurethanes for a variety of applications by means of epoxidation and ring-opening by methanol, ethylene glycol, or lactic acid; hydroformylation; or urethane reaction with isocyanates (Hai et al. 2020; Peyrton et al. 2020). Another attractive material that can be derived from microalgae is acrylonitrile, which is a monomer widely used in the production of a variety of plastics, rubbers, resins, acrylic fibers, and polyacrylonitrile (PAN) carbon fibers (Karp et al. 2017). Glycerol obtained from the transesterification of algal oils can be converted into acrylonitrile by direct ammoxidation in the gas phase (Guerrero-Pérez and Bañares 2008).

2.4.5 Oleochemicals

Although the lipids in microalgae are thought of mainly as precursors for biofuel production, they can also be used as feedstock to produce many oleochemicals. Oleochemicals are products derived from triglycerides, including fatty acids, fatty alcohols, methyl esters, and glycerin with a wide range of applications, from food and cosmetic additives to drilling fluids and lubricants. Most oleochemicals are derived from palm, soya, canola, coconut, and palm kernel oils (Parsons et al. 2020). The sustainability of the oleochemical industry, especially palm oil, has been the source of growing public concern due to its many negative environmental impacts (Rival and Levang 2014). Although biobased, the oleochemical industry has an elevated carbon footprint and its expansion through land-use conversion has resulted in permanent damage to the biodiversity of sensitive ecosystems along with the release of massive amounts of GHG (Parsons et al. 2020). Palm-driven land-use change in Southeast Asia emits nearly 0.5 Gt of CO_2 equivalent each year, roughly 1.4% of global net GHG emissions, and is responsible for extensive ecosystem degradation. The use of algal-derived lipids to produce oleochemicals will allow phasing out unsustainable feedstock.

2.5 Challenges and Recent Progress

The main challenge for scaling up CO_2 bio-capture using microalgal and cyanobacterial cultures is the low CO_2 diffusion rates from the gas phase into the liquid culture medium, which translates into reduced CO_2 capture efficiency (Lam et al. 2012; Yen et al. 2015). To enhance the CO_2 absorption and mass transfer some approaches have been suggested, such as improving the existing photobioreactors (PBR), designing new PBR systems, and evaluating the influence of several parameters (temperature, pH, mixing, culture type, culture density, and CO_2 concentration) in the CO_2 diffusion (Morales et al. 2018).

2.5.1 PBR Design

Microalgal and cyanobacteria cultivation can be done in open ponds or closed PBRs. Open ponds are low in capital and operating costs, which is beneficial for scaling up the production; however, biomass productivity is lower than in closed systems due to high CO_2 losses, evaporation, uncontrolled climate conditions, and contamination risk (Acién et al. 2017). On the other hand, closed systems provide a better-controlled environment and prevent CO_2 and evaporation losses, allowing to reach higher biomass productivity (Acién et al. 2017). The existing PBRs designs for CO_2 bio-capture and biomass cultivation are vertical column reactors (bubble columns or airlift), tubular reactors, flat-plate reactors, and stirrer tank reactors.

A key limiting factor in CO_2 bio-capture is the low photosynthetic conversion efficiency. Although PBRs are designed to provide a better light path than what is achieved in open ponds, light conversion efficiency is much lower than what can be theoretically achieved. Implementing new strategies to improve light penetration and delivery directly to cells minimizes energy losses and maximizes productivity. Light penetration can be increased by changing PBR orientation (horizontal, vertical, or tilted), using solar tracking devices to change the direction of the light coming to the PBR surface (Castrillo et al. 2018), optimizing light intensity and spectral distribution to prevent photoinhibition from excess light (Ooms et al. 2016), and maintaining heterogeneous light distribution to eliminate dark zones (Nwoba et al. 2019; De la Hoz Siegler 2022).

Several studies optimizing microalgae and cyanobacteria cultivation have focused on improving CO_2 gas–liquid mass transfer by improving reactor configurations to increase the contact between the gas and liquid phases. The initial bubble size is known to affect CO_2 mass transfer rate, with a smaller initial bubble size ($R = 0.98$ mm) resulting in increased CO_2 fixation. Hence, a small bubble is more suitable to be supplied in PBRs for the purpose of high CO_2 fixation (Barahoei et al. 2020). However, producing micro- or nanobubbles is a high energy-consuming process in which high-pressure devices are needed. Besides, high shear stress that is generated because of the bursting of small bubbles is damaging to algal cells.

Xu et al. (2020) developed a spiral-ascending CO_2 dissolver to enhance the CO_2 dissolution rate and prolong gas–

liquid contact time to improve microalgal growth in a horizontal tubular PBR. This cost-efficient and effective CO_2 dissolver reduced the bubble generation diameter by 23.4% and increased the CO_2 mass transfer rate by 69.2%. In another study, Gonçalves et al. (2021) designed an Oscillatory Flow Reactor with Smooth Periodic Constrictions (OFR-SPC) to improve CO_2 mass transfer without compromising fluid turbulence, which can negatively impact the most sensitive cells. This system promoted high gas–liquid mass transfer rates with low power consumption and controlled fluid turbulence. Therefore, it can be a promising technology to be used in microalgal cultivation, replacing the commonly used bubble-column and airlift PBRs.

2.5.2 Cultivation Parameters

The efficiency of the biological CO_2 capture process is also affected by CO_2 concentration, temperature, and pH. Temperature affects the solubility of CO_2 as well as the specific growth rates. CO_2 solubility decreases as temperature increases. Thus, to improve the solubility of CO_2, the culture medium must be maintained at cooler temperatures. Each microorganism has its own optimal growth temperature, with thermophilic strains typically having higher specific growth rate than mesophilic or psychrophilic organisms. For CO_2 bio-capture from stationary point sources, thermophilic microalgal and cyanobacterial strains can be used to tolerate the high temperature of flue gas without causing a decrease in cell growth. Varshney et al. (2018) reported isolation of two novel green algal strains, *Asterarcys quadricellulare* and *C. sorokiniana*, from water bodies that were near a steel plant in India. These strains had high specific growth rates of up to 0.06 h^{-1} and 0.1 h^{-1}, respectively. Furthermore, they were able to tolerate high temperatures up to 43 °C and high concentration of CO_2 and NO_x. In addition, they reported that when they eliminated NO coming from the flue gas, strains were able to accumulate lipids up to 44% to 46% of dry biomass. The tolerance for high temperature and CO_2 concentration and their ability to accumulate lipid make these strains very attractive for CO_2 bio-capture applications.

The pH of the culture media also has a direct impact on the dissolution of CO_2 and other inorganic molecules, as it affects the chemical equilibrium between HCO_3^- and CO_3^{2-}, precipitation of phosphates, volatilization of ammonia, and the solubility of trace elements. Moreover, it directly affects cell growth due to its effect on the activity of different enzymes. For many microalgae and cyanobacteria strains, the optimum pH value is between 7 and 9. However, some organisms are known to thrive in extreme conditions. For instance, *Spirulina* grows optimally in highly alkaline conditions (pH 9–11). Alkaline conditions, at pH above 11.0, can reduce the contamination risk from grazers, protozoa, and other competing algae. As discussed in Sect. 2.3.2, it also increases the CO_2 mass transfer rate and total dissolved inorganic carbon concentration.

Mixing is another key cultivation parameter that affects the gas–liquid mass transfer in PBRs. Poor mixing can cause dead areas that lack nutrients and CO_2, thus reducing overall reactor productivity. Increasing mixing rates through mechanical agitation or aeration can improve the CO_2 mass transfer, but it results in higher power consumption and introduce excessive shear stress and cellular damage.

2.5.3 Future Direction and Opportunities

Carbon capture technologies based on microalgae and cyanobacteria cultures are promising for a carbon-neutral future. However, for successful implementation of this technology, innovation in culture strategies, integration with other processes, and process optimization are needed.

Integration of other CO_2 capture processes with microalgae and cyanobacteria cultures can increase overall CO_2 capture and recovery. Chemical absorption technologies are based on the ability of different solvents, such as ionic liquids and alkaline solutions, to react with CO_2 (Vega et al. 2020). The successful integration of chemical absorption of CO_2 with *Spirulina* cultivation was demonstrated by De Rosa et al. (2015). Membrane separation uses a CO_2 permeable membrane to allow the CO_2 to pass through, while preventing other flue gasses from reaching the culture media. In this way, inhibition of the cell by toxic gasses is avoided (Cheng et al. 2021).

Recent studies have focused on the identification of highly efficient microalgae and cyanobacteria strains and on enhancing the efficiency of CO_2-fixing enzymes through genetic engineering (Barati et al. 2021). Increased CO_2 assimilation and biomass growth were reported by construction of new NADPH consumption pathways (Zhou et al. 2016), while the photosynthetic efficiency of *Nannochloropsis* sp. was improved by overexpression of RuBisCO activase (Wei et al. 2017). These achievements, however, need to be demonstrated at larger scale to ensure that the genetic constructs are stable over long cultivation periods.

2.6 Conclusion

Biological capture of CO_2 is a promising approach to mitigate CO_2 emissions and remove excess carbon from the atmosphere. Large-scale cultivation of microalgae and cyanobacteria provides environmental benefits with the possibility of producing a wide portfolio of valuable products for further a more sustainable and circular bioeconomy.

Although significant progress in bio-capture of CO_2 by microalgae and cyanobacteria has been achieved, additional research efforts are needed to improve CO_2 capture efficiency.

References

Acién FG, Molina E, Reis A, Torzillo G, Zittelli GC, Sepúlveda C, Masojídek J (2017) Photobioreactors for the production of microalgae. In: Gonzalez-Fernandez C, Muñoz R (eds) Microalgae-based biofuels and bioproducts from feedstock cultivation to end-products. Elsevier, Duxford, pp 1–44

Alesi WR, Kitchin JR (2012) Evaluation of a primary amine-functionalized ion-exchange resin for CO_2 capture. Ind Eng Chem Res 51(19):6907

Amorim ML, Soares J, Sélia J, Coimbra R, De M, Leite O, Fernando L, Albino T, Martins MA, Lopes Amorim M, Elia JS, De Oliveira LM, Ar M, Martins E (2020) Microalgae proteins: production, separation, isolation, quantification, and application in food and feed. Crit Rev Food Sci Nutr 61(12):1976–2002

Aslam A, Thomas-Hall SR, Mughal TA, Schenk PM (2017) Selection and adaptation of microalgae to growth in 100% unfiltered coal-fired flue gas. Bioresour Technol 233:271–283

Aslam A, Thomas-Hall SR, Manzoor M, Jabeen F, Iqbal M, uz Zaman Q, Schenk PM, Asif Tahir M (2018) Mixed microalgae consortia growth under higher concentration of CO_2 from unfiltered coal fired flue gas: Fatty acid profiling and biodiesel production. J Photochem Photobiol B Biol 179:126–133

Aziz MMA, Kassim KA, Shokravi Z, Jakarni FM, Lieu HY, Zaini N, Tan LS, Islam S, Shokravi H (2020) Two-stage cultivation strategy for simultaneous increases in growth rate and lipid content of microalgae: a review. Renew Sust Energ Rev 119:109621

Balaji S, Gopi K, Muthuvelan B (2013) A review on production of poly β hydroxybutyrates from cyanobacteria for the production of bio plastics. Algal Res 2(3):278–285

Balasubramaniam V, Gunasegavan RDN, Mustar S, Lee JC, Noh MFM (2021) Isolation of industrial important bioactive compounds from microalgae. Molecules 26(4):943

Barahoei M, Hatamipour MS, Afsharzadeh S (2020) CO_2 capturing by chlorella vulgaris in a bubble column photo-bioreactor; Effect of bubble size on CO_2 removal and growth rate. J CO2 Util 37:9–19

Barati B, Zeng K, Baeyens J, Wang S, Addy M, Gan SY, El-Fatah Abomohra A (2021) Recent progress in genetically modified microalgae for enhanced carbon dioxide sequestration. Biomass Bioenergy 145:105927

Barkia I, Saari N, Manning SR (2019) Microalgae for high-value products towards human health and nutrition. Mar Drugs 17(5):304

Barsanti L, Gualtieri P (2005) Algae: anatomy, biochemistry, and biotechnology, 2nd edn. CRC Press, Boca Raton, p 145

Bastiaens L, Van Roy S, Thomassen G, Elst K (2017) Biorefinery of algae: technical and economic considerations. In: Gonzalez-Fernandez C, Muñoz R (eds) Microalgae-based biofuels bioprod. From Feed. Cultiv. to End-Products. Elsevier, Duxford, pp 327–345

Beer C, Reichstein M, Tomelleri E, Ciais P, Jung M, Carvalhais N, Rödenbeck C, Arain MA, Baldocchi D, Bonan GB, Bondeau A, Cescatti A, Lasslop G, Lindroth A, Lomas M, Luyssaert S, Margolis H, Oleson KW, Roupsard O, Veenendall E, Viovy N, Williams C, Woodward FI, Papale D (2010) Terrestrial gross carbon dioxide uptake: global distribution and covariation with climate. Science 329(5993):834–838

Begum H, Yusoff FMD, Banerjee S, Khatoon H, Shariff M (2016) Availability and utilization of pigments from microalgae. Crit Rev Food Sci Nutr 56(13):2209–2222

Brennan L, Owende P (2010) Biofuels from microalgae—a review of technologies for production, processing, and extractions of biofuels and co-products. Renew Sust Energ Rev 14(2):557–577

Cabernard L, Pfister S, Oberschelp C, Hellweg S (2021) Growing environmental footprint of plastics driven by coal combustion. Nat Sustain 5(2):139–148

Caia M, Bernard O, Béchet Q (2018) Optimizing CO_2 transfer in algal open ponds. Algal Res 35:530–538

Camacho-Rubio F, Chamago FG, Fernández Sevilla JM, Chisti Y, Molina Grima E (2003) A mechanistic model of photosynthesis in microalgae. Biotechnol Bioeng 81:4

Canon-Rubio KA, Sharp CE, Bergerson J, Strous M, De la Hoz Siegler H (2016) Use of highly alkaline conditions to improve cost-effectiveness of algal biotechnology. Appl Microbiol Biotechnol 100:1611–1622

Carbajal Tejada EM, Martínez Hernández E, Fernández Linares L, Novelo Maldonado E, Limas Ballesteros R (2020) Techno-economic analysis of *Scenedesmus dimorphus* microalgae biorefinery scenarios for biodiesel production and glycerol valorization. Bioresour Technol Reports 12:100605

Casas-Arrojo V, Decara J, Arrojo-Agudo MÁ, Pérez-Manríquez C, Abdala-Díaz RT (2021) Immunomodulatory, antioxidant activity and cytotoxic effect of sulfated polysaccharides from *Porphyridium cruentum*. (S.F.Gray) Nägeli. Biomolecules 11(4)

Castrillo M, Díez-Montero R, Tejero I (2018) Model-based feasibility assessment of a deep solar photobioreactor for microalgae culturing. Algal Res 29:304–318

Cheah WY, Show PL, Chang JS, Ling TC, Juan JC (2015) Biosequestration of atmospheric CO_2 and flue gas-containing CO_2 by microalgae. Bioresour Technol 184:190–201

Chen CY, Zhao XQ, Yen HW, Ho SH, Cheng CL, Lee DJ, Bai FW, Chang JS (2013) Microalgae-based carbohydrates for biofuel production. Biochem Eng J 78:1–10

Cheng F, Cui Z, Mallick K, Nirmalakhandan N, Brewer CE (2018) Hydrothermal liquefaction of high- and low-lipid algae: mass and energy balances. Bioresour Technol 258:158–167

Cheng YW, Lim JSM, Chong CC, Lam MK, Lim JW, Tan IS, Foo HCY, Show PL, Lim S (2021) Unravelling CO_2 capture performance of microalgae cultivation and other technologies via comparative carbon balance analysis. J Environ Chem Eng 9(6):106519

Chew KW, Yap JY, Show PL, Suan NH, Juan JC, Ling TC, Lee DJ, Chang JS (2017) Microalgae biorefinery: high value products perspectives. Bioresour Technol 229:53–62

Chi Z, Xie Y, Elloy F, Zheng Y, Hu Y, Shulin C (2013) Bicarbonate-based integrated carbon capture and algae production system with alkalihalophilic cyanobacterium. Bioresour Technol 133:513–521

Chong JWR, Yew GY, Khoo KS, Ho SH, Show PL (2021) Recent advances on food waste pretreatment technology via microalgae for source of polyhydroxyalkanoates. J Environ Manag 293:112782

Chou HH, Su HY, Song XD, Chow TJ, Chen CY, Chang JS, Lee TM (2019) Isolation and characterization of *Chlorella* sp. mutants with enhanced thermo- and CO_2 tolerances for CO_2 sequestration and utilization of flue gases. Biotechnol Biofuels 12(1):1–14

Converti A, Casazza AA, Ortiz EY, Perego P, Del Borghi M (2009) Effect of temperature and nitrogen concentration on the growth and lipid content of *Nannochloropsis oculata* and *Chlorella vulgaris* for biodiesel production. Chem Eng Process Process Intensif 48(6): 1146–1151

Costa JAV, Lucas BF, Alvarenga AGP, Moreira JB, de Morais MG (2021) Microalgae polysaccharides: an overview of production, characterization, and potential applications. Polysaccharides 2(4): 759–772

Custódio L, Soares F, Pereira H, Rodrigues MJ, Barreira L, Rauter AP, Alberício F, Varela J (2015) *Botryococcus braunii* and *Nannochloropsis oculata* extracts inhibit cholinesterases and protect human dopaminergic SH-SY5Y cells from H_2O_2-induced cytotoxicity. J Appl Phycol 27(2):839–848

Daneshvar E, Wicker RJ, Show PL, Bhatnagar A (2022) Biologically-mediated carbon capture and utilization by microalgae towards sustainable CO_2 biofixation and biomass valorization – a review. Chem Eng J 427:130884

Darnoko D, Cheryan M (2000) Kinetics of palm oil transesterification in a batch reactor. J Am Oil Chem Soc 77(12):1263–1267

De Jesus Raposo MF, De Morais RMSC, De Morais AMMB (2013) Bioactivity and applications of sulphated polysaccharides from marine microalgae. Mar Drugs 11(1):233

De la Hoz Siegler H (2022) Process intensification for sustainable algal fuels production. In: El-Sheekh M, Abomohra A (eds) Handbook of algal biofuels: aspects of cultivation, conversion, and biorefinery. Elsevier, pp 503–521

De Rosa E, Checchetto V, Franchin C, Bergantino E, Berto P, Szabò I, Giacometti GM, Arrigoni G, Costantini P (2015) [NiFe]-hydrogenase is essential for cyanobacterium *Synechocystis* sp. PCC 6803 aerobic growth in the dark. Sci Reports 5(1):1–12

DiMario RJ, Machingura MC, Waldrop GL, Moroney JV (2018) The many types of carbonic anhydrases in photosynthetic organisms. Plant Sci 268:11–17

Duarte JH, Fanka LS, Costa JAV (2016) Utilization of simulated flue gas containing CO_2, SO_2, NO and ash for *Chlorella fusca* cultivation. Bioresour Technol 214:159–165

Fal S, Benhima R, El Mernissi N, Kasmi Y, Smouni A, El Arroussi H (2021) Microalgae as promising source for integrated wastewater treatment and biodiesel production. Int J Phytoremediation 24(1): 34–46

Fan LH, Zhang YT, Zhang L, Chen HL (2008) Evaluation of a membrane-sparged helical tubular photobioreactor for carbon dioxide biofixation by *Chlorella vulgaris*. J Memb Sci 325(1):336–345

Feron P, Cousins A, Jiang K, Zhai R, Garcia M (2020) An update of the benchmark post-combustion CO_2-capture technology. Fuel 273(1): 117776

Francavilla M, Kamaterou P, Intini S, Monteleone M, Zabaniotou A (2015) Cascading microalgae biorefinery: fast pyrolysis of *Dunaliella tertiolecta* lipid extracted-residue. Algal Res 11:184–193

García-Garibay M, Gómez-Ruiz L, Cruz-Guerrero AE, Bárzana E (2003) Single-cell protein | algae. In: Batt C, Tortorello ML (eds) Encyclopedia of food microbiology. Elsevier, London, pp 425–430

Geyer R, Jambeck J, Law K (2017) Production, use, and fate of all plastics ever made. Sci Adv 3(7):e1700782

Gonçalves AL, Simões M, Pires JCM (2014) The effect of light supply on microalgal growth, CO_2 uptake and nutrient removal from wastewater. Energy Convers Manag 85:530–536

Gonçalves AL, Almeida F, Rocha FA, Ferreira A (2021) Improving CO_2 mass transfer in microalgal cultures using an oscillatory flow reactor with smooth periodic constrictions. J Environ Chem Eng 9(6): 106505

González-González LM, Correa DF, Ryan S, Jensen PD, Pratt S, Schenk PM (2018) Integrated biodiesel and biogas production from microalgae: towards a sustainable closed loop through nutrient recycling. Renew Sustain Energy Rev 82:1137–1148

Gouda M, Tadda MA, Zhao Y, Farmanullah F, Chu B, Li X, He Y (2022) Microalgae bioactive carbohydrates as a novel sustainable and eco-friendly source of prebiotics: emerging health functionality and recent technologies for extraction and detection. Front Nutr 0: 391

Granado-Lorencio F, Herrero-Barbudo C, Acién-Fernández G, Molina-Grima E, Fernández-Sevilla JM, Pérez-Sacristán B, Blanco-Navarro I (2009) In vitro bioaccesibility of lutein and zeaxanthin from the microalgae *Scenedesmus almeriensis*. Food Chem 114(2):747–752

Gruber AV, Feiz L (2018) Rubisco assembly in the chloroplast. Front Mol Biosci 5:24

Guerrero-Pérez MO, Bañares MA (2008) New reaction: conversion of glycerol into acrylonitrile. ChemSusChem 1(6):511–513

Guo P, Zhang Y, Zhao Y (2018) Biocapture of CO2 by different microalgal-based technologies for biogas upgrading and simultaneous biogas slurry purification under various light intensities and photoperiods. Int J Environ Res Public Health 15(3)

Hai TAP, Neelakantan N, Tessman M, Sherman SD, Griffin G, Pomeroy R, Mayfield SP, Burkart MD (2020) Flexible polyurethanes, renewable fuels, and flavorings from a microalgae oil waste stream. Green Chem 22:3088–3094

Harun R, Davidson M, Doyle M, Gopiraj R, Danquah M, Forde G (2011) Technoeconomic analysis of an integrated microalgae photobioreactor, biodiesel and biogas production facility. Biomass Bioenergy 35(1):741–747

Hemalatha M, Sravan JS, Min B, Venkata Mohan S (2019) Microalgae-biorefinery with cascading resource recovery design associated to dairy wastewater treatment. Bioresour Technol 284:424–429

Hess SK, Lepetit B, Kroth PG, Mecking S (2018) Production of chemicals from microalgae lipids – status and perspectives. Eur J Lipid Sci Technol 120(1):1700152

Ho SH, Huang SW, Chen CY, Hasunuma T, Kondo A, Chang JS (2013) Bioethanol production using carbohydrate-rich microalgae biomass as feedstock. Bioresour Technol 135:191–198

Huang YT, Su CP (2014) High lipid content and productivity of microalgae cultivating under elevated carbon dioxide. Int J Environ Sci Technol 11(3):703–710

Hulatt CJ, Thomas DN (2011) Productivity, carbon dioxide uptake and net energy return of microalgal bubble column photobioreactors. Bioresour Technol 102(10):5775–5787

IEA (2021) Global Energy Review 2021, International Energy Agency, Paris, France. https://www.iea.org/reports/global-energy-review-2021

Issariyakul T, Dalai AK (2012) Comparative kinetics of transesterification for biodiesel production from palm oil and mustard oil. Can J Chem Eng 90:342–350

Jensen E, Clément R, Maberly SC, Gontero B (2017) Regulation of the Calvin–Benson–Bassham cycle in the enigmatic diatoms: biochemical and evolutionary variations on an original theme. Philos Trans R Soc B Biol Sci 372(1728)

Karp EM, Eaton T, Sanchez i Nogué V, Vorotnikov V, Biddy MJ, Tan ECD, Brandner DG, Cywar RM, Liu R, Manker LP, Michener WE, Gilhespy M, Skoufa Z, Watson MJ, Fruchey OS, Vardon DR, Gill RT, Bratis AD, Beckham GT (2017) Renewable acrylonitrile production. Science 358:1307–1310

Katayama N, Iijima H, Osanai T (2018) Production of bioplastic compounds by genetically manipulated and metabolic engineered cyanobacteria. Adv Exp Med Biol 1080:155–169

Kaufman AJ, Xiao S (2003) High CO_2 levels in the Proterozoic atmosphere estimated from analyses of individual microfossils. Nature 425(6955):279–282

Keith DW (2009) Why capture CO_2 from the atmosphere? Science 325(5948):1654–1655

Khoo KS, Chew KW, Yew GY, Leong WH, Chai YH, Show PL, Chen WH (2020) Recent advances in downstream processing of microalgae lipid recovery for biofuel production. Bioresour Technol 304:122996

Koch M, Berendzen KW, Forchhammer K (2020) On the role and production of Polyhydroxybutyrate (PHB) in the Cyanobacterium *Synechocystis* sp. PCC 6803. Life (Basel) 10(4):47

Koyande AK, Chew KW, Rambabu K, Tao Y, Chu DT, Show PL (2019) Microalgae: a potential alternative to health supplementation for humans. Food Sci Human Wellness 8(1):16–24

Kumar A, Ergas S, Yuan X, Sahu A, Zhang Q, Dewulf J, Malcata FX, van Langenhove H (2010) Enhanced CO_2 fixation and biofuel production via microalgae: recent developments and future directions. Trends Biotechnol 28(7):371–380

Lakatos GE, Ranglová K, Manoel JC, Grivalský T, Kopecký J, Masojídek J (2019) Bioethanol production from microalgae polysaccharides. Folia Microbiol (Praha) 64(5):627–644

Lam MK, Lee KT, Mohamed AR (2012) Current status and challenges on microalgae-based carbon capture. Int J Greenh Gas Control 10: 456–469

Larkum AWD, Ross IL, Kruse O, Hankamer B (2012) Selection, breeding and engineering of microalgae for bioenergy and biofuel production. Trends Biotechnol 30(4):198–205

Li FF, Yang ZH, Zeng R, Yang G, Chang X, Yan JB, Hou YL (2011) Microalgae capture of CO_2 from actual flue gas discharged from a combustion chamber. Ind Eng Chem Res 50:6496–6502

López Rocha CJ, Álvarez-Castillo E, Estrada Yáñez MR, Bengoechea C, Guerrero A, Orta Ledesma MT (2020) Development of bioplastics from a microalgae consortium from wastewater. J Environ Manag 263:110353

López G, Yate C, Ramos FA, Cala MP, Restrepo S, Baena S (2019) Production of polyunsaturated fatty acids and lipids from autotrophic, mixotrophic and heterotrophic cultivation of *Galdieria* sp. strain USBA-GBX-832. Sci Reports 9(1):1–13

Lupatini AL, Colla LM, Canan C, Colla E (2017) Potential application of microalga *Spirulina platensis* as a protein source. J Sci Food Agric 97(3):724–732

Markou G, Vandamme D, Muylaert K (2014) Microalgal and cyanobacterial cultivation: the supply of nutrients. Water Res 65: 186–202

McAdam B, Brennan Fournet M, McDonald P, Mojicevic M (2020) Production of Polyhydroxybutyrate (PHB) and factors impacting its chemical and mechanical characteristics. Polymers 12(12):2908

Moazami N, Ashori A, Ranjbar R, Tangestani M, Eghtesadi R, Nejad AS (2012) Large-scale biodiesel production using microalgae biomass of *Nannochloropsis*. Biomass Bioenergy 39:449–453

Moncada J, Tamayo JA, Cardona CA (2014) Integrating first, second, and third generation biorefineries: incorporating microalgae into the sugarcane biorefinery. Chem Eng Sci 118:126–140

Monlau F, Suarez-Alvarez S, Lallement A, Vaca-Medina G, Giacinti G, Munarriz M, Urreta I, Raynaud C, Ferrer C, Castañón S (2021) A cascade biorefinery for the valorization of microalgal biomass: biodiesel, biogas, fertilizers and high valuable compounds. Algal Res 59:102433

Montalvo GEB, Thomaz-Soccol V, Vandenberghe LPS, Carvalho JC, Faulds CB, Bertrand E, Prado MRM, Bonatto SJR, Soccol CR (2019) *Arthrospira maxima* OF15 biomass cultivation at laboratory and pilot scale from sugarcane vinasse for potential biological new peptides production. Bioresour Technol 273:103–113

Morales M, Sánchez L, Revah S (2018) The impact of environmental factors on carbon dioxide fixation by microalgae. FEMS Microbiol Lett 365(3):262

Moreira D, Pires JCM (2016) Atmospheric CO_2 capture by algae: negative carbon dioxide emission path. Bioresour Technol 215: 371–379

Morocho-Jácome AL, Ruscinc N, Martinez RM, de Carvalho JCM, Santos de Almeida T, Rosado C, Costa JG, Velasco MVR, Baby AR (2020) (Bio)Technological aspects of microalgae pigments for cosmetics. Appl Microbiol Biotechnol 104(22):9513–9522

Moroney JV, Ynalvez RA (2007) Proposed carbon dioxide concentrating mechanism in *Chlamydomonas reinhardtii*. Eukaryot Cell 6(8):1251–1259

Nagappan S, Devendran S, Tsai PC, Dahms HU, Ponnusamy VK (2019) Potential of two-stage cultivation in microalgae biofuel production. Fuel 252:339–349

Napan K, Teng L, Quinn JC, Wood BD (2015) Impact of heavy metals from flue gas integration with microalgae production. Algal Res 8: 83–88

Nwoba EG, Parlevliet DA, Laird DW, Alameh K, Moheimani NR (2019) Light management technologies for increasing algal photobioreactor efficiency. Algal Res 39:101433

Ooms MD, Dinh CT, Sargent EH, Sinton D (2016) Photon management for augmented photosynthesis. Nat Commun 7(1):1–13

Park J, Kumar G, Bakonyi P, Peter J, Nemestóthy N, Koter S, Kujawski W, Bélafi-Bakó K, Pientka Z, Muñoz R, Kim SH (2021) Comparative evaluation of CO_2 fixation of microalgae strains at various CO_2 aeration conditions. Waste Biomass Valorization 12(6):2999–3007

Parsons S, Raikova S, Chuck CJ (2020) The viability and desirability of replacing palm oil. Nat Sustain 3(6):412–418

Peyrton J, Chambaretaud C, Sarbu A, Avérous L (2020) Biobased polyurethane foams based on new polyol architectures from microalgae oil. ACS Sustain Chem Eng 8(32):12187–12196

Piiparinen J, Barth D, Eriksen NT, Teir S, Spilling K, Wiebe MG (2018) Microalgal CO2 capture at extreme pH values. Algal Res 32:321–328

Pires JCM, Gonçalves AL, Martins FG, Alvim-Ferraz MCM, Simões M (2014) Effect of light supply on CO_2 capture from atmosphere by *Chlorella vulgaris* and *Pseudokirchneriella subcapitata*. Mitig Adapt Strateg Glob Chang 19(7):1109–1117

Polat E, Altınbaş M (2020) Optimization of *Auxenochlorella protothecoides* lipid content using response surface methodology for biofuel production. Biomass Convers Biorefinery:1–15

Radmann EM, Camerini FV, Santos TD, Costa JAV (2011) Isolation and application of SO_x and NO_x resistant microalgae in biofixation of CO_2 from thermoelectricity plants. Energy Convers Manag 52(10):3132–3136

Rafa N, Ahmed SF, Badruddin IA, Mofijur M, Kamangar S (2021) Strategies to produce cost-effective third-generation biofuel from microalgae. Front Energy Res 9:517

Rahaman MSA, Cheng LH, Xu XH, Zhang L, Chen HL (2011) A review of carbon dioxide capture and utilization by membrane integrated microalgal cultivation processes. Renew Sust Energ Rev 15(8):4002–4012

Raheem A, Prinsen P, Vuppaladadiyam AK, Zhao M, Luque R (2018) A review on sustainable microalgae based biofuel and bioenergy production: Recent developments. J Clean Prod 181:42–59

Rahman A, Miller CD (2017) Microalgae as a source of bioplastics. Algal Green Chem Recent Prog Biotechnol:121–138

Rahman DY, Sarian FD, van Wijk A, Martinez-Garcia M, van der Maarel MJEC (2017) Thermostable phycocyanin from the red microalga Cyanidioschyzon merolae, a new natural blue food colorant. J Appl Phycol 29(3):1233–1239

Rajesh BJ, Preethi KS, Gunasekaran M, Kumar G (2020) Microalgae based biorefinery promoting circular bioeconomy-techno economic and life-cycle analysis. Bioresour Technol 302:122822

Ramkrishnan U, Bruno B, Swaminathan S (2014) Sequestration of CO_2 by halotolerant algae. J Environ Heal Sci Eng 12(1):1–7

Razzak SA, Hossain MM, Lucky RA, Bassi AS, De Lasa H (2013) Integrated CO_2 capture, wastewater treatment and biofuel production by microalgae culturing—a review. Renew Sustain Energy Rev 27: 622–653

Rival A, Levang P 2014. Palms of controversies oil palm and development challenges. Center for International Forestry Research. Bogor, Indonesia, 68p

Sánchez-Baracaldo P, Cardona T (2020) On the origin of oxygenic photosynthesis and Cyanobacteria. New Phytol 225(4):1440–1446

Santos A, Lamers P, Janssen M, Wijfels R (2013) Biomass and lipid productivity of Neochloris oleoabundans under alkaline–saline conditions. Algal Res 2:204–211

Sayre R (2010) Microalgae: the potential for carbon capture. BioSci 60(9):722–727

Senatore V, Buonerba A, Zarra T, Oliva G, Belgiorno V, Boguniewicz-Zablocka J, Naddeo V (2021) Innovative membrane photobioreactor for sustainable CO_2 capture and utilization. Chemosphere 273: 129682

Sepulveda C, Gómez C, El Bahraoui N, Acién G (2019) Comparative evaluation of microalgae strains for CO_2 capture purposes. J CO2 Util 30:158–167

Sharp CE, Urschel S, Dong X, Brady AL, Slater GF, Strous M (2017) Robust, high-productivity phototrophic carbon capture at high pH and alkalinity using natural microbial communities. Biotechnol Biofuels 10(1):1–13

Singh J, Dhar DW (2019) Overview of carbon capture technology: microalgal biorefinery concept and state-of-the-art. Front Mar Sci 29

Singh V, Tiwari A, Das M (2016) Phyco-remediation of industrial waste-water and flue gases with algal-diesel engenderment from micro-algae: a review. Fuel 173:90–97

Singh MK, Rai PK, Rai A, Singh S, Singh JS (2019) Poly-β-hydroxybutyrate production by the cyanobacterium *Scytonema geitleri* Bharadwaja under varying environmental conditions. Biomolecules 9(5)

Sobczuk TM, Camacho FG, Camacho Rubio F, Acién Fernández FG, Molina Grima E (2000) Carbon dioxide uptake efficiency by outdoor microalgal cultures in tubular airlift photobioreactors. Biotechnol Bioeng 67(4):45–475

Song C, Liu Q, Qi Y, Chen G, Song Y, Kansha Y, Kitamura Y (2019) Absorption-microalgae hybrid CO_2 capture and biotransformation strategy—a review. Int J Greenh Gas Control 88:109–117

Stern MC, Simeon F, Herzog H, Hatton TA (2013) Post-combustion carbon dioxide capture using electrochemically mediated amine regeneration. Energy Environ Sci 6(8):2505–2517

Thomas DM, Mechery J, Paulose SV (2016) Carbon dioxide capture strategies from flue gas using microalgae: a review. Environ Sci Pollut Res 23(17):16926–16940

Tokuşoglu Ö, Ünal MK (2003) Biomass nutrient profiles of three microalgae*: Spirulina platensis, Chlorella vulgaris, and Isochrisis galbana*. J Food Sci 68(4):1144–1148

Vadlamani A, Viamajala S, Pendyala B, Varanasi S (2017) Cultivation of microalgae at extreme alkaline pH conditions: a novel approach for biofuel production. ACS Sustain Chem Eng 5(8):7284–7294

Vadlamani A, Pendyala B, Viamajala S, Varanasi S (2019) High productivity cultivation of microalgae without concentrated CO2 input. ACS Sustain Chem Eng 7(2):1933–1943

Vale MA, Ferreira A, Pires JCM, Gonçalves AL (2020) CO_2 capture using microalgae. In: Rahimpour MR, Farsi M, Makerem MA (eds) Advances in carbon capture. Elsevier, Duxfor, pp 381–405

Varshney P, Beardall J, Bhattacharya S, Wangikar PP (2018) Isolation and biochemical characterisation of two thermophilic green algal species- *Asterarcys quadricellulare* and *Chlorella sorokiniana*, which are tolerant to high levels of carbon dioxide and nitric oxide. Algal Res 30:28–37

Vega F, Baena-Moreno FM, Gallego Fernández LM, Portillo E, Navarrete B, Zhang Z (2020) Current status of CO_2 chemical absorption research applied to CCS: towards full deployment at industrial scale. Appl Energy 260:114313

Venkata Mohan S, Modestra JA, Amulya K, Butti SK, Velvizhi G (2016) A circular bioeconomy with biobased products from CO_2 sequestration. Trends Biotechnol 34(6):506–519

Wang Y, Stessman DJ, Spalding MH (2015) The CO_2 concentrating mechanism and photosynthetic carbon assimilation in limiting CO_2: how *Chlamydomonas* works against the gradient. Plant J 82(3): 429–448

Wang Y, Tibbetts SM, McGinn PJ (2021) Microalgae as sources of high-quality protein for human food and protein supplements. Foods 10(12):1–18

Wei L, Wang Q, Xin Y, Lu Y, Xu J (2017) Enhancing photosynthetic biomass productivity of industrial oleaginous microalgae by overexpression of RuBisCO activase. Algal Res 27:366–375

Whitton R, Ometto F, Pidou M, Jarvis P, Villa R, Jefferson B (2015) Microalgae for municipal wastewater nutrient remediation: mechanisms, reactors and outlook for tertiary treatment. Environ Technol Rev 4(1):133–148

Wolf L, Cummings T, Müller K, Reppke M, Volkmar M, Weuster-Botz D (2021) Production of β-carotene with *Dunaliella salina* CCAP19/18 at physically simulated outdoor conditions. Eng Life Sci 21(3–4): 115–125

Xu J, Cheng J, Xin K, Xu J, Yang W (2020) Developing a spiral-ascending CO_2 dissolver to enhance CO_2 mass transfer in a horizontal tubular photobioreactor for improved microalgal growth. ACS Sustain Chem Eng 8(51):18926–18935

Yang X, Liu L, Yin Z, Wang X, Wang S, Ye Z (2020) Quantifying photosynthetic performance of phytoplankton based on photosynthesis – Irradiance response models. Environ Sci Eur 32:24

Yashavanth PR, Das M, Maiti SK (2021) Recent progress and challenges in cyanobacterial autotrophic production of polyhydroxybutyrate (PHB), a bioplastic. J Environ Chem Eng 9(4):105379

Yeh KL, Chang JS (2011) Nitrogen starvation strategies and photobioreactor design for enhancing lipid content and lipid production of a newly isolated microalga *Chlorella vulgaris* ESP-31: Implications for biofuels. Biotechnol J 6(11):1358–1366

Yen HW, Ho SH, Chen CY, Chang JS (2015) CO_2, NO_x and SO_x removal from flue gas via microalgae cultivation: a critical review. Biotechnol J 10(6):829–839

Zeng X, Danquah MK, Zhang S, Zhang X, Wu M, Chen XD, Ng IS, Jing K, Lu Y (2012) Autotrophic cultivation of *Spirulina platensis* for CO_2 fixation and phycocyanin production. Chem Eng J 183:192–197

Zhang S, Liu Z (2021) Advances in the biological fixation of carbon dioxide by microalgae. J Chem Technol Biotechnol 96(6): 1475–1495

Zhang W, Zhao C, Cao W, Sun S, Hu C, Liu J, Zhao Y (2020) Removal of pollutants from biogas slurry and CO_2 capture in biogas by microalgae-based technology: a systematic review. Environ Sci Pollut Res 27(23):28749–28767

Zhou J, Zhang F, Meng H, Zhang Y, Li Y (2016) Introducing extra NADPH consumption ability significantly increases the photosynthetic efficiency and biomass production of cyanobacteria. Metab Eng 38:217–227

Zhu C, Zhai X, Xi Y, Wang J, Kong F, Zhao Y, Chi Z (2020) Efficient CO_2 capture from the air for high microalgal biomass production by a bicarbonate Pool. J CO2 Util 37:320–327

Zuccaro G, Yousuf A, Pollio A, Steyer JP (2020) Microalgae cultivation systems. Microalgae Cultiv Biofuels Prod 11–29

Removal of Nitrogen and Phosphorus from Domestic Wastewater by Microalgal Cultures

3

María I. Ospina and Mohamed T. Darwich-Cedeño

Abstract

Domestic wastewater contains high concentrations of multiple organic and inorganic nutrient pollutants. When released without treatment, it can lead to a reduction in oxygen levels and eutrophication in water sources. Bioremediation utilizes the genetic and metabolic diversity of microorganisms in industrial processes, optimizing their biological capabilities. It is a promising method for approaching various contemporary issues such as climate change, remediation of contaminated environments, and the production of pharmaceutical, nutraceutical, or industrial molecules and biomass of interest. Microalgae, which are always associated with bacteria, can be used in bioreactors, treatment plants, or in-situ inoculations for wastewater remediation. Attached bacteria obtain oxidizable organic matter from the microalgae, which is released during photosynthesis. In turn, the bacteria provide factors that promote the growth of microalgae. Several studies have demonstrated the effectiveness of microalgae in the removal of nitrogen and phosphorus from wastewater. This chapter focuses on the removal of nitrogen and phosphorus from domestic wastewater using microalgae cultures from a metabolic perspective. These treatments can be carried out in open ponds, closed photobioreactors, or high-rate algal ponds (HRAP).

Keywords

Domestic wastewater · Microalgae · Nitrogen removal · Phosphorus removal · Bioremediation · HRAP

M. I. Ospina (✉)
Universidad del Valle, Cali, Colombia
e-mail: maria.ospina.gonzalez@correounivalle.edu.co

M. T. Darwich-Cedeño
Universidad Nacional de Colombia, Bogotá, Colombia
e-mail: mdarwich@unal.edu.co

Domestic wastewater contains high concentrations of organic and inorganic nutrients. If these waters are released without treatment, the receiving water sources such as rivers, lakes, and streams can have a reduction in oxygen levels, due to microbial oxidation of organic pollutants and ammonium (direct effect) and eutrophication (indirect effect) (Larsdotter 2006). Wastewater is a typical contaminant of anthropogenic activity, these wastes tend to promote high levels of eutrophication in underground and surface water bodies, generating ideal conditions for uncontrollable bacterial and microalgal growth, consuming the oxygen in the water column, altering pH, and generating toxic conditions.

In fact, a promising methodology to solve this problem is bioremediation, which focused on the use of genetic and metabolic diversity in microorganisms that are characterized to transform contaminants into innocuous products or less toxic and these products could be integrated into natural biogeochemical cycles. Thus, the aim is to carry out biodegradation processes with the use of natural algae in bioreactors, projecting them into treatment plants, fermentative lagoons in-situ inoculations, or even as wastewater remedial treatment in eutrophic conditions.

In natural or artificial, marine, or freshwater aquatic environments, microalgae are always associated with bacteria (Voltolina and Arredondedo 2015). These bacteria generally need a support to attach themselves to carry out their metabolic processes; these supports usually are made up of polysaccharides, DNA, and proteins. Mutualistic bacteria have been reported to take extracellular organic carbon (EOC) released during microalgae photosynthesis; they also obtain organic matter such as carbohydrates, nitrogen, fatty acids, lipids, and vitamins during the process. Microalgae use growth-promoting factors produced by heterotrophic bacteria; biotin, proline, and glycolic acid have also been reported for extracellular matrices excreted by *Chlorella sp.* (Giraldo 2012; Watanabe et al. 2005) or biofilms. The physiochemical characteristics of these biofilms are determined by the presence of photosynthetic organisms. The

A. de Jesús Martínez-Roldán (ed.), *Biotechnological Processes for Green Energy, and High Value Bioproducts by Microalgae, and Cyanobacteria Cultures*, Developments in Applied Phycology 13, https://doi.org/10.1007/978-3-031-43969-8_3

polysaccharides present in biofilms act as a nutritional factor and binding agent, forming a framework that retains other nutrients such as salts and biomass, allowing formation of new biofilms on top of existing ones (Giraldo 2012).

Several reports involve the *Chlorella* genus with various symbionts, such as paramecia, hydra, bacteria, where the binding and chelating properties of the exopolysaccharide gels produced during photosynthesis are the protagonists (Imase et al. 2008), although these topics are still under discussion, since most of the studies are focused on the same properties of the gel and not on its ecology as such (Imase et al. 2008). The symbiotic relationship between *Chlorella* sp. and various microorganisms depends almost exclusively on EOC under photoautotrophic conditions and this affects not only the associated microorganisms but also *Chlorella* sp. itself, given that axenic strains have been reported to decrease chlorophyll production in contrast to consortium strains due to a complete chain in the flow of nutrients and possibly phytohormonal factors in culture medium, an event that does not occur under axenic conditions (Watanabe et al. 2005).

Microalgae and cyanobacteria have been used as alternative biological systems in wastewater treatment, because of their ability to match their autotroph and heterotroph metabolism; in that way, they consume salts and ions in the presence of light, or consume organic molecules. Previous studies of algal–bacterial consortia in biochemical fields showed removal of chemical compounds, by considering important hormonal interactions, as reported by the works De-Bashan and Bashan (2003) and de-Bashan et al. (2008), and Márquez (2013), thus demonstrating how these interactions affect the metabolic pathways in nitrogen assimilation. Other studies suggest the implementation of algal biomass as a raw material in the bioenergy and food industries due to the caloric, nutritional, or removal properties announced for some microalgae of research interest (Table 3.1).

Recent experiments cultivated *Scytonema javanicum* cyanobacterium in artificial synthetic wastewater and determined the remotion potential of nitrogen and phosphates from this organism (Wu et al. 2018). Gorain et al. (2019) evaluated, using farm wastewaters, how nitrogen and phosphate were removed by *Anabaena sphaerica* and *A. variabilis* cyanobacteria for 30 days in a semi-continuous culture system; about 90% of nitrogen and phosphate was removed. Likewise, Sandoval Salazar et al. (2021) observed how *P. kessleri* and *Desmodesmus armatus* are efficient in wastewater nitrogen removal, no matter the initial concentration. *P. kessleri* removes 70, 67, and 61% nitrogen from water at 100, 50, and 25%, respectively.

3.1 Wastewater Treatment

There are different types of wastewater treatment. The primary treatment seeks the elimination of large particles in the wastewater through screens or sedimentation. Secondary treatment aims to reduce the biochemical oxygen demand (BOD) in wastewater through the oxidation of organic compounds and ammonium. Tertiary treatment seeks to remove plant nutrients, mainly nitrogen and phosphorus (Table 3.1) (Larsdotter 2006).

Nitrogen can be removed through nitrification and denitrification processes, where ammonium is oxidized to nitrate by nitrifying bacteria in aerobic reactors, and is subsequently recycled to an anoxic reactant, where it is converted to nitrogen gas (N_2) by denitrifying bacteria (Larsdotter 2006). For its part, phosphorus in wastewater is frequently removed through chemical precipitation with metal salts (aluminum or iron) to form aluminum or ferric phosphate. In addition to this process, the chemical removal of phosphorus can be achieved through adsorption on filter media containing positively charged surfaces (Larsdotter 2006).

In addition to chemical methods for phosphorus removal, there are biological methods. One of these methods is enhanced biological phosphorus removal (EBPR), where bacteria in the activated sludge accumulate excess phosphorus, responding to changes between aerobic and anaerobic conditions. In this way, when the environment is anaerobic, the bacteria use their phosphorus storage as an energy source to take up the substrate and release the phosphorus.

Conversely, when conditions are aerobic, cells take up more phosphorus than is released, and a net removal is achieved (Larsdotter 2006).

There are other biological processes for removing nutrients, including wetlands, hydroponics, and algae cultures, within the tertiary treatment processes. These methods are associated with the assimilation of nutrients in the biomass and other nutrient extraction processes (Larsdotter 2006).

3.2 Phosphorus

Phosphorus is an essential basic element in organisms, frequently found in water sources, mainly in dissolved form or in association with other elements (Mao et al. 2021). Phosphorus has a significant environmental impact. The excess of this element in the water accelerates the growth of algae and other types of microorganisms, generating an imbalance of nutrients in the water and causing eutrophication processes.

Table 3.1 Wastewater treatment by microalgae technologies

Specie	Contaminant	Removal percentage	Authors
Chlorella sacarophila, Chlamidomonas pseudococcum, Scenedesmus sp., *Neochloris oleoabundans*	Municipal wastewater	98	Hena et al. (2015)
Microalgae	Municipal Wastewater	99.1	Su et al. (2011)
Parachlorella kessleri	BTEX	40 Benzene, 30 Etil-Benzene, 63 Toluene 40 Xilene	Takáčová (2015)
Bostrychia caliptera	Heavy metals (Cr VI)	87	Rengifo et al. (2012)
Chlorella vulgaris, Scenedesmus quadricauda, Euglena gracilis, Ankistrodesmus convolutus, Chlorococcum oviforme	Leachate	91	Mustafa et al. (2012)
Chlorella sp.	Municipal wastewater	8	Halfhide et al. (2015)
Chlorella sp., *Chlamidomonas* sp., *Lyngbya* sp.	Heavy metals (Pb II)	86.16	Kumar and Goyal (2008)
Desmodesmus sp. *Chlorella* sp.	Domestic wastewater	*Desmodesmus sp.*96.5 NO^{3-} *Chlorella sp.*95 NO^{3-} Desmodesmus sp 88.3 $PO_4{}^{3-}$ Chlorella sp.69.3 $PO_4{}^{3-}$	Lopez Ponte et al. (2022)
Chlorella vulgaris, Botryococcus braunii, Ankistrodermus falcatus Tetradesmus obliquus	Municipal wastewater	>99% PO_4 62.4% PO_4	Lavrinovičs et al. (2021)
Chlorella pyrenoidosa	Digested anaerobic starch	61.9 Total Organic Carbon (TOC); 78.7 Total Nitrogen (TN) 97.2 Total Phosphorus (TP)	Chu et al. (2015)
Scenedesmus obliquus	Urban wastewater	86.8 TN 97.7 TP	Ruiz et al. (2013)
Scenedesmus sp.	Industrial wastewater	34.6 mg N/L/day 12.7 mg $PO_4{}^{3-}$/L/day	Jebali et al. (2015)
C. vulgaris	Aquaculture wastewater	86.1 TN, 82.7 TP	Gao et al. (2016)
Scenedesmus sp.	Muncipal wastewater	99 TN, 99 TP	McGinn et al. (2012)
Desmodesmus sp.	Digested piggery manure	94.2 TN; 88.7 $PO_4{}^{-}P$	Ji et al. (2015)
Poultry litter	S. platensis	$NH_4{}^{-}N > 99$ $PO_4{}^{-}P > 99$	Markou et al. (2012)

Some phosphorus-containing elements in water can become minerals and clog water supply lines and wastewater treatment facilities. In drainage systems, phosphorus comes mainly from domestic wastewater, and about 85% of the total phosphorus enters from domestic wastewater, containing human feces and detergents (Mao et al. 2021).

3.3 Removal of Phosphorus from Domestic Wastewater by Microalgal Cultures

Microalgae can be used in wastewater removal due to their great metabolic diversity and nutrient assimilation capacity. In the metabolic process of microalgae, oxygen is released that can be used to produce mechanical aeration and generate an aerobic environment suitable for the growth of bacteria that work in synergy with microalgae, so that the process is more efficient. Thus, the biomass generated from microalgae can be used to generate different commercial products (Ramírez Mérida 2020).

The species of the genera *Ankistrodesmus, Chlorella, Spirulina, Desmodesmus,* and *Phormidium* are involved in bioremediation. Some species of these genera have a high nutrient uptake rate, a high growth rate, and a high tolerance to contaminants and xenobiotics that are frequently found in wastewater. These microalgae can interact with bacteria, favoring the decontamination treatment system (Lopez Ponte et al. 2022).

Chlorella and *Desmodesmus* are green algae that dominate wastewater and have high removal rates of nutrients, nitrogen, and phosphorus. These algae grow rapidly in photoautotrophic cultures, demanding more nutrients from the medium, which makes their consumption capacity faster and more efficient. Furthermore, these species are very resistant to environmental changes, mechanical stress, and high concentrations of pollutants such as nutrients and heavy metals (Lopez Ponte et al. 2022).

In wastewater treatment by means of microalgae, these assimilate plant nutrients and support the bacteria with oxygen. The bacteria are degrading the organic matter in the wastewater, just like the process used in activated sludge. Microalgae help remove pollutants from wastewater and produce biomass that may have potential use in the production of value-added products and/or biogas through anaerobic digestion. Algae can also reduce the harmful effects of wastewater effluents and reduce eutrophication in water sources (Delgadillo-Mirquez et al. 2016; Larsdotter 2006).

The efficiency of phosphorus removal depends on the potential of microalgae to absorb, conserve, or degrade this nutrient in municipal wastewater (Mao et al. 2021. The efficiency of the algal-bacterial consortium for the absorption of nutrients is affected by the availability of nutrients and by the interactions between physical-chemical factors such as pH, temperature, light intensity, photoperiod, and biological factors (Delgadillo-Mirquez et al. 2016).

Microalgae treatment can be carried out in open ponds or in closed photobioreactors (PBRs) such as tubes or flat panels (Larsdotter 2006). Open systems refer to water-growing systems in the open air, i.e., water sources. The open system design is simple and easy to handle; however, nutrients can be diluted by contamination when exposed to the open environment for a long period. Closed systems are separated from the external environment, which allows them to protect the system from any interference from the external environment. PBRs are frequently used as closed systems for the cultivation of microalgae. These PBRs can be classified into pieces or flat plates, and can be made of glass or plastic, air, carbon dioxide (CO_2) and other gases. PBRs are highly controlled, which can be used to analyze the characteristics of microalgae and the effects on removal in wastewater under different conditions. In addition, it provides a culture system to improve the conditions for phosphorus absorption by microalgae in urban wastewater (Mao et al. 2021).

3.4 High-Rate Algae Ponds (HRAP)

When the algae-bacteria consortium is present, it is usually grown in high-rate algae ponds (HRAP) to manage municipal, industrial, and/or agricultural wastewater. This type of pond consists of a shallow racetrack reactor (between 0.3 and 0.4 m deep), with mixing, in which algae and bacteria are present. The treatment consists of introducing the wastewater into the HRAP and mixing it with the algae and bacteria culture inside the bioreactor. The agitation generates the growth of microalgae and prevents the sedimentation of biomass (Delgadillo-Mirquez et al. 2016).

High rates of nutrient removal efficiency have been achieved by HRPA. According to Kim et al. (2014), 93% of total nitrogen and 83% of total phosphorus were removed from untreated municipal wastewater in an area dominated by an algae consortium.

There are parameters that affect wastewater treatment in HRAP and biomass production, namely light intensity, temperature, pH, and photoperiod. In the case of photoperiod and temperature, these variables are dependent on seasonal variations that vary throughout the year, with an effect on the efficiency of wastewater treatment and biomass production. According to Lee et al. (2015), the photoperiod has a great impact on nutrient removal and biomass production, altering the population dynamics of algae and bacteria in a bioreactor used for municipal wastewater treatment (Fig. 3.1).

In a study using *Chlorella vulgaris* microalgae, Delgadillo-Mirquez et al. (2016) found that all phosphate ions were removed from the wastewater after approximately 100–150 hours. The highest rate of phosphate elimination occurred in cultures at 25 °C and 12 hours of photoperiod and with 15 °C and 18 hours of photoperiod. According to these conditions, the depletion rates of phosphorus as well as nitrogen were high and the system did not need organic carbon supply for the process. In this study, it was also found that the growth of microalgae slowed down and stopped when phosphate was no longer detected in the medium at 15 °C, which may suggest that phosphorus depletion may limit algal growth. However, in cultures at 25 °C, growth did not stop after phosphorus depletion, which may be related to the uncoupling of nutrient uptake and growth by algae, i.e., they can continue to grow after nutrient depletion. The mechanisms by which phosphate is depleted in a medium with algae are adsorption on the cell surface,

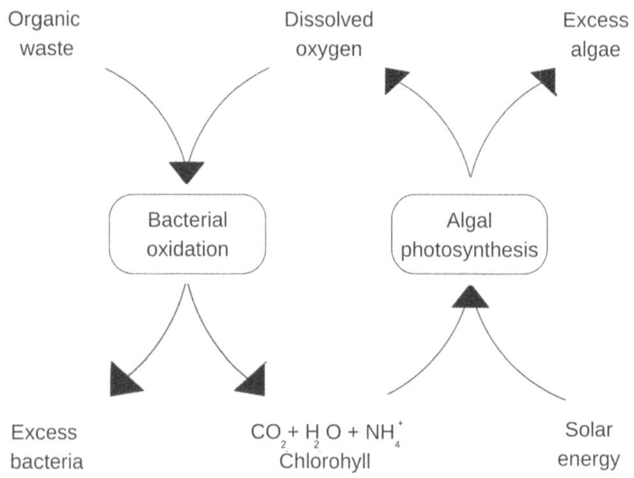

Fig. 3.1 The main processes involved in HRAP. (Adapted with permission from Larsdotter (2006))

assimilation by the biomass of part of the phosphate ions, and chemical precipitation (Delgadillo-Mirquez et al. 2016).

3.5 Phosphorus Removal Mechanism by Microalgae

In wastewater, inorganic phosphorus is present in various ionic states, depending on the pH (H_3PO_4, <2.15; $H_2PO_4^-$, 2.15 - 7.20; HPO_4^{2-}, 7.20 - 12.33; and PO_4^{3-}, >12.33). Inorganic phosphorus is considered the most bioavailable form of phosphorus, and algae preferentially assimilate HPO_4^{2-} and $H_2PO_4^-$. In microalgae, PO_4^{3-} enters the cell through active transport through a symporter channel with H^+ or Na^+ ions that generate the driving force, established by a plasma membrane H^+-ATPase pump. Soluble organic phosphorus compounds are a source of bioavailable phosphorus. These are made accessible to microalgae by the expression of free and extracellular membrane-bound phosphatases, which nonspecifically hydrolyze bound PO_4^{3-} groups (Mohsenpour et al. 2021).

Phosphorus is incorporated into organic compounds by phosphorylation of adenosine diphosphate (ADP). This reaction is endergonic with energy input obtained from the oxidation of respiratory substrates or from the photosynthetic electron transport chain. The adenosine triphosphate (ATP) produced allows the transfer of PO_4^{3-} groups to organic compounds at the substrate level, for example, in the conversion of glucose to glucose-6-phosphate in the glycolytic pathway (Fig. 3.2). When the environment is rich in phosphorus, microalgae can accumulate this nutrient more than their metabolic requirements and store it as insoluble polyphosphate in acid granules, which is called "luxury uptake," which only occurs without a previous starvation period (Mohsenpour et al. 2021).

3.6 Enhanced Algal Phosphorus Uptake (EAPU)

Enhanced algal phosphorus uptake (EAPU) and storage of excess phosphorus in algal cells can be achieved through manipulation with external availability of phosphorus. Known approaches that cause this phenomenon are luxury phosphorus uptake mechanism and phosphorus starvation, which are due to excess phosphorus availability and phosphorus-limited conditions, respectively. These biomass manipulation approaches can account for a ten-fold higher phosphorus uptake rate compared to regular consumption of biomass phosphorus (Lavrinovičs et al. 2021).

Lavrinovičs et al. (2021) found that *C. vulgaris, Botryococcus braunii,* and *Ankistrodermus falcatus* showed high nutrient removal rates, reaching removal of above 97% in nitrates and 91% in phosphate. In *C. vulgaris* and *B. braunii* almost complete removal (>99%) of PO_4 was achieved. After prior biomass exposure to phosphorus deficiency, *A. falcatus* showed almost complete removal of PO_4 after three days of phosphorus starvation. The growth rates, productivity, biomass increase, and response to phosphorus deficiency were higher in *C. vulgaris* and *B. braunii*, suggesting their great potential in nutrient removal in wastewater treatment. *B. braunii* is known as a producer of carbohydrates, with potential for bioenergy production. In this way, if a high biomass production is achieved, this species is suitable for the recovery of resources from wastewater. *A. falcatus* and *Tetradesmus obliquus* are species that can also be used in the removal of phosphorus in wastewater, and, in turn, have great potential for high economic performance due to their lipid productivity. But these two species, compared to *C. vulgaris* and *B. braunii*, have lower growth rate and productivity, so they are less suitable for post-treatment of wastewater with economic benefit.

3.7 Factors Affecting Phosphorus Removal from Municipal Wastewater by Microalgae

Among the factors that can affect the removal of phosphorus by microalgae are temperature, force, period of illumination, and pH, as well as hydraulic retention time, the ratio of nitrogen to phosphorus, the concentration of CO_2, microalgae species, and different types of wastewaters (Mao et al. 2021).

Regarding the nitrogen–phosphorous ratio (N/P), an adequate ratio generates a good growth environment for microalgae and increases the phosphorus removal performance of microalgae. The optimal range of N/P ratio in freshwater environments is 6.8–10, and in photobioreactors, the optimal range is 5–30 for total phosphorus removal (Mao et al. 2021).

Another factor that affects the efficiency of phosphorus removal is the concentration of CO_2. In this sense, an increase in CO_2 concentrations, in addition to improving the efficiency of phosphorus removal, also increases the biomass of microalgae. The higher the biomass of microalgae, the higher the performance in phosphorus removal. The CO_2 concentrations also improve the lipid content of microalgae cells, increasing their recovery value (Mao et al. 2021).

Regarding the species of microalgae, each species has optimal growth conditions, so species of microalgae suitable for the different initial concentrations of urban waste must be selected to achieve maximum benefit in terms of disposal of phosphorus. However, it is important to note that only one type of algae is rarely used for wastewater treatment. Generally, mixed microalgae are used, selecting species according to the specific characteristics of the wastewater to be treated (Mao et al. 2021).

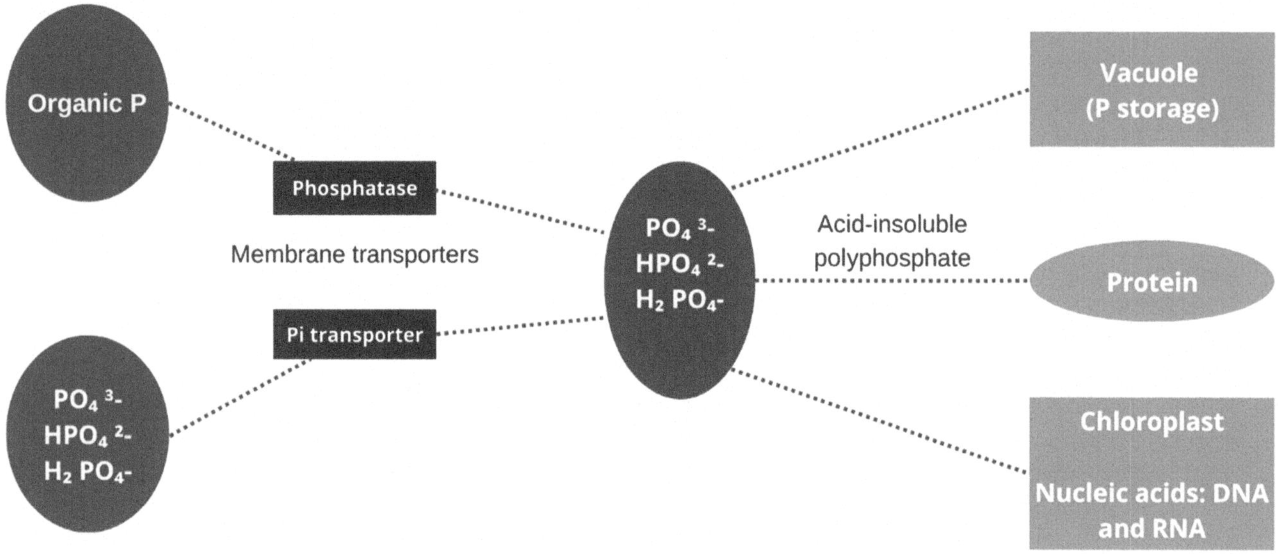

Fig. 3.2 Principal and minor routes in phosphorus assimilation.

According to the studies by Pachés et al. (2018), mixed microalgae culture could increase the rate of phosphorus removal and water productivity, by allowing the species to compete. Likewise, Devi et al. (2012) found that mixed cultures present a high degree of phosphorus removal in wastewater and concluded that the use of mixed cultures is one of the best methods for treating municipal wastewater and other low-toxicity wastewater.

3.8 Nitrogen

Nitrogen (N) can constitute as much as about 7–11% of the dry weight of microalgae cells (Goyena and Fallis 2019; Tebbani et al. 2014). Several microalgae such as cyanobacteria, Rhodophytes, and Chlorophytes, mainly use inorganic compounds (nitrate, dinitrogen, and ammonium) to supply their nitrogen requirements; in the same way urea or some amino acids can also be assimilated (Goyena and Fallis 2019; Bhaskar et al. 2015; Tebbani et al. 2014). Nitrogen is involved in the major metabolic pathways of microalgae. Therefore, increasing the concentration of nitrogen, up to an umbral value, leads to improved cellular and protein productivity and more significant chlorophyll synthesis. A deficiency of this element induces a significant accumulation of lipid reserves (polyunsaturated fatty acids) and a limitation of photosynthetic and cellular activities (Tebbani et al. 2014). Nitrogen is typically present in wastewater effluents as either dissolved inorganic nitrogen as ammonia (NH_3), ammonium (NH_4^+), nitrite (NO_2^-), and nitrate (NO_3^-), or as dissolved organic nitrogen as urea ($CO(NH_2)^2$) (Ross et al. 2018; Salbitani and Carfagna 2021); in addition, algae could consume dissolved organic nitrogen, such as amino acids, purines, pyrimidines, glycine betaine, yeast extract, peptone, and some peptides and proteins (Su 2020).

Nitrogen uptake depends on the concentration and typically follows a Michaelis–Menten type curve; nitrogen removal exhibits a linear trend while increasing external nitrogen concentration, and then reaches a plateau during uptake phase (Ross et al. 2018). The maximum removal nitrogen amount achieved directly depends on biological, physical, and chemical factors, which include the form of nitrogen in the media (Ross et al. 2018). The cornerstone of nitrogen metabolism process is played by glutamine synthetase/glutamate synthase (GS/GOGAT) cycle for ammonium assimilation (Fig. 3.3) (Goyena and Fallis 2019). Glutamate is itself a precursor for some amino acids and principal metabolites produced in this pathway but is not only the major nitrogen donor for the biosynthesis of other nitrogen-containing metabolites.

Fundamentally, mechanisms for nitrogen uptake are dependent on each nitrogen form present and all of these are geared toward the same objective: Utilization of different nitrogen sources involves transport from cell wall and membrane (Goyena and Fallis 2019; Ross et al. 2018). The membrane contains a special type of protein, known as "porins," which allow the passage of small molecules including amino acids and inorganic ions (Goyena and Fallis 2019) whereupon it is reduced to NH_4^+ and amino acids, before assimilation into macromolecules. Assimilation of NO_3^- occurs by sequentially reducing NO_3^- to NO_2^-, followed by NO_2^- to NH_4^+, via the action of nitrogen-reducing enzymes. Urea is converted into NH_4^+ by urease (Ross et al. 2018; Salbitani and Carfagna 2021) (Fig. 3.4). Microalgae such as plants prefer used NH_4^+ because the metabolic cost of reduction of the other nitrogen forms (NO_3^- and NO_2^-) is higher than

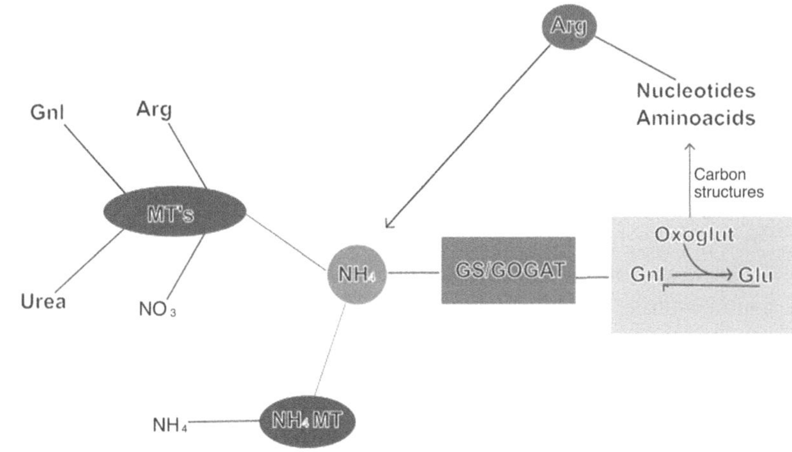

Fig. 3.3 Principal and minor routes in nitrogen assimilation. *MT* Membrane transporters, *GS* glutamine synthetase, *GOGAT* glutamate synthase

reducing ammonium to organic matter (Salbitani and Carfagna 2021). The convergence of ammonium levels and the GS/GOGAT cycle of several nitrogen assimilation pathways allows the operation of a regulatory network which helps the microalgae cell to enhance the nitrogen uptake process: ammonium inhibition by activity or/and expression of involved proteins in alternative nitrogen sources assimilation (Goyena and Fallis 2019; Salbitani and Carfagna 2021).

Many microalgae, such as cyanobacteria, can grow using atmospheric N_2 under aerobic conditions. Nitrogenase, the N_2 enzymatic fixation system, is irreversibly oxygen inactivated when an enzyme is extracted from any organism. Therefore, those organisms able to perform N_2 fixation in aerobic conditions have evolved protecting mechanisms for nitrogen-fixation machinery from the deleterious effects of oxygen. As photosynthesis releases oxygen, microalgae have to protect not only their nitrogen-fixation machinery from external oxygen income but also their internal release (Goyena and Fallis 2019).

In cyanobacteria and all nitrogen-fixing organisms, the enzyme complex nitrogenase catalyzes the ATP-dependent reduction of nitrogen to two molecules of ammonium (Kumar and Bera 2020); this reaction requires six electrons transference to nitrogen. Nitrogenase has proven to be very similar in all organisms tested. It consists of two different

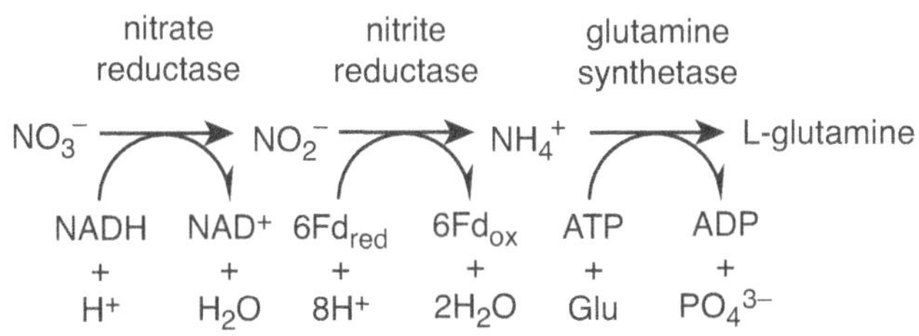

Fig. 3.4 Schematic nitrogen assimilation route. The upper NH_4^+ concentration inhibits the synthesis of nitrate and nitrite reductase. (Adapted from Cai et al. (2013))

protein components: dinitrogenase (Mo–Fe dependent protein) and dinitrogenase reductase (Fe-dependent protein), which mediates the ATP-dependent transference of electrons from external electron donors, such as ferredoxin or flavodoxin; then dinitrogenase binds through its Mo–Fe cofactors and catalyzes the reduction of two ammonia (Goyena and Fallis 2019).

3.9 Nitrate and Nitrite Uptake

Nitrate is probably the most abundant natural source of nitrogen (Goyena and Fallis 2019) for microalgae nutrition; its assimilation of nitrate (which is one of the major nitrogen oxides) by cyanobacteria involves nitrate uptake and reduction of intracellular nitrate to nitrite and followed by the formation of ammonium (Fig. 3.3) (Kumar and Bera 2020), which is the nitrogen form incorporated into organic compounds. Nitrite could also fulfill the nitrogen requirement of microalgae (Goyena and Fallis 2019). Reaction begins with the enzyme nitrate reductase; this enzyme needs Nicotinamide Adenine Dinucleotide (NADH) as electron donor for conducting the reduction process. This phenomenon contains an underlying intermediate reaction where NADH reduces a flavin protein which transfers the electron to a cytochrome component before finally reaching the nitrate reductase enzyme (Kumar and Bera 2020) evolving nitrite; in the same way iron-dependent nitrite reductase can reduce nitrite to ammonium with the help of ferredoxin, which acts as an electron donor and NADPH may act as a reducing agent only when ferredoxin is present. Finally, this ammonium becomes the principal protein-nitrogenated compound building block in the cell, and was observed to be strongly inhibited by ammonium in consumption of nitrate or nitrite, showing a self-regulating metabolic route (Kumar and Bera 2020).

Nitrite uptake by microalgae has two different ways: active transport and nitrous acid diffusion, whose contribution to the net uptake of nitrite decreases as the pH of the medium is raised. A competition for uptake between nitrate and nitrite is observed, showing that the same system operates the active transport of both substrates suggesting a competitive inhibition of the uptake route (Goyena and Fallis 2019) as is shown in Fig. 3.4. Vega (2020) describes four nitrate/nitrite transporter systems: nitrate transporter, nitrite transporter, bispecific transporter with a high affinity for nitrate and nitrite (once at time), and bispecific for both nitrate/nitrite. It is reported for some microalgae species that active transport of urea, purines, and amino acids involves Na^+, which is a urea, purines, and amino acids cotransport factor (Su 2020). Rees et al. (1980) used Li^+ and K^+ to substitute Na^+; they report disrupting the uptake of NO_3^- and urea and the presence of NH_4^+ ceases the NO_3^- uptake but there is no urea uptake.

3.10 Organic Nitrogen Uptake

In microalgae, the most common organic compounds found to serve as unique nitrogen sources are urea, amino acids (glutamine, arginine and asparagine), and uric acid (Goyena and Fallis 2019; Salbitani and Carfagna 2021). In case of urea, the most common metabolic way is by urease enzyme (Su 2020), and with Ni^+ as enzymatic cofactor, this urease transforms the urea in $2NH_4^+$ and CO_2 (Goyena and Fallis 2019; Ross et al. 2018). In amino acids, uptaking pathway is a little bit different, because each amino acid used by microalgae has a different uptake transporter. In the case of arginine, it uptakes by means of specific basic amino acid transporter; this kind of transporter in bacteria and cyanobacteria is a bimodular protein kinase; inside the cell, arginine is metabolized by arginase or arginine deiminase forming NH_4^+ and glutamate (Goyena and Fallis 2019).

For glutamine and glutamate uptake, microalgae have in general two types of membrane transporters; one of those transporters has high affinity for glutamine and the other one is, in general terms, a neutral amino acid transporter, which can uptake glutamate amino acid also (Goyena and Fallis 2019). Once inside of a plasmatic membrane, the glutaminase enzyme action releases NH_4^+ and the amino acid carbon backbone is used as a carbon source in intracellular carbon pathways (Goyena and Fallis 2019; Ross et al. 2018; Su 2020).

3.11 Nitrogen Removal from Domestic Wastewater by Microalgae Cultures

Municipal wastewater or agricultural wastewaters are commonly used to grow microalgae; overloads of nitrogen and phosphorus removal can be as high as 90% with short retention time, till the cultures reach stationary phase (McGinn et al. 2012). As Hongyang et al. (2011) shows the concentrations of all nitrogen forms found in wastewaters (TN, NH_4^+, NO_3, NO_2) rapidly decreased due to absorption and consumption of nutrients by *Chlorella* sp. At the end of the mixotrophic period, the concentrations of TN, NH_4^+, and TP decrease, with removal efficiencies of each nitrogen form as high as 70% in their experiments. This could be explained by the study of Ross et al. (2018) which used algal inoculum that has been adapted to low nitrogen media concentration, promoting the spending of their internal nutrient reserves. The new nitrogen addition in the media results in "surge uptake," where nitrogen uptake is promoted at a transiently enhanced rate due to the replenishment of internal nitrogen storage. The study also proved differences in nitrogen uptake preference for *C. parriaudii*, where this microalgae has a strong NH_4^+ removal rate and a higher degree than any other nitrogen source it was mixed with.

Nitrogen removal by nitrification–denitrification is commonly treated by electromechanical nutrient removal process, but it is costly and for a typical wastewater effluent nitrogen concentration of 3 mg N L requires additional aeration energy, assuming cost overruns for proper treatment. HRAP with CO_2 addition could provide energy-efficient tertiary-level nutrient removal for little additional energy cost (Viera Valencia and Garcia Giraldo 2019). Algal biomass can exhibit N:P ratios from 4:1 (nitrogen-depleted conditions) to 40:1. These N:P ratios correspond to algal biomass nitrogen and phosphorus composition percentages typically of 10% N and 1% P and 4% N and 0.4% P (under nitrogen-depleted conditions) (Viera Valencia and Garcia Giraldo 2019). In theory, the nutrient supply ratio only influences algal species composition, if one of the nutrients is, in fact, limiting the growth rate as long as irradiance levels and temperature are sufficient to support a higher growth rate (Yunes 2018; Ross et al. 2018). A change in the N:P ratio, for instance, may result in a substitution of algae species composition (Yunes 2018), making nitrogen removal a key to treating different wastewater sources given that phosphorus removal efficiency does not depend on biomass quantity but nitrogen concentrations (Bhaskar et al. 2015).

Some authors (Z. Chen et al. 2020; Papadopoulos et al. 2020) combine physics and biological strategies to remove nitrogenated compounds from wastewater. One promising methodology is electro-coagulation treatment. Papadopoulos et al. (2020) filtered municipal wastewater through a 0.45 μm membrane filter and then analyzed for NO_3^-, NH_4^-, NO_2^-, TP, chemical oxygen demand (COD), and total suspended solids (TSS) according to Standard Methods (APHA et al. 1999). Papadopoulos et al. (2020) used Fe electrodes which resulted in less radical NO_3^- reduction and NH_4^- production as Fe is oxidized partially to Fe_2 (Papadopoulos et al. 2020), having in consequence, a lower uptake of NO_3^- from algae because of the higher concentration of ammonia. The combined process proved very effective in terms of pollutant removal, i.e., 89.1%, 100%, 89.4%, 98.5%, 91.6%, and 100% removal of the initial NO_3^-, NH_4^-, TKN, TP, COD, and TSS concentrations, respectively (Papadopoulos et al. 2020).

Chlorella vulgaris–bacteria consortia shows acceptable efficiencies (Wang et al. 2020) compared to other consortia that include other microalgae such *Phomidium* sp., *Chlamydomonas* sp. (Sepehri et al. 2020) or *Desmodesmus* sp. with higher nitrogen and other contaminants removal rates. On the other hand, *Ettlia texensis*, *Ankistrodesmus falcatus*, *Scenedesmus obliquus* AS-6-1, *Chlorococcum* sp., *Tetraselmis suecica*, *C. vulgaris* JSC-7, and *Skeletonema marinoi* have autoflocculating properties, which is an additional advantage when it comes to recovering biomass (Z. Chen et al. 2020). Algae species as *Scenedesmus obliquus* and *Chlorella* sp. have been studied for wastewater treatment, showing promise in the removal of COD, phosphorus, and nitrogen, with approximate yields for *Desmodesmus* sp.—the removal rates are 92.9% for ammonium and 88.5% for total nitrogen, as well as removals of 40.8% for phosphorus and 61.9% for dissolved organic carbon (COD) in 14–17 days of treatment (Z. Chen et al. 2020). Also, Ferreira et al. (2017) experimented with the removal of nitrogen and phosphorus in brewery wastewater at different concentrations. They measure the nitrogen removal in terms of ammonia and total Kjeldahl nitrogen in a range of 71–93% and 73–89%, respectively. Likewise, *Chlorella* sp. removal yields for total nitrogen is 95–100%, phosphorus 85–95%, and COD 90% in 14–24 days of treatment (Z. Chen et al. 2020; Sepehri et al. 2020), while lipids production of both genera was emerging as promising raw material for production of biodiesel (Ferreira et al. 2017). However, one of the main limitations of the industrial application of this technology is the availability of light, since wastewater has high levels of turbidity caused by suspended solids that hinder photosynthesis (Papadopoulos et al. 2020). Therefore, the pretreatment process is essential for photosynthetic cultivation to be feasible in these effluents.

3.12 Conclusion

Microalgae are the primary producers per excellence in nature; they have one of most versatile and efficient metabolisms, optimizing resources available in natural environments to generate biomass. This implies the efficient oxygen generation and removal of ions from the environments, in which they are found, depending directly on the relationship in amounts of available nitrogen with respect to phosphorus for efficient production of biomass. These qualities have been used for design and implementation of multiple technologies that involve them in bioremediation and production processes where not only the use of nutrients in wastewater is optimized but also multiple value-added products are generated from biomass, along with the improvement in the quality of the treated water and a reduction in carbon footprint of anthropogenic activities.

Phosphorus is one of the main pollutants in wastewater. Microalgae can absorb this nutrient to remove it from wastewater; however, the effectiveness of microalgae to play this role depends, among other factors, on the potential of each species to absorb, conserve, or degrade phosphorus in this type of water. In addition to the use of microalgae for phosphorus removal, they produce biomass that can have potential use in the production of value-added products and/or biogas through aerobic digestion. Likewise, nitrogen is probably, may be after hydrogen and carbon, the largest constituent in organic matter and in the same way in wastewater; being part of nucleic acids and proteins, nitrogen could be taken by microalgae in almost all its forms "until organic matter to ionic inorganic forms," having different membrane transporters and metabolic routes to transform and store the nitrogen for its own catabolism and anabolism. This kind of metabolic versatility to uptake this nutrient is especially desirable and appreciated in wastewater treatment systems, particularly improving the efficiency and competitivity against traditional wastewater treatments. There is an extra benefit because of the biomass that could be used as raw material in diverse industrial processes supplying nutrients, biostimulants, biodegradable plastics, combustibles, drugs, and other industrial demands.

References

APHA, Water Environment Federation, & American Water Works Association. (1999). Standard methods for the examination of water and wastewater part 4000 inorganic nonmetallic constituents standard methods for the examination of water and wastewater. In Standard methods for the examination of water and wastewater

Bhaskar S, Kuldeep B, Faizal B (2015) Algae and environmental sustainability. Irrigation Drainage 66(3). https://doi.org/10.1002/ird.2161

Cai, T., Park, S. Y., & Li, Y. (2013). Nutrient recovery from wastewater streams by microalgae: Status and prospects. Renew Sust Energ Rev, 19, 360–369. https://doi.org/10.1016/j.rser.2012.11.030

Chen Z, Shao S, He Y, Luo Q, Zheng M, Zheng M, Chen B, Wang M (2020) Nutrients removal from piggery wastewater coupled to lipid production by a newly isolated self-flocculating microalga Desmodesmus sp. PW1. Bioresour Technol 302(January):122806. https://doi.org/10.1016/j.biortech.2020.122806

Chu HQ, Tan XB, Zhang YL, Yang L, Bin ZFC, Guo J (2015) Continuous cultivation of Chlorella pyrenoidosa using anaerobic digested starch processing wastewater in the outdoors. Bioresour Technol 185:40–48. https://doi.org/10.1016/j.biortech.2015.02.030

De-Bashan L, Bashan Y (2003) Bacterias promotoras de crecimiento de microalgas: una nueva aproximación en el tratamiento de aguas residuales. Rev Colomb Biotecnol 5:85–90

De-Bashan L, Antoun H, Bashan Y (2008) Involvement of indole-3-acetic acid produced by the growth-promoting Bacterium azospirillum spp. in promoting growth of Chlorella vulgaris. J Phycol 44(4):938–947

Delgadillo-Mirquez L, Lopes F, Taidi B, Pareau D (2016) Nitrogen and phosphate removal from wastewater with a mixed microalgae and bacteria culture. Biotechnol Reports 11:18–26. https://doi.org/10.1016/j.btre.2016.04.003

Devi MP, Subhash GV, Mohan SV (2012) Heterotrophic cultivation of mixed microalgae for lipid accumulation and wastewater treatment during sequential growth and starvation phases: effect of nutrient supplementation. Renew Energy 43:276–283

Ferreira, A., Ribeiro, B., Marques, P. A. S. S., Ferreira, A. F., Dias, A. P., Pinheiro, H. M., Reis, A., & Gouveia, L. (2017). Scenedesmus obliquus mediated brewery wastewater remediation and CO2 biofixation for green energy purposes. J Clean Prod, 165, 1316–1327. https://doi.org/10.1016/j.jclepro.2017.07.232

Gao F, Li C, Yang ZH, Zeng GM, Feng L, Liu J, Zhi J, Liu M, Cai HW (2016) Continuous microalgae cultivation in aquaculture wastewater by a membrane photobioreactor for biomass production and nutrients removal. Ecol Eng 92:55–61. https://doi.org/10.1016/j.ecoleng.2016.03.046

Giraldo M (2012) Aislamiento y caracterización de microalgas formadoras de tapetes microbianos asociados a un cultivo hidropónico de plantas halófitas. Universidad de Las Palmas de Gran Canaria. http://acceda.ulpgc.es/bitstream/10553/6792/4/0654092_00000_0000.pdf

Gorain PC, Paul I, Bhadoria PS, Pal R (2019) An integrated approach towards agricultural wastewater remediation with fatty acid production by two cyanobacteria in bubble column photobioreactors. Algal Res 42:101594. https://doi.org/10.1016/j.algal.2019.101594

Goyena R, Fallis A (2019) The molecular biology of cyanobacteria. J Chem Inf Model 53(9). https://doi.org/10.1017/CBO9781107415324.004

Halfhide T, Dalrymple OK, Wilkie AC et al (2015) Growth of an indigenous algal consortium on anaerobically digested municipal sludge centrate: photobioreactor performance and modeling. Bionergy Res 8(1):249–258. https://doi.org/10.1007/s12155-014-9513-x

Hena S, Fatimah S, Tabassum S (2015) Cultivation of algae consortium in a dairy farm wastewater for biodiesel production. Water Resources and Industry 10:1-14. https://doi.org/10.1016/j.wri.2015.02.002

Hongyang S, Yalei Z, Chunmin Z, Xuefei Z, Jinpeng L (2011) Cultivation of Chlorella pyrenoidosa in soybean processing wastewater. Bioresour Technol 102(21):9884–9890. https://doi.org/10.1016/j.biortech.2011.08.016

Imase M, Watanabe K, Aoyagi H, Tanaka H (2008) Construction of an artificial symbiotic community using a Chlorella-symbiont association as a model. FEMS Microbiol Ecol 63(3):273–282. https://doi.org/10.1111/j.1574-6941.2007.00434.x

Jebali A, Acién FG, Gómez C, Fernández-Sevilla JM, Mhiri N, Karray F, Dhouib A, Molina-Grima E, Sayadi S (2015) Selection of native Tunisian microalgae for simultaneous wastewater treatment and biofuel production. Bioresour Technol 198:424–430. https://doi.org/10.1016/j.biortech.2015.09.037

Ji F, Zhou Y, Pang A, Ning L, Rodgers K, Liu Y, Dong R (2015) Fed-batch cultivation of Desmodesmus sp. in anaerobic digestion wastewater for improved nutrient removal and biodiesel production. Bioresour Technol 184:116–122. https://doi.org/10.1016/j.biortech.2014.09.144

Kim BH, Kang Z, Ramanan R, Choi JE, Cho DH, Oh HM, Kim K (2014) Nutrient removal and biofuel production in high rate algal pond using real municipal wastewater. J Microbiol Biotechnol 24:1123–1132

Kumar A, Bera S (2020) Revisiting nitrogen utilization in algae: a review on the process of regulation and assimilation. Bioresource Technol Reports 12(October):100584. https://doi.org/10.1016/j.biteb.2020.100584

Kumar R, Goyal D (2008) Comparative biosorption of Pb2+ by live algal consortium and immobilized dead biomass from aqueous solution. Indian J Exp Biol 47(8):690–694

Larsdotter K (2006) Microalgae for phosphorus removal from wastewater in a Nordic climate. PhD Thesis, School of Biotechnology, Royal Institute of Technology, Stockholm, Sweden, ISBN: 91-7178-288-5

Lavrinovičs A, Murby F, Zīverte E, Mežule L, Juhna T (2021) Increasing phosphorus uptake efficiency by phosphorus-starved microalgae for municipal wastewater post-treatment. Microorganisms 9(8):1598. https://doi.org/10.3390/microorganisms9081598. PMID: 34442678; PMCID: PMC8399584

Lee CS, Lee SA, Ko SR, Oh HM, Ahn CY (2015) Effects of photoperiod on nutrient removal, biomass production, and algal-bacterial population dynamics in lab-scale photobioreactors treating municipal wastewater. Water Res 68:680–691

Lopez Ponte WM, Zamora Talaverano N, Oscanoa Huaynate A, Cafferata EA, Cervantes Gallegos M (2022) Efficiency of microalgae cultures for nutrient removal from domestic wastewater. Adv Environ Technol 1:73–81

Mao Y, Xiong R, Gao X, Jiang L, Peng Y, Xue Y (2021) Analysis of the status and improvement of microalgal phosphorus removal from municipal wastewater. PRO 9:1486. https://doi.org/10.3390/pr9091486

Markou G, Chatzipavlidis I, Georgakakis D (2012) Cultivation of Arthrospira (Spirulina) platensis in olive-oil mill wastewater treated with sodium hypochlorite. Bioresour Technol 112:234–241. https://doi.org/10.1016/j.biortech.2012.02.098

Márquez BM (2013) "EFECTO DEL ÁCIDO INDOL-3-ACÉTICO PRODUCIDO POR Azospirillum brasilense EN LAS ENZIMAS DE ASIMILACIÓN DE AMONIO EN Chlorella vulgaris, BAJO CONDICIONES DE CO- INMOVILIZACIÓN.", *Centro de investigacionesbiológicas del norte*, http://cybertesis.cibnor.mx:8080/sdx/cibnor/2013/meza_b.pdf

McGinn PJ, Dickinson KE, Park KC, Whitney CG, MacQuarrieb SP, Black FJ, Frigon JC, Guiot SR, O'Leary SJB (2012) Assessment of the bioenergy and bioremediation potentials of the microalga Scenedesmus sp. AMDD cultivated in municipal wastewater effluent in batch and continuous mode. Algal Res 1(2):155–165. https://doi.org/10.1016/j.algal.2012.05.001

Mohsenpour SF, Hennige S, Willoughby N, Adelove A, Gutierrez T (2021) Integrating micro-algae into wastewater treatment: a review. Sci Total Environ 752:142168. https://doi.org/10.1016/j.scitotenv.2020.142168

Mustafa EM, Phang SM, Chu WL (2012) Use of an algal consortium of five algae in the treatment of landfill leachate using the high-rate algal pond system. J Appl Phycol 24(4):953–963

Pachés M, Martínez-Guijarro R, González-Camejo J, Seco A, Barat R (2018) Selecting the most suitable microalgae species to treat the effluent from an anaerobic membrane bioreactor. Environ Technol 41:267–276

Papadopoulos KP, Economou CN, Tekerlekopoulou AG, Vayenas DV (2020) Two-step treatment of brewery wastewater using electrocoagulation and cyanobacteria-based cultivation. J Environ Manag 265:110543. https://doi.org/10.1016/j.jenvman.2020.110543

Ramírez Mérida (2020) Agrobiología. Una visión general y sus aplicaciones. Mérida Publisher CC-BY

Rees TAV, Cresswell RC, Syrett PJ (1980) Sodium-dependent uptake of nitrate and urea by a marine diatom. Biochim Biophys Acta 596(1):141–144

Rengifo A, Pena E, Benítez N (2012) Efecto de la asociación alga-bacteria Bostrychia calliptera (Rhodomelaceae) en el porcentaje de remoción de cromo en laboratorio. Biología Tropical 60:1055–1064

Ross ME, Davis K, McColl R, Stanley MS, Day JG, Semião AJC (2018) Nitrogen uptake by the macro-algae Cladophora coelothrix and Cladophora parriaudii: influence on growth, nitrogen preference and biochemical composition. Algal Res 30:1–10. https://doi.org/10.1016/j.algal.2017.12.005

Ruiz J, Álvarez-Díaz PD, Arbib Z, Garrido-Pérez C, Barragán J, Perales JA (2013) Performance of a flat panel reactor in the continuous culture of microalgae in urban wastewater: prediction from a batch experiment. Bioresour Technol 127:456–463. https://doi.org/10.1016/j.biortech.2012.09.103

Salbitani G, Carfagna S (2021) Ammonium utilization in microalgae: a sustainable method for wastewater treatment. Sustainability 13(2): 17. https://doi.org/10.3390/su13020956

Sandoval Salazar IA, Darwich-Cedeño MT, Castañeda MDR, Torres Munar WA, Montenegro Ruiz LC (2021) Aproximación al tratamiento de aguas residuales del lavado del café con las microalgas Parachlorella kessreli y Desmodesmus armatus. Revista Mutis 11(2):32–43. https://doi.org/10.21789/22561498.1755

Sepehri A, Sarrafzadeh MH, Avateffazeli M (2020) Interaction between Chlorella vulgaris and nitrifying-enriched activated sludge in the treatment of wastewater with low C/N ratio. J Clean Prod 247. https://doi.org/10.1016/j.jclepro.2019.119164

Su Y (2020) Revisiting carbon, nitrogen, and phosphorus metabolisms in microalgae for wastewater treatment. Sci Total Environ 144590. https://doi.org/10.1016/j.scitotenv.2020.144590

Su Y, Mennerich A, Urban B (2011) Municipal wastewater treatment and biomass accumulation with a wastewater-born and settleable algal-bacterial culture. Water Res 45(11):3351–3358

Takáčová A (2015) Degradation of btex by microalgae P. kessleri. Petroleum & Coal 57(2):101–107

Tebbani S, Lopes F, Filali R, Dumur D, Pareau D (2014) CO_2 biofixation by microalgae: modeling, Estimation and Control In Co (Vol. 9781848215986). https://doi.org/10.1002/9781118984475

Vega JM (2020) Nitrogen and sulfur metabolism in microalgae and plants: 50 years of research. in: Progress in botany Vol. 81, (Eds.) F.M. Cánovas, U. Lüttge, C. Leuschner, M.-C. Risueño, Springer. Cham, 1–40

Viera Valencia LF, Garcia Giraldo D (2019) Algae for biofuels and energy book 2013. Angew Chem Int Ed 6(11):951–952

Voltolina D, Arredondedo B (2015) Métodos y herramientas analíticas en la evaluación de la biomasa microalgal. In D Voltolina & B Arredondedo (eds.); 1st ed., Issue September). http://www.researchgate.net/publication/265198034

Wang Y, Wang S, Sun L, Sun Z, Li D (2020) Screening of a Chlorella-bacteria consortium and research on piggery wastewater purification. Algal Res 47(October 2019):101840. https://doi.org/10.1016/j.algal.2020.101840

Watanabe K, Takihana N, Aoyagi H, Hanada S, Watanabe Y, Ohmura N, Saiki H, Tanaka H (2005) Symbiotic association in Chlorella culture. FEMS Microbiol Ecol 51(2):187–196. https://doi.org/10.1016/j.femsec.2004.08.004

Wu L, Zhu Q, Yang L, Li B, Hu C, Lan S (2018) Nutrient transferring from wastewater to desert through artificial cultivation of desert cyanobacteria. Bioresour Technol 247:947–953. https://doi.org/10.1016/j.biortech.2017.09.127

Yunes JS (2018) Cyanobacterial toxins. In Cyanobacteria: from basic science to applications. Elsevier. https://doi.org/10.1016/B978-0-12-814667-5.00022-2

Wastewater Treatment by Microalgae and Cyanobacteria. The Reduction of the Contaminant Potential of Real Domestic Wastewater: A Case Study

4

Alfredo de Jesús Martínez-Roldán, Rebeca Paola Villanueva-Garcia, María Dolores Josefina Rodríguez Rosales, Sergio Valle Cervantes, and Hugo Virgilio Perales Vela

Abstract

Microalgae can consume high amounts of inorganic nutrients, such as nitrogen and phosphorus, from culture media. This characteristic can be exploited to eliminate nitrogen and phosphorus present in wastewater, and the amount of these contaminants can vary depending on the source. For domestic wastewater, the amounts of nitrogen, phosphorus, and organic matter are low, and its treatment by microalgal cultures can reach high efficiencies. This chapter discusses the treatment of domestic wastewater from Durango City, Mexico, using cultures of *Synechocystis aquatilis* and *Stigeoclonium nanum*. The results showed an efficient consumption of nitrogen and phosphorus and a reduction of organic matter (as chemical oxygen demand) and pathogenic bacteria; improvements in pH, conductivity, color, and odor were also observed. Therefore, the use of microalgae and cyanobacteria cultures to treat wastewater from small cities is proposed because of the reduced pollutant potential of the effluent achieved through a simple process.

Keywords

Microalgae · Municipal wastewater · *Stigeoclonium nanum* · *Synechocystis aquatilis* · Phycoremediation

4.1 Introduction

Wastewater (WW) contains diverse types of contaminants, many of which are toxic and can pose health risks. To eliminate these pollutants, technologies have been developed based on their characteristics, degradability, and toxicity (Droste and Gehr 2018; Lum et al. 2013; Rawat et al. 2011). The treatment of WW can be carried out using three kind of processes according to the strategy of removing or degrading the contaminants: physical, chemical, and biological. Physical and chemical technologies are efficient in reducing the chemical oxygen demand (COD) and removing compounds, such as heavy metals and colorants, but are less effective in eliminating nitrogen and inorganic phosphorus. Technologies that use live organisms such as bacteria, fungi, yeast, plants, and macro- or microalgae are preferred for the removal of nitrogen and phosphorus (Abdel-Raouf et al. 2012; Droste and Gehr 2018; Lum et al. 2013) because of the high requirements of these biogenic elements.

Microalgae are efficient microorganisms for the removal nitrogen, regardless of the oxidation state (ammonium, nitrate, urea, or organic nitrogen), and phosphorus; they are also capable of removing other pollutants such as organic matter or metal ions (Al, Fe, Ca, Mg, Mn, etc.) (Abdel-Raouf et al. 2012; Flórez-Miranda et al. 2017; Franco Martínez et al. 2017; Lum et al. 2013; Mohamed Ramli et al. 2017; Wang et al. 2009). N-NH_3 can be efficiently and rapidly eliminated by microalgal cultures, reaching a volumetric rate of 5–11 mg L^{-1} d^{-1} (Franco Martínez et al. 2017; Rawat et al. 2011; Wang et al. 2009). N-NO_3 can also be quickly eliminated; however, generally only when the concentration of N-NH_3 is low (Xie et al. 2013), because of the energy consumption by the enzymatic reduction of N-NO_3 before it

A. de Jesús Martínez-Roldán (✉)
Investigador Nacional Nivel I, Catedrático CONAHCYT - TecNM/ITD, Departamento de Ingenierías Química y Bioquímica, División de Estudios de Posgrado, Maestría en Sistemas Ambientales, Durango, Mexico

Tecnológico Nacional de México/IT de Durango, Maestría en Sistemas Ambientales, División de Posgrado e Investigación, Durango, Dgo, Mexico
e-mail: adjmartinezro@conacyt.mx

R. P. Villanueva-Garcia · M. D. J. Rodríguez Rosales · S. Valle Cervantes
Tecnológico Nacional de México/IT de Durango, Maestría en Sistemas Ambientales, División de Posgrado e Investigación, Durango, Dgo, Mexico

H. V. Perales Vela
Facultad de Estudios Superiores Iztacala, UNAM, Laboratorio de Bioquímica, Unidad de Morfología y Función, Tlalnepantla, Mexico

A. de Jesús Martínez-Roldán (ed.), *Biotechnological Processes for Green Energy, and High Value Bioproducts by Microalgae, and Cyanobacteria Cultures*, Developments in Applied Phycology 13, https://doi.org/10.1007/978-3-031-43969-8_4

enters the cell. In the case of P-PO_4, the consumption rate is only 1–2 mg L^{-1} d^{-1} (Franco Martínez et al. 2017), since the amount of phosphorus in the biomass is lower (approximately 1%). Nevertheless, the cell consumes high amounts of phosphorus and the excess from polyphosphate in the cytosol accumulates, which is a strategy to deal with periods of phosphorus deficiency without affecting metabolic functioning (Jácome-Pilco et al. 2009; Perales-Vela et al. 2007).

4.2 Microalgae in the Tertiary Treatment of Wastewater and the Elimination of Specific Pollutants

4.2.1 Microalgae Versus Nitrogen and Phosphorus

Microorganisms such as microalgae require macro- and micronutrients for their growth, including carbon as a macronutrient that can be obtained from inorganic (carbon dioxide) or organic sources depending on the active metabolic route (autotrophic or heterotrophic) (Martínez-Roldán and Cañizares-Villanueva 2022). For nitrogen, many sources with different oxidation states can be used, such as nitrate, ammonia, nitrite, or organic compounds, whereas phosphorus is used in the form of phosphate (Franco Martínez et al. 2017). In WW treatment, bacteria and fungi are effective for the removal of organic matter, but other contaminants such as nitrogen and phosphorus are not greatly affected; however, microalgae can reduce these elements, and under some conditions, can completely eliminate them from the effluent (Franco Martínez et al. 2017; Martínez-Roldán and Cañizares-Villanueva 2017).

4.2.2 Microalgae in the Tertiary Treatment of Urban Wastewater

The metabolic diversity of microalgae is an important factor to consider in the phycoremediation of WW, because these organisms use autotrophic metabolism (fixing CO_2), organic carbon compound consumption (heterotrophy), or a mixture of both (mixotrophy). This metabolic flexibility allows microalgae to remove organic matter in the treatment of WW (Perez et al. 2015). In addition, microalgae can increase the efficiency of aerobic processes in facultative or aerobic lagoons because they can generate symbioses with aerobic bacteria. Bacteria produce CO_2 that is used by microalgae during photosynthesis, and microalgae produce oxygen, stimulating the metabolism of the bacteria; this symbiosis increases the efficiency of the lagoons (Oswald et al. 1957; Oswald and Golueke 1960; Oswald William and Gotaas 1957; Shelef et al. 1984). Symbiosis with aerobic bacteria and the ability to consume organic matter and remove nitrogen and phosphorus potentiate microalgae as useful tools for the treatment of different types of WW, especially if the organic matter concentration is not high (Abdel-Raouf et al. 2012; Franco Martínez et al. 2017; Wu et al. 2014). In addition, microalgae have a specific niche in the tertiary treatment of WW, as they eliminate contaminants without affecting the anaerobic and aerobic processes based on bacteria. Some contaminants that can be eliminated are nitrogen and phosphorus in the polish phase from domestic effluents previously treated to remove organic matter. Table 4.1 shows some studies focused on the elimination of these pollutants from WW.

The employment of microalgae and cyanobacteria reduces organic matter, COD, nitrate, ammonia, phosphorus, and other compounds. Microalgae metabolism also increases the pH and dissolved oxygen in the effluent, which reduces the concentration of pathogens, and can completely eliminate coliforms (Franco Martínez et al. 2017; Martínez-Roldán and Ibarra-Berumen 2019). The interaction of the cells with the effluent favors the elimination of other contaminants, including metallic ions, healthcare products, medicaments, and hormones, through adsorption/absorption processes or mineralization, although in some cases, biotransformation is required (Franco Martínez et al. 2017; Moondra et al. 2020; Rani et al. 2021).

WW can include contaminants that microalgae are not able to degrade, but adsorption (by a superficial interaction with the charge of the cell wall) or absorption (into the cell and detoxification by different strategies) can be used to remove these compounds. These contaminants include heavy metal ions, colorants, and molecules used for healthcare called emerging contaminants (Escapa et al. 2015; Keen et al. 2014).

Metallic ions can be present in WW from different sources, such as industry and old domestic pipes, and they can reach high concentrations in WW plants, principally in water and sludge (Abou 2011). Microalgae are useful for the elimination of heavy metals because their cellular walls are negatively charged, which allows them to interact with positive metallic ions, favoring adsorption. This process employs both living and dead biomass, occurs rapidly, and removes large amounts of metals (Jácome-Pilco et al. 2009). Live biomass allows the absorption process to increase the number of ions that can be eliminated (Cañizares-Villanueva 2000). Nevertheless, the absorption process is slower, and the heavy metals are translocated inside the cell for neutralization in vacuoles, chloroplasts, nuclei, or the cytoplasm by the production of metallothionein and phytochelatins, where they remain chelated, reducing their toxicity (Goswami et al. 2021). Adsorption and absorption occur simultaneously in living cells, but the amount of metal removed by adsorption is many times higher than that by absorption (Cañizares-

Table 4.1 Tertiary treatment of effluents employing microalgae and cyanobacteria

Effluent type	Benefits	Operational conditions
Tertiary effluent	Reduction of N-NO_3 (53%) and COD (70%)	8 d HRT
Domestic secondary WW	Reductions in N-NO_3, P-PO_4, and N-NH_4	Immobilized in alginate Different HRT
Municipal WW	Removal of nitrate and ammonia, reduction of COD (80%) Removal of *Salmonella* and *Shigella*	Microalgae + sludge
Secondarily treated water	Daily: 2.86 g COD/day, 0.12 g PO_4-P/day, 0.82 g NH_4-N/day, 0.13 g NO_3-N/day, 0.88 g total nitrogen/day	*Spirulina platensis* Open raceway pond
Synthetic WW simulating a secondary treatment effluent	Daily removal rates of 11.7 mg N gVSS^{-1} d^{-1} and 3.04 mg P gVSS^{-1} d^{-1}	150-L closed tubular photobioreactor
Domestic WW	Removal of COD and ammonia higher than 80%	Different HRT

Villanueva 2000; Jácome-Pilco et al. 2009). The capability of microalgae to remove metallic ions from WW has been demonstrated in many studies. Table 4.2 lists some reports related to the elimination of metals from liquid effluents.

In wastewater, there are other types of contaminants that cannot be removed by aerobic or anaerobic processes because they are recalcitrant or because the degradation products are more toxic than the original; these compounds are colorants. These molecules have the capability to absorb a fraction of the light and reflect the rest, which confers color on the surface. The chemical composition of colorants is diverse, although all have high molecular sizes and double chemical bonds or aromatic rings, which hinder their mineralization (Ferreira et al. 2022). The removal and degradation of colorants is possible using microalgae and cyanobacteria; some examples are listed in Table 4.3.

Recently, the presence of additional pollutants has been reported because their concentrations have increased and it is now possible to quantify them. The concentrations of these compounds are still low, but their presence in urban wastewater has been demonstrated; they are referred to as emerging contaminants (Peña-Guzmán et al. 2019). The existence of these contaminants in wastewater is caused by the consumption of chemical products such as plasticizers, perfluorinated surfactants, pesticides, surfactants, pharmaceuticals, personal care products, drugs, food additives, artificial sweeteners, flame retardants, and off-flavor compounds, and more recently disinfection byproducts, such as benzothiazoles and benzotriazoles (Pal et al. 2014; Peña-Guzmán et al. 2019). These pollutants are present in wastewater at very low concentrations, with the highest at approximately 100 μg L^{-1}, although their accumulation in water bodies can cause different and severe problems (Pal et al. 2014). Microalgae have demonstrated their potential to eliminate these types of contaminants, and the results of some investigations are listed in Table 4.4.

Wastewater that contains high amounts of emerging pollutants cannot be treated by biodegradation (as mineralization) or removal (adsorption or absorption), but only by biotransformation. This creates a challenge because certain products can cause problems in the environment or create higher toxicity than that of the original molecule. This is observed mainly with colorants (aniline derivatives) or medical compounds such as antibiotics, hormones, antivirals, antifungal, and endocrine disrupters (Min et al. 2018). It is important to track these products from their biotransformation by employing a sensitive apparatus such as nuclear magnetic resonance or mass spectrometer (Jain et al. 2022; Rempel et al. 2022).

4.3 The Reduction of the Contaminant Potential of Real Domestic Wastewater: A Case Study

This case study shows the results obtained using a microalgae-based process to eliminate nitrogen and phosphorus from real domestic wastewater in Durango City, Mexico. A green microalga (*Stigeoclonium nanum*) and a cyanobacterium (*Synechocystis aquatilis*) were used. The objective of this study was to evaluate the potential of both microalgae to reduce COD, eliminate pollutants such as nitrate, ammonia, and phosphate, and decrease the number of coliforms present in wastewater that has had no previous biological treatment. The experiments were conducted under laboratory conditions using an airlift photobioreactor (PBR). The results demonstrated the potential of both microalgal strains to be employed in the treatment of domestic wastewater with substantial elimination of nitrogen and phosphorus from the effluents and important reductions in COD and coliforms.

4.4 Materials and Methods

Native strains of *Stigeoclonium nanum* (ITD001) and the cyanobacterium *Synechocystis aquatilis* (ITD005) were isolated from the clarifier of a wastewater treatment plant (WWTP) in Durango City, Mexico. The inoculum was grown in mineral medium without an organic carbon source (BG11) (CCAP 2019). In the first phase, the growth of both

Table 4.2 Removal of metallic ions by microalgae and cyanobacteria (Goswami et al. 2021; Leong and Chang 2020; Pavithra et al. 2020)

Species/Genera	Metal	Removal efficiency
Ulothrix cylindricum	Arsenic	2.45 mg g^{-1} maximal sorption 98% removal efficiency
Spirulina sp.	Arsenic	365 mg g^{-1} maximal sorption 60.2% removal efficiency
Spirulina platensis	Chromium (VI)	56.9 mg g^{-1} maximal sorption
Phormidium sp.	Lead	2.3 mg/g^{-1} maximal sorption 92% removal efficiency
Spirogyra sp.	Mercury	0.25 mg g^{-1} maximal sorption 76% removal efficiency
Scenedesmus almeriensis	Arsenic	5 mg g^{-1} maximal sorption 41.7% removal efficiency
Scenedesmus quadricauda	Chromium (VI)	100% removal efficiency (initial concentration of 10 mg L^{-1})
Scenedesmus sp.	Zinc Iron	84% removal efficiency 65% removal efficiency
Chlorella vulgaris	Arsenic	3.89 mg g^{-1} maximal sorption 32.4% removal efficiency
Chlorella vulgaris	Chromium (VI)	140 mg g^{-1} maximal sorption 56% removal efficiency
Chlorella vulgaris	Mercury	17.49 mg g^{-1} maximal sorption 72.9% removal efficiency
Chlorella vulgaris	Nickel (II)	
Chlorella minutissima	Cadmium	303 mg g^{-1} maximal sorption
Chlorella minutissima	Copper Manganese Cadmium Zinc	83.68% removal efficiency 83.68% removal efficiency 74.34% removal efficiency 62.05% removal efficiency
Chlorella sp.	Lead	10.4 mg g^{-1} maximal sorption 78% removal efficiency
Chlorella sp.	Mercury	3.3 mg g^{-1} maximal sorption 12.5% removal efficiency
Chaetoceros sp.	Lead	8 mg g^{-1} maximal sorption 60% removal efficiency
Chaetoceros sp.	Lead Cadmium Copper	90% removal efficiency 62% removal efficiency 83% removal efficiency
Desmodesmus sp.	Cadmium	>58% removal efficiency
Mixed culture (*Nannochloropsis, Pavlova, Tetraselmis, Chaetoceros*)	Municipal leachate and saline effluent	95% removal efficiency

Table 4.3 Elimination of colorants by microalgae and cyanobacteria (Carbajo Arteaga et al. 2018; Chu and Phang 2019; Hernández-Zamora et al. 2015; Premaratne et al. 2021)

Species/Genus	Dye	Initial concentration	Efficiency
Chlorella vulgaris	Aniline blue dye	100 mg L^{-1}	Removal ~60%
Chlorella vulgaris	Congo red	5–25 mg L^{-1}	Removal 58–83%
Chlorella pyrenoidosa	Textile wastewater	Not reported	Decolorization 100%
Chlorella sp.	Malachite green	2–20 mg L^{-1}	Biosorption capacity 9.45–33.7 mg g^{-1}
Scenedesmus quadricauda	Indigo blue	12–75 mg L^{-1}	Decolorization 100%
Micractinium sp.	Textile wastewater	Not reported	Decolorization 70%
Chlamydomonas sp.	Textile dye effluent	Not reported	Decolorization 100%
Microspora sp.	Methylene blue	20–2500 mg L^{-1}	Biosorption capacity 107–139 mg g^{-1}
Pithophora sp.	Malachite green	20–100 mg L^{-1}	Biosorption capacity 117.6 mg g^{-1}
Spirulina platensis	Methylene blue	30–200 mg L^{-1}	> 80% dye removal
Spirogyra sp.	Azo dye	5–15 mg L^{-1}	Removal 35.3–64%
Ulothrix zonata (pyrolyzed biomass)	Malachite green	500 mg L^{-1}	Biosorption capacity 5306 mg g^{-1}
Mixed microalgal culture	Not reported	Not reported	60–72% Decolorization of the effluent

Table 4.4 Elimination of emerging contaminants by microalgae

Emerging contaminant	Microalgae	Main conclusion
Diclofenac	*Scenedesmus* sp. *Synechocystis* sp.	Adsorption capacity of 28 and 22 mg g^{-1}
Salicylic acid Paracetamol	*Chlorella sorokiniana*	Removal 67% for paracetamol and 73% for salicylic acid
Carbamazepine	*Chlamydomonas mexicana* *Scenedesmus obliquus*	Degradation 21–35% Degradation 13–28%
Metropolol	*Selenastrum capricornutum* *Chlorella vulgaris* *Scenedesmus quadricauda* *Haematococcus pluvialis*	Important removal
Diclofenac Ibuprofen Paracetamol Trimethoprim Carbamazepine	*Chlorella sorokiniana*	Removal of 60–40% 100% 100% 60% 30%
Ciprofloxacin Sulfadiazine	*Chlamydomonas* sp.	Removal of 65% 17%
Progesterone	*Scenedesmus obliquus*	Removal of 95%
Caffeine	*Chlamydomonas acidophila*	Removal of 99%
Erythromycin	*Chlamydomonas acidophila*	Removal 93–65%
Clarithromycin	*Chlamydomonas acidophila*	Removal 64–50%
Levofloxacin	*Chlorella vulgaris*	2.8–92.3%

strains and their consumption of $N\text{-}NO_3$ and $P\text{-}PO_4$ occurred under laboratory conditions by employing an airlift PBR constructed in acrylic (Osuna Salazar et al. 2016) for eight days. The cultures were inoculated with an exponential-growing inoculum at 15% and exposed to white light (OSRAM DULUX L Daylight, Germany) at an irradiance of 125 μE m^{-2} s^{-1}. The cultures were bubbled with air without CO_2 enrichment at 0.5 vvm and maintained at 25 °C. The culture was traced by daily determination of dry weight (DW) and total chlorophyll, $N\text{-}NO_3$, and $P\text{-}PO_4$ concentrations. DW was determined by filtering a specific volume of culture through a fiberglass Whatman® filter that was previously weighed (Martínez-Roldán et al. 2014). Total chlorophyll was extracted with methanol at 100 °C and the concentration was calculated using the equations reported by Wellburn (Wellburn 1994). Using the cell-free supernatant, the concentrations of $N\text{-}NO_3$ and $P\text{-}PO_4$ were estimated by spectrophotometric determination (Franco Martínez et al. 2017; Zhang and Fischer 2006).

In the second phase, the growth and nutrient consumption of both species were evaluated using WW from the WWTP of Durango City. The WW was obtained from primary treatment (sedimentation process) and was passed through a paper filter to remove larger suspended solids. These experiments were performed using a PBR (Osuna Salazar et al. 2016) under the same operational conditions as those used in the first phase. The cultures were traced using total chlorophylls, as well as $N\text{-}NO_3$ and $P\text{-}PO_4$ concentrations using the previously mentioned techniques. $N\text{-}NH_3$ was quantified using the phenate method (Franco Martínez et al. 2017). The initial and final estimations of total coliforms, pH, conductivity, and COD were performed according to standard methods (APHA, AWWA,, and WEF 1999). Statistical analyses were performed using GraphPad Prism 5 (GraphPad Software, Inc., USA). First, a two-way ANOVA was implemented, followed by a Bonferroni post hoc analysis.

4.5 Results and Discussion

4.5.1 Characterization of the Growth of *S. aquatilis* and *S. nanum* in a Mineral Medium

BG11 medium has high concentrations of nutrients (nitrogen and phosphorus), which allows the microalgae and cyanobacteria to grow rapidly, reaching biomass concentrations above 0.500 g L^{-1}. In the case of DW (Fig. 4.1), both strains grew until day eight of culture. *S. aquatilis* was initiated with a biomass concentration of 0.08 ± 0.016 g L^{-1} and reached a final concentration of 0.49 ± 0.083 g L^{-1}, which represents a biomass gain of 0.410 g L^{-1} during eight days of culture. Using the biomass gain and the time required to attain this gain, the biomass productivity was calculated at 0.050 g $L^{-1}d^{-1}$; this value is similar to that reported by other authors for *Synechocystis* species in the range of 40 to 120 mg L^{-1} d^{-1} (Cai et al. 2013; Gonçalves et al. 2016; Touloupakis et al. 2016).

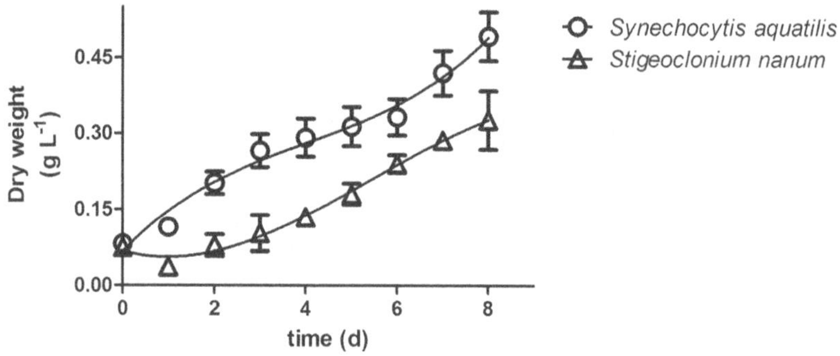

Fig. 4.1 Dry weight concentration during growth in the BG11 medium

The *S. nanum* culture started with a biomass concentration of 0.08 ± 0.007 g L^{-1} and reached a final concentration of 0.33 ± 0.100 g L^{-1}, representing a biomass gain and biomass productivity of 0.251 g L^{-1} and 0.032 g L^{-1} d^{-1}, respectively. Mohamed Ramli et al. (2017) reported final biomass concentrations close to 0.200 g L^{-1} after 12 days of culture, whereas on the eighth day, the concentration was 0.100 g L^{-1}. The biomass obtained on the eighth day was 150% higher than the value reported by Ramli et al. because the medium used was basal bold medium, which had lower nutrient concentrations than that of the BG11 medium used here (CCAP 2019; Mohamed Ramli et al. 2017).

The biomass concentration during culture was used to calculate the specific growth rates (μ) of 0.279 d^{-1} and 0.247 d^{-1} for *S. aquatilis* and *S. nanum*, respectively. The difference between these values is not large; however, the final biomass concentration of *S. aquatilis* was 20% higher than that of *S. nanum*. This reduction was due to the decreased biomass concentration of *S. nanum* during the first 24 h. Internal loop airlift PBRs (as used in this study) are useful for the production of unicellular and coenobium-forming microorganisms; however, their use in the culture of filament microalgae such as *S. nanum* can cause cellular lysis by shear stress (A. J. Martínez-Roldán et al. 2019). Nevertheless, the stress produced in the PBR was not high, and after an adaptation period, *S. nanum* achieved substantial biomass accumulation, similar to reported values for this genus (Mohamed Ramli et al. 2017).

The behavior of chlorophyll concentration (Fig. 4.2) was very similar to that observed for DW, and higher values were observed with *S. aquatilis* than *S. nanum*. The final chlorophyll concentrations were 9.52 ± 1.652 and 6.39 ± 2.035 mg_{Chl} L^{-1} for *S. aquatilis* and *S. nanum*, respectively. The value achieved by *S. nanum* was 30% lower than that of *S. aquatilis*, and a similar reduction was observed with DW. This similarity between biomass and chlorophyll concentrations relates to the direct proportionality of the pigment to the biomass prior to exponential growth (Franco Martínez et al. 2017). The calculation of the specific chlorophyll content was performed by considering the biomass and chlorophyll concentrations during the entire experiment. The total chlorophyll content in *S. aquatilis* reached 19.44 ± 1.492 mg_{Chl} $g_{DW}{}^{-1}$ ($R^2 = 0.872$), whereas *S. nanum* showed higher values of 23.07 ± 1.606 mg_{Chl} $g_{DW}{}^{-1}$ ($R^2 = 0.892$). This was expected, since *S. nanum* is a green alga and usually has a specific content close to 22 mg_{Chl} $g_{DW}{}^{-1}$, while cyanobacteria have levels less than 20 mg_{Chl} $g_{DW}{}^{-1}$ (Franco Martínez et al. 2017; Touloupakis et al. 2016).

The behavior of the concentrations of nitrogen and phosphorus in the mineral medium is shown in Figs. 4.3 and 4.4. The initial concentration of N-NO_3 in both cultures was high at approximately 260 mg N-NO_3 L^{-1} (251.7 ± 6.17 and 288.5 ± 2.75 mg N-NO_3 L^{-1} for *S. aquatilis* and *S. nanum*, respectively). Considering the final concentration, the total consumption of N-NO_3 was calculated as 48.5 and

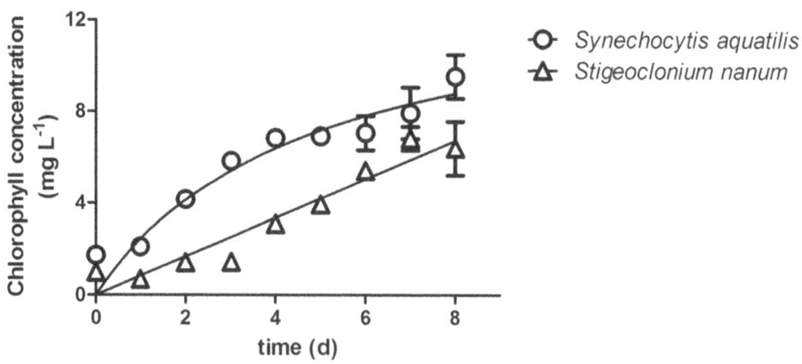

Fig. 4.2 Total chlorophyll concentration during growth in the BG11 medium

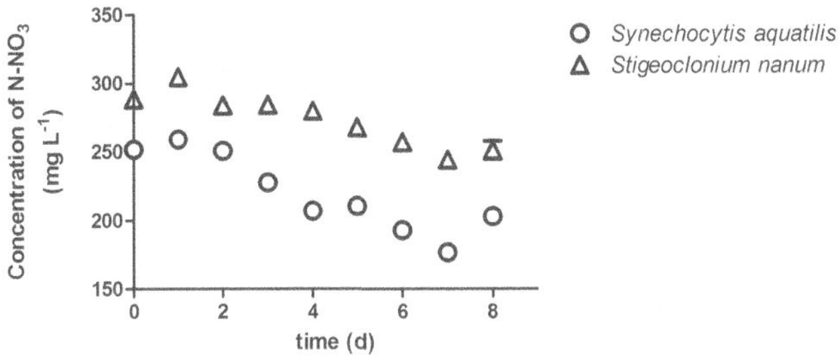

Fig. 4.3 N-NO3 concentration during growth in the BG11 medium

37.3 mg N-NO_3 L^{-1} for *S. aquatilis* and *S. nanum*, respectively. The consumption of this nutrient was directly linked to biomass production because the highest consumption amount corresponded to the culture that reached the higher biomass (*S. aquatilis*). The difference in the nitrogen absorbed was approximately 25%, and the final biomass concentration was 30% higher in *S. aquatilis*. Considering the days of culture, the volumetric consumption rate for nitrogen was calculated as 9.6 ± 0.93 and 6.7 ± 0.63 mg N-NO_3 L^{-1} d^{-1} for *S. aquatilis* and *S. nanum*, respectively. These values are similar to those reported for nitrate consumption under laboratory conditions, ranging from 5 to 10 mg N-NO_3 L^{-1} d^{-1} (Franco Martínez et al. 2017).

The amount of phosphorus consumed was lower than that of N. The initial concentration of P-PO_4 was 6.4 ± 1.07 and 6.5 ± 0.69 mg P-PO_4 L^{-1} for *S. aquatilis* and *S. nanum*, respectively; both strains reached P-PO_4 concentrations lower than 1.5 mg P-PO_4 L^{-1}, but at different times, whereby *S. aquatilis* had a minimal concentration of 0.96 ± 0.080 mg P-PO_4 L^{-1} on the fourth day; the minimum for *S. nanum* was 1.3 ± 0.19 mg P-PO_4 L^{-1} on the eighth day. Using P-PO_4 consumption and the time-lapse employed, the consumption rates were calculated as 1.6 ± 0.18 and 0.7 ± 0.03 mg P-PO_4 L^{-1} d^{-1}, for *S. aquatilis* and *S. nanum*, respectively. These values correspond to those reported for different species of microalgae under diverse culture conditions (Franco Martínez et al. 2017; Touloupakis et al. 2016) (Fig. 4.5).

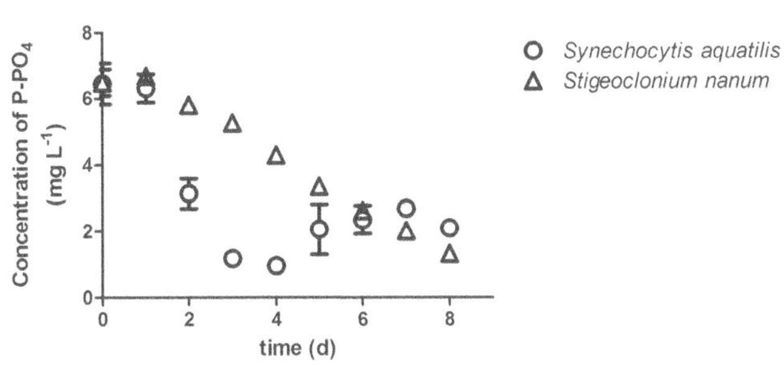

Fig. 4.4 P-PO4 concentration during growth in the BG11 medium

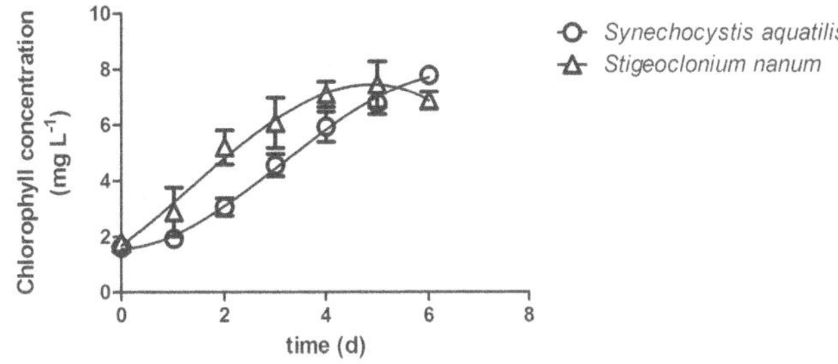

Fig. 4.5 Total chlorophyll concentration during growth in real wastewater

4.5.2 Characterization of *S. aquatilis* and *S. nanum* Growth in Real Wastewater from a WWTP in Durango City

In the next stage, the mineral medium was replaced with real WW from a WWTP, and water was collected from the sedimentation unit (primary treatment). Although only the WW was filtered, the concentration of contaminants was low because the majority of the WW that passed through the drain was domestic. Some of its parameters such as COD, coliforms, alkalinity, N-NO_3, and P-PO_4 are listed in Table 4.5. The COD, N-NO_3, N-NH_3, and P-PO_4 values permit its classification as domestic WW with low pollutant loads (McCarty et al. 2011; Rawat et al. 2011). Although the concentrations of nitrogen and phosphorus were low, they were sufficient to maintain significant growth of both species for at least five days. The presence of suspended solids made it difficult to determine the DW, and therefore, the tracing of the biomass was performed only by the quantification of the chlorophyll concentration (Fig. 4.5) based on the direct relationship observed between the biomass and chlorophyll in the first phase (Figs. 4.1 and 4.2).

Both strains began with a chlorophyll concentration close to 1.6 mg_{Chl} L^{-1} (7.9 ± 0.25 and 6.9 ± 0.44 mg_{Chl} L^{-1} for *S. aquatilis* and *S. nanum*, respectively); however, significant differences between the strains were observed on the second and third days of culture ($p < 0.05$) when the values for *S. nanum* were higher than that of *S. aquatilis*. Nevertheless, from the fourth day to the end of the experiment, the chlorophyll concentrations in both cultures were not significantly different ($p > 0.05$). The final Chl values were similar to those obtained in the experiments performed with the mineral medium (Fig. 4.2). These similar values prove that regardless of the low concentrations of nitrogen and phosphorus in the WW (Table 4.5), it was sufficient to allow for growth. In the experiments with BG11, a limitation of nutrients might have occurred because there were no differences between the values observed in WW and those in the mineral medium, and perhaps the limiting nutrient was light. Some authors affirm that cultures with Chl concentrations of 7–8 mg $_{Chl}$ L^{-1} are light-

Table 4.5 Initial and final characterization of the effluent

Parameter	Initial value	Final value	
		S. aquatilis	*S. nanum*
COD (mg L^{-1})	305.5 ± 94.89	94.3 ± 15.91	156 ± 35.00
pH	7.86 ± 0.10	11.0 ± 0.04	9.7 ± 0.01
Alkalinity (mg_{CaCO_3} L^{-1})	294.2 ± 6.24	68.3 ± 1.48	161.7 ± 2.12
Conductivity (μS cm^{-1})	808.3 ± 12.45	873.6 ± 15.49	857.4 ± 12.62
N-NO_3 ($mg_{N\text{-}NO3}$ L^{-1})	0.3 ± 0.03	20.1 ± 2.79	27.7 ± 2.67
N-NH_3 ($mg_{N\text{-}NH3}$ L^{-1})	11.7 ± 1.53	nd	nd
P-PO_4 ($mg_{P\text{-}PO4}$ L^{-1})	7.9 ± 0.17	1.7 ± 0.06	3.4 ± 0.12
Total coliforms (CFU (100 mL)$^{-1}$)	>2.4*10^8	1.1*10^3	9.36*10^2

nd Not detected, values below the quantification limit of the technique (0.1 mg L^{-1})

limited if the irradiance is lower than 300 μE m^{-2} s^{-1} (Zou and Richmond 2000).

The initial characterization of WW showed a low concentration of N-NO_3, probably due to the presence of anaerobic conditions in the drain. Anaerobic conditions in water usually favor the accumulation of N-NH_3 and the reduction of N-NO_3 (Metcalf and Eddy et al. 2002). However, during the growth experiments in WW, high N-NO_3 concentrations were observed (38.2 ± 0.54 and 37.1 ± 0.08 mg N-NO_3 L^{-1} for *S. aquatilis* and *S. nanum,* respectively) (Fig. 4.6). The presence of N-NO_3 was due to the inocula in BG11 medium, which contained a high amount of this nutrient (approximately 247 mg N-NO_3 L^{-1}) (CCAP 2019). The N-NO_3 concentrations during the experiments are shown in Fig. 4.6, and present a similar behavior to that observed in the cultures in the mineral medium (Fig. 4.3). This nutrient was consumed quickly until the end of the culture, and reached a final concentration of 20.1 ± 2.80 and 25.2 ± 3.46 mg N-NO_3 L^{-1} for *S. aquatilis* and *S. nanum,* respectively; the consumption of N-NO_3 (deference between initial and final values) by the strains was 18.2 and 11.8 mg N-NO_3 L^{-1}, respectively. Notably, a significant reduction occurred in the first days of culture; for *S. aquatilis,* the N-NO_3 concentration decreased by 27.5% in 48 h, whereas for *S. nanum*, the reduction was 31.1% for the same period. From day two until the end of the experiment, the N-NO_3-concentration decreased, but at a slower rate. At the end of the culture experiment, the reductions in the N-NO_3 concentration were 47.5% and 31.9%, respectively. This atypical behavior was compared with that shown in a mineral medium, with the consumption rate calculated using the total consumption and time of the culture. The values reached were 3.03 mg N-NO_3 L^{-1} d^{-1} for *S. aquatilis*, and 1.97 mg N-NO_3 L^{-1} d^{-1} for *S. nanum*. These results were significantly lower than the values reached in mineral medium (68.3% and 70.5% lower for *S. aquatilis* and *S. nanum*, respectively). The reduction in the consumption rate of this nutrient was due to the presence of the most reduced form of nitrogen (in this case, N-NH_3); the microalgae prefer the use of N-NH_3 because no energy is required for its incorporation into the cell, whereas N-NO_3 must be reduced to N-NH_3 before its use (Franco Martínez et al. 2017; Xin et al. 2010).

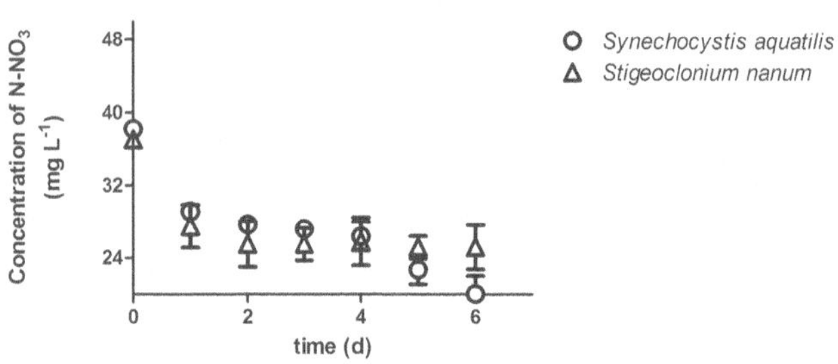

Fig. 4.6 N-NO_3 concentration during growth in real wastewater

The total consumption of N-NH_3 (Fig. 4.7) occurred within four days of culture; however, the major fraction was consumed in the first 24 h in a manner similar to that of N-NO_3. The consumption rate for this nutrient was 1.89 and 2.18 mg N-NH_3 L^{-1} d^{-1} for *S. aquatilis* and *S. nanum,* respectively; however, these values are lower than those reported for microalgae growing in mineral media with N-NH_3 as the sole nitrogen source. Franco Martínez et al. (2017) reported that the N-NH_3 consumption rate depends on the initial concentration; at a lower concentration of 22 mg N-NH_3 L^{-1}, the microalgae reached a volumetric consumption rate of 5.38 mg N-NH_3 L^{-1} d^{-1}. This value was higher than that obtained in this study; however, the initial concentration was lower.

The behavior of P-PO_4 during the experiments with WW is shown in Fig. 4.8, and was similar to that observed in the mineral medium. The initial concentrations were approximately 7 mg P-PO_4 L^{-1} (6.5 ± 0.35 and 7.6 ± 1.26 for *S. aquatilis* and *S. nanum*, respectively) without significant differences ($p > 0.05$). Consumption was consistently rapid during the first 36 h when the cultures reached the limit of detection by the technique. From the third day until the end of the experiment, the P-PO_4 concentration remained constant, without a significant difference between the values ($p > 0.05$). The reduction in the volumetric consumption rate was calculated using the amount consumed and the

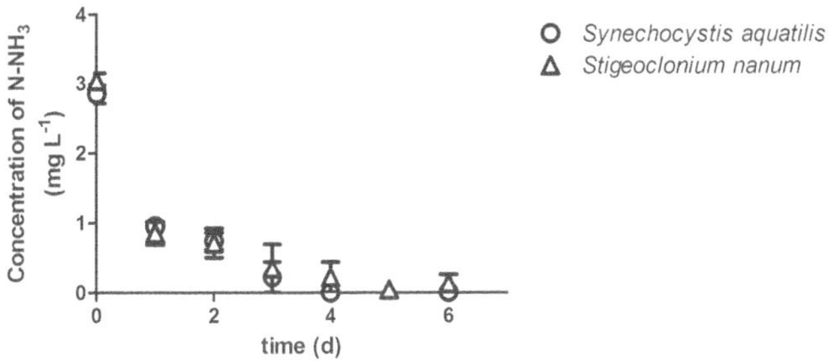

Fig. 4.7 N-NH_3 concentration during growth in real wastewater

duration of the experiment; and the values reached 1.26 and 1.42 mg P-PO_4 L^{-1} d^{-1} for *S. aquatilis* and *S. nanum*, respectively. These values are similar to those obtained in the experiments in the mineral medium of 0.7–1.6 mg P-PO_4 L^{-1} d^{-1} and those reported for different microalgae under various culturing conditions (Franco Martínez et al. 2017; A. J. Martínez-Roldán et al. 2014). There was no correlation between the P-PO_4 consumption rate and the operational conditions because phosphorus is consumed in excess regardless of the cellular necessities, and the excess accumulates in the cytosol as polyphosphate inclusions (Franco Martínez et al. 2017; Jácome-Pilco et al. 2009; Perales-Vela et al. 2007).

4.5.3 Effect of Microalgal Growth on the Final Characteristic of the Effluent

The WW used in this study was not treated; nevertheless, the concentrations of pollutants, such as nitrogen and phosphorus, and COD were low because they were of domestic origin. The characterization results are presented in Table 4.5. The concentration of COD was low (305.5 ± 94.89 mg L^{-1}) and similar to the values reported for minimally contaminated domestic WW (Lum et al. 2013; Metcalf and Eddy et al. 2002; Rawat et al. 2011). The COD decreased at the end of the experiment by 60% for *S. aquatilis* and 48% for *S. nanum*. Two different phenomena can cause a reduction in COD: (1) the capability of microalgae to metabolize small amounts of organic matter and (2) the stimulation

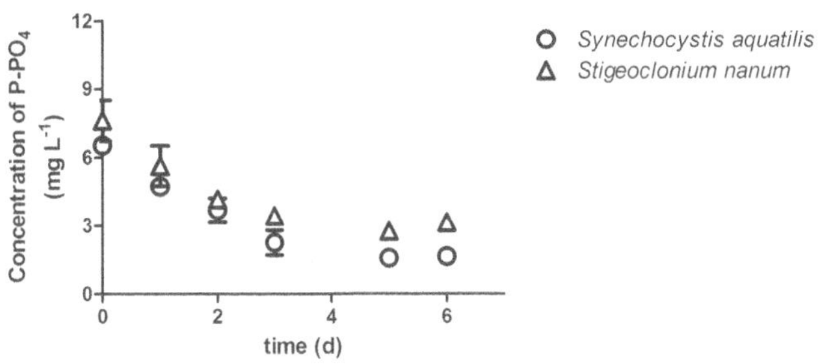

Fig. 4.8 P-PO4 concentration during growth in real wastewater

of the aerobic bacteria population by an increase in the dissolved oxygen concentration (Larsdotter 2006). The pH value increased from 7.8 to >9 at the end of the experiment; this increase was due to the bicarbonate ion consumption and photosynthetic activity. (Larsdotter 2006). An increase in pH contributes to the elimination of coliforms; however, it is not convenient to have a pH value >9 if the treated water empties into a water body (Metcalf and Eddy et al. 2002; Ruiz-Marin et al. 2010). The alkalinity decreased due to an imbalance in the carbonate–bicarbonate–carbonic acid–CO_2 system by the consumption of bicarbonate ions from the culture medium (Larsdotter 2006). The total coliforms were reduced by five orders of magnitude; the initial concentrations were approximately 10^8 colony-forming units (CFU) 100 mL^{-1} and the final values were 10^3 and 10^2 CFU 100 mL^{-1} for *S. aquatilis* and *S. nanum*, respectively. The reduction in coliforms occurred in relation to three factors, namely the increase in the pH, the rise in the dissolved oxygen concentration (by photosynthetic activity), and the ultraviolet (UV) radiation from the lamps. UV radiation affected the coliforms because the cultures were illuminated by fluorescent lamps positioned 5 cm from the PBR. At this distance, the lights emit radiation with wavelengths in the range of 200–400 nm, which corresponds to that of short, medium, and long UV wavelengths (Khazova and O'Hagan 2008), and this type of radiation has germicidal activity. The rise in the pH demonstrated that the pathogens and coliforms were susceptible to pH values over 8.5 (Larsdotter 2006; Schumacher et al. 2003). Some authors have reported a reduction higher than 97% in coliform concentrations when the pH was >11 (Ruiz-Marin et al. 2010). For both strains, the pH values were similar and > 9. Photosynthetic activity causes an increase in oxygen concentration, and cultures can reach three or four times the saturation value, which can considerably reduce pathogens and coliforms (Ruiz-Marin et al. 2010).

Domestic WW treatments are usually performed using aerobic processes. However, microalgae are becoming more commonly used because they can improve the quality of the effluents produced. WW treated with microalgae generally has low pathogen and coliform concentrations, and high pH and dissolved oxygen concentrations, which substantially reduces the risk of eutrophication of the receptor water body (Abdel-Raouf et al. 2012; Larsdotter 2006; Rawat et al. 2011). However, the process with microalgae has the highest operational cost compared to those with bacteria that include stabilization and oxidation ponds. This increased cost can be remedied by developing the process into a biorefinery concept (Pittman et al. 2011). This concept has not only the ecological benefit of producing treated water with high quality but also the advantage of generating biomass with multiple uses, such as energy production (by the fermentation or digestion of the biomass), livestock nutrition, and high-value product acquisition such as those of pigments. The economic contribution of these sub-products can favor the cost-effectiveness of the process, making it attractive to investors (Cherubini 2010).

The results obtained in this investigation demonstrate the potential of *S. aquatilis* and *S. nanum* for the removal of contaminants (mainly nitrogen and phosphorus) from domestic WW. Some of these contaminants can reach values lower than the allowable limits stipulated in the Mexican Regulation for Wastewater. The potential of *S. nanum* for WW treatment is higher than that of *S. aquatilis*, because *S. nanum* can auto-flocculate, making it easier to recuperate biomass without energy consumption or the addition of inorganic chemicals (coagulants). Nevertheless, it is necessary to evaluate the characteristics of the biomass to include the WW treatment process with the biorefinery concept and to generate economic benefits.

References

Abdel-Raouf N, Al-Homaidan AA, Ibraheem IBM (2012) Microalgae and wastewater treatment. Saudi J Biol Sci 19:257–275

Abou RAEAI (2011) Bioremediation: new approaches and trends. In: Khan MS, Zaidi A, Goel R, Musarrat J (eds) Biomanagement of metal-contaminated soils. Springer, pp 65–94

APHA, AWWA, & WEF (1999) Standard methods for the examination of water and wastewater

Cai T, Ge X, Park SY, Li Y (2013) Comparison of *Synechocystis* sp. PCC6803 and *Nannochloropsis salina* for lipid production using artificial seawater and nutrients from anaerobic digestion effluent. Bioresour Technol 144:255–260

Cañizares-Villanueva RO (2000) Biosorption of heavy metals by microorganisms. Rev Latin Microbiol 42:131–143

Carbajo Arteaga L, Ponce Zavaleta M, Moreno Eustaquio W, Mendoza Bobadilla J (2018) Removal of aniline blue dye using live microalgae *Chlorella vulgaris*. J Energy Environ Sci 2:6–12

CCAP (2019) Culture Collection of Algae and Protozoa (CCAP). Retrieved from https://www.ccap.ac.uk/

Cherubini F (2010) The biorefinery concept: using biomass instead of oil for producing energy and chemicals. Energy Convers Manag 51: 1412–1421

Chu WL, Phang SM (2019) Biosorption of heavy metals and dyes from industrial effluents by microalgae. In: Alam A, Wang Z (eds) Microalgae biotechnology for development of biofuel and wastewater treatment. Springer, Singapore, pp 599–634

Droste RL, Gehr RL (2018) Theory and practice of water and wastewater treatment, 2nd edn. Wiley, p 992

Escapa C, Coimbra RN, Paniagua S, García AI, Otero M (2015) Nutrients and pharmaceuticals removal from wastewater by culture and harvesting of *Chlorella sorokiniana*. Bioresour Technol 185: 276–284

Ferreira JT, Bortoleti KCDA, Motta LS, Gavazza S, Brasileiro-Vidal AC, Bezerra RP (2022) Microalgae-based remediation approaches in textile dye removal. In: Khadir A, Muthu SS (eds) Biological approaches in dye-containing wastewater. sustainable textiles: production, processing, manufacturing & chemistry. Springer, Singapore, pp 107–127

Flórez-Miranda L, Cañizares-Villanueva RO, Melchy-Antonio O, Martínez-Jerónimo F, Flores-Ortíz CM (2017) Two stage heterotrophy/photoinduction culture of *Scenedesmus incrassatulus*: potential for lutein production. J Biotechnol 262:67–74

Franco Martínez ML, Rodríguez Rosales MDJ, Moreno Medina CU, Martínez-Roldán AJ (2017) Tolerance and nutrients consumption of *Chlorella vulgaris* growing in mineral medium and real wastewater under laboratory conditions. Open Agric 2:394–400

Gonçalves AL, Pires JCM, Simões M (2016) Biotechnological potential of *Synechocystis salina* co-cultures with selected microalgae and cyanobacteria: nutrients removal, biomass and lipid production. Bioresour Technol 200:279–286

Goswami R K, Agrawal K, Shah M P, Verma P (2021) Bioremediation of heavy metals from wastewater: a current perspective on microalgae-based future. Lett Appl Microbiol 1–17

Hernández-Zamora M, Cristiani-Urbina E, Martínez-Jerónimo F, Perales-Vela HV, Ponce-Noyola T, Montes-Horcasitas MC, Cañizares-Villanueva RO (2015) Bioremoval of the azo dye Congo Red by the microalga *Chlorella vulgaris*. Environ Sci Pollut Res 22: 10811–10823

Jácome-Pilco CR, Cristiani-Urbina E, Flores-Cotera LB, Velasco-García R, Ponce-Noyola T, Cañizares-Villanueva RO (2009) Continuous Cr(VI) removal by *Scenedesmus incrassatulus* in an airlift photobioreactor. Bioresour Technol 100:2388–2391

Jain M, Khan SA, Sharma K, Jadhao PR, Pant KK, Ziora ZM, Blaskovich MAT (2022) Current perspective of innovative strategies for bioremediation of organic pollutants from wastewater. Bioresour Technol 344:126305

Keen OS, Bell KY, Cherchi C, Finnegan BJ, Mauter MS, Parker AM, Rosenblum JS, Stretz HA (2014) Emerging pollutants – Part II: Treatment. Water Environ Res 86:2036–2096

Khazova M, O'Hagan JB (2008) Optical radiation emissions from compact fluorescent lamps. Radiat Prot Dosim 131:521–525

Larsdotter K (2006) Wastewater treatment with microalgae – a literature review. Vatten 62:31–38

Leong YK, Chang JS (2020) Bioremediation of heavy metals using microalgae: recent advances and mechanisms. Bioresour Technol 303:122886

Lum KK, Kim J, Lei X (2013) Dual potential of microalgae as a sustainable biofuel feedstock and animal feed. J Anim Sci Biotechnol 4:53

Martínez-Roldán AJ, Cañizares-Villanueva RO (2017) The effect of the nitrogen-source modification and the addition of CO_2 on the growth and composition of lipids in *Nannochloropsis*. In: Jan M, Kazik P (eds) Nannochloropsis: biology, biotechnological potential and challenges. Nova Science Publishers

Martínez-Roldán AJ, Cañizares-Villanueva RO (2022) Carbon dioxide capture and its use to produce microalgae-based fuels. In: Jacob-Lopes E, Queiroz Zepka L, Manzoni Maroneze M (eds) 3rd Generation biofuels disruptive technologies to enable commercial production Woodhead publishing series in energy. Woodhead Publishing, pp 135–151

Martínez-Roldán AJ, Ibarra-Berumen J (2019) Employment of wastewater to produce microalgal biomass as a biorefinery concept. In: Alam A, Wang Z (eds) Microalgae biotechnology for development of biofuel and wastewater treatment. Springer, Singapore, pp 487–504

Martínez-Roldán AJ, Perales-Vela HV, Cañizares-Villanueva RO, Torzillo G (2014) Physiological response of *Nannochloropsis* sp. to saline stress in laboratory batch cultures. J Appl Phycol 26: 115–121

Martínez-Roldán AJ, Rodriguez-Rosales MDJ, Gómez-Lozano BP, Díaz-Ramírez MA, García-Ruíz MA, Villanueva-García RP (2019) Los fotobiorreactores como herramientas para el desarrollo de bioprocesos a partir de microalgas. In: Extenso en Memoria del Congreso Internacional de Investigación Academia Journals Celaya 2019. Academia Journals, pp 2110–2115

McCarty PL, Bae J, Kim J (2011) Domestic wastewater treatment as a net energy producer–can this be achieved? Environ Sci Technol 45: 7100–7106

Metcalf & Eddy I, Tchobanoglous G, Burton F, Stensel H (2002) Wastewater engineering: treatment and reuse, 4th edn. McGraw-Hill, Inc

Min L, Lo C, Hsu CC, Wang Y, Chiang YR, Sun Q, Wu Y, Li Y, Chen L, Yu CP (2018) Identification of enantiomeric byproducts during microalgae-mediated transformation of metoprolol by MS/MS spectrum based networking. Front Microbiol 9:2115

Mohamed Ramli N, Verdegem M, Yusoff F, Zulkifely M, Verreth J (2017) Removal of ammonium and nitrate in recirculating aquaculture systems by the epiphyte *Stigeoclonium nanum* immobilized in alginate beads. Aquac Environ Interac 9:213–222

Moondra N, Jariwala ND, Christian RA (2020) Sustainable treatment of domestic wastewater through microalgae. Int J Phytoremediation 22: 1480–1486

Osuna Salazar NK, Pacheco Fernández MA, Gómez Lozano BP, Martínez-Roldán AJ (2016) Design and construction of a bubbled column for microalgae culture. 3rd Biotechnology Summit 2016, 436–440

Oswald WJ, Golueke CG (1960) Biological transformation of solar energy. Adv Appl Microbiol 2:223–262

Oswald William J, Gotaas HB (1957) Photosynthesis in sewage treatment. Trans Am Soc Civil Eng 122:73–105

Oswald WJ, Gotaas HB, Golueke CG, Kellen WR, Gloyna EF, Hermann ER (1957) Algae in waste treatment. Sewage and industrial wastes. Wiley

Pal A, He Y, Jekel M, Reinhard M, Gin KYH (2014) Emerging contaminants of public health significance as water quality indicator compounds in the urban water cycle. Environ Int 71:46–62

Pavithra KG, Kumar PS, Jaikumar V, Vardhan KH, SundarRajan PS (2020) Microalgae for biofuel production and removal of heavy metals: a review. Environ Chem Lett 18:1905–1923

Peña-Guzmán C, Ulloa-Sánchez S, Mora K, Helena-Bustos R, Lopez-Barrera E, Alvarez J, Rodriguez-Pinzón M (2019) Emerging pollutants in the urban water cycle in Latin America: a review of the current literature. J Environ Manag 237:408–423

Perales-Vela HV, González-Moreno S, Montes-Horcasitas C, Cañizares-Villanueva RO (2007) Growth, photosynthetic and respiratory responses to sub-lethal copper concentrations in *Scenedesmus incrassatulus* (Chlorophyceae). Chemosphere 67:2274–2281

Perez M, Nolasco NA, Vasavada A, Johnson M, Kuehnle A (2015) Algae-mediated valorization of industrial waste streams. Ind Biotechnol 11:229–234

Pittman JK, Dean AP, Osundeko O (2011) The potential of sustainable algal biofuel production using wastewater resources. Bioresour Technol 102:17–25

Premaratne M, Nishshanka GKSH, Liyanaarachchi VC, Nimarshana PHV, Ariyadasa TU (2021) Bioremediation of textile dye wastewater using microalgae: current trends and future perspectives. J Chem Technol Biotechnol 96:3249–3258

Rani S, Chowdhury R, Tao W, Nedbalova L (2021) Microalga-mediated tertiary treatment of municipal wastewater: removal of nutrients and pathogens. Sustainability 13:9554

Rawat I, Kumar RR, Mutanda T, Bux F (2011) Dual role of microalgae: phycoremediation of domestic wastewater and biomass production for sustainable biofuels production. Appl Energ 88:3411–3424

Rempel A, Gutkoski JP, Nazari MT, Biolchi GN, Biduski B, Treichel H, Colla LM (2022) Microalgae growth with a high concentration of emerging pollutants and phytotoxicity evaluation of cultivation wastewater. J Water Process Eng 46:102616

Ruiz-Marin A, Mendoza-Espinosa LG, Stephenson T (2010) Growth and nutrient removal in free and immobilized green algae in batch and semi-continuous cultures treating real wastewater. Bioresour Technol 101:58–64

Schumacher G, Blume T, Sekoulov I (2003) Bacteria reduction and nutrient removal in small wastewater treatment plants by an algal biofilm. Water Sci Technol 47:195–202

Shelef G, Sukenik A, Green M (1984) Microalgae harvesting and processing: a literature review. Solar Energy Research Institute, Golden, Colorado, p 71

Touloupakis E, Cicchi B, Benavides AMS, Torzillo G (2016) Effect of high pH on growth of *Synechocystis* sp. PCC 6803 cultures and their contamination by golden algae (Poterioochromonas sp.). Appl Microbiol Biotechnol 100:1333–1341

Wang L, Yecong Li MM, Chen P, Chen Y, Liu Y, Wang Y, Ruan R (2009) Cultivation of green algae Chlorella sp. in different wastewaters from municipal wastewater treatment plant. Appl Biochem Biotechnol 162:1174–1186

Wellburn AR (1994) The spectral determination of Chlorophylls a and b, as well as Total Carotenoids, using various solvents with spectrophotometers of different resolution. J Plant Physiol 144: 307–313

Wu Y, Xia L, Yu Z, Shabbir S, Kerr PG (2014) In situ bioremediation of surface waters by periphytons. Bioresour Technol 151:367–372

Xie Y, Ho SH, Chen CNN, Chen CY, Ng IS, Jing KJ, Chang JS, Lu Y (2013) Phototrophic cultivation of a thermo-tolerant *Desmodesmus* sp. for lutein production: effects of nitrate concentration, light intensity and fed-batch operation. Bioresour Technol 144:435–444

Xin L, Hong-ying H, Ke G, Ying-xue S (2010) Effects of different nitrogen and phosphorus concentrations on the growth, nutrient uptake, and lipid accumulation of a freshwater microalga *Scenedesmus* sp. Bioresour Technol 101:5494–5500

Zhang J-Z, Fischer CJ (2006) A simplified resorcinol method for direct spectrophotometric determination of nitrate in seawater. Mar Chem 99:220–226

Zou N, Richmond A (2000) Light-path length and population density in photoacclimation of *Nannochloropsis* sp. (Eustigmatophyceae). J Appl Phycol 12:349–354

Phycoremediation of Industrial Wastewater

5

Mariana Jiménez-Veuthey, Luz Marina Zapata, and María Luján Flores

Abstract

The constant increase in the contamination of wastewater from the activities of different industries, constitutes a growing environmental problem, mainly due to the limitations of many treatment plants for such waters that do not allow effective elimination. When contaminants persist in the water resources above their allowed limits it may cause some harmful and toxic effects on aquatic organisms and humans, such as eutrophication, mutagenic, and carcinogenic risks. Bioremediation is a biological process where organisms can naturally treat pollutants from wastewater to release them into the environment in compounds with less or no toxicity. In this way, it is important to promote the use of the nutrients present in industrial wastewater for the development of microalgae cultures, which will allow, on the one hand, to produce biomass of species of interest and, on the other, to remove contaminants, generating an economically profitable process. In this phycoremediation process, the efficiency that can be achieved is fundamentally related to the microalgae species, the bioreactor design, the growth conditions, and the composition of the residual wastewater. In this context, this chapter highlights the main factors to take into account to achieve effectiveness in the process of phycoremediation and production of microalgae. In addition, a case study of our research groups is presented based on the use of the native microalga *Scenedesmus obliquus* isolated from the Embalse Salto Grande (Entre Ríos, Argentina), which has shown significant effectiveness in the treatment of effluents from the poultry industry.

Keywords

Bioreactor · Contamination · Industrial wastewater · Microalga · Nutrients removal · Phycoremediation · *Scenedesmus obliquus*

5.1 Water Pollution Due to Industrial Effects

The significant growth of the world population, which exceeds 7.9 billion according to 2022 data from the United States Census Bureau (Zerrouki and Henni 2019), consequently generates an increase in the use of water. The demand for water comes mainly from industries, agricultural activity, and municipal and domestic areas in general. Currently, agricultural activity demands around 70% of fresh water, and the industrial sector 20% (UNW-DPAC 2015; United Nations 2021).

On the other hand, the quality of the water is being affected due to the strong contamination that occurs in the main rivers of Africa, Asia, and Latin America, being the amount of pathogenic nutrients one of the main sources of contamination. About 80% of the wastewater from industries and municipalities in the world is discharged into the environment without any prior treatment, which has a negative

M. Jiménez-Veuthey (✉)
Laboratorio de Análisis de Metales en Alimentos y Otros Sustratos (LAMAS), Facultad de Ciencias de la Alimentación, Universidad Nacional de Entre Ríos (UNER), Instituto de Ciencia y Tecnología de los Alimentos de Entre Ríos (ICTAER), Concordia, Entre Ríos, Argentina

CONICET, Consejo Nacional de Investigaciones Científicas y Técnicas, Buenos Aires, Argentina
e-mail: mariana.jimenezveuthey@uner.edu.ar

L. M. Zapata
Laboratorio de Análisis de Metales en Alimentos y Otros Sustratos (LAMAS), Facultad de Ciencias de la Alimentación, Universidad Nacional de Entre Ríos (UNER), Instituto de Ciencia y Tecnología de los Alimentos de Entre Ríos (ICTAER), Concordia, Entre Ríos, Argentina

M. L. Flores
Farmacognosia, Grupo de Investigación Química Bioactividad y Metabolismo de Recursos Naturales Patagónicos (GQBMRNP) y Area de Análisis Instrumental (AAI) – Centro Regional de Investigación y Desarrollo Científico Tecnológico (CRIDECIT), Facultad de Ciencias Naturales y Ciencias de la Salud (FCNyCS), Universidad Nacional de la Patagonia San Juan Bosco (UNPSJB), Chubut, Argentina

A. de Jesús Martínez-Roldán (ed.), *Biotechnological Processes for Green Energy, and High Value Bioproducts by Microalgae, and Cyanobacteria Cultures*, Developments in Applied Phycology 13, https://doi.org/10.1007/978-3-031-43969-8_5

impact on ecosystems and human health. This situation is even more unfavorable in less developed countries, whose wastewater treatment systems are extremely deficient (WWAP 2016; United Nations 2021).

This situation generates a significant lack of water that affects more than 40% of the population in the most disadvantaged places, which is increasing given that other factors are also added, for example climate change. According to the United Nations (2021), declining supplies of drinking water are affecting all continents. Thus, in the 2030 Agenda for Sustainable Development, Sustainable Development Goal 6 points to "clean water and sanitation" (United Nations Development Program 2022).

Therefore, water is one of the issues considered a priority in the world, highlighting the economic and security consequences of its improper use, in the United Nations reports on the development of water resources in the world (Cosgrove and Loucks 2015).

One of the matters to be addressed is the discharge of effluents into natural resources. Discharges from industrial establishments occur in surface watercourses or in the urban sewage network. Many times, industrial waste reaches watercourses directly or indirectly, with or without prior treatment; which depends on the regulations in each jurisdiction, as well as the respect for these, and the degree of commitment of the industrialists to population's life quality and the environment (United Nations Development Program 2022).

The main contaminants present in industrial, agricultural, and domestic effluents are nitrogen, phosphorous, heavy metals, organic matter, and bacteria from fecal contamination. Its discharge into bodies of water constitutes a threat of eutrophication, degrading water resources with the consequent environmental and human damage (Tao et al. 2005; Renuka et al. 2015; Zerrouki and Henni 2019).

The discharge of wastewater containing high amounts of nitrogen and phosphorus alters natural processes such as photosynthesis, respiration, nitrogen fixation, evaporation and precipitation, causing a deterioration in water quality, the constitution of ecosystems, and the functionality of aquatic systems. In addition, it seriously damages freshwater reserves for human consumption. As a mitigation measure, the removal of nutrients must be carried out prior to the discharge of wastewater into aquatic environments. However, the removal of such nutrients is generally inadequate, being the secondary municipal effluents from the wastewater treatment plants the most frequent sources of eutrophication since they contain residues of NO_2, NH_4^+, PO_4^{-3} and are discharged into the natural channels (Rawat et al. 2011; Beltrán-Rocha et al. 2017; Gonçalves et al. 2017).

According to OXFAM International (2021), the poor treatment of effluents is the cause of numerous deaths worldwide due to diseases related to water pollution, in many cases exceeding other causes, including wars. This clearly generates the urgent need to optimize the treatment of domestic and industrial effluents to prevent health and environmental problems and reduce the causes of related deaths. For this, it is important to promote the development and use of technologies that are more efficient (Renuka et al. 2015; Acién Fernández et al. 2017, 2018).

For this reason, many countries have environmental regulations that are increasingly demanding, but at the same time dynamic to adapt to situations that arise. In this sense, the regulations for wastewater discharge of the European Union set the discharge limits for total phosphorus and nitrogen for wastewater from domestic and industrial activities at 1 mg L^{-1} and 10 mg L^{-1}, respectively (Directive 91/271/EEC 2003). In Argentina, each province has authority over its natural resources, which is why it establishes its own regulations in terms of water quality and wastewater discharges. An example is the province of Buenos Aires, which sets the discharge limits for total phosphorus and nitrogen at 1 mg L^{-1} and 35 mg L^{-1}, respectively (Resolution No. 336/03, Argentina, 2003). In the province of Entre Ríos, the discharge limits for these nutrients are 30 mg L^{-1} and 105 mg L^{-1}, respectively (Law 6260 of 1992; Resolution No. 554/15 of 2015).

In this context, it is necessary to generate strategies that make it possible to comprehensively plan and manage water sources, particularly effluents from industries and domestic chores, both at the national and municipal levels. This requires that wastewater become an integral part of urban planning and the integrated management of basins and coasts, linking the population that makes use of water in anthropogenic and industrial activities. These populations are who must have conceptual bases that allow rational and efficient use of water (ONU 2015; Acién Fernández et al. 2018; Montenegro-Gómez and Osorio 2019).

5.2 Wastewater Treatment Methods

The increase in urbanization and industrial activities worldwide has led to an annual wastewater production of around 300 km^3 (Zhou et al. 2022). Their discharge into water bodies triggers serious environmental challenges (Rawat et al. 2011; Verma 2022).

Physical, chemical, and biological methods are used to remove contaminants from wastewater. In physical methods mechanical forces are applied; examples are crushing, flow equalization, sedimentation, flotation, filtration, adsorption, membrane technology, activated carbon, sorption, and volatilization. Chemical processes involve chemical reactions that cause some type of change. Operations associated with chemical treatments consist of adsorption, disinfection, dechlorination, oxidation processes, photodegradation, ozonation, and photocatalysis. For biological treatment, microorganisms are used, which produce the decomposition of wastewater pollutants (Zerrouki and Henni 2019). Added to this, this type of treatment contributes to the production of value-added compounds for various uses. In the biological treatment unit, the aerated lagoon, rotating biological contactors, anaerobic digestion, membrane bioreactors, microalgae-based treatment, and enzyme-assisted biodegradation stand out (Rawat et al. 2011; Bilal et al. 2019); Gomes et al. 2019; Ahmed et al. 2021; Rout et al. 2021).

For the treatment of wastewater, physical, chemical, and biological methods are combined, called primary, secondary, and tertiary treatments. In the primary treatment, heavy solids are separated by sedimentation and lighter solids by flotation. Settled and floating materials are removed and the liquid can be discharged for secondary treatment; in the latter, dissolved and suspended biological matter is removed. Secondary treatment involves biological and/or chemical mechanisms. The biological ones use native microorganisms, which are subsequently eliminated or inactivated before going on to tertiary treatment; in chemicals, photodegradation, sorption, volatilization, among others, are used. In tertiary treatment, before discharge to natural resources, the water is chemically or physically disinfected by oxidation, activated carbon adsorption (Rawat et al. 2011; Rout et al. 2021). In tertiary treatment, before discharge to natural resources, the water is chemically or physically disinfected by oxidation, adsorption with activated carbon (Rawat et al. 2011; Rout et al. 2021). According to Rout et al. (2021), in wastewater treatment plants, the removal efficiency of emerging contaminants varies from 20 to 50% in primary treatment, from 30 to 70% in secondary treatment, and exceeds 90% in tertiary treatment.

Conventional treatment plants in general tend to have a high degree of mechanization and high energy costs for pumping; they also require highly trained labor for their management (García-Einschlag 2011; Rawat et al. 2011). In contrast, wastewater treatments using unconventional technologies such as bioremediation require less expensive technologies (Huang and Lee 2015).

5.3 Phycoremediation Process as Wastewater Tertiary Treatment

As described in previous paragraphs, wastewater treatment has become an important challenge for the sustainability of human activities (Acién Fernández et al. 2017). These waters must be treated to avoid health and environmental problems, but this entails a cost and the need for increasingly efficient technologies (Zhao et al. 2022).

Discharge of untreated wastewater into the environment causes eutrophication problems, seriously damaging ecosystems and, therefore, freshwater reserves for human consumption (Acién Fernández et al. 2017; Larronde-Larretche and Jin 2022).

For this reason, it is essential to optimize the treatment of wastewater in order to eliminate contaminants and discharge it safely (Acién Fernández et al. 2017). In this context, phycoremediation constitutes a potential solution to tertiary treatments. In this process, macro or microalgae are used for the removal or biotransformation of contaminants from wastewater, mainly nutrients, xenobiotics, and CO_2. For their development, microalgae can use large amounts of carbon, nitrogen, and phosphorus; they also act as disinfectants due to the increase in pH that they produce during photosynthesis and have the capacity to remove heavy metals and emerging compounds (Olguín 2012). Phycoremediation represents a cost-effective and environmentally compatible method, also incorporating the benefits of recovery and recycling of resources. The combination of microalgae production with wastewater treatment dates back to the 1960s when it was proposed by Oswald (Acién Fernández et al. 2013, 2017). However, the real application of this type of process on a commercial scale is still scarce today (Craggs et al. 2012). This may be due to different reasons, the most relevant being the low efficiency of current technologies due to the fact that they require very long hydraulic retention times (HRT) (up to 10 days), and therefore large surfaces. To increase this efficiency, it is necessary to improve the operating conditions as well as the use of new photobioreactor designs such as thin-layer reactors (Acién Fernández et al. 2017).

The removal of nutrients from effluents using microalgae on a laboratory scale is the subject of different investigations. In relation to the absorption range of nitrogen and phosphorus in different agro-industrial wastewaters, it ranges from 24 to 100% and from 25 to 100%, respectively. On the other hand, microalgae can also reduce the COD in a percentage that oscillates between 23 and 95%. N is found in general in agro-industrial effluents, constituting nitrates, nitrites, and ammonia. Microalgae absorb N fundamentally in the plasmatic membrane, where $N\text{-}NO_3^-$ is reduced to $N\text{-}NO_2^-$; subsequently, a part of the $N\text{-}NO_2^-$ produced is released into the aqueous phase, and the other part is reduced to form $N\text{-}NH_4^+$ in the chloroplast and finally to form amino acids (Zhou et al. 2022). Total phosphorus in agro-industrial effluents is found mainly as orthophosphate and dissolved organic phosphorus. During metabolism, $P\text{-}PO_4^{-3}$ is assimilated as $H_2PO_4^-$ and HPO_4^{-2} and is incorporated into organic compounds through phosphorylation (Zhou et al. 2022). In this way the cells obtain energy. In addition, microalgae can accumulate minerals such as Mo, Fe, Ni, Cu, Zn, Co, B, and Mn, which are present in agro-industrial effluents and thus decrease their amount in discharge water (Zhou et al. 2022).

Lau et al. (2014) reported a removal of N and P of up to 90% and 80%, respectively, when carrying out the treatment of wastewater with *Chlorella vulgaris*. Oliveira et al. (2018) published the results obtained when using *Scenedesmus obliquus* in the removal of nutrients from wastewater from the poultry industry, with a removal efficiency of 97% for both inorganic N and phosphorus. Pereira et al. (2019) reached a removal efficiency of TN, TP, and COD of up to 94, 96.6, and 90%, respectively; in addition, they achieved a 99% elimination of *Escherichia coli* when treating municipal wastewater with *Chlorella* sp.

In this way, for the phytoremediation process in the open air on an industrial scale, is important to take into account various issues, fundamentally technical and economics. Among them, Zerrouki and Henni (2019) highlighted that suitable microalgae strains must be found for the process to be carried out, evaluate the N/P ratio contained in the wastewater, determine the presence of toxicants and/or biological inhibitors, establish the optimal operating conditions of the bioreactor and to know the environmental conditions. In order to reach the maximum growth potential of the microalgae strain and to achieve a high contaminant removal rate, it is necessary to design balanced formulas. In this sense, agro-industrial effluents are an appropriate option to be used as an artificial medium for the production of microalgae, where the composition and characteristics of untreated effluents depend on the source and location (Zerrouki and Henni 2019).

In addition to what is described in previous paragraphs, the biomass of microalgae generated from the phycoremediation process constitutes a potential source for its use as animal feed, as a biofertilizer, biostimulant, and/or as raw material for the extraction of antioxidant products, compounds with different biological activities, fatty acids, pigments, amino acids, among others (Yangüez et al. 2015; Renuka et al. 2016; Beltrán-Rocha et al. 2017).

5.3.1 Microalga Species Used in the Phycoremediation Process

Microalgae species that are used in the phycoremediation must have certain conditions; among them: (a) high growth rate, (b) high tolerance to seasonal and diurnal variations, and (c) good capacity to form aggregates for their subsequent separation (Park et al. 2011). In addition, to avoid problems related to the introduction of foreign species to the ecosystem, as well as to achieve efficiency in the phycoremediation process, it is crucial to find strains that are adapted to culture in open ponds by selecting native species adapted to the natural environment (Zerrouki and Henni 2019). *Chlorella*, *Scenedesmus*, *Oscillatoria*, and *Nitzschia* species, and some cyanobacteria, have been successful in the tertiary treatment of water from agricultural activities, as well as municipal and industrial wastewater, given their high tolerance to environmental conditions (Table 5.1).

The phycoremediation process using microalgae has been extensively studied and applied at a laboratory scale. However, on a larger scale, much less has been done. Environmental factors such as solar radiation, temperature, and seasonal changes can be a significant challenge to achieve scaling (Zerrouki and Henni 2019). In the world, in recent years, a few successful pilot plants have been built, located in places of high irradiation (Rawat et al. 2013; Zhu et al. 2016; Zerrouki and Henni 2019).

5.3.2 Production of Microalgae in Industrial Photobioreactors

The production of microalgae outdoors at industrial scale from laboratory scale cultures implies the adjustment and optimization of parameters such as nutrient absorption and growth kinetics of the microalga strain. Biomass productivity and photosynthetic activity. Biomass productivity and photosynthetic activity of microalgae depend on the solar radiation on the culture's surface, the dilution rate (D), the N/P ratio, the pH, and the temperature (Cabello et al. 2015; Mahdy 2016; Zerrouki and Henni 2019). On a laboratory level, many strains have been identified and high biomass productivity have been achieved.

Diurnal environmental conditions and seasonal conditions, such as fluctuations in temperature and light, can cause a strain to grow adequately under particular laboratory conditions (room temperature and relatively low light intensities) and not grow in outdoor photobioreactors. Sometimes, it is difficult to extrapolate the results obtained at laboratory scale to industrial scale open photobioreactors (Zerrouki and Henni 2019).

5.3.2.1 Microalgae Production at Laboratory Scale

At laboratory scale, the culture process starts from axenic microalgal colonies maintained on a solid medium with which small volume liquid cultures are inoculated (usually up to 250 mL). These cultures are kept as "stock," which will serve as inoculum for the production systems. The advantage of laboratory production is that it allows the temperature, radiation, and the light:dark cycles to be kept constant.

Starting from a stock culture, it is brought to a larger volume and cultivated under optimal growth conditions until sufficient quantity and quality of biomass to start a crop.

In scaling, normally, the inoculum is 10–20% of the final volume. In this way, production is increased using different sizes of recipients, as shown in Fig. 5.1, until reaching a biomass concentration of around 1 g L^{-1}.

Table 5.1 Wastewater composition and microalgae species used in phytoremediation process

Wastewater sources	Origin	TN (mg L^{-1})	TP (mg L^{-1})	COD (mg L^{-1})	N/P ratio	Microalage species	References
Industrial	Brewery	9–480	5–45	547–6730	1.44–10.67	*Chlorella vulgaris*	Zheng et al. (2018)
	Molasses	458	67	514.000	6.84	*Scenedesmus* sp.	Ma et al. (2017)
	Soybean processing	70–73	16.8–17.5	2990–3750	4.17	*Chlorella pyrenoidosa*	Hongyang et al. (2011)
	Winery	7.54-7.80	21.53-22.03	–	0.35	*Diplosphaera* sp.	Liu et al. (2016)
	Paper Mill	8.06-11.8	16.97-33.53	–	0.35-0.47	Mixed culture of two *Scenedesmus* sp.	Usha et al. (2016)
	Fish processing	46–50	2.7–10.7	–	4.67-17.03	*Oocystis* sp.	Riaño et al. (2011)
Agricultural	Dairy manure	3305	266	38.230	12.42	*Chlorella* sp.	Wang et al. (2010)
	Digested pig	1405–1519	164–620	2746–4157	2.21–7.43	*Spirulina*	Olguin et al. (2003)
	Poultry (diluted)	76–152	6–12	–	12.67	*Chlorella minutissima, Chlorella sorokiniana, Scenedesmus bijuga*	Singh et al. (2011)
	Swine	75.4-97.4	19.2-21.2	–	3.93-4.59	*Scenedesmus acutus, Scenedesmus spinosus, Scenedesmus quadricauda*	Kim et al. (2007)
	Dairy	10.66–86.65	2.52–9.50	–	4.23-9.12	*Scenedesmus quadracauda, Tetraselmis suecica*	Daneshvar et al. (2019)
Municipal	Secondary	23.1–35.4	3.8–4.9	103–190.9	6.07-7.22	Fil. blue-green algae	Su et al. (2011)
	Secondary	34.5	2.5	–	13.8	*Scenedesmus obliquus, Chlorella vulgaris*	Ruiz-Marin et al. (2010)
	Centrate	116.1–132.3	212.0–215.1	2304.0–2389.5	0.55-0.62	*Chlorella* sp.	Li et al. (2011)
	Sewage	27.59-27.81	1.56-1.62	–	17-18	*Chlorella vulgari, Scenedesmus* sp.	Zhang et al. (2013)
	Secondary treated sewage	6.3–130	0.04–0.88	–	147.73-157.5	*Botryococcus braunii*	Sawayama et al. (1992)
	Digested sludge diluted 10×	39.5	5.3	178.4	7.45	*Coelastrum microporum*	Lee et al. (2015)

5.3.2.2 Large-Scale Microalgae Production Systems

The next step in the scaling involves large-scale microalgae production. The selection place for microalgae production is one of the main factors to consider. Microalgae can be grown photoautotrophically, photoheterotrophically, heterotrophically, or under mixotrophic conditions (Fig. 5.2). During photosynthesis, microalgae assimilate inorganic carbon in the form of CO_2 from soluble carbonates (HCO_3^-) in the aqueous medium, both atmospheric CO_2 and that generated by industry. Under photoautotrophic growth conditions, microalgae use light as an energy source; which is used to convert inorganic carbon to organic carbon. However, in the absence of CO_2, microalgae experience heterotrophic processes of metabolism, which are characterized by consuming only organic compounds as a carbon source, such as glycerol, acetate, glucose, and other organic compounds (Zhou et al. 2022).

When microalgae use light as an energy source, they are known as photoheterotrophs; while if they obtain energy through the oxidation of organic compounds, they are called heterotrophs. On the other hand, when microalgae assimilate organic and inorganic carbon for their growth, they are said to have a mixotrophic metabolism (Dragone 2022). However, photoautotrophic production is the only one feasible to carry out on a large scale due to its profitability and technical feasibility (Brenan and Owende 2010).

To achieve high efficiency in tertiary wastewater treatment, another important aspect to consider is the bioreactor design. These can be classified into open systems and closed

Fig. 5.1 Scale-up of microalgae production from laboratory to pilot scale

systems. The first consists of artificial lagoons and raceways; while closed systems are plastic bags, or automated systems, such as bioreactors. Open culture systems are a relatively simple and cheap technology, but they have the drawback that the culture can be contaminated by other species of microalgae and/or microorganisms (Fig. 5.3). While one of the main advantages of closed culture systems is the greater ease of maintaining a monoculture, without contamination by other species, providing microalgal biomass of purity suitable for processing in the pharmaceutical or food industry. These types of systems, more expensive and complex, consist of transparent photobioreactors (acrylic or glass) of different geometries: tubular, cylindrical, or flat (Fig. 5.4).

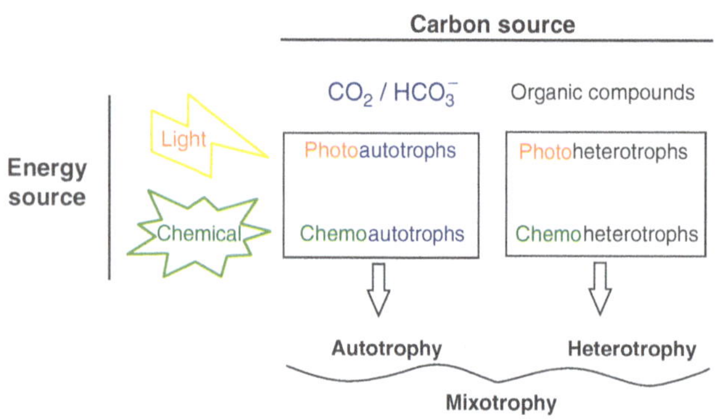

Fig. 5.2 Nutritional modes of microalgae (adapted from Dragone 2022)

At a commercial scale, massive microalgae cultures in open outdoor ponds are viable to obtain low-value products or services such as biofuels, wastewater treatment, biofertilizers, biostimulants, among others. While the use of closed photobioreactors is justified for the production of biomass or products such as carotenoids, polyunsaturated fatty acids, pharmaceutical products, cosmetics, food and feed production, besides other compounds of high commercial value. Table 5.2 shows the different microalgae production systems used in the tertiary treatment of agro-industrial effluents, and the most used microalgae species and their productivities.

Worldwide, more than 90% of microalgae production is carried out in open reactors (Acién Fernández et al. 2017). There are numerous examples around the world of

Fig. 5.3 Open systems of microalgae production. Raceway reactors (**a** y **b**). Thin-layer reactors (**c** y **d**)

[Photo: M. Jiménez Veuthey in IFAPA facilities, Almería, Spain (A, C and D), and LAS CAMELIAS S.A. facilities (B)]

microalgae production facilities using raceway reactors. Above all, in processes related to wastewater treatment or in the biofertilizers production, which are currently the ones that are arousing the most interest (Park et al. 2011; Chisti 2013).

5.4 Case Study: Tertiary Treatment of Poultry Effluent with Microalgae *Scenedesmus Obliquus*

After the anaerobic digestion of poultry effluents, the removal of contaminants from the biologically treated sludge in the secondary treatments is insufficient; therefore, the digestate obtained has a high content of nitrogen and phosphorus. To reduce their concentrations, it is necessary to recirculate the effluent within the effluent treatment plant, to reduce the chances of eutrophication when wastewater is disposed of into natural water courses (Rosso et al. 2008; Jiménez-Veuthey et al. 2019).

The high energy consumption of these treatments causes increases between 45 and 75% in production costs. Poultry effluents can be used as a nutrient source in microalgae production (Rosso et al. 2008; Uggetti et al. 2014). Usually, to obtain high biomass productivity of freshwater microalgae strains, ammonia nitrogen concentrations in the effluent are required between 65 and 200 mg L^{-1} and phosphate concentrations of 10–100 mg L^{-1} (McGinn et al. 2011; Li et al. 2012). Effluent may also present compounds that are potentially toxic and capable of inhibiting microalgae growth, as for example pesticides, phenols, organic acid, and urea. Consequently, high concentrations of these compounds in effluents could limit the potential of microalgae for industrial use (Djelal et al. 2014). It is necessary, in each case, to study the optimal dilution rate and effluent percentage that can be mixed with freshwater for microalgae production (Jiménez-Veuthey et al. 2019).

5.4.1 Effluents from Poultry Industry in Argentina

According to Servicio Nacional de Sanidad y Calidad Agroalimentaria (SENASA), in Argentina there are 60 slaughter plants authorized, reaching an average population of 139 million. Entre Ríos is the main poultry meat

Fig. 5.4 Closed microalgae production systems. Photobioreactors: Flat plate (**a**), bubble columns (**b**), and tubular (**c**)

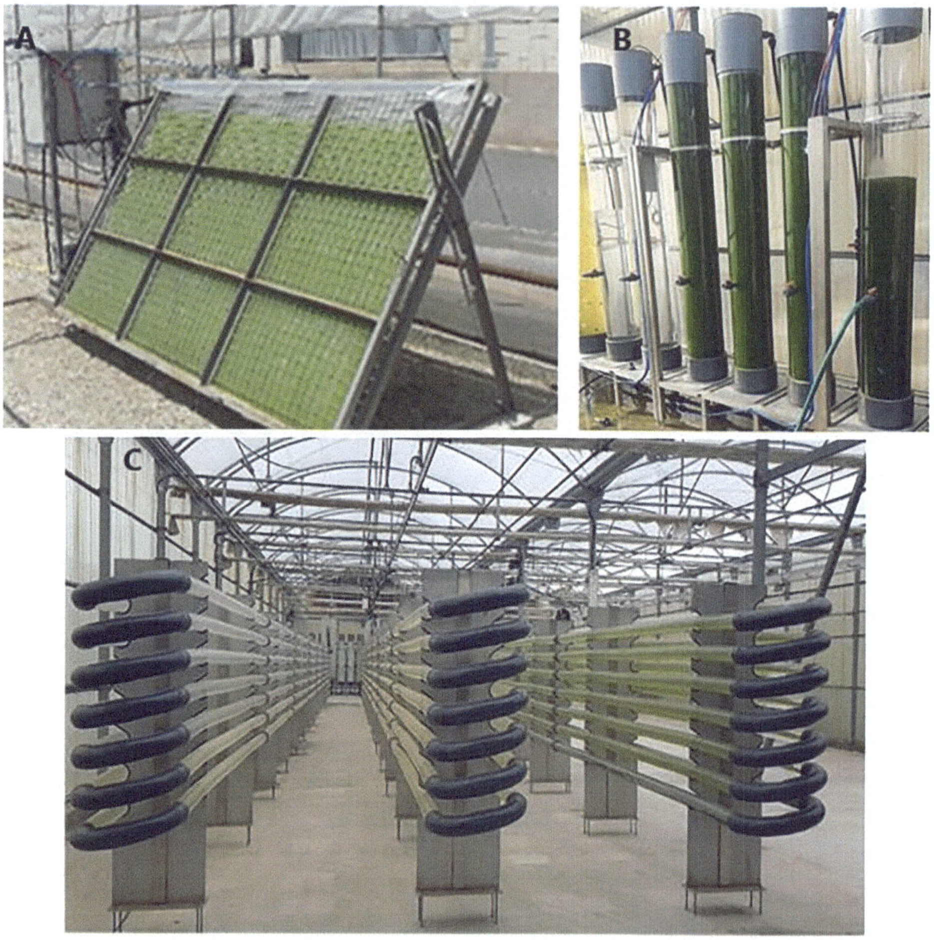

(Photo: M. Jiménez Veuthey in IFAPA facilities, Almería, Spain)

Table 5.2 Microalgae production systems used in agro-industrial wastewater treatment

Bioreactor	Wastewater	Microalgae species	Productivity (g L^{-1} day)	References
Raceway	Municipal wastewater	*Scenedesmus*, *Micratinium*, *Chlorella*	0.02	Woertz et al. (2009)
Turf scrubber	Dairy manure	Mixed culture	8.3- 25.1	Mulbry et al. (2008)
High-Rate Algal Pond (HRAP)	Secondary effluent	*Scenedemus obliquus*	8.7–13.6	Arbib et al. (2017)
Raceway	Domestic wastewater	*Pediastrum* sp.	0.10	Park and Craggs (2010)
Bubble columns	Centrate from an urban wastewater	*Nannochloropsis gaditana*	0.60	Villegas et al. (2017)
Airlift circulation photobioreactor	Starch processing wastewater	*Chlorella pyrenoidosa*	0.33-0.35	Chu et al. (2015)
Tubular photobioreactors	Anaerobic digestion of municipal wastewater	*Nannochloropsis gaditana*	0.48	Ledda et al. (2015)
Open tanks	Treated wastewater (post-clarifier)	*Chlorella*, *Scenedesmus*, *Spirulina*	–	Bhattacharjee and Siemann (2015)
Open photobioreactors	Piggery wastewater diluted 10- and 20-fold	*Chlorella vulgaris*	–	González-Fernández et al. (2016)
Raceway	Anaerobic digestate of piggery effluent	Microalgal consortium	0.02	Ayre et al. (2017)
Enclosed photobioreactors	Urban wastewater	*Galdieria sulphuraria*	–	Henkanatte-Gedera et al. (2017)
High Rate Algal Pond (HRAP)	Municipal wastewater	Indigenous algal consortium	0.47–0.53	Kim (2014)

producing province; it concentrates 51% of production and 53% of fattening farms and has 16 authorized slaughter industries (SENASA 2019). These industries are particularly concentrated in the departments of Colón, Concepción del Uruguay, and Gualeguay. In this type of industry, water is used in different stages of the slaughter process. The average water consumption varies between 8 and 15 L/slaughtered birds (Matsumura and Mierzwa 2008). According to the indicated data, the average consumption of water used in the process is between 1,112,000 and 2,085,000 m^3 per year, so this valuable resource becomes wastewater for the poultry industry (SENASA 2019). Frequently, effluents from poultry slaughter plants have high concentrations of chemical oxygen demand (COD), organic matter, nitrogen, phosphorus, fats, suspended solids, and coliform bacteria. The effluents from slaughterhouses located in the province of Entre Ríos, after primary and secondary treatments, are discharged into the Río Uruguay. A study carried out by the Laboratorio de Análisis de Metales en Alimentos y Otros Sustratos (LAMAS) of the Facultad de Ciencias de la Alimentación of the Universidad Nacional de Entre Ríos (workplace of the authors of this study), where it was observed that poultry effluents from activated sludge had 120 mg of nitrogen L^{-1} of effluent; 82 mg of phosphorus L^{-1} of effluent; a COD of 559 mg L^{-1}; 2.20 × 10^4 CFU of total coliform bacteria mL^{-1} of effluent and 3.40 × 10^4 CFU of *Escherichia coli* mL^{-1} of effluent (Jiménez-Veuthey et al. 2019). These values widely exceed the discharge limits to watercourses established in Law 6260 of the province of Entre Ríos and in Resolution N° 336/03 of the province of Buenos Aires.

Regarding sanitation, one of the biggest problems observed in the waters of the Río Uruguay coast is the presence of significant concentrations of coliform bacteria. When these bacteria come into contact with humans, they can cause multiple conditions, from fever, diarrhea, nausea, and vomiting, including Hemolytic Uremic Syndrome (HUS). In Argentina, the mortality rate for those suffering from HUS is 3–5%. In this country, between 300 and 350 cases of HUS per year have been reported in children under 5 years old (Ministerio de Salud 2014).

With the purpose of offering alternatives for the management and handling of effluents aimed at mitigating the environmental and social impacts caused by effluents from poultry industries, the research group of the LAMAS proposed phycoremediation technology to these industries.

To reach the goal, these researchers worked on the isolation of native microalgae able to remediation the poultry effluent, and then they carried out laboratory-scale studies to evaluate the effectiveness of the isolated microalgae in phycoremediation and, finally, scale up the technology to industry.

5.4.2 Isolation of Native Microalgae

Water samples from Embalse Salto Grande, located in the North-East of Argentina, were collected monthly between June and September 2016. The reservoir is in a warm tropical area with an average annual temperature of 23, 16.8, and 31 °C minimum and maximum temperatures, respectively (INTA 2017). The sample was collected at 30° 59′ 19.07″ South latitude and 57° 54′ 19.24″ West latitude.

Water samples were obtained by a five-minute horizontal dragging on the dam surface. A plankton net with a 20 μm mesh size was used (Hidro-Bios©, D-2300 Kiel). Once collected, the samples were placed in 250 mL sterile flasks and maintained at 6 ± 1 °C until they were carried to the laboratory.

Each sample was observed with an optical microscope, and in those where *Scenedesmus obliquus* was present, centrifugation was carried out. The precipitate was cultivated in Allen & Arnon culture medium enriched with a nitrogen source, which was monitored microscopically. Once a greater number of microalgae cells were detected, these cells were placed in sterile distilled water and a series of dilutions were made. Next, the same steps were repeated until only microalgae cells were observed. Microalga was identified microscopically using taxonomic keys (Edmondson 1959; Bourrelly 1966) and with molecular techniques through ARNr 18S ITS region analysis. Results of alignment indicated a 99% identity correspondence with *Scenedesmus obliquus*.

In the microscopic, the microalgae cells with an intense green color were observed. Cells were ovoid, ellipsoid, fusiform, or spherical in shape with a well-defined apical crest, a narrowing central section in small individuals and very small or no apical crest in larger cells (Fig. 5.5a–c). Some cells were found as single cells (2–7 μm wide and 8–5 μm long) and also as 4–8 celled coenobium of 16–21 μm wide and 17–27 μm long, and 24–34 μm wide and 36–52 μm long, respectively. Cells were arranged in single series, connected to each other by three-fourth wall while 8-celled coenobium were arranged into two alternating 4 cell rows (Fig. 5.5d–i). Intermediate cells were arranged rhomboid shaped. However, distal cells were flat convex, with an outward concavity. Each cell contained a central nucleus, with a cup-shaped chloroplast surrounded by a membrane and an eyespot. The large chloroplast had a simple pyrenoid embedded in the basal region of the cell.

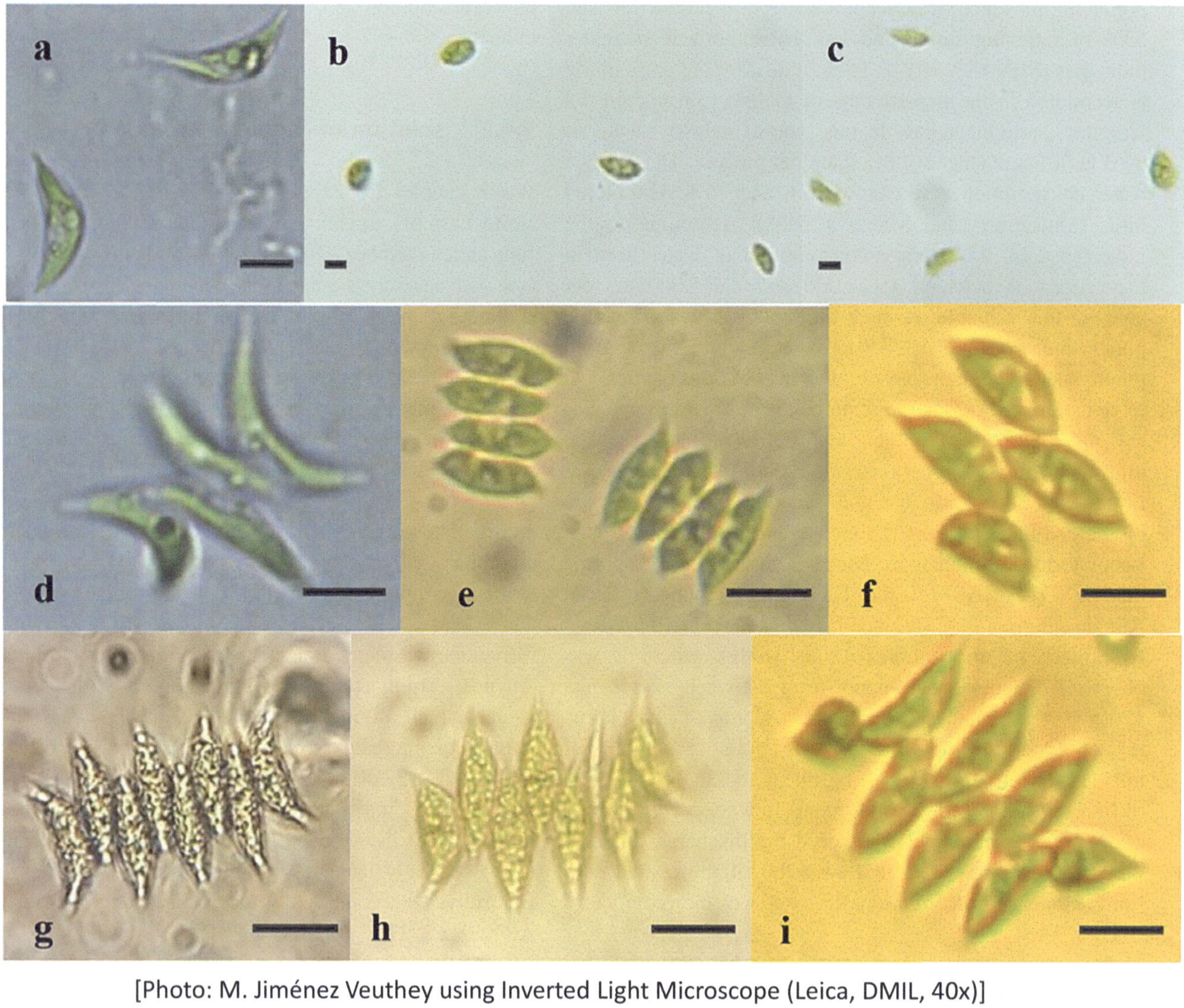

Fig. 5.5 *Scenedesmus obliquus* cell images under an inverted microscope (Leica, DMIL; 40x): **a–c**, single cell; **d–f**, 4-celled coenobium; **g–i**, 8-celled coenobium. Scale bars: Fig. 5.2 a = 30 μm; Fig. 5.2 b–c = 10 μm; Fig. 5.2 d–i = 50 μm

5.4.3 Poultry Effluent Treatment Using Microalgae

5.4.3.1 Poultry Effluent as Culture Media

Poultry effluent was used as culture media. This effluent came from an active sludge treatment plant of "Las Camelias" company, located in San José (Entre Ríos, Argentina) (Jiménez-Veuthey et al. 2019).

Different assays using poultry effluent and microalgae were carried out in bubble-column bioreactors in order to study the behavior and growth of microalgae in this media. Three bioreactors were filled with 200 mL of poultry effluent and 50 mL of microalgae; they were placed in a culture chamber (400H-MGC, China) with automatic control of temperature, humidity, and photoperiod. The temperature, humidity, and photoperiod inside the culture chamber were kept at 25 °C, 65% and 12 h light and 12 h dark, respectively. The irradiance inside the culture was measured using a spherical SQS- LI 192SA sensor (LICOR, USA), this parameter was maintained in 870 μmol photons m^{-2} s^{-1} using eight 36 W fluorescent tubes (Jiménez-Veuthey et al. 2019).

Bioreactors were operated in batch mode until the stationary phase was achieved. pH was maintained at 8.0 ± 0.1 with CO_2 injection on-demand and bioreactors were shaken using an air flow of 0.20 v v^{-1} min.

5.4.4 Study of Dilution Rate at Laboratory Scale

Four dilution rates of poultry effluent (0.20; 0.25; 0.30, 0.35 day^{-1}) were assayed. Thus, each dilution rate was performed in triplicate. The 12 bubble-column bioreactors of 250 mL were performed at a laboratory scale. These were inoculated with secondary undiluted poultry effluents and 20% of microalgae inoculum. All bioreactors were placed in a culture chamber (400H-MGC, China) with automatic control of temperature, relative humidity, and photoperiod. The photoperiod (12:12 h light/dark) was controlled with 8 fluorescent tubes of 36 W each one, while the irradiance was maintained at 870 μmol photons m^{-2} s^{-1}, using a spherical sensor SQS-LI 192SA, USA). In addition, bioreactors were aerated with an air's flow of 0.20 v v^{-1} min and the pH was set at 8.0 ± 0.1 using CO_2 injection on-demand.

In the beginning, all experiments were operated in batch mode (around 4 days) and then in a semi-continuous mode until the stationary phase (where the biomass concentration remained constant for at least 3 days). Semi-continuous mode was performed by harvesting 20–35% of culture daily and replacing it with fresh secondary effluent. Finally, the microalgae culture was centrifuged and the supernatant was used for treatment efficiency determination (Jiménez-Veuthey et al. 2019).

5.4.4.1 Laboratory-Scale Results

The highest biomass productivity (0.57 $g_{biomass}L^{-1}$ day^{-1}) was achieved at a dilution rate of 0.30 day^{-1}. With a dilution rate of 0.20 day^{-1} the highest nutrient removal efficiency was obtained. Thus, the N and P removal efficiencies were up to 80% and up to 99% for *E. coli*. However, the COD removal efficiency was above 90% at all dilution rates tested.

In this sense, the use of poultry effluents as a culture medium for *Scenedesmus obliquus* production can be adjusted according to the aim, i.e., a dilution rate of 0.30 day^{-1} is appropriate for the highest biomass production and 0.20 day^{-1} for highest nutrient removal (Jiménez-Veuthey et al. 2019).

5.4.5 Bioreactors and Operational Conditions at Pilot Scale

Experiments were carried out at a large-scale raceway of 60 m^3, located in the poultry industry "Las Camelias S.A." (San José, Entre Ríos, Argentina). Previously, this bioreactor was inoculated with culture from a smaller raceway (6 m^3).

The large-scale raceway of 1.36 m^2 of illuminated surface and a depth of 0.40 m was built of concrete with two 43.4 m length channels, 3.8 m width connected by 180° bends at each end (Jiménez-Veuthey et al. 2019). In addition, the culture mixing speed was set at 0.2 m s^{-1} provided by a six-bladed paddle wheel driven by a 0.1-kW electric motor. Initially, the large-scale raceway was filled with poultry effluent from anaerobic digester and 10% of *Scenedesmus obliquus*'s inoculum from the smaller raceway. This bioreactor was operated in batch mode for 1 week to allow the stabilization of microalgae. After that, started the semi-continuous experiment, in this case, the poultry effluent flow rate was set at 12 m^3 per day using an automatically controlled valve (Cepex L10, Spain), which was equivalent to a hydraulic retention time (HRT) of 5 days. The dilution rate of 0.20 day^{-1} was maintained throughout the year. Likewise, fresh medium daily was added to the reactor to offset the evaporation.

The same operational conditions, in the four seasons, were maintained until a stable steady state was achieved (i.e., constant dry biomass weight). Solar radiation and external temperature were obtained from the local meteorological station "Estación Meteorológica Automática San José." While the pH was controlled using a pH-meter AD1030 (Adwa, China) (Jiménez-Veuthey et al. 2019).

5.4.5.1 Results

There are several aspects to be considered to achieve a successful phycoremediation process; among them: (a) select a robust strain with high tolerance to nutrients and external conditions, (b) use an appropriate bioreactor, and (c) use an optimal dilution rate of the fresh media (Jiménez-Veuthey et al. 2022). Thus, in the present study, *Scenedesmus obliquus* was the strain selected which is native to the Uruguay River (Salto Grande region, Entre Ríos, Argentina) and this is an advantage for the strain's adaptation at the external conditions where the raceways are located. The Salto Grande region presents a mild climate where the mean annual temperature is 19.0 ± 1.0 °C, the relative humidity is 72 ± 2% and solar radiation is 921 ± 5 Wm^{-2}.

Microalgae requires different nutrients, among them, macro and micronutrients, however, the most important nutrients for their growth are nitrogen and phosphorus. In effluents from the poultry industry, nitrogen and phosphorus are present, mainly, as ammonia and phosphate, respectively (Jiménez-Veuthey et al. 2019). Authors such as Rawat et al. (2011) and Renuka et al. (2015) have shown that microalgae are efficient in removing nutrients and bacteria from wastewater. However, it is important to study their behavior and removal efficiency when the season and effluent composition change.

In order to determine the removal efficiency of phycoremediation process, Table 5.3 shows the poultry effluent's composition at the inlet and outlet, as well as the removal percentage of each one of them (Jiménez-Veuthey et al. 2022). At the end of phycoremediation process, a removal of up to 90% for nitrogen, phosphorous, *Escherichia coli*, calcium, iron, and COD was observed. Thus, the nutrient removal from effluent by microalgae involves several metabolic pathways present in the cell, where the components C, N, and P are the most important for the growth of microalgae.

This study demonstrated that microalga *Scenedesmus obliquus* is capable of efficient nutrient removal in 5 days

Table 5.3 Poultry effluent composition and removal percentages after the phycoremediation process

Poultry effluent composition	Inlet concentration (mg L^{-1})	Outlet concentration (mg L^{-1})	Removal (%)
Total nitrogen	132.00	13.00	90.15
Total Kjeldahl nitrogen	53.45	3.22	93.98
Ammonia nitrogen	48.00	0.64	98.67
Organic nitrogen	5.53	2.58	53.35
Nitrate	3.00	0.65	78.33
Chemical oxygen demand	597.00	38.00	93.63
Total phosphorus	11.30	0.37	96.73
Sodium	37.36	19.23	48.53
Potassium	46.00	45.69	0.67
Calcium	9.49	14.53	90.15
Magnesium	5.00	4.57	8.60
Zinc	0.61	0.32	47.54
Iron	1.70	0.54	98.15
Escherichia coli (CFU mL^{-1})	3.40E+04	14.00	99.94

due to its robustness and, therefore, it is promising for large-scale phytoremediation processes. Therefore, it is technically feasible to treat effluents from the poultry industry without the need to dilute them. This work demonstrates the efficiency of the phycoremediation process for agro-industrial wastewater treatment, which can be coupled with current systems as tertiary treatment.

The results achieved at laboratory scale motivate the scale up of the process, in order to reduce the impact on the environment and improve the population's quality of life quality. Therefore microalgae-based systems constitute a viable alternative to conventional systems in terms of sustainability and costs.

5.5 Conclusions

Microalgae biotechnology can contribute positively to the circular economy through multiple processes from food and feed production to agro-industrial wastewater treatment. Thus, the tolerance of microalgae to different environmental conditions and their high productivity make them suitable for multiple applications. In addition, it is possible to couple the agro-industrial wastewater treatment with microalgae production. This constitutes a viable alternative to conventional systems in terms of sustainability and costs.

However, as the production capacity is still low and production costs are high, one of the short-term aims is to improve the production and harvesting systems, mainly in the development of new technologies.

Acknowledgments Authors thank the financial support for their research from Universidad Nacional de Entre Ríos, Secretaría de Políticas Universitarias (SPU, Argentina), Secretaría de Modernización del Estado, Ciencia y Tecnología of the Entre Ríos (PID Novel N° 8091, SPU N° 9312 projects and Programa de Fortalecimiento de I + D, respectively) and "Las Camelias SA" (Entre Ríos). Also, they acknowledge the Asociación Universitaria Iberoamericana de Postgrado (AUIP), Universidad Nacional de la Patagonia San Juan Bosco (Argentina), and the Universidad de Almería (España) for the MJV mobility scholarship and collaboration in research.

References

Acién Fernández FG, Fernández-Sevilla E, Grima M (2013) Photobioreactors for the production of microalgae. Rev Environ Sci Biotechnol 12:131–151

Acién Fernández FG, Fernández Sevilla JM, Molina Grima E (2017) Microalgae: the basis of mankind sustainability, 123–140. https://doi.org/10.5772/67930

Acién Fernández FG, Fernández-Sevilla JM, Molina-Grima E (2018) Contribución de las microalgas al desarrollo de la bioeconomía. Mediterráneo Económico 31:309–332

Ahmed SF, Mofijur M, Nuzhat S, Chowdhury AT, Rafa N, Uddin MA, Inayat A, Mahlia TMI, Ong HC, Chia WY, Show PL (2021) Recent developments in physical, biological, chemical, and hybrid treatment techniques for removing emerging contaminants from wastewater. J Hazard Mater 416:125912

Arbib Z, de Godos I, Ruiz J, Perales JA (2017) Optimization of pilot high rate algal ponds for simultaneous nutrient removal and lipids production. Sci Total Environ 589:66–72

Ayre JM, Moheimani NR, Borowitzka MA (2017) Growth of microalgae on undiluted anaerobic digestate of piggery effluent with high ammonium concentrations. Algal Res 24:218–226

Beltrán-Rocha JC, Guajardo-Barbosa C, Barceló-Quintal ID, López-Chuken UJ (2017) Biotratamiento de efluentes secundarios municipales utilizando microalgas: Efecto del pH, nutrientes (C, N y P) y enriquecimiento con CO_2. Rev Biol Mar Oceanogr 52:417–427

Bhattacharjee M, Siemann E (2015) Low algal diversity systems are a promising method for biodiesel production in wastewater fed open reactors. Algae 30:67–79

Bilal M, Adeel M, Rasheed T, Zhao Y, Iqbal HM (2019) Emerging contaminants of high concern and their enzyme-assisted biodegradation–a review. Environ Int 124:336–353

Bourrelly P (1966) Les algues d'eau douce. Initiation à la systématique. Tome I. Les algues vertes. Boubée & Cie, París. 511 pp

Brenan M, Owende P (2010) Biofuels from microalgae – a review of technologies for production, processing, and extraction of biofuels and co-products. Renew Sust Energ Rev 14:557–577

Cabello J, Toledo-Cervantes A, Sanchez L, Revah S, Morales M (2015) Effect of the temperature, pH and irradiance on the photosynthetic activity by *Scenedesmus obtusiusculus* under nitrogen replete and deplete conditions. Bioresour Technol 181:128–135

Chisti Y (2013) Raceways-based production of algal crude oil. Green 3: 195–216

Chu H, Tan X, Zhang Y, Yang L, Zhao F, Guo J (2015) Continuous cultivation of *Chlorella pyrenoidosa* using anaerobic digested starch processing wastewater in the outdoors. Bioresour Technol 185:40–48

Cosgrove WJ, Loucks DP (2015) Water management: current and future challenges and research directions. Water Resour Res 51:4823–4839

Craggs R, Sutherland D, Campbell H (2012) Hectare-scale demonstration of high rate algal ponds for enhanced wastewater treatment and biofuel production. J Appl Phycol 24:329–337

Daneshvar E, Zarrinmehr MJ, Koutra E, Kornaros M, Farhadian O, Bhatnagar A (2019) Sequential cultivation of microalgae in raw and recycled dairy wastewater: microalgal growth, wastewater treatment and biochemical composition. Bioresour Technol 273:556–564

Directive 91/271/CEE (2003) Tratamiento de las aguas residuales urbanas. Manual de interpretación y elaboración de informes. Retrieved from https://www.miteco.gob.es/es/agua/publicaciones/03_Manual_Directiva_91_271_CEE_tcm30-214069.pdf on 6 April 2022

Djelal H, Tahrani L, Fathallah S, Cabrol A, Ben Mansour H (2014) Treatment process and toxicities assessment of wastewater issued from anaerobic digestion of household wastes. Environ Sci Pollut Res 21:2437–2447

Dragone G (2022) Challenges and opportunities to increase economic feasibility and sustainability of mixotrophic cultivation of green microalgae of the genus *Chlorella*. Renew Sust Energ Rev 160: 112284

Edmondson WT (ed) (1959) Fresh-water biology. Wiley, New York, pp 115–170

García-Einschlag FS (ed) (2011) Wastewater management, waste water – evaluation and management, vol 20. InTech, Cham, 379

Gomes J, Lincho J, Domingues E, Quinta-Ferreira RM, Martins RC (2019) N–TiO_2 photocatalysts: a review of their characteristics and capacity for emerging contaminants removal. Water 11:373

Gonçalves AL, Pires JCM, Simões M (2017) A review on the use of microalgal consortia for wastewater treatment. Algal Res 24:403–415

González-Fernández C, Mahdy A, Ballesteros I, Ballesteros M (2016) Impact of temperature and photoperiod on anaerobic biodegradability of microalgae grown in urban wastewater. Int Biodeterior Biodegradation 106:16–23

Henkanatte-Gedera SM, Selvaratnam T, Karbakhshravari M, Myint M, Nirmalakhandan N, Van Voorhies W, Lammers PJ (2017) Removal of dissolved organic carbon and nutrients from urban wastewaters by *Galdieria sulphuraria*: laboratory to field scale demonstration. Algal Res 4:450–456

Hongyang S, Yalei Z, Chunmin Z, Xuefei Z, Jinpeng L (2011) Cultivation of *Chlorella pyrenoidosa* in soybean processing wastewater. Bioresour Technol 102:9884–9890

Huang L, Lee DJ (2015) Membrane bioreactor: a mini review on recent R&D works. Bioresour Technol 194:383–388

INTA (Instituto Nacional de Tecnología Agropecuaria) (2017) Sistema de Información y Gestión Agrometeorológica (SIGA). Retrieved from http://siga.inta.gob.ar/#/ on 30 August 2017

Jiménez-Veuthey M, Zapata LM, Flores ML (2019) Phycoremediation and microalgae production using poultry effluent from anaerobic digestion as a culture medium. J Environ Biotechnol Res 8:52–64

Kim H (2014) Nutrient removal and biofuel production in high rate algal pond using real municipal wastewater. J Microbiol Biotechnol 24: 1123–1132

Kim MK, Park JW, Park CS, Kim SJ, Jeune KH, Chang MU, Acreman J (2007) Enhanced production of *Scenedesmus* spp. (green microalgae) using a new medium containing fermented swine wastewater. Bioresour Technol 98:2220–2228

Larronde-Larretche M, Jin X (2022) The influence of forward osmosis module configuration on nutrients removal and microalgae harvesting in osmotic photobioreactor. Membranes 12(9):892. https://doi.org/10.3390/membranes12090892

Lau KY, Pleissner D, Sze C, Lin K (2014) Recycling of food waste as nutrients in *Chlorella vulgaris* cultivation. Bioresour Technol 170: 144–151

Law 6260 (1992). Retrieved from http://www.entrerios.gov.ar/ambiente/ambiente_flujograma/descargas/Ley_6260.pdf on: 6 April 2022

Ledda C, Villegas GIR, Adani F, Fernández FGA, Grima EM (2015) Utilization of centrate from wastewater treatment for the outdoor production of *Nannochloropsis gaditana* biomass at pilot-scale. Algal Res 12:17–25

Lee CS, Lee SA, Ko SR, Oh HM, Ahn CY (2015) Effects of photoperiod on nutrient removal, biomass production, and algal-bacterial population dynamics in lab-scale photobioreactors treating municipal wastewater. Water Res 68:680–691

Li Y, Chen YF, Chen P, Min M, Zhou W, Martinez B, Zhu J, Ruan R (2011) Characterization of a microalga *Chlorella* sp. well adapted to highly concentrated municipal wastewater for nutrient removal and biodiesel production. Bioresour Technol 102:5138–5144

Li Y, Zhou W, Hu B, Min M, Chen P, Ruan RR (2012) Effect of light intensity on algal biomass accumulation and biodiesel production for mixotrophic strains *Chlorella kessleri* and *Chlorella protothecoide* cultivated in highly concentrated municipal wastewater. Biotechnol Bioeng 109:2222–2229

Liu C, Subashchandrabose S, Ming H, Xiao B, Naidu R, Megharaj M (2016) Phycoremediation of dairy and winery wastewater using *Diplosphaera* sp. MM1. J Appl Phycol 28:3331–3341

Ma C, Wen H, Xing D, Pei X, Zhu J, Ren N, Liu B (2017) Molasses wastewater treatment and lipid production at low temperature conditions by a microalgal mutant *Scenedesmus* sp. Z-4. Biotechnol Biofuels 10:111

Mahdy AA (2016) Impact of temperature and photoperiod on anaerobic biodegradability of microalgae grown in urban wastewater. Int Biodeterior Biodegradation 106:16–23

Matsumura E, Mierzwa J (2008) Water conservation and reuse in poultry processing plant – a case study. Resour Conserv Recy 52: 835–842

McGinn PJ, Dickinson KE, Bhatti S, Frigon JC, Guiot SR, O'Leary SJB (2011) Integration of microalgae cultivation with industrial waste remediation for biofuel and bioenergy production: opportunities and limitations. Photosynth Res 109:231–247

Ministerio de Salud (2014) Síndrome Urémico Hemolítico (SUH) en Argentina, 2010–2013. Extracto del Boletín Integrado de Vigilancia. Retrieved from http://www.msal.gob.ar/images/stories/bes/graficos/0000000799cnt-201408_Informe

Montenegro-Gómez SP, Osorio JA (eds) (2019) Servicios ecosistémicos: Un enfoque introductorio con experiencias del occidente colombiano. UNAD, p 146

Mulbry W, Kondrad S, Buyer J (2008) Treatment of dairy and swine manure effluents using freshwater algae: fatty acid content and composition of algal biomass at different manure loading rates. J Appl Phycol 20:1079–1085

Olguín EJ (2012) Dual purpose microalgae–bacteria-based systems that treat wastewater and produce biodiesel and chemical products within a biorefinery. Biotechnol Adv 30:1031–1046

Olguin EJ, Galicia S, Mercado G, Perez T (2003) Annual productivity of *Spirulina* (*Arthrospira*) and nutrient removal in a pig wastewater recycling process under tropical conditions. J Appl Phycol 15:249–257

Oliveira AC, Barata A, Batista AP, Gouveia L (2018) *Scenedesmus obliquus* in poultry wastewater bioremediation. Environ Technol 40:3735–3744

ONU (2015) Nota informativa sobre Implementación de mejoras para la calidad del agua y la protección de servicios ecosistémicos, Agua y Desarrollo Sostenible: de la visión a la acción. Conferencia Anual 2015 de ONU, Zaragoza. Retrieved from http://www.un.org/spanish/waterforlifedecade/waterandsustainabledevelopment2015/information_briefs.shtml on 6 April 2022

OXFAM International (2021) Water and sanitation. Retrieved from https://www.oxfam.org/en/what-we-do/issues/water-and-sanitation on 6 April 2022

Park JBK, Craggs RJ (2010) Wastewater treatment and algal production in high rate algal ponds with carbon dioxide addition. Water Sci Technol 61:633–639

Park JBK, Craggs RJ, Shilton AN (2011) Wastewater treatment high rate algal ponds for biofuel production. Bioresour Technol 102:35–42

Pereira MV, Dassoler AF, Antunes PW, Gonçalves RF, Cassini ST (2019) Indigenous microalgae biomass cultivation in continuous reactor with anaerobic effluent: effect of dilution rate on productivity, nutrient removal and bioindicators. Environ Technol 14:1–13

Rawat I, Ranjith Kumar R, Mutanda T, Bux F (2011) Dual role of microalgae: phycoremediation of domestic wastewater and biomass production for sustainable biofuels production. Appl Energy 88: 3411–3424

Rawat I, Ranjith Kumar R, Mutanda T, Bux F (2013) Biodiesel from microalgae: a critical evaluation from laboratory to large scale production. Appl Energy 103:444–467

Renuka N, Sood A, Prasanna R, Ahluwalia AS (2015) Phycoremediation of wastewaters: a synergistic approach using microalgae for bioremediation and biomass generation. IJSR 12: 1443–1460

Renuka N, Prasanna R, Sood A, Ahluwalia AS, Bansal R, Babu S, Singh R, Shivay YS, Nain L (2016) Exploring the efficacy of wastewater-grown microalgal biomass as a biofertilizer for wheat. ESPR 23:6608–6620

Resolution N° 554 (2015). Retrieved from https://www.entrerios.gov.ar/ambiente/userfiles/files/archivos/Normativas/Res_%20554-15.pdf on 6 April 2022

Riaño B, Molinuevo B, García-Gonzalez MC (2011) Treatment of fish processing wastewater with microalgae-containing microbiota. Bioresour Technol 102:10829–10833

Rosso D, Larson LE, Stenstrom MK (2008) Aeration of large-scale municipal wastewater treatment plants: state of the art. Water Sci Technol 57:973–978

Rout PR, Zhang TC, Bhunia P, Surampalli RY (2021) Treatment technologies for emerging contaminants in wastewater treatment plants: a review. Sci Total Environ 753:141990

Ruiz-Marin A, Mendoza-Espinosa LG, Stephenson T (2010) Growth and nutrient removal in free and immobilized green algae in batch and semi-continuous cultures treating real wastewater. Bioresour Technol 101:58–64

Sawayama S, Minowa T, Dote Y, Yokoyama S (1992) Growth of the hydrocarbon-rich microalga *Botryococcus braunii* in secondarily treated sewage. Appl Microbiol Biotechnol 38:135–138

SENASA (Servicio Nacional de Sanidad y Calidad Agroalimentaria) (2019) Aves sector primario. Retrieved from https://www.argentina.gob.ar/senasa/mercados-y-estadisticas/estadisticas/animal-estadisticas/aves on 10 April 2022

Singh M, Reynolds DL, Das KC (2011) Microalgal system for treatment of effluent from poultry litter anaerobic digestion. Bioresour Technol 102:10841–10848

Su Y, Mennerich A, Urban B (2011) Municipal wastewater treatment and biomass accumulation with a wastewater-born and settleable algal-bacterial culture. Water Res 45:3351–3358

Tao F, Yokozawa M, Hayashi Y, Lin E (2005) A perspective on water resources in China: interactions between climate change and soil degradation. Clim Chang 68:169–197

Uggetti E, Sialve B, Latrille E, Steyer JP (2014) Anaerobic digestate as substrate for microalgae culture: the role of ammonium concentration on the microalgae productivity. Bioresour Technol 152:437–443

United Nations (2021) The United Nations World Water Development Report. Valuing Water. Retrieved from https://unesdoc.unesco.org/ark:/48223/pf0000375724 on 6 April 2022

United Nations Development Program (2022) Sustainable Development Goals, Goal 6: Clean Water and Sanitation. Retrieved from https://www.undp.org/sustainable-development-goals#clean-water-and-sanitation on 2 April 2022

UNW-DPAC (2015) Agua y desarrollo sostenible. Retrieved from https://www.un.org/spanish/waterforlifedecade/waterandsustainabledevelopment2015/pdf/03_sustainable_development_esp.pdf on 6 April 2022

Usha MT, Chandra TS, Sarada R, Chauhan VS (2016) Removal of nutrients and organic pollution load from pulp and paper mill effluent by microalgae in outdoor open pond. Bioresour Technol 214: 856–860

Verma P (ed) (2022) Micro-algae: next-generation feedstock for biorefineries. Cultivation and refining processes. Springer, Singapore. pp 250. https://doi.org/10.1007/978-981-19-0793-7

Villegas GIR, Fiamengo M, Fernández FGA, Grima EM (2017) Outdoor production of microalgae biomass at pilot-scale in seawater using centrate as the nutrient source. Algal Res 25:538–548

Wang L, Li Y, Chen P, Min M, Chen Y, Zhu J, Ruan RR (2010) Anaerobic digested dairy manure as a nutrient supplement for cultivation of oil-rich green microalgae *Chlorella* sp. Bioresour Technol 101:2623–2628

Woertz I, Feffer A, Lundquist T, Nelson Y (2009) Algae grown on dairy and municipal wastewater for simultaneous nutrient removal and lipid production for biofuel feedstock. J Environ Eng 135:1115–1122

WWAP (United Nations World Water Assessment Programme) (2016). The United NationsWorld Water Development Report 2016: Water and Jobs. UNESCO, Paris

Yangüez K, Lovazzano C, Contreras-Porcia L, Ehrenfeld N (2015) Response to oxidative stress induced by high light and carbon dioxide (CO_2) in the biodiesel producer model *Nannochloropsis salina* (Ochrophyta, Eustigmatales). Rev Biol Mar Oceanogr 50: 163–175

Zerrouki D, Henni A (2019) Outdoor microalgae cultivation for wastewater treatment. In: Gupta SK, Bux F (eds) Application of microalgae in wastewater treatment. Springer, Cham. https://doi.org/10.1007/978-3-030-13913-1_5

Zhang TY, Wu YH, Zhu S, Li FM, Hu HY (2013) Isolation and heterotrophic cultivation of mixotrophic microalgae strains for domestic wastewater treatment and lipid production under dark condition. Bioresour Technol 149:586–589

Zhao S, Ren T, Ma L et al (2022) Multi-period planning of municipal solid waste management: a case study in Qingdao. Process Integr. Optim. Sustain. https://doi.org/10.1007/s41660-022-00279-7

Zheng H, Liu M, Lu Q, Wu X, Ma Y, Cheng Y, Addy M, Liu Y, Ruan R (2018) Balancing carbon/nitrogen ratio to improve nutrients removal and algal biomass production in piggery and brewery wastewaters. Bioresour Technol 249:479–486

Zhou Q, Sun H, Jia L, Wu W, Wang J (2022) Simultaneous biological removal of nitrogen and phosphorus from secondary effluent of wastewater treatment plants by advanced treatment: a review. Chemosphere 296:134054

Zhu LD, Xu ZB, Qin L, Wang ZM, Hiltunen E, Li ZH, Xu ZB, Qin L, Wang ZM, Hiltunen E, Oil ZHL (2016) Oil production from pilot-scale microalgae cultivation: an economics evaluation. Energy Sources B Econ Plan Policy 11:11–17

Treatment of Industrial Wastewater with Microalgae

6

Mónica Cristina Rodríguez-Palacio, Cruz Lozano-Ramírez, and Marisol Martínez-Hernández

Abstract

For some time, microalgae cultures have been studied in municipal wastewater bioremediation processes. It has been shown that they are an efficient alternative to remove pollutants from them, in this type of effluent the main pollutants are nitrogen and phosphorus, which are assimilated by microalgae for their growth, in the form of ammonium and orthophosphates, respectively. However, the treatment of industrial wastewater represents a difficult challenge to overcome due to the complexity of the effluents that vary according to the industry. This means that in some cases the effluents cannot be treated using microalgae cultures directly, but a previous treatment is needed to allow them to grow. Some investigations show us that the use of microalgae cultures in free state presents limitations in the treatment of industrial wastewater, due to the inherent toxicity of the effluent as well as the competition between native and foreign populations of microalgae as well as other microorganisms, whereas immobilized cultures on different supports provide the microorganisms with protection against the toxic effect of the substances present in the water being treated and with protection against predation by other microorganisms. In addition to this, the immobilization of the microalgae facilitates the harvesting process, reducing costs and facilitating the subsequent use of the harvested biomass. In this chapter, we will analyze the bioremediation capacity of microalgae in the treatment of industrial wastewater of different origins using free and immobilized cultures. The bioremediation process of industrial wastewater using microalgae should be managed as a virtuous circle where environmental service, biomass production, and the production of bioproducts with high-added value are considered.

M. C. Rodríguez-Palacio (✉) · C. Lozano-Ramírez · M. Martínez-Hernández
Laboratory of Applied Phycology, Hydrobiology Department, Metropolitan Autonomous University, Mexico City, Mexico
e-mail: mony@xanum.uam.mx

Keywords

Microalgae culture · Industrial wastewater · Bioremediation · Immobilized cultures

6.1 Introduction

One of the main problems of contamination that we confront today is contamination from aquifers, due to poor management and the treatment of wastewater. The wastewater are defined as waters of varied composition arising from uses such as urban public, domestic, commercial, from services, agricultural, livestock, from treatment plants and, in general, of any use, as well as the mixture of these. These wastewater are classified as municipal wastewater and as industrial wastewater. Municipal wastewater are those that are generated in homes and communities as the product of daily use, from the washing of food, floors, and organic wastes; however, due to urban growth and the proximity of industries, other contaminants are also present. The most common components in municipal wastewater are solids in suspension, biodegradable organic materials, pathogens, nutrients, refractory organic material, medicaments, heavy metals, and dissolved organic solids (Tchobanoglous et al. 2003; García et al. 2006; Zheng et al. 2011).

The content of total solids is the most important physical characteristic of wastewater; these are compounds that comprise floating material and material in suspension, in colloidal dispersion, and in dissolution. Among the pathogens present in the effluent, we find the following highlighted: bacteria; parasites (protozoans and helminths), and viruses. Some of the bacteria found, as well as the diseases to which they give rise, include *Salmonella* (salmonellosis), *Shigella* (bacillary dysentery), *Vibrio cholerae* (cholera), and *Campylobacter jejuni* (bacterial diarrhea). The eggs and larvae of helminths include the stomach parasite *Ascaris lumbricoides*, *Taenia saginata* (beef tapeworm), *Taenia solium* (pork tapeworm), the whipworm *Trichuris trichuria*, and the hookworm

A. de Jesús Martínez-Roldán (ed.), *Biotechnological Processes for Green Energy, and High Value Bioproducts by Microalgae, and Cyanobacteria Cultures*, Developments in Applied Phycology 13, https://doi.org/10.1007/978-3-031-43969-8_6

Ancylostoma duodenale. The most important enteric viruses and those that can be found in wastewater include enteroviruses (polio, echo, and coxsackie), Norwalk virus, rotavirus, rhinovirus, adenovirus, and the hepatitis A virus (Araujo et al. 1997; Campos 2003; Hanan et al. 2003; Tchobanoglous et al. 2003).

Industrial wastewater are wastewater that are that dumped from premises utilized for carrying out any commercial or industrial activity, that are not domestic wastewater, nor rainfall runoff waters. These waters contain compounds that are not eliminated by conventional treatments, due to two possible factors: their high concentration, or their chemical nature. Many organic and inorganic compounds found in industrial wastewater are the object of special regulation because of their long-term toxic effects or their bacterial effect (Araujo et al. 1997; Hanan et al. 2003; Tchobanoglous et al. 2003).

Among the principal components in industrial wastewater, we find arsenic, cadmium, chlorides, chromium, copper, cyanides, fluorides, lead, mercury, nickel, organic tin compounds, phosphorus, zinc, benzene, toluene, ethylbenzene, xylenes, chloroalkanes, halogenated organics, among others. The chemical demand for oxygen and the biological demand for oxygen are indicators that, from the ecological point of view, are very important in the discharges of industrial wastewater; at the international level, there are established limits for these chemical parameters (Tchobanoglous et al. 2003; García et al. 2006; Zheng et al. 2011).

6.1.1 Problematic of Treatments for Industrial Wastewater

The emerging technologies for the treatment of these affluents include advanced oxidation, a treatment applied to the compounds or to the oxidable contaminating material and the membranes, and physicochemical factors such as the gradients of concentration and pressure.

Industrial wastewater by nature also present a great amount of highly recalcitrant contaminants, for which there exist the so-called advanced oxidation treatments, which are utilized in textile industry wastewater (Fig. 6.1) and with the presence of detergents (Duarte et al. 2013; Gil-Pavas 2005; Lapertot et al. 2006), in leachates deriving from sanitary landfills (Vedrenne et al. 2012), and also in wastewater contaminated with pesticides and phenols (Pulgarin 2015; Papoutsakis et al. 2015). Filtration processes with the use of membranes are considered as advanced processes of industrial effluent purification and, although there are processes for improving the quality of these, the wastewater that this generates contains a high content of metallic salts, which could be taken advantage of through electrochemical processes (Tchobanoglous et al. 2003; De Florio et al. 2005; Benitez et al. 2009; Escobar et al. 2011).

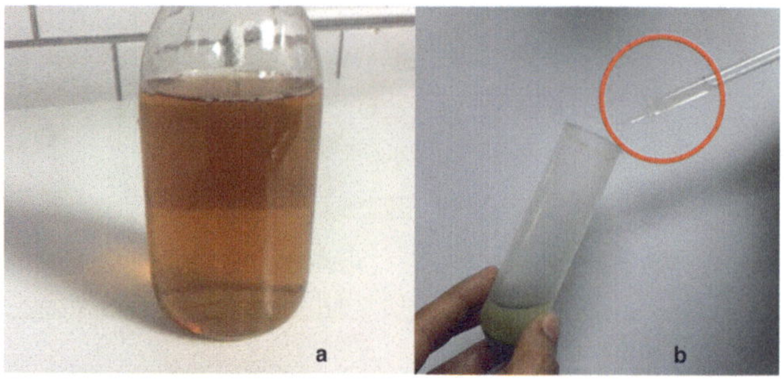

Fig. 6.1 (**a**) Wastewater from the textile industry. (**b**) Color removal after treatment with the microalgae *Lagerhemia* sp.

Some of the most contaminating industries are the following: farms breeding birds that discard feces and feathers; fish farms that release nutrient-enriched waters employed for the growth and fattening of fish; meat- and dairy-producing cattle ranches that not only consume great amounts of water, but that also release nutrient-enriched; the textile manufacturing industry that releases dyes and substances that alter the pH; the tanning industry, which releases detergents, acids, and sulfurs; the manufacture of paper that releases chelating substances and dissolved lignin chlorates; agriculture, which releases a great amount of nitrogenated products; the petroleum transformation and extraction industry, which releases crude oil, as well as the diverse chemicals used in its transformation, and mining, which releases cyanide, sulfuric acid, among various others (Choudri and Baawain 2014, 2015, 2016a, b).

Most industrial wastewater treatment plants eliminate soluble organic material; however, some compounds remain, such as ammonium, nitrates, phosphates, among others. Primary and secondary treatment processes are employed in many places with the purpose of eliminating easily sedimented particles (primary), and the oxidation of organic material (secondary). The final result of each of these treatment stages comprises an apparently clean effluent, but one containing a high concentration of nitrogen and phosphorus, which can cause the eutrophication of the receptor bodies and other problems in the long term due to the presence of compounds difficult to degrade (De la Noüe et al. 1992; Escorihuela et al. 2007).

The use of microalgae for the treatment of diverse types of wastewater has been studied over the last years (Phang et al. 2000; Jiang et al. 2011; Renuka et al. 2015; Powell et al. 2008), and these are employed in numerous parts of the world due to their capacity to eliminate from the water, nitrogen, as well as phosphorus, among others, demonstrating that the elimination of phosphorus can be as efficient as that of the conventional chemical treatment (Hoffman 1998). Their main advantages are a lower cost, in that chemical products are no longer necessary, and the recovery of nutrients in algal biomass form that can be employed as a biofertilizer (De Bashan and Bashan 2004), as biofuels (Bhatnagar et al. 2010; Li et al. 2011), and as other added-value products.

6.2 Microalgae and Cyanobacteria

Algae comprise a non-natural group of organisms. This means that they do not derive from a common ancestor. Historically, botanists have approached their study in that these share characteristics with plants, principally in their being photosynthetic. This group is composed of a miscellany of organisms with very dispersed characteristics (Keeling 2004; Lewis and McCourt 2004). Microalgae and cyanobacteria are present in soil, fresh water, and salt water. They are considered responsible for the organic material found in aquatic ecosystems, as well as for 40% of the total photosynthesis of the planet (Abalde et al. 1995), and they constitute the base of the trophic chains that permit life in the oceans, which is equivalent to approximately 70% of the world biomass. The adaptive characteristics that the microalgae present are related to their resistance to sinking, with the ability to absorb nutrients rapidly when these are found available or stored intracellularly for when they become scarce. The life cycles are short and include the formation of resistance spores or resting cells that help them to survive in unfavorable situations.

In the field of Biotechnology, advances have been made in the types and systems of the culture of these microorganisms utilized in bioremediation, in such a way that the produced algal biomass can be employed in turn as a producer source of fertilizers, animal food (poultry farming, pig farming, cattle ranching) and aquiculture, energy, bioplastics, in addition to other applications in the chemical, Biomedicine, and Pharmacology industries (Abalde et al. 1995; Chisti 2007; Raja et al. 2008; Hernández-Reyes et al. 2012; Sahua et al. 2012; Tarkashvand et al. 2014; Vanthoor-Koopmans et al. 2014; Gangl et al. 2015; Pienkos et al. 2011; Drexler et al. 2014; Morales-Avedaño et al. 2014).

6.3 Bioremediation

Bioremediation can be defined as the capacity of an organism to remove or destroy contaminants, if possible, or at least transform them into less harmful substances (Sharma 2012; Kumar et al. 2011), and can be carried out in soil, air, or water. With respect to the treatment of residual waters, phycoremediation has been utilized for some years now; it has been defined as serving for the use of aquatic plants, as well as for macro- or microalgae for the removal or biotransformation of contaminants (Olguín 2003, 2012). Microalgae can be used in wastewater rich in nutrients, metals, pharmacological residues, hydrocarbons, acids, and salts of different types, and even to desalinate water (Rao et al. 2011; Amezaga et al. 2014).

Bioremediation processes with algae are interesting due to the collateral benefits of culturing them. Not only is the removal of the contaminant obtained but also a great amount of biomass can be generated, which is susceptible to being used for many other purposes. There are many investigations focused on the use of microalgae for the treatment of wastewater due to the capacity of adaptation of these microorganisms to this type of effluent, observing rapid growth and a high capacity for removing nutrients. It was more than 50 years ago that Caldwell (1946), and later Oswald et al. (1953), suggested the utilization of the massive culture of microalgae for the treatment of wastewater and the simultaneous production of protein (Oswald 2003). However, recently this has been acquiring more relevance, because of the problems of contamination, climate change, and demographic growth, which render the increase of discharges of untreated wastewater. Therefore, there are increasingly fewer bodies of clean or pristine water.

Microalgae can be utilized for the tertiary treatment of wastewater due to their capacity to assimilate nutrients from the effluent, and once the cultures are established, the pH increases as a consequence of the algal growth, the latter favoring the precipitation of phosphorus and the volatilization of the ammonium. Microalgae cultures are a more cost-effective way to reduce biochemical oxigen demand, the pathenogenic agents, and the N and P of activated muds (Green et al. 1996). In terms of the consumption of nutrients, the forms of N that prefer microalgae include ammonium ($NH4^+$) and NO_3^- nitrates. In addition to these N compounds, urea (NH_2CONH_2) and nitrogen dioxide (NO_2^-) can be utilized as sources of N (Becker 1994). However, ammonium is found at concentrations higher than 20 mg per liter, which is toxic for these (Becker 1994; Larsdotter 2006).

Phosphorus is another nutrient that is essential for the growth of algae, and these assimilate it preferentially as inorganic phosphate. Its uptake comprises an active process that requires oxygen, energy, compounds such as adenosine and triphosphate, and nucleotide co-enzymes play a key role in photosynthesis. Microalgae are capable of accumulating phosphorus, which is stored in the granules of polyphosphates. These reserves can be sufficient for prolonged growth in the absence of available phosphorus. Thus, the growth rate of the algae cannot respond immediately to the changes in the external concentration of phosphorus, different from the immediate responses to light and temperature (Larsdotter 2006).

Microorganisms can capture heavy metals by absorption, and this biomass can be used for other purposes, whenever it does not surpass the sanitary norms permitted for heavy metals (Borde et al. 2002).

The bioremediation of wastewater using microalgal cultures presents advantages as important as the improvement of the quality of the effluent, through a mechanism of a low energy cost, in which nutrients that were being disposed of are now incorporated into the biomass, which can have a high commercial value (Pellon et al. 2003; Mahapatra et al. 2013; Mehrabadi et al. 2015).

6.4 Culture Types and Systems

Wastewater treatment installations with tanks can generally be relatively deep, but these are not as efficient in favoring photosynthesis (Kauser et al. 2014). Therefore, if microalgae are to be utilized for bioremediation in these wastewater, they can be cultured in photobioreactors or in open raceway-type tanks.

There are many discussions and dissertations on the advantages and disadvantages of open and closed culture systems, but in general, both of these are required to achieve a task at the industrial and pilot scale, it being necessary, in the first stages of the culture, to work in closed culture systems, thus ensuring the quality of the biomass and the innocuousness of the inoculums. Therefore, when these are seeded in industrial wastewater, they can adapt positively to the effluent (Fig. 6.2). In addition, at a higher scale, it is recommended to work always in open, not very deep, tanks, which facilitate the manipulation of the algae, optimize maximal biomass production, and require less investment for this production phase.

Depending on the nature of the effluent, physiochemical analyses should be conducted of same, and also, bioassays at laboratory scale, in which dilutions are prepared for testing the adaptation of the microorganisms to a high load of contaminants. On achieving success at laboratory scale, we can perform the bioassays at pilot scale and later, work with the effluent at a higher scale.

On increasing the scale of the cultures, we face a new challenge: the harvest. How will we separate high amounts of the effluent's algal biomass once it is clean? We can resort to industrial centrifuges, which would increase the operational cost; the most common method implies the addition of coagulants and flocculants, followed by separation of the microalgae. In both cases, the harvesting cost can minimize the viability of the process or, in the case of the flocculants, can change the quality of the algal biomass.

An alternative that minimizes harvesting costs is working with the immobilized cultures in supports that allow the microalgae to carry out the bioremediation of the effluent and facilitate the separation of the already treated algal biomass. Immobilized microalgae cultures consists of confining

Fig. 6.2 Closed culture systems, to work at laboratory scale and pilot plant. (**a**) 250 mL flasks, (**b**) plastic reactors with recycled pet bottles, (**c**) 16 L glass photobioreactors, (**d**) Technified photobioreactors with sensors and LED light of 80 L

the entirety of the cells through natural or artificial means, thus limiting their movement to any part of the aqueous phase of the system (Tampion and Tampion 1987), this therefore differing from the free culture, in which cells are dispersed in the liquid medium.

The use of free microalgal industrial cultures presents limitations due to the presence of toxins in the medium that can inhibit growth, as well as the competition between the native and exogenous populations (Fantroussi and Agathos 2005). In immobilized cell cultures, immobilization confers protection on the cells against the toxic effect of the substances present in the medium and predation on the part of other microorganisms. In addition, the use of immobilized cells in contaminant removal processes achieve increasing the global rate of biodegradation, due to the high cellular densities reached (Mishra et al. 2001).

At present, a great variety of biotechnological industrial processes are being conducted from immobilized cells, such as in the production of amino acids, fatty acids, antibiotics, steroids, and enzymes (Najafpour 2015; Linhardt et al. 1987). In addition, immobilization also favors the increase in the stability and tolerance of the microorganisms to toxic compounds, in the case of the encapsulation of gels, or of polymers, in that these function as a physical barrier (Godjevargova et al. 2003; Hanaki et al. 1994).

Six immobilization types have been defined: covalent bonds; the affinity of immobilization; adsorption; confinement in liquid–liquid emulsions; capture behind semipermeable membranes, and entrapment in polymers (Mallick 2002). These immobilization types can be grouped into passive, which are those that utilize the properties of some microorganisms to adhere to surfaces and grow on them, and active, where flocculant agents, chemical adhesion, and gel entrapment are utilized (Moreno 2008).

Immobilization in a porous matrix is one of the most utilized processes. It can arise in synthetic polymers (resins, acrylamides, polyurethanes), proteins (gelatin, collagen, egg albumin), or natural polysaccharides (agars, carrageenans, or alginates) (Moreno 2008). This consists of cell immobilization within a matrix, which impedes the diffusion of the cells in the medium but permits the passage of metabolites and nutrients (Kourkoutas et al. 2004).

With regard to the immobilization of microalgae, Grizeau and Navarro (1986) studied immobilization in alginate in a hypersaline medium of the species *Dunaliella tertiolecta*, and these authors reported a significant production of glycerol. Labeau et al. (2000), by means of the immobilized culture of the diatom *Haslea ostrearia*, reported increased production of a blue pigment called Marenin, which is used for commercial oyster culture.

Immobilization can include more than one species. Such is the case of the immobilization of two species, *Rhodobactor sphaeroides* (bacteria) and *Chorella sorokiniana* (microalgae), in the same culture, which removed acetate, propionate, ammonium, ammonium nitrate, and phosphate in synthetic wastewater; however, separately, the performance in the removal of contaminants decreased considerably (Ogbonna et al. 2000).

De la Noüe and Proulx (1988) in their study registered the use of chitosan as support for immobilizing a culture of *Phormidium* sp.; later, we utilized these immobilized cultures as biological material for the treatment of effluent of urban wastewater, reporting the removal of 95% of the organic nitrogen after 6 h, and the phosphate levels were reduced by 87% after 24 h. As a support matrix, chitosan offers better yields in the removal of anions such as nitrates, nitrites, and phosphates, while the agars and alginates resulted in more efficiency in the removal of cations (Mallick and Rai 1994).

The immobilization of microalgae also entertains important applications in the purification of wastewater contaminated with heavy metals. Although the dead biomass can absorb heavy metals, it does this in a lesser proportion than the live biomass (Moreno-Garrido et al. 2002); in this respect, the immobilization of microalgae for this purpose offers a greater advantage.

In the UAMI Applied Phycology Laboratory, we have worked with different supports for the immobilization of microalgae to facilitate the harvesting process. For this, we have utilized dry corn husk, loufah (*Luffa Aegyptiaca*), polyethylene pearls, and calcium alginate (Fig. 6.3). These supports have demonstrated to be efficient with differences due to the species that is being immobilized and the type of effluent-to-treat.

Microalgae can be immobilized and the efficiency of the supports can be compared in terms of the growth of the algae, the bioremediation of the effluent and, later, in harvesting, carrying out the analysis of the quality of the biomass, in order to, based on this, find the possible uses of the latter. One of the most utilized genera in the bioremediation of effluents is *Chlorella*. Many species of this genus are cosmopolitan and adapt to different media and culture conditions. The species *Chlorella miniata* has been scarcely studied for this purpose, and it presents good results with regard to the removal of ammonium and orthophosphates from wastewater (Table 6.1) in experiments at laboratory scale, making it a good candidate for working with it at pilot scale.

The values that we obtained of nutrient removal utilizing this microalga are above those reported by Hernández-Reyes et al. (2012).

Another interesting chlorophyte microalga and one with high biotechnological potential is the microalga *Coelastrella* sp., in that it produces, under conditions of stress, high concentrations of beta-carotene and lipids that can be utilized for the production of biodiesel and for the treatment of different types of industrial effluents (Fig. 6.4). It is important

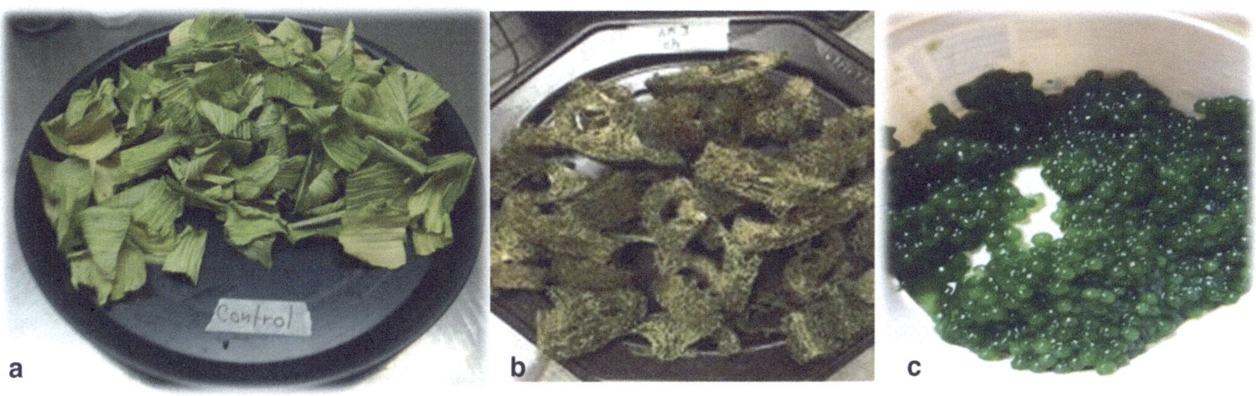

Fig. 6.3 Different supports with immobilized algae. (**a**) Corn leaf, (**b**) Loofah, and (**c**) Calcium alginate beads

Table 6.1 Removal of nutrients and quality of the algal biomass of the microalga *Chlorella miniata* in the different supports

Support	Culture media	% PO_4 removal	% NH_4 removal	% Lípids	% Proteins	% Carbohydrates
Corn leaf	Municipal wastewater	84.34	76.77	16.48	27.31	4.95
	Control	89.79	98.51	18.54	23.20	9.7
Loofah	Municipal wastewater	70	99	19	38	15
	Control	93	98	17	31.1	35
Alginate	Municipal wastewater	90	99	10	37	38
	Control	89	99	10	32	38

to consider that there is an enormous diversity of species among microalgae, and most biotechnological studies are circumscribed to the utilization of only a few of these. Therefore, the potential of investigation and applications is still under development phase and there remains much work to be done.

In works with *Coelastrella* sp. for the treatment of porcine wastewater, removal values of 90–100% are reported of the ammonium content (Luo et al. 2016) and, in comparison with a conventional medium such as BG11, the species grew more rapidly in all of the samples of swineherd wastewater than in the BG11 medium. In the laboratory, we have probed with different effluents of the food industry, of the liquor industry, and of municipal wastewater, and it has yielded good results in bioremediation (Table 6.2).

The results obtained with this microalga reflect an advance toward carrying out pilot-scale studies for its utilization and the treatment of effluents, above all of the most aggressive of these, such as *nejayote* (calcium-rich *nixtamal* wastewater generated during maize cooking and tortilla production), which is generated in great amounts in Mexico and that is dumped, without previous treatment, into the environment. Species under culture conditions can provide us with environmental services, and biomass has many potential uses.

Fig. 6.4 (**a**) Photobioreactor installed on the esplanade of the Iztapalapa Metropolitan Autonomous University, Mexico. (**b** and **c**) Free microalgae culture. (**d** and **e**) Culture of microalgae immobilized in calcium alginate beads

The designs of external photobioreactors can go hand in hand with the environment in which these are installed; in addition, they fulfill a landscaping function, one of lively shadows and of improving the environment, and they capture CO_2, the main greenhouse effect gas and produce oxygen, ideal for improving air quality in large cities. Every day, interesting novel models appear, such as this umbrella-type photobioreactor (Fig. 6.5), in which cultures provide places for the microalga *Coelastrella* sp. for bioremediation and improvements in air quality.

Algae collected in impacted, contaminated zones or in the effluent-to-be-purified are ideal alternatives to be exploited and utilized in the bioremediation process of the latter. Such is the case of the species *Lagerheimia* spp., which was worked on by Martínez-Hernández et al. (2018); this species was isolated from the textile industry effluent in the industrial zone of Altamira, Tamaulipas state, Mexico. On testing its culture in the contaminated effluent of the same zone of study, good results were yielded in terms of growth, nutrient consumption, and lipid accumulation (Table 6.1).

Likewise, the species *Verrucodesmus verrucosus*, which was isolated from temporary rain puddles in Cd. Azteca, State of Mexico, Mexico, has demonstrated high biotechnological potential, above all in the bioremediation processes of contaminated effluents. The species *Verrucodesmus verrucosus* was shown to adapt to organic leachates, animal excreta leachates, and municipal wastewater, in free and immobilized cultures. The latter suggests that microalgae have the potential to grow in alternative culture media and to serve as a biological mechanism for the treatment of effluents, avoiding or modulating mainly eutrophication and other types of contamination. This species has been marginally studied in terms of bioremediation processes and is an alga with high potential, regardless of the effluent (Arenas et al. 2016) and because it is already being worked in free or in immobilized systems.

Some species of *Neochloris oleoabundans* were studied by Band-Schmidt (1997); the culture utilized conventional media, obtaining a maximal density of 6.67×10^6 cells/L in 17 days, and 6.77×10^6 cells/L in 5 days employing the F/2

Table 6.2 Percentages of removal and nutritional content of the biomass of the microalgae *Coelastrella* sp. in different culture media

Culture media	% PO_4 removal	% NH_4 removal	% COD removal	% Proteins	% Lipids	% Carbohydrates
F2	19.25	94.68	96.62	2.4	10.94	26.8
Bayfolan forte	49	98	14	14.6	11.8	24.6
Triple 17	–	–	–	4.3	14.02	25.2
Municipal wastewater	53.42	91.02	94.75	24.2	12.33	28.0
Nejayote at 50%	90	96	36	16.96	46	15
Nejayote at 20%	80	75	17	14.64	23	27.2

medium. Similarly, Elvira-Antonio et al. (2011) reported a maximal cellular density, during an 8-day treatment period, of 28.1×10^6 cells/mL in culture utilizing artificial wastewater. On the other hand, Olguín (2012), on using dilutions of anaerobic effluents from porcine waste matter, with the species *N. oleoabundans*, obtained removals of nearly 98%. *N. oleoabundans* is one of the most studied microalgae for the production of biofuels, due to its capacity to store great amounts of lipids (Arenas et al. 2017; Chisti 2007; Elvira-Antonio et al. 2011). Therefore, if it does indeed grow in a wastewater effluent, it is somewhat promising, in that it lowers production costs and generates an environmental benefit.

Investigations mainly report on the genera *Chlorella* and *Scenedesmus*, as well as on some cyanobacteria, for the treatment of different types of municipal or industrial wastewater, or on the effluents of other types (Hernández-Reyes et al. 2012; Hongyang et al. 2011; Jebali et al. 2015; León and Chaves 2010; Li et al. 2011; Matamoros et al. 2016; Sundaramoorthy et al. 2016). The effectiveness of the system will depend on the culture medium (Acién et al. 2012; Chen et al. 2011) and on the purpose of the culture. If we think along the lines of a purpose for the culture, for example, in the production of biofuels, microalgae provide different types of renewable biofuels, the methane produced by the anaerobic digestion of the algal biomass, the biodiesel derived from the microalgal oil, and biohydrogen (Chisti 2007; Escobar et al. 2009; Lei et al. 2015; Jebali et al. 2015; Milano et al. 2016). The idea of utilizing microalgae as a source of fuel is not new, but now it is being taken up again due to the emerging concern about global warming, which is associated with the combustion of fossil fuels (Chisti 2007; Brennan and Owende 2010).

Fig. 6.5 (**a**) Bioremediation of residual water from the cheese production industry, (**b** and **c**) wastewater from the traditional liquor production industry (pulque). Both with the microalgae *Coelastrella* sp.

Economically, the production of biodiesel from microalgae would imply a low production cost accompanied by environmental benefits, in that microalgae consume a great amount of the CO_2 responsible for global warming, and this would achieve the creation of new sources of employment, supporting the economy of the country (Loera-Quezada and Olguín 2010; Milano et al. 2016; Singh et al. 2011).

The algal biomass harvested from the bioremediation process can also be utilized for the generation of biofertilizers; microalgae store in their structure a great amount of nutrients that can be assimilable by atmospheric nitrogen-fixing plants in order to render these bioavailable for same; in this manner, low-cost, energy and environmentally biofertilizers are generated (Morales-Avedaño et al. 2014).

6.5 Bioenergy and Other Bioproducts

The use of wastewater as a source of nutrients for the protection of microalgae with energy potential is described by diverse authors. Jiang et al. (2011) studied the production of lipids by means of the culture of *Nannochloropsis* sp. in municipal wastewater under nitrogen limitations and under high-luminosity conditions. Abou-Shanab et al. (2013) reported 33% in the lipid production of *Chlorella* spp., which was cultured in wastewater from pig raising farms; Arora et al. (2016) studied the production of lipids in *Chlamydomonas debaryana* utilizing wastewater from different sources (domestic, municipal, the paper industry, and from dairy farms), reporting a lipid production in the range of 40–45% in dry biomass.

Xin et al. (2010) reported a lipid content ranging from 31 to 33% in a dry biomass of *Scenedesmus* spp. cultured in a secondary effluent of wastewater; the following species were also utilized for studying the yield in the content of lipids utilizing wastewater as culture medium: *Desmodesmus quadricauda* (27.36%); *Chlorella ellipsoidea* (43%), and *Chlorella* sp. (20–43%) (Han et al. 2015; Yang et al. 2011; Cheirsilp et al. 2017).

Another of the application in which biomass can be used in order to take advantage of its composition in terms of carbohydrates is the obtention of biofuels such as ethanol. Ho et al. (2013) studied the production of carbohydrates in *Chlorella vulgaris*, reaching up to 51% and applying an enzymatic hydrolysis treatment and a process coupled with saccharification fermentation, these authors obtained a concentration of 11.7 g/L of ethanol with yield during the process of its obtention of 87.6%.

Bchir et al. (2016) described the following species as yielding a high production of carbohydrates: *Porphyridium cruentum* (40–57%); *Chlorococum humicola* (32–52%); *Spirogyra* spp. (33–64%), and *Tetradesmus dimorphus* (21–52%). Control of the variables could increase the production of carbohydrates utilizing culture strategies based on the intensity of the light, diminution of the nitrogen source, variation of the temperature, the pH, and the addition of CO_2; among the latter, the limitation of nitrogen would aid not only in the production of carbohydrates but also in that of lipids (Chen et al. 2013).

Once the bioremediation process of an effluent is finalized, quality analyses of the algal biomass can be conducted (Table 6.3) to search for a use of or an application for the latter. Taking advantage of the microalgal biomass as a sub-product deriving from residual industrial wastewater treatments offer potential in terms of the production of energy, prime matter for biodiesel or ethanol, through fermentation of the biomass.

Microalgal cultures in wastewater or in media elaborated from wastewater offers us results of very different yields from those in the literature because there are different variables that are linked with cellular growth. In the first place, the species can be a local species (as in this case), or a species in international collections, and there can be even genetically modified species. Another important aspect includes the characteristics of the culture medium in terms of its nature: artificial, commercial, natural, wastewater effluents, leachable, and others, and the third aspect, the conditions of the culture, include the conditions (autotrophic, heterotrophic, mixotrophic), the type of light (artificial or natural), the light cycle, the intensity of the light, the volume of the reactor, the type of reactor, the transfer of the mass (shaking, gasification), the temperature, the pH, the concentration of CO_2, and the concentration of O_2. Thus, it is very difficult to establish points of comparison. However, microalgae of the *Chlorophyta* class are the most reported in the literature for this type of treatment (Arenas et al. 2016, 2017; Chisti 2007; Rodríguez-Palacio et al. 2018).

The reported values of the efficiency and the removal of nitrogen and phosphorus also vary depending on the composition of the medium and the culture conditions, mainly the initial concentrations of nutrients, the intensity of the light, the cycles of light–darkness and, of course, the species type.

The harvesting process to separate the algal biomass from the already treated water is difficult at times, due to the very nature of the algae. The most economical method could be that of sedimentation. This process reduces operational costs and depends on the type of species, the stem, the density, the motility of the cells, and whether, if the flocculants are added to accelerate the process, this can contaminate the clean effluent (Pezzolesi et al. 2015; Chatsungnoen and Chisti 2016; Singh and Patidar 2018). Therefore, by immobilizing the cells in sodium alginate, we render the harvesting process more environmentally friendly and easier, leaving the effluent clean and recovering the biomass for the elaboration of added-value products, or even for their utilization in a new process of bioremediation. Notwithstanding this, it must be borne in mind that their efficiency can be greater under laboratory conditions with respect to cultures on a greater scale.

We can affirm that microalgae isolated in bodies of water or in a natural environment could adapt better and present

Table 6.3 Quality analysis of the microalgal biomass of some species cultivated in residual water

Microalgae	Lipids	Carbohydrates	Proteins
Chorella miniata	37.3	4.78	34.4
Coelastrella sp.	46.9	16.2	33.0
Desmodesmus quadricauda	19.20	11.72	15.64
Neochloris oleoabundans	37.5	18.7	33.16
V. verrucosus (free cell culture)	24.61	18.9	15.5
V. verrucosus (immobilized cell culture)	10.47	22.3	33.1
Lagerhemia sp. (free cell culture)	31.1	15	–
Lagerhemia sp. (immobilized cell culture)	7.55	22	–

good results in the nutrient removal of contaminated effluents, under specific conditions. In addition to this, exotic species are not introduced, and knowledge is increased concerning the potential of the use of our resources. Thus, the capacity demonstrated of native microalgae to remove N and P from municipal wastewater and from other effluents, because this efficiency does not diminish on working with an immobilized system, thus facilitating the harvesting of the biomass and the release of the effluent, this being an efficient bioremediation system. Investigation of microalgae presents us with alternatives to energy, the bioremediation of wastewater, and basic and applied investigation (Crisóstomo-Vázquez et al. 2016; Rodríguez-Palacio et al. 2011, 2018, 2022). However, although most of the country's bodies of water have been identified and analyzed, alternatives have not in turn been analyzed with regard to taking economic advantage of microalgae as a natural resource. It is necessary to conduct a systematized study in which these opportunities are analyzed and that, at a future date, permit the establishment of mechanisms of the transfer of technology for a sustainable exploitation of the resource.

Acknowledgments The autors thank the Project: Cultivation of microalgae for potential uses. Caribbean and Gulf of Mexico. Metropolitan Autonomous University.

References

Abalde J, Cid A, Fidalgo P, Torres E, Herreno C (1995) Microalgas: Cultivo y Aplicaciones. Universidad de Coruña, España, p 210

Abou-Shanab R, Ji M, Kim H, Paeng KJ, Jean BH (2013) Microalgal species growing on piggery wastewater as a valuable candidate for nutrient removal and biodiesel production. J Environ Manag 115: 257–264

Acién FG, Fernández JM, Magán JJ, Molina E (2012) Production cost of a real microalgae production plant and strategies to reduce it. Biotechnol Adv 30:1344–1353

Amezaga JM, Amtmann A, Biggs CA, Bond T, Gandy CJ, Honsbein A, Karunakaran E, Lawton L, Madsen MA, Minas K, Templeton MR (2014) Biodesalination: a case study for applications of photosynthetic bacteria in water treatment. Plant Physiol 164(4):1661–1676

Araujo RM, Puig A, Lasobras J, Lucena F, Jofre J (1997) Phages of enteric bacteria in fresh water with different levels of faecal pollution. J Appl Microbiol 82(3):281–286

Arenas EG, Rodríguez-Palacio MC, Juantorena AU, Sebastian PJ (2016) GC-MS and FTIR spectroscopic determination of lipids and fatty acid profile of microalgae *Verrucodesmus verrucosus* under different residual culture media. The 6th International Conference on Algal Biomass, Biofuels and Bioproducts. Paradise Point, San Diego, USA. 26 al 29 June

Arenas EG, Rodríguez-Palacio MC, Juantorena AU, Sebastian PJ (2017) Microalgae as a potential source for biodiesel production: techniques, methods and other challenges. Int J Energy Res 41(6): 761–789. https://doi.org/10.1002/er.3663

Arora N, Patel A, Sartaj K, Pruthi P, Pruthi V (2016) Bioremediation of domestic and industrial wastewater integrated with enhanced biodiesel production using novel oleaginous microalgae. Environ Sci Pollut Res 23:20977–21007

Band-Schmidt CJ (1997) Generación biotecnológica para la producción de microalgas. Ciencia y mar 1:23–30

Bchir F, Falleh A, Ghabbarou E, Hamdi M (2016) 3rd generation bioethanol production from microalgae isolated from slaughterhouse wastewater. Waste Biomass Valor 7:1041–1046

Becker EW (1994) Microalgae biotechnology and microbiology. Cambridge studies in biotechnology. Cambridge Press, p 293

Benitez FJ, Acero LJ, Leal IA, Gonzalez M (2009) The use of ultrafiltration and nanofiltration membranes for the purification of cork processing wastewater. Filtration Separation 162:1438–1445

Bhatnagar A, Bhatnagar M, Chinnasamy S, Das K (2010) *Chlorella minutissima*, a promising fuel alga for cultivation in municipal wastewaters. Appl Biochem Biotechnol 161:523–536

Borde X, Guieysse B, Delgado O, Muñoz R, Hatti-Kaul R, Nugier-Chauvin C, Patin H, Mattiasson B (2002) Synergistic relationships in algal-bacterial microcosmos for the treatment of aromatic pollutants. Bioresour Technol 86(3):293–330

Brennan L, Owende P (2010) Biofuels from microalgae. A review of technologies for production, processing, and extraction of biofuels and co-products. Renew Sust Energ Rev 14:557–577

Caldwell DH (1946) Sewage oxidation ponds: performance, operation and design. Sewaje Works J 18:433–458

Campos C (2003) Indicadores de contaminación fecal en aguas. In: Agua potable para comunidades rurales, reuso y tratamientos avanzados de aguas residuales domésticas. RYDA-CYTED / CIRA-UAMEX, Toluca, pp 224–229

Chatsungnoen T, Chisti Y (2016) Harvesting microalgae by flocculation–sedimentation. Algal Res 13:271–283. https://doi.org/10.1016/ELSEVIER_CM_POLICY

Cheirsilp B, Tippayut J, Romprom P, Prasertsan P (2017) Phytoremediation of secondary effluent from palm oil by using oleaginous microalgae for integrated lipid production and pollutant removal. Waste Biomass Valor 8:2889–2897

Chen CY, Yeh KL, Aisyah R, Lee DJ, Chang JS (2011) Cultivation, photobioreactor design and harvesting of microalgae for biodiesel production: a critical review. Bioresour Technol 102:71–81

Chen CY, Zhao XQ, Yen HW, Ho SH, Cheng CL, Lee DJ, Bai FW, Chang JS (2013) Microalgae-based carbohydrates for biofuel production. Biochem Eng J 78:1–10

Chisti Y (2007) Biodiesel from microalgae. Biotechnol Adv 25:294–306

Choudri BS, Baawain M (2014) Effects of pollution on freshwater organisms. Water Environ Res 86(10):1832–1868

Choudri BS, Baawain M (2015) Effects of pollution on freshwater organisms. Water Environ Res 87(10):1679–1717

Choudri BS, Baawain M (2016a) Bioenergy from biofuel residues and wastes. Water Environ Res 88(10):1446–1466

Choudri BS, Baawain M (2016b) Effects of pollution on freshwater organisms. Water Environ Res 88(10):1672–1692

Crisóstomo-Vázquez L, Alcocer-Morales C, Lozano-Ramírez C, Rodríguez-Palacio MC (2016) Fitoplancton de la Laguna del Carpintero Tampico, Tamaulipas, México. Interciencia 41(2): 103–109

De Bashan L, Bashan Y (2004) Recent advances in removing phosphorus from wastewater and it's future use as fertilizer (1993–2003). Water Res 38:4222–4246

De Florio L, Giordano A, Mattioli D (2005) Nanofiltration of low-contaminated textile rinsing effluents for on-site treatment and reuse. Desalination 181:283–292

De la Noüe J, Proulx D (1988) Biological tertiary treatment of urban wastewaters with chitosan-immobilized *Phormidium*. Appl Microbiol Biotechnol 29:292–297

De la Noüe J, Laliberеté G, Proulx D (1992) Algae and wastewater. J Appl Phycol 4:247–254

Drexler I, Caryssa Joustra LC, Prieto A, Bair R, Yeh DH (2014) AlgaSim: a model for integrated algal biofuel production and wastewater treatment. Water Environ Res 86(2):163–176

Duarte FM, Maldonado-Hódar FJ, Madeira LM (2013) Influence of the iron precursor in the preparation of heterogeneous Fe/activated carbon Fenton-like catalysts. Appl Catal A Gen (0) https://doi.org/10.1016/j.apcata.2013.03.030

Elvira-Antonio N, Ruíz MA, Pérez RLJ, Reyes GZ, Canedo LY, Campos GSC, Sabido PMY, Cerón BG, Cordova QAV, Moreno LJH (2011) Densidad celular y acumulación de lípidos en cultivos libres de *Chlorella vulgaris* y *Neochloris oleoabundans* a diferentes concentraciones de nitrógeno y carbonato de sodio. Unacar Tecnociencia 5(1):56–71

Escobar JC, Lora ES, Venturini OJ, Yanez EE, Castillo EF, Almazan O (2009) Biofuels: environment, technology and food security. Renew Sust Energ Rev 13:1275–1287

Escobar J, Muro C, Castellanos J (2011) Optimización del proceso de ultrafiltración de efluentes de una industria de cereales. Afinidad LXVIII 552:117–123

Escorihuela A, Nuñez M, Rosales N, Morales E (2007) Microalgas presentes en una laguna para pulimento de efluentes de una planta de tratamiento de aguas residuales urbanas. Revista de la Facultad de Agronomía de La Universidad del Zulia 24(1):225–230. ISSN-e 2477-9407

Fantroussi S, Agathos S (2005) Is bioaugmentation a feasible strategy for pollutant removal and site remediation? Curr Opin Microbiol 8: 268–275

Gangl D, Zedler JAZ, Rakakumar PD, Ramos Martínez EM, Riseley A, Włodarczyk A, Purton S, Sakuragi Y, Howe CJ, Jensen PE, Robinson C (2015) Biotechnological exploitation of microalgae. J Exp Bot 66(22):6975–6990

García IM, Betancort Rodríguez JR, Salas Rodríguez JJ, Peñate Suárez B, Pidre Bocardo JR, Sardón Martín N (2006) Guía sobre tratamientos de aguas residuales urbanas para pequeños núcleos de población Mejora de la calidad de los efluentes. ICREW, CENTRA. ISBN: 84-689-7604-0. 128p

Gil-Pavas E (2005) Oxidación fotocatalfítica de aguas residuales de la industria de lavandería utilizando TiO_2 como catalizador y luz UV. Ingeniería y Ciencia 1(2):25–40

Godjevargova T, Ivanova D, Alexieva Z, Dimov N (2003) Biodegradation of toxic organic components from industrial phenol production waste waters by free and immobilized *Trichosporon cutaneum* R57. Process Biochem 38:915–920

Green FB, Bernstone LS, Lundquist TJ, Oswald WJ (1996) Advanced integrated wastewater pond systems for nitrogen removal. Water Sci Technol 33(7):207–217

Grizeau D, Navarro JM (1986) Glycerol production by *Dunaliella tertiolecta* immobilized within Ca-alginate beads. Biotechnol Lett 8:261–264

Han L, Pei H, Hu W, Jiang L, Ma G, Zhang S, Han F (2015) Integrated campus sewage treatment and biomass production. Bioresour Technol 175:262–268

Hanaki K, Hirunmasuwana S, Matsuo T (1994) Protection of methanogenic bacteria from low pH and toxic materials by immobilization using polyvinyl alcohol. Water Res 28:877–885

Hanan M, Munir, Mohammad J (2003) Survival and accumulation of microorganisms in soils irrigated with secondary treated wastewater. J Basic Microbiol 43(1):47–55

Hernández-Reyes BM, Rodríguez-Palacio MC, Lozano-Ramírez C, Castilla-Hernández P (2012) Remoción de nutrientes por tres cultivos de microalgas libres e inmovilizados. Revista Latinoamericana de Biotecnología Algal y Ambiental 3(1):80–94

Ho S, Huang S, Chen C, Hasunuma T, Kondo A, Chang J (2013) Bioethanol production using carbohydrate-rich microalgae biomass as feedstock. Bioresour Technol 135:191–198

Hoffman JP (1998) Wastewater Treatment with suspended and nonsuspended algae. J Phycol 34:757–763

Hongyang S, Yalei Z, Chunmin Z, Xuefei Z, Jinpeng L (2011) Cultivation of *Chlorella pyrenoidosa* in soybean processing wastewater. Bioresour Technol 102:9884–9890

Jebali A, Acién FG, Gómez C, Fernández-Sevilla JM, Mhiri N, Karray F, Dhouib A, Molina-Grima E, Sayadi S (2015) Selection of native Tunisian microalgae for simultaneous wastewater treatment and biofuel production. Bioresour Technol 198:424–430

Jiang L, Lou S, Fan X, Yang Z, Gou R (2011) Biomass and lipid production of marine microalgae using municipal wastewater and high concentration of CO_2. Appl Energy 28(10):3336–3341

Kauser J, Hoque S, Ahmed T (2014) Activated sludge and other aerobic suspended culture processes. Water Environ Res 86(10):1026–1069

Keeling PJ (2004) Diversity and evolutionary history of plastids and their hosts. Am J Bot 91(10):1481–1493

Kourkoutas Y, Bekatorou A, Banat I, Marchan R, Koutinas A (2004) Immobilization technologies and support materials suitable in alcohol beverages production: a review. Food Microbiol 21:377–397

Kumar A, Bisht BS, Joshi VD, Dhewa T (2011) Review on bioremediation of polluted environment: a management tool. Int J Environ Sci 1(6):1079–1093

Labeau T, Gaudin P, Junter GA, Mignot L, Robert J (2000) Continuous marennnin production by agar-entrapped *Halsea ostrearia* using a tubular photobioreactor with internal illumination. Appl Microbiol Biotechnol 54:847–850

Lapertot M, Pulgarín C, Fernández P, Maldonado M, Leonidas PL, Oller I, Gernjak W, Malato S (2006) Enhancing biodegradability of priority substances (pesticides) by solar photo-Fenton. Water Res 40(5):1086–1094

Larsdotter K (2006) Wastewater treatment with microalgae. A literature review. Vatten 62:31–38

Lei X, Chen Y, Shao Z, Chen Z, Li Y, Zhu H, Zhang J, Zheng W, Zheng T (2015) Effective harvesting of the microalgae *Chlorella vulgaris* via flocculation–flotation with bioflocculant. Bioresour Technol 198: 922–925

León C, Chaves D (2010) Tratamiento de residual vacuno utilizando microalgas, la lenteja de agua *Lemna aequinoctiales* y un humedal subsuperficial en Costa Rica. Revista Latinoamericana de Biotecnología Ambiental y Algal 1(2):155–177

Lewis LA, McCourt RM (2004) Green algae and the origin of land plants. Am J Bot 91(10):1535–1556

Li Y, Chen YF, Chen P, Min M, Zhou W, Martinez B, Zhu J, Ruan R (2011) Characterization of a microalga *Chlorella* sp. well adapted to highly concentrated municipal wastewater for nutrient removal and biodiesel production. Bioresour Technol 102:5138–5143

Linhardt RJ, Amotz S, Rugh S, Markussen EK, Thomsen K, Bigwood MP, Naples JO, Blattler WA, Lambert JM, Senter PD (1987) Patents and literature immobilized biocatalysts. Appl Biochem Biotechnol 14:121–145

Loera-Quezada M, Olguín E (2010) Las microalgas oleaginosas como fuente de biodiesel: retos y oportunidades. Revista Latinoamericana de Biotecnología Ambiental y Algal 1(1):91–116

Luo L, He H, Yang C, Wen S, Zeng G, Wu M, Zhou Z, Lou W (2016) Nutrient removal and lipid production by *Coelastrella* sp. in anaerobically and aerobically treated swine wastewater. Bioresource Technol 216:135–141

Mahapatra D, Chanakya H, Ramachandra T (2013) Treatment efficacy of algae-based sewage treatment plants. Springer Sci Environ Monit Assess 185(9):7145–7164

Mallick N (2002) Biotechnological potential of immobilized algae for wastewater N,P and metal removal: a review. Biometals 15:377–390

Mallick N, Rai L (1994) Removal of inorganic ions from wastewater by immobilized microalgae. World J Microb Biotechnol 10:439–443

Martínez-Hernández M, Suastes-Acosta SL, Lozano-Ramírez C, Rodríguez-Palacio MC (2018) Aislamiento y perfil lipídico de la microalga *Lagerheimia* sp. proveniente de aguas residuales industriales de Tamaulipas, México. Avances en Ciencias e Ingeniería. Chile 9(1):25–33

Matamoros V, Uggetti E, García J, Bayona JM (2016) Assessment of the mechanisms involved in the removal of emerging contaminants by microalgae from wastewater: a laboratory scale study. J Hazard Mater 301:197–205

Mehrabadi A, Craggs R, Farid M (2015) Wastewater treatment high rate algal ponds (WWT HRAP) for low-cost biofuel production. Bioresour Technol 184:202–214

Milano J, Ong HC, Masjuki HH, Chong WT, Lam MK, Loh PK, Vellayan V (2016) Microalgae biofuels as an alternative to fossil fuel for power generation. Renew Sust Energ Rev 58:180–197

Mishra S, Jyot J, Kuhad R, Lal B (2001) Evaluation of inoculum addition to stimulate it in situ bioremediation of oily-sludge-contaminated soil. Appl Environ Microbiol 67:1675–1681

Morales-Avedaño E, Martínez-Pérez R, Suárez-Rodríguez G (2014) Aislamiento, cultivo, viabilidad y evaluación de un consorcio cianobacteria-microalga como acondicionador de suelos. Revista Ecuatoriana de Medicina y Ciencias Biológicas 1 y 2:41–71

Moreno I (2008) Microalgae immobilization current techniques and uses. Biosource Technolog 99:3949–3964

Moreno-Garrido I, Codd GA, Gadd GM, Lubian L (2002) Cu and Zn accumulation by calcium alginate immobilized marine microalgal cells of *Nannochloropsis gaditana* (eustigmatophyceae). Cienc Mar 28:107–119

Najafpour G (2015) Biochemical engineering and biotechnology. Elsevier Science. http://www.totalboox.com/book/id-8147351353007806007

Ogbonna JC, Yoshizawa H, Tanaka H (2000) Treatment of high strength organic wastewater by a mixed culture of photosynthetic microorganisms. J Appl Phycol 12:227–284

Olguín EJ (2003) Phycoremediation: key issues for cost-effective nutrient removal processes. Biotechnol Adv 22(1–2):81–91

Olguín EJ (2012) Dual purpose microalgae–bacteria-based systems that treat wastewater and produce biodiesel and chemical products within a Biorefinery. Biotechnol Adv

Oswald WJ (2003) My sixty years in applied algology. J Appl Phycol 15:99–106

Oswald WJ, Gotaas HB, Ludwig HF, Lynch V (1953) Algae symbiosis in oxidation ponds. Photosynthetic oxygenation. Sewage Indust Wastes 25(6):692–705

Papoutsakis S, Afshari Z, Malato S, Pulgarin C (2015) Elimination of the iodinated contrast agent iohexol in water, wastewater and urine matrices by application of photo-Fenton and ultrasound advanced oxidation processes. J Environ Chem Eng 3(3):2002–2009

Pellon A, Benitez F, Frades J, Garcia L, Cerpa A, Alguacil F (2003) Empleo de microalga *Scenedesmus obliquos* en la eliminación de Cromo presente en aguas residuales galvánicas. Rev Metal 39:9–16

Pezzolesi L, Samorì C, Pistocchi R (2015) Flocculation induced by homogeneous and heterogeneous acid treatments in *Desmodesmus communis*. Algal Res 10:145–151

Phang S, Miah M, Yeoh G, Hashim M (2000) Spirulina cultivation in digested sago starch factory wastewater. J Appl Phycol 12:395–400

Pienkos P, Laurens TL, Aden A (2011) Making biofuel from microalgae: so much potential coexists with so many scientific environmental and economic challenges. Am Sci 99(6):474–481

Powell N, Shilton A, Pratt S, Chisti Y (2008) Factors influencing luxury uptake of phosphorus by microalgae in waste stabilization ponds. Environ Sci Technol 42(16):5958–5962

Pulgarin C (2015) Effect of advanced oxidation processes on the micropollutants and the effluent organic matter contained in municipal wastewater previously treated by three different secondary methods. Water Res 84:295–306

Raja R, Hemaiswaryz S, AshokKumar N, Sridhar S, Rengasamy R (2008) A perspective on the biotechnological potential of microalgae. Crit Rev Microbiol 34:77–88

Rao PH, Ranjith Kumar R, Raghavan BG, Subramanian VV, Sivasubramanian V (2011) Is phycovolatilization of heavy metals a probable (or possible) physiological phenomenon? An in situ pilot scale study at a leather. Processing Chemical Industry. Water Environ Res 83(4):291–297

Renuka N, Sood A, Prasanna R (2015) Phycoremediation of wastewaters: a synergistic approach using microalgae for bioremediation and biomass generation. Int J Environ Sci Technol 12:1443–1460

Rodríguez-Palacio MC, Crisóstomo-Vázquez L, Álvarez-Hernández S, Lozano-Ramírez C (2011) Strains of toxic and harmful microalgae, from wastewater, marine, brackish and freshwater. Food Addit Contam Part A, DO 1:10

Rodríguez-Palacio MC, Cabrera Cruz RBE, Rolón Aguilar JC, Lozano Ramírez C, Galeana Hurtado LJ, Avendaño EDM (2018) Comparative study on the removal of N and P from municipal waste waters and leached vermicomposting using five microalgae strains. Desalin Water Treat 131:180–186. https://doi.org/10.5004/dwt.2018.23035

Rodríguez-Palacio MC, Cruz RBEC, Aguilar JCR, Tobías-Jaramillo R, Hernández MM, Ramírez CL (2022) Culture of five species of microalgae and their potential for biodiesel production. Energy Sustainabil Soc 12:10. https://doi.org/10.1186/s13705-022-00337-5

Sahua D, Priyadarshani I, Rath B (2012) Cyanobacteria as Potential biofertilizer. J Microbiol 1(2–3):20–26

Sharma S (2012) Bioremediation: features, strategies and applications. Asian J Pharm Life Sci 2(2):202–213

Singh G, Patidar SK (2018) Microalgae harvesting techniques: a review. J Environ Manag 217:499–508

Singh A, Nigam PS, Murphy JD (2011) Renewable fuels from algae: an answer to debatable land based fuels. Bioresour Technol 102:10–16

Sundaramoorthy B, Thiagarajan K, Mohan S, Mohan S, Rao PR, Ramamoorthy S, Chandrasekaran R (2016) Biomass characterization and phylogenetic analysis of microalgae isolated from estuaries: role in phycoremediation of tannery effluent. Algal Res 14:92–99

Tampion J, Tampion MD (1987) Immobilized cells. Principles and applications. Cambridge University Press, Cambridge

Tarkashvand M, Alikhani HA, Pourbabae AA (2014) Nitrogen fixation ability of some cyanobacteria isolated of maize field (case Study Iran). Int J Res Agric Sci 1:2348–3997

Tchobanoglous G, Burton FL, Stensel HD (2003) Wastewater engineering: treatment, disposal and reuse. Ed. Metcalf & Eddy INC. 4 edition. New York: Mc Graw Hill. 18046p

Vanthoor-Koopmans M, Cordoba-Matson MV, Arredondo-Vega BO, Lozano-Ramírez C, García Trejo JF, Rodríguez-Palacio MC (2014) Microalgae and cyanobacteria production for food and food supplements. In: Guevara-Gonzalez R, Torres-Pacheco I (eds) Biosystems engineering: biofactories for food production in XXI century, vol 8. Springer, pp 253–275

Vedrenne M, Vazquez M, Prato G, Frontana U, Ibañez JC (2012) Characterization and detoxification of a mature landfill leachate using a combined coagulation-flocculation/photo fenton treatment. Hazard Mater 205:208–215

Xin L, Hong-Ying H, Ke G, Jia Y (2010) Growth and nutrient removal properties of a freshwater microalga *Scenedesmus* sp. LX1 under different kinds of nitrogen sources. Ecol Eng 36:379–381

Yang J, Li X, Hu H, Zhang X, Yu Y, Chen Y (2011) Growth and lipid accumulation properties of a freshwater microalga, *Chlorella ellipsoidea* YJ1, in domestic secondary effluents. Appl Energy 88(10):3295–3299

Zheng Chunli, Ling Zhao, Xiaobai Zhou, Zhimin Fu, An Li. (2011) Treatment technologies for organic wastewater. In: Water treatment Edited by Walid Elshorbagy and Rezaul Kabir Chowdhury. https://www.intechopen.com/chapters/41953. https://doi.org/10.5772/52665

Biofertilization by Nitrogen-Fixing Cyanobacteria, Nutrient Supplementation, and Growth Promotion

7

Rosa Olivia Cañizares-Villanueva, Citlally Ramírez-López, Pablo A. López-Pérez, and Dulce J. Hernández-Melchor

Abstract

The indiscriminate use of agrochemicals to control pests and fertilize crops has been used to satisfy the population's demand, causing environmental pollution problems. The use of non-harmful and environmentally friendly fertilizers such as biofertilizers has been promoted. Biofertilizers consist of the preparation of living soil microorganisms that add, conserve, and mobilize nutrients to the rhizosphere or to the interior of plants, which encourage their growth; some of them also can fix nitrogen and produce growth-promoting (cytokinins, auxins, and gibberellins), in addition to maintaining the soil structure in arable fields. In this way, biofertilizers based on cyanobacteria are a vital element in developing sustainable agriculture that seeks to enhance the productivity of crops. For these photosynthetic microorganisms to be exploited on a large scale to improve culture technology and develop new economically feasible protocols for producing quality inoculums, it is necessary to implement strategies that allow controlling abiotic factors that limit their growth in photobioreactors. An example of biofertilization technology in wheat crops using a photosynthetic microbial consortium produced in a photobioreactor is presented.

R. O. Cañizares-Villanueva
Centro de Investigación y de Estudios Avanzados del Instituto Politécnico Nacional, Departamento de Biotecnología y Bioingeniería, Ciudad de México, Mexico

C. Ramírez-López
Universidad del Mar, campus Puerto Ángel, Instituto de Ecología, San Pedro Pochutla, Oaxaca, Mexico

P. A. López-Pérez
Universidad Autónoma del Estado de Hidalgo, Escuela Superior Apan, Apan, Hgo., Mexico

D. J. Hernández-Melchor (✉)
Colegio de Postgraduados, campus Montecillo, Departamento de Edafología e Hidrociencias, Estado de México, Mexico

Keywords

Biofertilizer · Cyanobacteria · Nitrogenase · Crops · Photobioreactors

7.1 Introduction

Since man began with agriculture, he has used fertilizers to increase the yield of the land, such as manure, animal bones, wood ashes, vegetable waste, guano, fish, and lime, among others. As the demand for agricultural products increased rapidly, these fertilizers were insufficient, forcing man to seek methods to produce substances that would provide the nutrients required by crops. In this way, the chemical fertilizer industry was born in the mid-nineteenth century, a time in which the chemists Juntus Von Liebig (German) and Thomas Green Glemson (American) demonstrated that it is essential to supply N, P, and K to plants in order to obtain better results in the harvests (Ramírez-López 2011; Meena and Busi 2019). However, the use of these inputs has not been adequate. There have been environmental pollution problems such as soil acidification, damage to health (blue baby syndrome), eutrophication of water bodies, and global warming (Brusseau and Artiola 2019).

Therefore, it is possible to generate a sustainable agricultural system through the use of environmentally friendly technologies, such as the use of transgenic crops and biofertilizers (Bhattacharyya et al. 2020). Biofertilizers are classified according to their action in the plant growth process, either as phosphate solubilizing microorganisms (PO_4^{-3}), growth-promoting rhizobacteria, mycorrhizae, and N_2-fixing microorganisms (Díaz et al. 2020).

Cyanobacteria are used as biofertilizers in agricultural crops due to their ability to fix nitrogen (N_2), in addition to increasing crop productivity and root growth through the production of substances that promote plant growth and active substances that they protect the plant from soil pathogens (Rai et al. 2019; Sarma and Prakash 2020).

A. de Jesús Martínez-Roldán (ed.), *Biotechnological Processes for Green Energy, and High Value Bioproducts by Microalgae, and Cyanobacteria Cultures*, Developments in Applied Phycology 13, https://doi.org/10.1007/978-3-031-43969-8_7

These microorganisms provide a large amount of biological nitrogen since per year it is estimated that the fixation of N_2 by cyanobacteria provides from 0.23 to 75.5 kg of N ha^{-1} $year^{-1}$, so their use as biofertilizers, it would help reduce the application of chemical fertilizers in agricultural soils by between 25 and 35% (Heimann and Cirés 2015).

For photosynthetic microorganisms, specialized systems have been designed in which closed photobioreactors (PBR) and open ponds are used more frequently. In closed photobioreactors, microbial cultures have little or no contact with the atmosphere, thus allowing us to keep microorganisms free of external contaminants and at the same time control the appropriate operating conditions for their growth, which favors the quantity and quality of the inocula produced (Lu et al. 2020). The most used PBR's are usually flat-panel, bubble columns, and tubular loops (Acién et al. 2017).

7.2 Cyanobacteria General Description

Cyanobacteria (Class Cyanophyceae), prokaryotic microorganisms also known as blue-green algae, carry out oxygenic photosynthesis and a large number of them are capable of fixing nitrogen in aquatic and terrestrial environments, so they are ecologically relevant (Venkataraman 1981; Kumar et al. 2019; Pandey et al. 2020; Heggerud et al. 2022). Cyanobacteria have unicellular and filamentous forms, individual or forming colonies; we can find them as free-living microorganisms, or in symbiotic associations, in addition, due to their photosynthetic nature, they have photosynthetic pigments such as chlorophyll a and phycobiliproteins (Alvarez et al. 2021). They have an extensive distribution in nature, on the ground, and in marine habitats. They are generally more tolerant of extreme environments than eukaryotic algae and are often the only photosynthetic organisms present in hot springs, saline lakes, and other extreme environments (Sarma and Prakash 2020).

7.2.1 Microbiome

Cyanobacteria are part of ecosystems since they have the ability to interact with plants and be part of their microbiome (Lee and Ryu 2021). A microbiome is defined as a characteristic microbial community that is part of a habitat with specific physicochemical properties. The microbiome not only involves the microorganisms present but also their microbial interactions and activities, giving rise to the formation of ecological niches (Berg et al. 2020).

The participation of cyanobacteria as part of the soil microbiome was demonstrated by Davey and Clarke (1991) who found that the Oscillatoriaaceae family was dominant at the soil surface and down to a depth of 8 cm. It has also been shown that cyanobacteria together with other microorganisms can be present in places with extreme climatic conditions, such as semi-arid steppes, hot deserts, and Polar Regions, forming a biological crust on the soil (Lee and Ryu 2021). Examples of the presence of these microorganisms in the microbiome is that mentioned by Nelson et al. (2021) who described the exchange of metabolites between biocrust communities from dryland soils. These communities form a diazotroph-rich "cyanosphere" where the cyanobacterium *Microcoleus vaginatus* is present; which participates in the exchange of nitrogen and carbon between photoautotrophic and heterotrophic microorganisms; thus demonstrating a symbiotic relationship between cyanobacteria and the microorganisms that make up the cyanosphere. Likewise, the benefit of the inoculation as biofertilizers of some cyanobacteria such as *Calothrix* to improve the yield of crops of agricultural importance such as rice, interacting in turn synergistically with the endophytic microbiome of rice seedlings (Priya et al. 2015).

7.2.2 Nitrogen-Fixing Cyanobacteria

The utilization of atmospheric nitrogen is called "nitrogen fixation" and is a property possessed only by some bacteria and cyanobacteria. The basic requirements for biological nitrogen fixation are an enzyme complex called nitrogenase, a reducing agent, ATP, and low O_2 tension (Brock et al. 2004; Sarma and Prakash 2020).

7.2.2.1 Nitrogenase

The reduction of N_2 to NH_4^+ is catalyzed by the nitrogenase enzyme complex, which is very similar in composition and function in all prokaryotic organisms. This complex consists of the iron protein (Fe-protein) and the molybdo iron protein (MoFe-protein). The Fe-protein has two main functional characteristics: it is binding to ATP and the second is a functional site that is the 4Fe:4S group. The available evidence indicates that the Fe-protein is involved in the hydrolysis of ATP, necessary for the transfer of electrons in the nitrogen fixation process. Substrate binding and nitrogen reduction reactions occur in MoFe-protein (Heimann and Cirés 2015).

7.2.2.2 Heterocytes

Filamentous cyanobacteria of the genera *Anabaena* and *Nostoc* protect their nitrogenase complex by differentiating approximately 5–10% of their vegetative cells called heterocytes (Rai et al. 2019). Heterocytes are distributed along the filament or at the end of it. Its function is to create an oxygen-free environment (anaerobic) for the proper functioning of the nitrogenase complex, in addition to providing combined nitrogen to the rest of the filament (Heimann and Cirés 2015). Heterocytes have a low concentration of phycobilin (photosynthetic accessory pigment) and lack photosystem II, which is the one that produces oxygen. Due to the reductive nature of nitrogen fixation and the sensitivity of the nitrogenase enzyme complex to oxygen, it seems likely that the heterocyte, by maintaining an anaerobic environment, makes it possible to stabilize the nitrogen-fixing system in organisms that are not only aerobic but also oxygen producers. It is also possible that some non-heterocytic filamentous cyanobacteria produce nitrogenase and fix nitrogen in normal vegetative cells if grown anaerobically (Brock et al. 2004; Nozue et al. 2017; Mącik et al. 2020).

7.3 Nutrient Supplementation

The main contributor to the presence of cyanobacteria in the soil is nitrogen due to its ability to fix this element, however, they are also capable of benefiting the plant and enriching the soil by facilitating the availability of other nutrients such as phosphorus (Pandey et al. 2020).

7.3.1 Phosphorus-Solubilizing

One of the limiting nutrients in agriculture apart from nitrogen is phosphorus, since it participates in various metabolic activities that affect plant growth, and therefore the yield of agricultural crops. Phosphorus can be solubilized by cyanobacteria from insoluble phosphate minerals present in the soil such as $FePO_4$, $AlPO_4$, hydroxyapatite $[Ca_5(PO_4)_3OH]$, and $(Ca)_3(PO_4)_2$, by action of the phosphatase enzyme; thus improving soil fertility and quality (Sarma and Prakash 2020). Mandal et al. (1999) proposed a series of chemical reactions to explain the phosphorus solubilization process by cyanobacteria, through the synthesis of Ca^{2+} chelators that change the pH, or with the release of carbonic acids:

$$Ca_{10}(OH)_2(PO_4)_6 \leftrightarrow 10Ca^{2+} + 2OH^- + 6PO_4{}^{3-} \ (Ca^{2+} \text{ chelators})$$

$$Ca_3(PO_4)_2 + 2H_2CO_3 \leftrightarrow 2CaHPO_4 + Ca(HCO_3)_2 \ (\text{carbonic and organic acids})$$

Nostoc carneum, *Nostoc piscinola*, *Anabaena torulosa*, and *Anabaena doliolum* are examples of cyanobacteria that have the ability to mineralize phosphorus (Joshi et al. 2020; Pandey et al. 2020).

7.4 Phenolic Compounds and Phytohormones of Importance for Agriculture

Within the phenolic compounds of importance for agriculture we can mention phenolic acids, flavonoids, and phenylpropanoids. These compounds have antimicrobial, fungicidal, and antioxidant properties, so their presence in agricultural crops confers protection against pathogens and biotic or abiotic stress (Pan et al. 2019). Microalgae and cyanobacteria have the ability to produce various phenolic compounds as part of the secondary metabolites of their growth, which participate in their reproduction process and protection against various stress conditions (Goiris et al. 2012).

Phytohormones are signaling molecules produced by plants that participate in their growth and development; some microorganisms are also capable of producing them as secondary metabolites. According to their chemical structure and functionality, we can classify these compounds as auxins, cytokinins, gibberellins, and abscisic acid (Alvarez et al. 2021). Zarezadeh et al. (2020) report the production of plant growth-promoting compounds such as indole-3-acetic, indole-3-propionic, and indole-3-butyric acids by the inoculation of the *Wollea vaginicola* ISB89, *Nostoc punctiforme* ISB90, and *Nostoc carneum* ISB88 cyanobacteria in chamomile (*Matricaria chamomilla* L.) cultures.

Cyanobacteria have the ability to produce auxins such as indole-3-acetic acid (IAA) and indole-3-butyric acid (IBA), responsible for regulating plant growth and development (Lu and Xu 2015). *Anabaena vaginalicola*, *Nostoc calcicola*, and *Scytonema bohneri* are examples of cyanobacteria that produce IBA as a secondary metabolite (Gayathri et al. 2017).

Abscisic acid participates in the response of plants to saline, osmotic, oxidative, drought, and nutritional stress. A response to the activation of the abscisic acid pathway in cyanobacteria is the production of carotenoids, which act as protective agents against oxidative stress (Han et al. 2018). Lu and Xu (2015) report the synthesis of ethylene and abscisic acid in *Arthrospira platensis*, *Synechococcus* sp., *Anabaena* sp., and *Nostoc* sp.

Cytokinins mediate several processes in plants, such as cell division, growth, biogenesis, chloroplast differentiation, and improve their tolerance to adverse environmental conditions. The synthesis of these compounds in microalgae and cyanobacteria stimulates cell division and photosynthetic activity, in addition to protecting cells from environmental stress (Lu and Xu 2015).

Gibberellin production in plants and cyanobacteria has similar functions in organisms related to growth rates and the accumulation of pigments and proteins (Lu and Xu 2015;

Han et al. 2018). As examples of gibberellin-producing microalgae and cyanobacteria, we can mention *Chlorella ellipsoidea*, *Chlorella pyrenoidosa*, *Chlorella vulgaris*, *Microcystis aeruginosa*, and *Arthrospira* (*Spirulina*) *platensis* (Pan et al. 2019; Bayona-Morcillo et al. 2020).

Compounds such as auxins, cytokinins, and gibberellins produced by cyanobacteria are summarized in Table 7.1.

7.5 Siderophores

In addition to the ability of cyanobacteria to produce phytohormones, they can produce siderophores, defined as ferric iron chelating organic compounds, which make this element (iron) available to other microorganisms and plants, thus affecting the productivity and yield of agricultural crops (Chakraborty et al. 2019).

The *Anabaena*, *Lyngbya*, *Nodularia*, *Synechococcus*, and *Trichormus* cyanobacteria have the ability to produce siderophores to chelate Fe or Cu, catalysts of enzymatic processes such as photosynthesis, respiration, and nitrogen fixation (Toribio et al. 2020). Among the siderophores that can be produced by cyanobacteria, we find schizoquine, anachelin, and sinecobactin; its synthesis is regulated by the iron concentration, the nitrogen source, the pH, and the temperature present in the medium (Singh et al. 2016a; Chakraborty et al. 2019).

The application of cyanobacteria in agricultural crops as biofertilizers increases the concentration of iron in the root zone by the action of siderophores, which facilitates the absorption and translocation of iron in the plant, in addition to other elements such as zinc (Santini et al. 2021). Anitha et al. (2016) evaluated the production of siderophores of *Spirulina platensis* when inoculating it in cultures of *Amaranthus gangeticus*, *Phaseolus aureus*, and tomato. An increase in zinc levels was observed due to the presence of siderophores in agricultural crops, which have the capacity to absorb chelated zinc ions from the soil and transport them to the plants.

Another important application of the use of siderophores produced by cyanobacteria is in the bioremediation of heavy metals, since currently, due to the effect of anthropogenic activities of man, the presence and persistence of these metals in soil and bodies of water has increased, therefore that

Table 7.1 Promoting plant growth compounds produced by cyanobacteria in different crops

Cyanobacteria	Agricultural crop	Promoting plant growth compounds	References
Anabaena flos-aquae and *Nostoc muscorum*	*Lupinus termis*	Indole-3-acetic acid, cytokinin, and gibberellin	Tantawy and Atef (2010)
Anabaena sp., *Oscillatoria* sp., *Phormidium* sp., *Chroococcidiopsis* sp., and *Synechocystis* sp.	Wheat	Indole-3-acetic acid and cytokinins	Hussain and Hasnain (2011)
Leptolyngbya sp., *Phormidium* sp., *Chroococcidiopsis* sp., and *Synechocystis* sp.	Wheat	Indole-3-acetic acid	Mazhar et al. (2013)
Anabaena vaginicola, *Cylindrospermum Michailovskoense*, and *Nostoc calcicola*	Peppermint	Indole-3-acetic acid and indole-3-propionic acid	Shariatmadari et al. (2015)
Anabaena, *Chlorogloeopsis*, and *Calothrix*	Radish and tomato	Cytokinins	Singh et al. (2016b)
Anabaenopsis and *Cylindrospermum*	Rice	Gibberellins	Kumar et al. (2019)
Anabaena variabilis and *Nostoc calcicola*	Corn, rice, and sorghum	Indole-3-acetic acid	Suresh et al. (2019)
Anabaena vaginicola and *Cylindrospermum michailovskoense*	*Plantago major* L.	Indole-3-acetic acid and indole-3-propionic acid	Chookalaii et al. (2020)
Wollea vaginicola, *Nostoc punctiforme*, and *Nostoc carneum*	Chamomile	Indole-3-acetic acid, indole-3-propionic acid, and indole-3-butyric acid	Zarezadeh et al. (2020)
Anabaena, *Trichormus*, *Leptolyngbya*, and *Nostoc*	Cucumber	Cytokinins	Toribio et al. (2020)
Nostoc kihlmani and *Anabaena cylindrica*	Wheat	Indole-3-acetic acid, cytokinin, and gibberellin	Alvarez et al. (2021)
Chroococcidiopsis sp., *Synechocystis* sp., and *Leptolyngbya* sp.	Wheat and bean	Indole-3-acetic acid	Lee and Ryu (2021)

looking for environmentally friendly alternatives to solve this problem is imperative. Siderophores can form complexes with these heavy metals, which allows the environment to detoxify these elements; examples of these processes are mentioned below. Singh et al. (2016a) evaluated the ability of the cyanobacterium *Anabaena oryzae* to produce siderophores and tolerate high cadmium concentrations in rice fields. The authors demonstrated that *Anabaena oryzae* has the ability to produce siderophores exposed to a cadmium concentration of 1.0 μM under iron-depleted conditions; therefore, it is suggested that these compounds can act as sequestering agents for cadmium metal ions, thus improving the productivity of rice crops. Additionally, Rashmi et al. (2013) studied the ability of a marine cyanobacterium to bioaccumulate uranium through the production of siderophores. The results show that *Synechococcus elongatus* has the ability to form a uranium–siderophore complex, so the use of this cyanobacterium to sequester uranium can be considered as a bioremediation method.

7.6 Biofertilizer in Crops

Biofertilizers are a set of microorganisms, macro and micronutrients that, when added to the soil, favor the growth and productivity of plants. The microorganisms present in these biofertilizers have the ability to fix nitrogen, solubilize minerals, and synthesize plant growth promoters and substances for the biocontrol of pathogens (García de Salamone et al. 2019; Joshi et al. 2020).

Cyanobacteria can fix 20–40 kg N ha^{-1} and are widely used as biofertilizers in rice cultivars (Mącik et al. 2020). Among the genera of cyanobacteria that we can find in the rice growing area are: nitrogen-fixing cyanobacteria such as *Gloeotrichia*, *Nostoc linkia*, *Anabaena variabilis*, *Aulosira fertilisima*, *Calothrix* sp., *Tolypothrix* sp., and *Scytonema* sp., as well as another group of cyanobacteria such as *Microcystis*, *Chroococcus*, *Oscillatoria*, *Lyngbya*, and *Phormidium* (Fernández-Valiente and Quesada 2004). We can also use cyanobacteria to biofertilize other crops of agronomic importance such as corn, wheat, sugarcane, sorghum, and tomato (Suresh et al. 2019; Sarma and Prakash 2020; Lee and Ryu 2021; Prasanna et al. 2021; Gulia et al. 2022). Bidyarani et al. (2016) demonstrated that the inoculation of *Anabaena laxa* as fresh biomass in chickpea plants improved soil parameters (available nitrogen and phosphorus), in addition to increasing dry weight, the number of pods per plant and yield (50% higher) of the crop. Prasanna et al. (2015) evaluated the inoculation of *Anabaena* and *Calothrix* as fresh biomass combined with 50% nitrogen fertilizer in cotton plants, resulting in an increase in the available nitrogen content in the soil and in plant height.

Some research groups have implemented cyanobacterial-based microbial consortia to biofertilizer different crops. Ramírez-López et al. (2019) used a nitrogen-fixing photosynthetic microbial consortium to fertilize wheat under greenhouse conditions, achieving a 75% reduction in the recommended chemical fertilization (ammonium sulfate). Additionally, beneficial effects on soil parameters (pH, nitrogen, phosphorus, electrical conductivity) were observed.

Despite the recognized usefulness of cyanobacteria as biofertilizers in various crops of agricultural importance, several authors have sought strategies to improve their interaction with the plant and permanence in the soil through the formation of consortia or *biofilms*, as is the case of Prasanna et al. (2016) who evaluated combinations of consortia and biofilm (*Anabaena-Azotobacter*, *Anabaena-Trichoderma biofilms* and the *Anabaena* + *Providencia* consortium), in order to increase soil microbial activities and the production of maize hybrid crops. The use of the *Anabaena-Trichoderma biofilm* gave the highest values both for the growth of the hybrid maize crop and for the microbiological parameters.

7.7 Bioprocess for the Biofertilizer Production

The development of bioprocesses to transfer a study at the laboratory level to its commercialization depends on the availability of the strain under study and having defined the bioproducts of interest. Reactor design and bioprocess optimization are essential steps for the new product to be sold (Acién et al. 2017). For the production of cyanobacterial biofertilizers to be implemented on a large scale, several steps must be taken into account that must be optimized depending on the farmer's needs, the crop where it will be applied, and the strain to be used. Figure 7.1 shows a general scheme of the bioprocess for obtaining a biofertilizer based on cyanobacteria.

7.7.1 Isolation

According to nutritional requirements, it is possible to isolate a wide variety of photosynthetic microorganisms from water and soil samples, which can be selected for use as biofertilizers based on their nitrogen fixation capacity, nutrient mineralization, and production hormones (auxins, gibberellins, and cytokinins); characteristics that enhance the growth of plants from their inoculation (Heimann and Cirés 2015; Toribio et al. 2020). These microorganisms can be isolated as pure strain, mixed culture, consortia, or *biofilm*, being crucial for their successful implementation in the study of their compatibility and interaction capacity of different microorganisms growing together. Among the cyanobacteria,

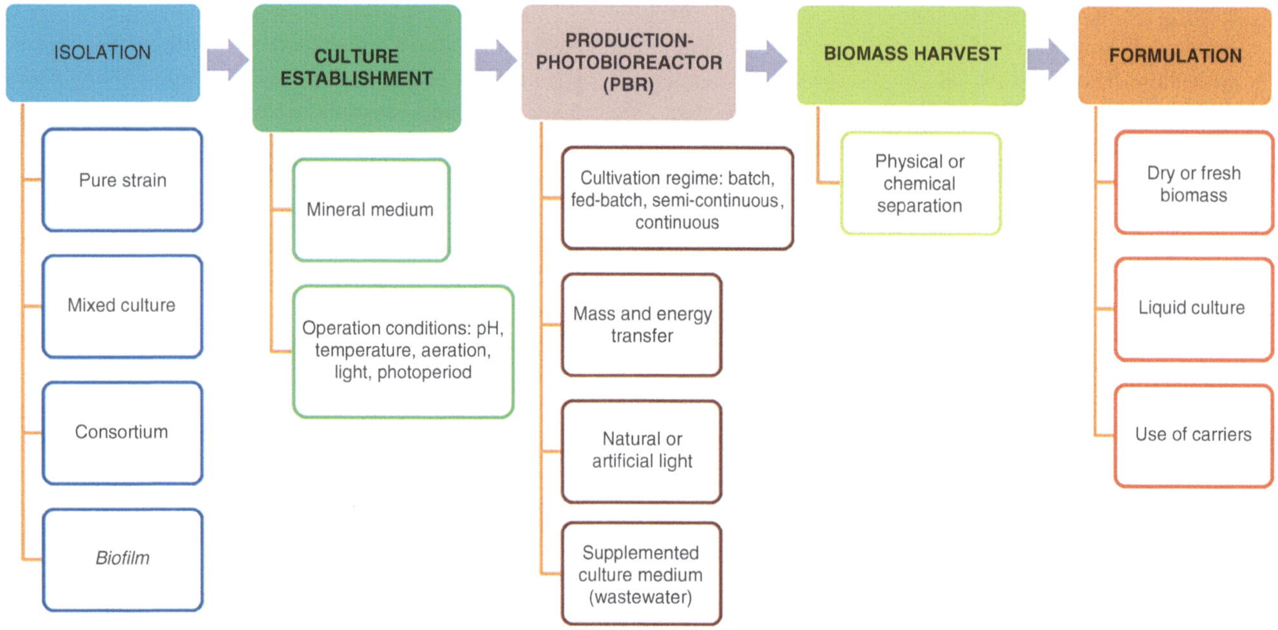

Fig. 7.1 Bioprocess steps for obtaining a biofertilizer based on cyanobacteria

more used as pure strains are *Nostoc*, *Anabaena*, *and Tolypothrix* for fixing nitrogen (Sarma and Prakash 2020; Song et al. 2022).

Mixed culture: A combination of two or more pure strains with different characteristics of agronomic importance (Renuka et al. 2018). Manjunath et al. (2016) demonstrated that the inoculation of the mixed culture of *Anabaena* sp.–*Providencia* sp. in the okra plant (*Abelmoschus esculentus* (L.) Moench), it increases the availability of nitrogen and phosphorus in the soil, and the growth of the plant in comparison with the inoculation of pure strains.

Biofilms: A set of microorganisms, of a single species or of multiple species, adhered to the biotic or abiotic surface of an extracellular polymeric matrix (Li and Tian 2016). Algae biofilms are made up of microalgae and other microorganisms that flourish in bodies of water or any terrestrial surface with a large amount of moisture (e.g., submerged stones, tree bark, among others) (Rana et al. 2020). Prasanna et al. (2020) observed that inoculation with *Anabaena torulosa* biofilms in wheat seeds favored germination (95–100%) over the control without inoculation.

Consortium: Consists of a system of two or more members where each organism does something for the benefit of another (Brock et al. 2004; Zhang et al. 2020). Ramírez-López et al. (2019) used a nitrogen-fixing photosynthetic microbial consortium to fertilize wheat under greenhouse conditions, achieving a 75% reduction in the recommended chemical fertilization (ammonium sulfate), as well as benefits on plant growth and performance, and an increase in soil parameters (pH, nitrogen, phosphorus, electrical conductivity).

7.7.2 Culture Establishment

Cyanobacteria have three types of nutrition (Lee 1999):

1. Facultative chemoheterotrophs. Some cyanobacteria can grow using an organic carbon source in the dark, but they can grow phototrophically with light too.
2. Obligate phototrophs. They can grow only in an inorganic medium and with light; some cyanobacteria are auxotrophic, since they require a small amount of an organic compound, invariably some vitamin, which is not used as a carbon source.
3. Photoheterotrophs. In the presence of light, cells can consume organic compounds as a carbon source but not in the dark.

When cyanobacteria are isolated from their natural environments and adapted to laboratory conditions, specific nutritional and environmental needs are required for their optimal growth and maintenance, among which the following stand out (Barsanti and Gualtieri 2014):

Temperature. To have good cell growth, a stable and constant temperature is required. It has been observed that most cyanobacteria can be grown at temperatures between 15 and 35 °C, with the optimal temperature being between 20 and 25 °C.

Mixed. Mixing is necessary to keep the medium homogeneous and guarantee the availability of nutrients throughout the geometry of the culture system. One way to keep the culture system mixed is by aeration and the fact that the cyanobacteria can use the CO_2 contained in the supplied air, which is used for photosynthesis.

pH. For optimal growth, cyanobacteria generally prefer growth at a pH range of 6.7–7.3 (neutral to slightly alkaline pH) (Pandey et al. 2020).

Illumination. Lighting is necessary for photosynthesis, in photoautotrophic and mixotrophic crops. The light is supplied by photoperiod to prevent photoinhibition of microorganism growth. The photoperiod necessary for proper growth is usually 16:8, 14:10, or 12:12, hours of light and hours of darkness, respectively (Xiong et al. 2013).

Nutrients. For the proper establishment and growth of cyanobacteria, a mineral medium containing macronutrients (such as N, P, K, Ca, S, Mg, Fe, and Na) and micronutrients (B, Mn, Zn, Mo, Cu, Co), which are provided by mineral salts in aqueous solution, in addition to CO_2 as carbon source (Chittora et al. 2020).

7.7.3 Production

Photobioreactor (PBR) is a system where the culture has little or no contact with the atmosphere; these systems have been evaluated in various configurations concerning their productivity potential and economic feasibility (Lu et al. 2020). The culture systems used are closed, open, and hybrid systems. Closed photobioreactors can be classified as large bags (clear plastic bags supported by a metal mesh frame) or tanks, towers (vertical cylinders made of fiberglass or acrylic), plate reactors, or tubular (glass, clear Teflon o clear PVC tubing) reactors. Open ponds are cheaper in construction and operation, offering many advantages for certain selected species. Open ponds can be classified as shallow unmixed ponds, raceway ponds with a paddle wheel, circular ponds mixed with a rotating arm, and sloping thin-layer cascade systems. Hybrid culture systems combine a closed photobioreactor and an open raceway. Closed photobioreactors produce a high-density inoculum for the open ponds, optimizing biomass production in less time (Ramírez-López and Fernández-Linares 2015).

Microalgae biotechnology can achieve high cell density cultures if an adequate reactor design is used and the process is optimized. For this reason, it is important to emphasize the most critical parameters for scaling and operation, such as mass transfer, shear stress, mixing, and light. These parameters correspond to each other and determine the productivity and efficiency of the culture system (Reyna-Velarde et al. 2010). Hernández-Melchor et al. (2017) carried out the scale-up, mass transfer properties, and the characterization hydrodynamics (mixing time (tm), gas hold-up (ε), and Reynolds number (Re)) of a flat-panel airlift photobioreactor (PBR) of 150 L. This 150 L PBR was adequate for biomass production of a microbial consortium (MC) with nitrogen fixation capacity and potential use as a biofertilizer. The better parameters hydrodynamic ($\varepsilon = 0.048$, $tm = 103.8$ s, $Re = 1761$) and the mass transfer ($k_La = 14.4\ h^{-1}$) were found to operate the PBR in a superficial velocity of gas of $0.0068\ m\ s^{-1}$. Likewise, this workgroup evaluated the biomass production capacity of the MC in two flat-panel airlift PBR (50 and 150 L) outdoors in batch cultivation, in both FBRs semicontinuous cultivation in the higher volume PBR. The maximum biomass concentration in batch culture was obtained in the PBR of 150 L, with a value of 600 mg L^{-1}. The semicontinuous MC culture was carried out for three cycles, increasing biomass productivity after each cycle, with values of 1.86, 1.94, and 2.94 mg $L^{-1}\ h^{-1}$, respectively (Hernández-Melchor et al. 2017b).

In addition, it is important to consider the production costs, so the use of nutrient-rich wastewater is considered an alternative for supplying nutrients for cyanobacteria growth (Alvarez et al. 2021). Sánchez-Contreras et al. (2021) evaluated the growth of carbohydrate-rich biomass of the filamentous cyanobacteria *Geitlerinema* sp., which grew in a stirred tank photobioreactor, operated in a semicontinuous regime using the effluent from a wastewater treatment plant as substrate. The maximal carbohydrate accumulation was achieved on day 8 (with an N:P ratio of 22:1), the constant carbohydrate content of 57% without any additional carbon source and biomass productivity of 0.033 g $L^{-1}\ d^{-1}$.

7.7.4 Biomass Harvest

Biomass harvesting of cyanobacteria is difficult due to the low concentration of biomass in the culture medium and the small size of the cells, for which sedimentation is not an option (Iasimone et al. 2020).

The most widely used methods on a small scale for the biomass harvesting of cyanobacteria are centrifugation and chemical flocculation. These cannot be applied at large scales due to centrifugation being a high-energy consuming process, while chemical flocculation generates secondary pollution in the liquid effluent (Iasimone et al. 2020). An alternative to these methods is the autoflocculation and/or bioflocculation processes. Flocculation of the suspended biomass around a nucleus of inorganic compounds generates autoflocculation, whereas organic secretions from cells can induce spontaneous flocculation called bioflocculation (Ummalyma et al. 2017). Iasimone et al. (2021) studied the bioflocculation process for *Leptolyngbya* sp. and *Pseudanabaena* sp. (filamentous cyanobacteria). Their cultivation was carried out in batch mode using the synthetic aqueous solution and urban wastewater as culture media. Results showed that bioflocculation and sedimentation processes occur efficiently only in synthetic aqueous solution. In contrast, the specific characteristics of urban wastewater prevent bioflocculation.

7.7.5 Formulation

The appropriate formulation is an important factor for biofertilizers containing living microbial cells since it affects the quality of biological agents. A suitable formulation effectively introduces microorganisms to the target site and enhances their activity after inoculation on the host plant (Mącik et al. 2020). Biofertilizers based on cyanobacteria involve both direct inoculations into the soil: the dry biomass addition (pellets, powder, granules) or liquid culture and its formulation with a suitable carrier (material that carries microorganisms and keeps them in a good physiological state: peat, lignite, charcoal). Commonly, liquid formulations contain a high cell concentration and the application of a lower quantity of inoculum is necessary (Bhattacharyya et al. 2020; Mącik et al. 2020). Gulia et al. (2022) evaluated the ability to enhance the growth of tomato with *Calothrix elenkinii, Anabaena laxa*, and *biofilm—Anabaena torulosa—Trichoderma viride* using as carrier perlite, vermiculite, and cocopeat (P, V, C) in different combinations. The results of leaf, soil, and nutrients suggested that the combination of C:V:P (3,1:1) + *A. laxa* was the best.

7.8 Study Case 1: Photosynthetic Microbial Consortium as Biofertilizer in Wheat Crops

Reyna-Velarde et al. (2012) isolated a photosynthetic Microbial Consortium (MC) able to fix nitrogen, from a rice field located at Alpuyeca, state of Morelos, México. The MC was maintained at BG-11_0 medium (without nitrogen source) and progressively scaled up at 0.5 L, 1 L, 1.9 L, and 11 L working volume. The Nitrogen fixation capacity of MC in the 1.9 L photobioreactor was 3794 nmol Ethene g^{-1} dry weight h^{-1}.

The microorganisms that make up the MC were identified morphologically and phylogenetically to establish, theoretically, the probable interaction between its components and explore its possible biotechnological application. In the MC were founded 21 different microorganisms belonging to the *phyla* Bacteroidetes (*Flavobacterium* sp. and *F. aquatile*), Proteobacteria (*Rhodobacter* sp., *Devosia insulae*, *Pedomicrobium americanum*, *Alpha proteobacteria*, *Aquaspirillum delicatum*, *Methylibium petroleiphilum*, and *Nannocystis* sp.), Chlorophyta (*Monoraphidium* sp., *Chlorella* sp.), Heterokontophyta (*Cyclotella meneghiniana*, *Melosira varians*, *Cocconeis placentula*, *Achnanthidium chlidanos*, *Navicula radiosa*, *Fragilaria ulna*, and *Nitzschia* sp.), Cyanobacteria (*Aphanizomenon aphanizomenoides*, *Leptolyngbya* sp., and *Anabaena oscillarioides*). A greater relative abundance of cyanobacteria was founded (Hernández-Melchor et al. 2016).

From the above, the biomass production of the MC was carried out in a photobioreactor of 11 L in batch culture. The biomass produced was subjected to two drying methods and later used in wheat seed germination tests.

7.8.1 MC Cultivation in a Photobioreactor

The propagation of the microbial consortium (MC) was carried out in an 11-L bubble column photobioreactor, evaluating its biological performance in batch culture for 14 days by determining the amount of dry biomass (Tredici et al. 1991) and nitrogenase activity by acetylene reduction assay (Venkataraman 1981). The culture conditions were 80 μmol photons m^{-2} s^{-1} (light intensity), aeration of 11 L min^{-1} (1 vvm), and an inoculum of 4 L measuring a total volume of 11 L with mineral medium, without nitrogen source BG-11_0 (Rippka et al. 1979).

7.8.2 Drying Process

The methods used were lyophilization and spray-drying. For the lyophilization process, the biomass obtained from the 11 L photobioreactor batch culture was concentrated by centrifugation and placed in 125 mL Erlenmeyer flasks that were placed in dry ice and acetone for freezing, constantly rotating the flask to obtain homogeneity in the freezing of the biomass, later the flasks were placed in the lyophilizer (Labconco 2.5 Free Zone) for 5 h. The spray-drying process was realized according to Zhang et al. (2022).

7.8.3 Germination Tests

Wheat seeds variety Barcenas S2002 were used, which were previously disinfected with a bleach solution (Cloralex®) and distilled water in a ratio of 1:4, respectively, for 15 minutes (Ramírez-López et al. 2019). Subsequently, they were rinsed with distilled water until the total elimination of the disinfectant solution. The seeds were placed in square transparent plastic boxes with lids, which were previously washed and sterilized with UV radiation, and a sterile cotton bed was placed inside. Eight treatments were applied to the seeds, including different presentations of the MC, to evaluate the percentage of germination. Each treatment was carried out in duplicate, placing 8 seeds per box. The treatments were designated as follows: (A) Distilled water (5 mL), (B) Sterile BG-11_0 medium (5 mL), (C) MC biomass in suspension (1.25 mL), (D) MC biomass in suspension (2.5 mL), (E) MC biomass in suspension (5 mL), (F) MC biomass in suspension (7.5 mL), (G) Lyophilized MC + distilled water (5 mg + 5 mL), (H) Spray-drying MC + distilled water (5 mg + 5 mL). Every third day, 5 mL of sterile distilled water was added to the wheat seeds to maintain the necessary humidity for their development.

The germination boxes were maintained under the following working conditions: photoperiod 12:12 (light: dark), 80 μmol photons m^{-2} s^{-1} (light intensity), and temperature of 23 ± 2 °C for seven days. After the time, the percentage of germination per treatment was evaluated; in addition, the height of the seedling, flag leaf length, and the number and length of roots were measured.

7.8.4 Statistical Analysis

An analysis of variance (ANOVA) and the multiple mean comparison test (Tukey, $P \leq 0.05$), with the statistical SAS package for Windows version 6.2.9200 was used to assess significant differences among germination treatments.

7.9 Results

7.9.1 Biomass Production

The biomass used for the drying process (lyophilization and spray-drying) and in the germination tests as biomass in suspension was obtained from the 11-L photobioreactor at the end of the batch culture (day 14), from which was obtained an amount of dry biomass of 305 mg L^{-1}, with a nitrogenase activity of 10,294 nmol Ethene g^{-1} dry weight h^{-1}. The latter value is approximately five times higher than that reported in the literature for *Anabaena variabilis* Sao (2678 nmol Ethene g^{-1} dry weight h^{-1}) and *Nostoc muscorum* DOH (2237 nmol Ethene g^{-1} dry weight h^{-1}) (Kannaiyan et al. 1997; Kannaiyan 2002), thus highlighting its potential use as a MC biofertilizer.

7.9.2 Germination Tests

The seeds of all treatments began to germinate on the third day of exposure; however, the germination percentages varied in the different treatments until day seven of the experiment. Figure 7.2 shows the germinated wheat seeds

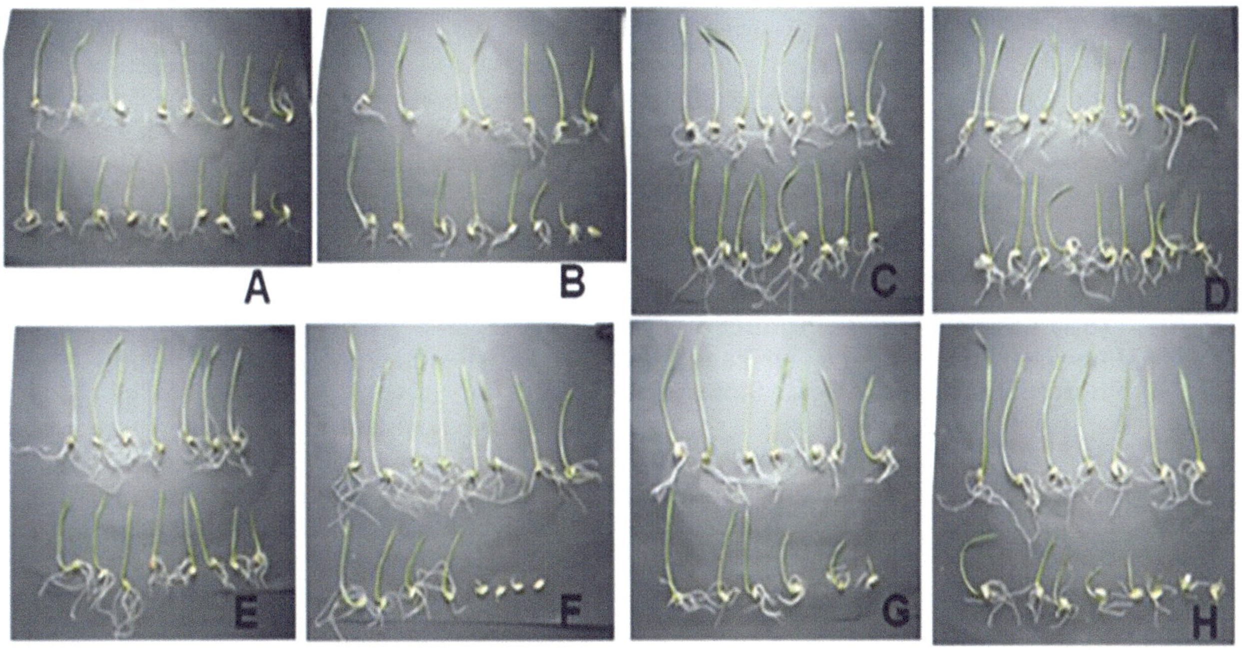

Fig. 7.2 Seedlings of wheat variety Barcenas S2002 on the seventh day after inoculation. (**a**) Distilled water (5 mL), (**b**) Sterile BG-11_0 medium (5 mL), (**c**) MC biomass in suspension (1.25 mL), (**d**) MC biomass in suspension (2.5 mL), (**e**) MC biomass in suspension (5 mL), (**f**) MC biomass in suspension (7.5 mL), (**g**) Lyophilized MC + distilled water (5 mg + 5 mL), (**h**) Spray-drying MC + distilled water (5 mg + 5 mL)

Table 7.2 Global results of the germination of wheat seeds variety Barcenas S2002 on day 7

Treatment	% germination	Seedling height (cm)	Flag leaf length (cm)	Number of roots	Root length (cm)
(A) Distilled water	94.38 b ± 0.078	5.38 de ± 0.071	3.18 d ± 0.072	4 a	3.63 e ± 0.071
(B) Sterile BG-11$_0$ medium	83.38 d ± 0.078	5.84 c ± 0.072	3.65 b ± 0.071	6 a	4.31 cd ± 0.071
(C) MC biomass in suspension (1.25 mL)	83.38 d ± 0.078	7.17 a ± 0.071	4.70 a ± 0.072	5 a	5.36 a ± 0.072
(D) MC biomass in suspension (2.5 mL)	99.99 a ± 0.007	5.84 c ± 0.072	3.44 bcd ± 0.071	6 a	4.70 b ± 0.072
(E) MC biomass in suspension (5 mL)	83.38 d ± 0.078	6.80 b ± 0.071	4.45 a ± 0.071	5 a	5.31 a ± 0.072
(F) MC biomass in suspension (7.5 mL)	88.94 c ± 0.078	5.43 d ± 0.072	3.60 bc ± 0.071	5 a	4.88 b ± 0.071
(G) Lyophilized MC + distilled water	94.38 b ± 0.078	5.13 e ± 0.071	3.22 d ± 0.072	5 a	4.07 d ± 0.035
(H) Spray-drying MC + distilled water	83.38 d ± 0.078	5.37 de ± 0.072	3.35 cd ± 0.072	5 a	4.41 c ± 0.072

Mean values (± standard errors) in a column followed by different letters are significantly different (Tukey; $p \leq 0.05$)

per treatment on day seven. Table 7.2 shows the global results of the treatments tested.

All the treatments tested produced a germination percentage higher than 80% (Table 7.2); however, the treatments where low concentrations of biomass in suspension (1.25, 2.5, and 5 mL), and they showed significant differences (Tukey, $\alpha = 0.05$) in the percentage of germination and physiological parameters of seedling growth (seedling height, flag leaf length, number and length of roots), compared with the control and even concerning the use of lyophilized or spray-drying MC biomass. The above-mentioned indicates a greater contribution of nutrients and growth promoters by the MC to the wheat seedlings. It is also important to consider the viability of the microorganisms that make up the MC for the drying processes by which it was subjected since the temperatures that these processes use can damage the cellular apparatus of the microorganisms present in the MC, thus changing their dynamics and population integrity. According to the results mentioned above, applying the biomass in the suspension of the MC as a biofertilizer is suggested.

7.10 Study Case 2: Biomass Production of a Photosynthetic MC in a Flat-Panel Airlift Photobioreactor for Use as Biofertilizer

According to the results obtained and analyzed in study case 1, the photosynthetic MC is of interest due to its high nitrogen fixation capacity compared to pure strains, in addition to promoting the germination of wheat seeds. Therefore, it is imperative to develop a strategy to produce large amounts of biomass, for which a 16-L flat-panel airlift photobioreactor (PBR) was designed, built, and characterized.

For the design of the PBR, the geometry and the A/V ratio (illuminated area/operation volume) of 7.3 m^{-1} described by Reyna-Velarde et al. (2010) were applied. The PBR was elaborated with acrylic sheets and has the following specifications: 0.41 m length, 0.34 m height, 0.14 m width (light-path), 16 L working volume, and a ratio of the cross-sectional area of the downcomer zone with respect to the riser zone (Ad/Ar) equal to 1. Air flow entered the PBR through a stainless steel tube placed along the base of the PBR and it was controlled with a Carboloy-float rotameter (Cole-Parmer, USA). Once the 16 L PBR was built, the hydrodynamic and mass transfer characterization was carried out in a biphasic system (air-water) at 21 ± 2 °C, with the atmospheric pressure of Mexico City (585 mmHg), evaluating the parameters of volumetric mass transfer coefficient (k_La), mixing time and gas hold-up according to the techniques described by Reyna-Velarde et al. (2010).

After the characterization of the PBR, a kinetics per batch culture was carried out for 14 days under controlled conditions of light and temperature, to evaluate biomass production on dry weight (Tredici et al. 1991), chlorophyll *a* (Brock et al. 2004), and nitrogenase activity by acetylene reduction assay (Venkataraman 1981) of the MC. The culture conditions were 80 μmol photons m^{-2} s^{-1} (light intensity), temperature 21 ± 2 °C, aeration of 10 L min^{-1} (equivalent to a superficial gas velocity of 0.0029 m s^{-1}) and an inoculum of 5 L measuring a total volume of 16 L with mineral medium without nitrogen source BG-11$_0$ (Rippka et al. 1979).

7.11 Results

The better parameters hydrodynamic and the mass transfer were found to operate the PBR in a superficial velocity of gas equal to 0.0029 m s^{-1} with values equal to mixing time = 52.5 s, Gas hold-up = 0.025 and k_La = 10.62 h^{-1}.

Figure 7.3 shows the dry weight and chlorophyll *a* concentration in batch culture of MC in the 16 L PBR. The dry weight starts with a concentration equal to 226 mg L^{-1} reaching the maximum biomass with 323 mg L^{-1} on day four, this value corresponds to the end of the exponential phase, after this point an establishment phase is observed until day 8, after this point the culture begins a phase of cell death, the productivity obtained at the end of the exponential phase was 1 mg L^{-1} h^{-1} with a specific growth rate of 0.092 d^{-1}. The concentration of

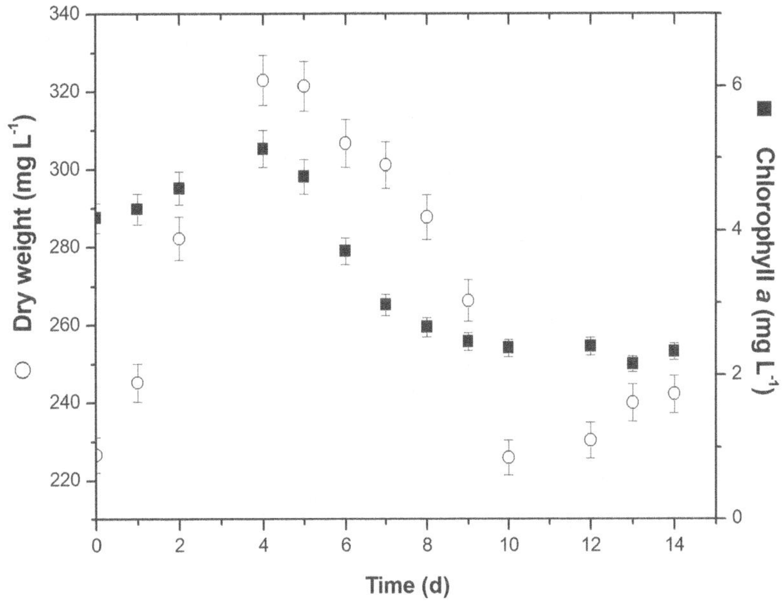

Fig. 7.3 Dry weight and Chlorophyll *a* concentration *vs* time in the 16-L PBR

chlorophyll *a* increased over time, obtaining the highest value on day 4 with 5.13 mg L^{-1}, after this day a decrease in the content of chlorophyll in the MC was observed, which coincides with the kinetics of biomass production.

The maximum value of nitrogenase activity was obtained on day four with a value equal to 81,851 nmol Ethene g^{-1} dry weight h^{-1}, the day on which the highest value of biomass was obtained. This nitrogenase activity value is approximately 9–23 times higher than that reported in the literature for *Leptolyngbya valderiana* DBU20041 (9000 nmol Ethene g^{-1} dry weight h^{-1}) (Prabaharan et al. 2010) and *Anabaena azollae* SK6 (3571 nmol Ethene g^{-1} dry weight h^{-1}) (Kannaiyan et al. 1997), respectively.

According to the results obtained in the 16-L PBR, it is possible to use this design for the production of MC at different scales and thus obtain a high concentration of biomass without affecting its nitrogen fixation capacity, parameters necessary for its application as a biofertilizer.

7.12 Conclusions and Future Perspectives

In conclusion, the nitrogen-fixing rates obtained in the 11-L photobioreactor were five times higher than reported by other nitrogen-fixing cultures at the laboratory scale. According to the morphological analysis, together with the phylogenetic analysis, it was possible to identify cyanobacteria nitrogen-fixing filamentous, and proteobacteria as the main organisms that make up the MC; this ensures the existence of an interaction between them and an exchange of metabolites between all the organisms that make up the MC, which allows them to survive in a different environment. In addition, the application of MC as biomass in suspension favors the germination of wheat seeds, which suggests the presence of growth promoters, so it is recommended that the MC application be isolated in these investigations as a biofertilizer for different crops.

Actually, chemical fertilizers and pesticides become a limitation for the farmers because of their cost-function ineffectiveness, also proven to be harmful to the land. On the other hand, biofertilizers represent a cost-effective alternative compared with the agrochemicals available (Bhattacharyya et al. 2020). Therefore, cyanobacteria are an excellent alternative as biofertilizers because of their photosynthesis and nitrogen fixation capability, making soil enriched in carbon and nitrogen and enhancing phosphorus bioavailability to plants. However, its application in agricultural fields has not yet become widespread due to technological limitations and lack of knowledge about its use. It is still necessary to identify potential cyanobacteria strains and combinations of strains and identify bioactive compounds promoting plants. For those above to be economically feasible on a large scale, it is necessary to design and optimize biomass production systems for cyanobacteria and their by-products through the design of photobioreactors, bioseparation, and product recovery methods.

References

Acién FG, Molina E, Reis A, Torzillo G, Zittelli GC, Sepúlveda C, Masojídek J (2017) Photobioreactors for the production of microalgae. In: Gonzalez-Fernandez C, Muñoz R (eds) Microalgae-based biofuels and bioproducts. Woodhead Publishing, Sawston, pp 1–44

Alvarez AL, Weyers SL, Goemann HM, Peyton BM, Gardner RD (2021) Microalgae, soil and plants: a critical review of microalgae as renewable resources for agriculture. Algal Res 54:102200. https://doi.org/10.1016/j.algal.2021.102200

Anitha L, Bramari GS, Kalpana P (2016) Effect of supplementation of *Spirulina platensis* to enhance the zinc status in plants of *Amaranthus gangeticus*, *Phaseolus aureus* and tomato. Adv Biosci Biotechnol 7(6):289–299

Barsanti L, Gualtieri P (2014) Algae: anatomy, biochemistry, and biotechnology, 2nd edn. CRC Press, Boca Raton, p 362

Bayona-Morcillo PJ, Plaza BM, Gómez-Serrano C, Rojas E, Jiménez-Becker S (2020) Effect of the foliar application of cyanobacterial hydrolysate (*Arthrospira platensis*) on the growth of *Petunia x hybrida* under salinity conditions. J Appl Phycol 32:4003–4011

Berg G, Rybakova D, Fischer D, Cernava T, Vergès MCC, Charles T, Chen X, Cocolin L, Eversole K, Corral GH, Kazou M, Kinkel L, Lange L, Lima N, Loy A, Macklin JA, Maguin E, Mauchline T, McClure R, Mitter B, Ryan M, Sarand I, Smidt H, Schelkle B, Roume H, Kiran GS, Selvin J, de Souza RSC, van Overbeek L, Singh BK, Wagner M, Walsh A, Sessitsch A, Schloter M (2020) Microbiome definition re-visited: old concepts and new challenges. Microbiome 8(1):1–22

Bhattacharyya C, Roy R, Tribedi P, Ghosh A, Ghosh A (2020) Biofertilizers as substitute to commercial agrochemicals. In: Prasad MNV (ed) Agrochemicals detection, treatment and remediation. Butterworth-Heinemann, Oxford, pp 263–290

Bidyarani N, Prasanna R, Babu S, Hossain F, Saxena AK (2016) Enhancement of plant growth and yields in Chickpea (*Cicer arietinum* L.) through novel cyanobacterial and biofilmed inoculants. Microbiol Res 188–189:97–105

Brock JT, Madigan MT, Martinko JM, Parker J (2004) Biología de los microorganismos, 10th edn. Pearson Education, Madrid, p 1200

Brusseau ML, Artiola JF (2019) Chemical contaminants. In: Brusseau ML, Pepper IL, Gerba CP (eds) Environmental and pollution science. Academic Press, New York, pp 175–190

Chakraborty S, Verma E, Singh SS (2019) Cyanobacterial siderophores: ecological and biotechnological significance. In: Mishra AK, Tiwari DN, Rai AN (eds) Cyanobacteria. Academic Press, Cambridge, pp 383–397

Chittora D, Meena M, Barupal T, Swapnil P, Sharma K (2020) Cyanobacteria as a source of biofertilizers for sustainable agriculture. Biochem Biophys Rep 22:100737. https://doi.org/10.1016/j.bbrep.2020.100737

Chookalaii H, Riahi H, Shariatmadari Z, Mazarei Z, Hashtroudi MS (2020) Enhancement of total flavonoid and phenolic contents in *Plantago major* L. with plant growth promoting cyanobacteria. J Agric Sci Technol 22:505–518

Davey M, Clarke K (1991) The spatial distribution of microalgae on Antarctic fellfield soils. Antarct Sci 3:257–263

Díaz IFC, Molina LXZ, Cárdenas CIC, Anaya ER, Ramírez SR, de los Santos Villalobos S (2020) Consideraciones sobre el uso de biofertilizantes como alternativa agro-biotecnológica sostenible para la seguridad alimentaria en México. Rev Mexicana Cienc Agric 11(6):1423–1436

Fernández-Valiente E, Quesada A (2004) A shallow water ecosystem: rice-fields. The relevance of cyanobacteria in the ecosystem. Limnetica 23:95–108

García de Salamone IE, Esquivel-Cote R, Hernández-Melchor DJ, Alarcón A (2019) Manufacturing and quality control of inoculants from the paradigm of circular agriculture. In: Singh D, Gupta V, Prabha R (eds) Microbial interventions in agriculture and environment. Springer, Singapore, pp 37–74

Gayathri M, Shunmugam S, Thajuddin N, Muralitharan G (2017) Phytohormones and free volatile fatty acids from cyanobacterial biomass wet extract (BWE) elicit plant growth promotion. Algal Res 26:56–64

Goiris K, Muylaert K, Fraeye I, Foubert I, De Brabanter J, De Cooman L (2012) Antioxidant potential of microalgae in relation to their phenolic and carotenoid content. Environ Boil Fishes 24:1477–1486

Gulia U, Shukla J, Nishanth S, Kokila V, Bharti A, Singh AK, Shivay YS, Prasanna R (2022) Fortifying nursery soil-less media with cyanobacteria for enhancing the growth of tomato. S Afr J Bot 146:564–572

Han X, Zeng H, Bartocci P, Fantozzi F, Yan Y (2018) Phytohormones and effects on growth and metabolites of microalgae: a review. Fermentation 4(2):25. https://doi.org/10.3390/fermentation4020025

Heggerud CM, Wang H, Lewis MA (2022) Coupling the socio-economic and ecological dynamics of cyanobacteria: single lake and network dynamics. Ecol Econ 194:107324. https://doi.org/10.1016/j.ecolecon.2021.107324

Heimann K, Cirés S (2015) N_2-fixing cyanobacteria: ecology and biotechnological applications. In: Kim SK (ed) Handbook of marine microalgae. Academic Press, Cambridge, pp 501–515

Hernández-Melchor DJ, Carmona Jiménez J, Hidalgo Lara ME, Dendooven L, Marsch Moreno R, Cañizares Villanueva RO (2016) Identificación morfológica y filogenética de un consorcio microbiano fotosintético de posible interés biotecnológico. Hidrobiologica 26(2):311–321

Hernández-Melchor DJ, Cañizares-Villanueva RO, Terán-Toledo JR, López-Pérez PA, Cristiani-Urbina E (2017) Hydrodynamic and mass transfer characterization of flat-panel airlift photobioreactors for the cultivation of a photosynthetic microbial consortium. Biochem Eng J 128:141–148

Hernández-Melchor DJ, Urbina EC, Cerrato RF, García FJE, Villanueva ROC (2017b) Producción a la intemperie de un consorcio microbiano fotosintético fijador de nitrógeno en cultivo por lote y semicontinuo en un fotobiorreactor air-lift de cara plana. Rev Int Contam Ambient 33:73–81

Hussain A, Hasnain S (2011) Phytostimulation and biofertilization in wheat by cyanobacteria. J Ind Microbiol Biotechnol 38:85–92

Iasimone F, Seira J, Desmond-Le Quemener E, Panico A, De Felice V, Pirozzi F, Steyer JP (2020) Bioflocculation and settling studies of native wastewater filamentous cyanobacteria using different cultivation systems for a low-cost and easy to control harvesting process. J Environ Manag 256:109957. https://doi.org/10.1016/j.jenvman.2019.109957

Iasimone F, Seira J, Panico A, De Felice V, Pirozzi F, Steyer JP (2021) Insights into bioflocculation of filamentous cyanobacteria, microalgae and their mixture for a low-cost biomass harvesting system. Environ Res 199:111359. https://doi.org/10.1016/j.envres.2021.111359

Joshi H, Shourie A, Singh A (2020) Cyanobacteria as a source of biofertilizers for sustainable agriculture. In: Singh PK, Kumar A, Singh VK, Shrivastava AK (eds) Advances in cyanobacterial biology. Academic Press, Cambridge, pp 385–396

Kannaiyan S (2002) Biofertilizers for sustainable crop production. In: Kannaiyan S (ed) Biotechnology of biofertilizers. Narosa Publishing House, New Delhi, pp 9–49

Kannaiyan S, Aruna SJ, Merina PKS, Hall DO (1997) Immobilized cyanobacteria as a biofertilizer for rice crops. J Appl Phycol 9: 167–174

Kumar J, Singh D, Tyagi MB, Kumar A (2019) Cyanobacteria: applications in biotechnology. In: Mishra AK, Tiwari DN, Rai AN (eds) Cyanobacteria. Academic Press, Cambridge, pp 327–346

Lee RE (1999) Phycology, 3rd edn. Cambridge University Press, Cambridge, p 561

Lee SM, Ryu CM (2021) Algae as new kids in the beneficial plant microbiome. Front Plant Sci 12:91. https://doi.org/10.3389/fpls.2021.599742

Li YH, Tian XL (2016) Quorum sensing and bacterial social interactions in biofilms: bacterial cooperation and competition. In: de Bruijn FJ (ed) Stress and environmental regulation of gene expression and adaptation in bacteria. Wiley, pp 1195–1205. https://doi.org/10.1002/9781119004813

Lu Y, Xu J (2015) Phytohormones in microalgae: a new opportunity for microalgal biotechnology? Trends Plant Sci 20:273–282

Lu W, Alam MA, Liu S, Xu J, Saldivar RP (2020) Critical processes and variables in microalgae biomass production coupled with bioremediation of nutrients and CO_2 from livestock farms: a review. Sci Total Environ 716:135247. https://doi.org/10.1016/j.scitotenv.2019.135247

Mącik M, Gryta A, Frąc M (2020) Biofertilizers in agriculture: an overview on concepts, strategies and effects on soil microorganisms. Adv Agron 162:31–87

Mandal B, Vlek PLG, Mandal LN (1999) Beneficial effects of blue-green algae and Azolla, excluding supplying nitrogen, on wetland rice fields: a review. Biol Fertil Soils 28:329–342

Manjunath M, Kanchan A, Ranjan K, Venkatachalam S, Prasanna R, Ramakrishnan B, Hossain F, Nain L, Shivay YS, Rai AB, Singh B (2016) Beneficial cyanobacteria and eubacteria synergistically enhance bioavailability of soil nutrients and yield of okra. Heliyon 2(2):e00066. https://doi.org/10.1016/j.heliyon.2016.e00066

Mazhar S, Cohen JD, Hasnain S (2013) Auxin producing non-heterocystous cyanobacteria and their impact on the growth and endogenous auxin homeostasis of wheat. J Basic Microbiol 53: 996–1003

Meena H, Busi S (2019) Use of microbial biofertilizers technology in agro-environmental sustainability. In: Singh JS (ed) New and future developments in microbial biotechnology and bioengineering. Elsevier, Ámsterdam, pp 99–211

Nelson C, Giraldo-Silva A, Garcia-Pichel F (2021) A symbiotic nutrient exchange within the cyanosphere microbiome of the biocrust cyanobacterium, *Microcoleus vaginatus*. ISME J 15(1):282–292

Nozue S, Katayama M, Terazima M, Kumazaki S (2017) Comparative study of thylakoid membranes in terminal heterocysts and vegetative cells from two cyanobacteria, *Rivularia* M-261 and *Anabaena variabilis*, by fluorescence and absorption spectral microscopy. Biochim Biophys Acta-Bioenerg 1858(9):742–749

Pan S, Jeevanandam J, Danquah MK (2019) Benefits of algal extracts in sustainable agriculture. In: Hallmann A, Rampelotto P (eds) Grand challenges in marine biotechnology. Springer, Berlin, pp 501–534

Pandey SN, Verma I, Kumar M (2020) Cyanobacteria: potential source of biofertilizer and synthesizer of metallic nanoparticles. In: Singh PK, Kumar A, Singh VK, Shrivastava AK (eds) Advances in cyanobacterial biology. Academic Press, Cambridge, pp 351–367

Prabaharan D, Kumar DA, Uma L, Subramanian G (2010) Dark hydrogen production in nitrogen atmosphere-an approach for sustainability by marine cyanobacterium *Leptolyngbya valderiana* BDU 20041. J Hydrogen Energ 35:10725–10730

Prasanna R, Babu S, Bidyarani N, Kumar A, Triveni S, Monga D, Mukherjee AK, Kranthi S, Gokte-Narkhedkar N, Adak A, Yadav K, Nain L, Saxena AK (2015) Prospecting cyanobacteria-fortified composts as plant growth promoting and biocontrol agents in cotton. Exp Agric 51:42–65

Prasanna R, Kanchan A, Ramakrishnan B, Ranjan K, Venkatachalam S, Hossain F, Shivay YS, Krishnan P, Nain L (2016) Cyanobacteria-based bioinoculants influence growth and yields by modulating the microbial communities favourably in the rhizospheres of maize hybrids. Eur J Soil Biol 75:15–23

Prasanna R, Gupta H, Yadav VK, Gupta K, Buddhadeo R, Gogoi R, Bharti A, Mahawar H, Nain L (2020) Prospecting the promise of cyanobacterial formulations developed using soil-less substrates as carriers. Environ Technol Innov 18:100652. https://doi.org/10.1016/j.eti.2020.100652

Prasanna R, Hossain F, Saxena G, Singh B, Kanchan A, Simranjit K, Shivay YS (2021) Analyses of genetic variability and genotype x cyanobacteria interactions in biofortified maize (*Zea mays* L.) for their responses to plant growth and physiological attributes. Eur J Agron 130:126343. https://doi.org/10.1016/j.eja.2021.126343

Priya H, Prasanna R, Ramakrishnan B, Bidyarani N, Babu S, Thapa S, Renuka N (2015) Influence of cyanobacterial inoculation on the culturable microbiome and growth of rice. Microbiol Res 171: 78–89

Rai AN, Singh AK, Syiem MB (2019) Plant growth-promoting abilities in cyanobacteria. In: Mishra AK, Tiwari DN, Rai AN (eds) Cyanobacteria. Academic Press, Cambridge, pp 459–476

Ramírez-López C (2011) Fertilización de trigo (Triticum aestivum) variedad Bárcenas empleando un consorcio microbiano fotosintético fijador de nitrógeno. Master Thesis, Centro de Investigación y de Estudios Avanzados del Instituto Politécnico Nacional, CDMX México

Ramírez-López C, Fernández-Linares L (2015) Technologies for the production of biodiesel from microalgae: biomass production the main challenge. In: Torres-Bustillos LG (ed) Microalgae and other phototrophic bacteria: culture, processing, recovery and new products. Nova Publishers, New York, pp 197–218

Ramírez-López C, Esparza-García FJ, Ferrera-Cerrato R, Alarcón A, Cañizarez-Villanueva RO (2019) Short-term effects of a photosynthetic microbial consortium and nitrogen fertilization on soil chemical properties, growth, and yield of wheat under greenhouse conditions. J Appl Phycol 31:3617–3624

Rana KL, Kour D, Yadav AN, Yadav N, Saxena AK (2020) Agriculturally important microbial biofilms: biodiversity, ecological significances, and biotechnological applications. In: Yadav MK, Singh BP (eds) New and future developments in microbial biotechnology and bioengineering: microbial biofilms. Elsevier, Ámsterdam, pp 221–265

Rashmi V, ShylajaNaciyar M, Rajalakshmi R, D'Souza SF, Prabaharan D, Uma L (2013) Siderophore mediated uranium sequestration by marine cyanobacterium *Synechococcus elongatus* BDU 130911. Bioresour Technol 130:204–210

Renuka N, Guldhe A, Prasanna R, Singh P, Bux F (2018) Microalgae as multi-functional options in modern agriculture: current trends, prospects and challenges. Biotechnol Adv 36:1255–1273

Reyna-Velarde R, Cristiani-Urbina E, Hernández-Melchor DJ, Thalasso F, Cañizares-Villanueva RO (2010) Hydrodynamic and mass transfer characterization of a flat-panel air-lift photobioreactor with high light path. Chem Eng Process Process Intensif 49(1): 97–103

Reyna-Velarde R, Hernández-Melchor DJ, Vázquez-Hernández M, Cañizares-Villanueva RO (2012) Isolation of a photosynthetic microbial consortium formed mainly by cyanobacteria with capacity to fix nitrogen. Int J Environ Eng 4(3–4):183–196

Rippka R, Deruelles J, Waterbury JB, Herdman M, Stanier RY (1979) Generic assignments, strain histories and properties of pure cultures of cyanobacteria. J Gen Microbiol 111: 1–61

Sánchez-Contreras MI, Morales-Arrieta S, Okoye PU, Guillén-Garcés RA, Sebastian PJ, Arias DM (2021) Recycling industrial wastewater for improved carbohydrate-rich biomass production in a semi-continuous photobioreactor: effect of hydraulic retention time. J Environ Manag 284:112065. https://doi.org/10.1016/j.jenvman.2021.112065

Santini G, Biondi N, Rodolfi L, Tredici MR (2021) Plant biostimulants from cyanobacteria: an emerging strategy to improve yields and sustainability in agriculture. Plan Theory 10(4):643. https://doi.org/10.3390/plants10040643

Sarma RS, Prakash P (2020) Physiological aspects of cyanobacterial nitrogen fixation and its applications in modern sciences. In: Singh PK, Kumar A, Singh VK, Shrivastava AK (eds) Advances in cyanobacterial biology. Academic Press, Cambridge, pp 205–217

Shariatmadari Z, Riahi H, Abdi M, Hashtroudi MS, Ghassempour AR (2015) Impact of cyanobacterial extracts on the growth and oil content of the medicinal plant *Mentha piperita* L. J Appl Phycol 27:2279–2287

Singh A, Kaushik MS, Srivastava M, Tiwari DN, Mishra AK (2016a) Siderophore mediated attenuation of cadmium toxicity by paddy field cyano-bacterium *Anabaena oryzae*. Algal Res 16:63–68

Singh JS, Kumar A, Rai AN, Singh DP (2016b) Cyanobacteria: a precious bio-resource in agriculture, ecosystem, and environmental sustainability. Front Microbiol 7:529. https://doi.org/10.3389/fmicb.2016.00529

Song X, Zhang J, Li D, Peng C (2022) Nitrogen-fixing cyanobacteria have the potential to improve nitrogen use efficiency through the reduction of ammonia volatilization in red soil paddy fields. Soil Tillage Res 217:105274. https://doi.org/10.1016/j.still.2021.105274

Suresh A, Soundararajan S, Elavarasi S, Oscar FL, Thajuddin N (2019) Evaluation and characterization of the plant growth promoting potentials of two heterocystous cyanobacteria for improving food grains growth. Biocatal Agric Biotechnol 17:647–652

Tantawy STA, Atef NM (2010) Growth responses of *Lupinus termis* to some plant growth promoting cyanobacteria and bacteria as biofertilizers. J Food Agric Environ 8:1178–1183

Toribio AJ, Suárez-Estrella F, Jurado MM, López MJ, López-González JA, Moreno J (2020) Prospection of cyanobacteria producing bioactive substances and their application as potential phytostimulating agents. Biotechnol Rep 26:e00449. https://doi.org/10.1016/j.btre.2020.e00449

Tredici MR, Carlozzi P, Chini-Zitelli G, Materassi R (1991) A vertical alveolar panel (VAP) for outdoor mass cultivation of microalgae and cyanobacteria. Bioresour Technol 38:153–159

Ummalyma SB, Gnansounou E, Sukumaran RK, Sindhu R, Pandey A, Sahoo D (2017) Bioflocculation: an alternative strategy for harvesting of microalgae – an overview. Bioresour Technol 242: 227–235

Venkataraman GS (1981) Blue-green algae for rice production: a manual for its promotion (No. 46). Food and Agriculture Organization of the United Nations, Rome. 102 pp

Xiong W, Yang Z, Zhai H, Wang G, Xu X, Ma W, Tang R (2013) Alleviation of high light-induced photoinhibition in cyanobacteria by artificially conferred biosilica shells. Chem Comm 49(68): 7525–7527

Zarezadeh S, Riahi H, Shariatmadari Z, Sonboli A (2020) Effects of cyanobacterial suspensions as bio-fertilizers on growth factors and the essential oil composition of chamomile, *Matricaria chamomilla* L. J Appl Phycol 32:1231–1241

Zhang B, Li W, Guo Y, Zhang Z, Shi W, Cui F, Lens PNL, Tay JH (2020) Microalgal-bacterial consortia: from interspecies interactions to biotechnological applications. Renew Sust Energ Rev 118: 109563. https://doi.org/10.1016/j.rser.2019.109563

Zhang H, Gong T, Li J, Pan B, Hu Q, Duan M, Zhang X (2022) Study on the effect of spray drying process on the quality of microalgal biomass: a comprehensive biocomposition analysis of spray-dried *S. acuminatus*. Biomass Bioenergy Res 15:320–333

Biodiesel Production by Microalgal Biomass and Strategies to Improve Its Quality

8

Martha Trinidad Arias Peñaranda

Abstract

Microalgal oil has been described as the raw material with the greatest potential for third-generation biodiesel production, without competing with arable land suitable for food production. The fuel quality of biodiesel depends on the raw materials, the reagents used, and the values of the operational variables in the transesterification.

ASTM D6751 and EN 14214 standards describe the limits of biodiesel properties values such as Cetane Number (CN), Oxidation Stability (OS), Kinematic Viscosity (KV), Acid Value (AV), Iodine Value (IV), and Cold Flow Properties. These properties are highly influenced by the chain length and degree of unsaturation of the Fatty Acid Methyl Esters (FAMEs) that constitute biodiesel. The CN represents the ignition quality of biodiesel in the engine, high cetane value is the indicator of better combustion and low emission of nitrous oxide. In general, the presence of long-chain FAMEs with a high proportion of saturated fatty acids (SFA) increases CN and improves oxidative stability. In contrast, biodiesel with a relatively high content of monounsaturated fatty acids (MUFA) or polyunsaturated fatty acids (PUFA) tends to perform better in cold climates.

Microalgae have a significantly higher yield of lipids, compared to conventional crops for biodiesel production; however, the fatty acid composition varies with the species of microalgae and may differ in the size of the carbon chain and/or in the presence and amount of unsaturation. Thus, to choose the most suitable microalgae for biodiesel production, other factors such as the fatty acid composition must be evaluated.

This chapter gives detailed information about parameters to consider for improving microalgae biodiesel properties. In addition, updated research, challenges, and strategies to improve its quality are also presented.

Keywords

Biodiesel quality · Microalgae · Nutrient stress · Wastewater · Coculture

8.1 Introduction

Government policies aimed at achieving energy security and mitigating greenhouse gas pollution have promoted global interest in the search for raw materials and the development of technologies to produce liquid biofuels such as bioethanol and biodiesel. Biodiesel is known as an environmentally friendly biological fuel, its production in the world began in the 1990s and has had an annual growth of 10% in the last 10 years, reaching 53 billion liters in 2019, exceeding expectations of 41.4 billion liters projected for 2025 (Mesa 2020).

Biodiesel is produced by transesterification of triglycerides (TAG) contained in vegetable oils or animal fats, generally with methanol, to obtain a mixture of fatty acid methyl esters (FAME). Given that first-generation biodiesel derived from edible vegetable oils such as, palm, rapeseed, soybean, and sunflower oil has brought great controversy due to the threat to food security, the use of arable land and the impact on the environment associated with fertilizer consumption and water requirements; Second-generation biodiesel has emerged is that it is made from raw materials that are not used for food and are grown on non-agricultural or marginal land, and third-generation biodiesel is produced from lipids of microorganisms such as bacteria, fungi, and algae. In this context, in recent decades, microalgae, omnipresent microorganisms in various natural habitats and easily isolated have been studied as a raw material with high potential to produce third-generation biodiesel.

M. T. Arias Peñaranda (✉)
Programa de Química Industrial, Grupo de Investigación GIQUIBA, Universidad Francisco de Paula Santander, Cúcuta, Colombia
e-mail: marthatrinidadap@ufps.edu.co

A. de Jesús Martínez-Roldán (ed.), *Biotechnological Processes for Green Energy, and High Value Bioproducts by Microalgae, and Cyanobacteria Cultures*, Developments in Applied Phycology 13, https://doi.org/10.1007/978-3-031-43969-8_8

Microalgae oil as a feedstock to produce biodiesel has gained worldwide attention in recent decades, due to its multiple advantages compared to feedstocks used to obtain first- and second-generation biodiesel. However, due to the high cost of production, the commercialization of biodiesel from microalgae is not yet profitable; Therefore, obtaining lipids from microalgae with the appropriate characteristics to obtain quality biodiesel presents several challenges that range from strain selection to lipid extraction (Arias-Peñaranda et al. 2013; Chandra et al. 2021; Gour et al. 2020).

The quality of biodiesel is determined by the variables of the transesterification process and the characteristics of the oil used as raw material. Generally, the raw materials used in each country depend on their local availability; The fatty acid profile of the oil used as raw material determines some of the properties of biodiesel such as cetane number (CN), density, kinematic viscosity (KV), oxidation stability (OS), cold flow properties, higher calorific value, acid number (AV), and iodine value (IV) which are briefly described below.

8.2 Biodiesel Quality Characteristics

The quality of biodiesel for vehicular use affects the life of an engine and its operation, the main characteristics that define the quality of biodiesel have been issued by the European Committee for Standardization (EN 14214 2012) and the American Society for Testing and Materials (ASTM D6751 2012), which have served as a starting point for biodiesel specifications developed in other countries, such as the Colombian Technical Standard (NTC 5444, 2016) biodiesel specifications for use in diesel engines, the resolution of the National Agency of Petroleum, Natural Gas and Biofuels (RANP45, 2014) in Brazil and biodiesel specification, Indian Bureau of Standards (IS:1560, 2006) (Rao et al. 2010).

Properties of biodiesel can be grouped conveniently into product, process, and performance properties, viz.:

- Physicochemical properties of the product: density, kinematic viscosity, oxidative stability, Cold Flow properties, acid number, saponification number, copper corrosion, thermal degradation, Flash point, etc.
- Process properties: Methyl ester content, Methanol content, free/total glycerin content, total Contamination.
- Performance properties: Cetane number (CN), Heating value, Lubricity.

The limits specified by the standards for some of these characteristics are listed in Table 8.1.

Previous studies have shown that some of the properties listed in Table 8.1 are largely determined by the chain length and degree of unsaturation of the fatty acid methyl esters (FAME) that make up biodiesel, a profile that in turn corresponds to that of the oil or fat from which it was produced (Knothe and Razon 2017; Moreno-Garcia et al. 2021; Ramos et al. 2009). Some of the critical fuel properties that depend on the fatty acid composition of the feedstock are described below.

8.2.1 Cetane Number

Cetane Number (NC) measures the self-ignition ability of a fuel and influences combustion quality, noise level, engine performance and smoke emissions; a low NC produces engine noise and prolonged ignition delay, while a high NC is correlated with higher combustion efficiency and lower emissions of oxides of nitrogen, carbon, and sulfur, although a small increase in the presence has been observed. of soot (Ramos et al. 2009; Sajjadi et al. 2016).

The CN can be affected by the structure and composition of the fatty acids present in the raw material, especially by the presence of double bonds, the length and the branching of the fatty acid chain; a raw material rich in unsaturated fatty acids such as oleic, linoleic and linolenic acids has a low CN while the presence of saturated fatty acids such as palmitic and stearic acids confer a high CN; if the double bonds are more toward the ends than toward the middle of the chain, the CN is higher and vice versa. For saturated fatty acid alkyl esters as the length of the fatty acid chain increases the greater the NC and if the branching of the chain increases the CN is reduced, again also depending on the position of the bifurcation, if the position of the branch is toward the center of the molecule, the CN decreases (Knothe and Razon 2017).

Biodiesel has shown higher CN than petrodiesel, European and Colombian standards (EN14214 and NTC 5444) establish a minimum CN of 47 for biodiesel, while the United States and India (ASTM D6751 and IS:15607) establish a minimum CN of 51.

8.2.2 Density

The density of a fuel is directly related to its thermal energy content and directly influences fuel atomization and consumption, engine efficiency, and emissions (Çakmak and Özcan 2020; Knothe and Steidley 2005). The injection system must deliver precise amounts of fuel in such a way that adequate combustion occurs, since the amount of fuel injected into an engine is measured and controlled by its volume, at a higher density, the same volume would deliver a greater amount of fuel than necessary, which would be more difficult to atomize and would result in higher emissions and engine deposits (Huang et al. 2010). Density correlates with other biodiesel properties, such as cetane number, viscosity, and higher heating value (HHV), which

Table 8.1 Biodiesel quality standards

Specification	EN 14214:2012	ASTM D6751-12	NTC 5444:2006	RANP 45/2014	IS:15607 2006
Cetane number	≥51	≥47	≥47	Report	≥51
Density at 15 °C (kg/m^3)	860–900	–	860–900	850–900	860–900
Kinematic viscosity at 40 °C (mm^2/s)	3.5–5.0	1.9–6.0	1.9–6.0	3.0–6.0	2.5–6.0
Water content (max) mg/kg	500	500	500	200	500
Flash point (min) °C	101	130	120	100	120
Oxidation stability, 110 °C (h)	≥8	≥3	≥6	≥8	≥6
Cloud point (°C)	CS[a]	Report[b]	Report	19	–
Pour point (°C)	–	Report[b]	Report	–	–
Cold filter plugging point (°C)	−13	Report[b]	Report	–	–
Higher calorific value (MJ/kg)	35	39–41	–	–	–
Iodine value *IV* ($gI_2/100$ *g*)	–	<120	<120	Report	Report
Acid value, mg KOH/g, máx.	0.5	0.5	0.5	0.5	–
Free glycerol, max wt%	0.02	0.02	0.02	0.02	0.02
Total glycerol (max) wt%	0.25	0.24	0.25	0.25	0.25
Linolenic acid methyl ester wt%	<12	–	<12	–	–
Polyunsaturated methyl esters, wt%	<1	–	–	–	–

[a]Country specific
[b]Low-temperature properties are not strictly specified, they depend on geography and time of year, but must be agreed upon by the fuel supplier or purchaser

increase with decreasing biodiesel density (Alptekin and Canakci 2008). The density of biodiesel is influenced by the profile of methyl esters, it has been reported that long saturated chains lead to a lower density of biodiesel (Knothe and Razon 2017).

The ASTM D6751 standard does not specify a value for the density of biodiesel, while biodiesel from India, Colombia, and Europe must have a density at 15 °C between 860 and 900 kg/m^3 and for Brazil at 20 °C between 850 and 900 kg/m^3.

8.2.3 Kinematic Viscosity

Kinematic viscosity is one of the main properties of biodiesel, along with density they control injection characteristics and affect engine performance and emissions. At low temperatures, it builds up and can cause excessive fuel injection pressure during engine warm-up and because the fuel flows more slowly, the engine may have starvation problems, resulting in incomplete combustion and poor performance. Poor engine. Engine (Çakmak and Özcan 2020; Rao et al. 2010). Also, high viscosity values can cause carbon deposits and damage fuel injectors; therefore, it is required to maintain the viscosity at an acceptable level to avoid negative impacts on the performance of the injection system (Knothe and Razon 2017; Meher et al. 2006; Moreno-Garcia et al. 2021).

Kinematic viscosity, like the other properties of biodiesel, is influenced by the structure and chain length of the alkyl esters; increases with chain length and with the degree of saturation, while unsaturation reduces it. However, branching has little influence on viscosity. Transesterification with short-chain alcohols reduces the viscosity of the oil used as feedstock for biodiesel production (Knothe and Razon 2017; Sajjadi et al. 2016).

Biodiesel has higher kinematic viscosity than petrodiesel, the European standard establishes a narrow range of kinematic viscosity between 3.5 and 5.0 mm^2/s for biodiesel, while in the United States and Colombia the range is wider between 1.9 and 6.0 mm^2/s; for their part, the standards of Brazil and India allow viscosities of 3–6.0 and 2.5–6.0 mm^2/s, respectively.

8.2.4 Oxidative Stability of Biodiesel

Although biodiesel is thermodynamically stable, in contact with oxygen present in the environment it can oxidize. Oxidative stability is a measure of how long a fuel can resist oxidation when exposed to air, so its determination is important for storing biodiesel for extended periods of time. Additionally, factors such as exposure to temperatures above room temperature, to air and/or to light, can have a catalytic effect on oxidation (Knothe and Razon 2017; Lanjekar and Deshmukh 2016).

Several authors have observed that the speed of the biodiesel oxidation reaction is proportional to the degree of unsaturation and especially to the position of the double bonds; Polyunsaturated chains containing an allylic site (CH2 adjacent to the unsaturated carbon atom) and bis-allylic sites (CH2 located directly between two double bonds) are prone to oxidation since the double bonds and adjacent allylic sites are susceptible to oxidation (McCormick et al. 2007). Therefore, linolenic acid (C18:3) is more susceptible to oxidation than linoleic acid (C18:2) and biodiesel with high proportions of saturated methyl esters has better oxidation stability (Knothe and Razon 2017; Lanjekar and Deshmukh 2016).

The minimum time for oxidative stability measured at 110 °C varies in different standards. The EN 14214 standard requires a minimum time of 8 hours, the same as the Brazilian standard RANP 45/2014, while the NTC 5444 and ISIS: 15607 standards establish a time of 6 hours and the ASTM D6751-12 of only 3.

8.2.5 Cold Flow Weather Properties

Cold flow properties are associated with operational problems and are of great importance in regions with low ambient temperatures, as they can limit their use due to the tendency of fuels to crystallize or gel at low temperatures, resulting in clogged filters and fuel lines (Knothe and Razon 2017; Lanjekar and Deshmukh 2016).

Cold flow properties can be characterized by various parameters: The cloud point (CP) is the temperature at which crystals generally form a cloudy suspension when cooling a fuel; The pour point (PP) is the temperature at which the fuel cannot be pumped, that is, the lowest temperature at which the fuel will continue to flow and the cold filter plugging point (CFPP) is defined as the temperature at which fuel is unable to pass through the filter due fuel has crystalized, solidified or gelled enough to plug the fuel filter (Lown et al. 2014). This behavior is attributed to the presence of saturated esters that have high melting points and tend to precipitate and clog the fuel filter when the biodiesel cools; while branched esters with a high proportion of PUFA improve low-temperature properties; CP generally occurs at a higher temperature than PP and CFPP (Knothe and Razon 2017).

Biodiesel quality standards do not establish limits for cold flow properties, each country must define these temperature limits depending on the climatic conditions in which the fuel is used. For example, Spain established 0 °C in summer and −10 ° C in winter; while tropical countries like Colombia, Venezuela, Brazil, Singapore, and Indonesia among others do not have such problems.

8.2.6 Iodine Value

The iodine value is the amount of iodine (in grams) necessary to saturate 100 g of oil sample, providing information on the degree of unsaturation of its esters, independently of the relative proportion of mono-, di-, tri-, and polyunsaturated within a mix. This indicates the tendency of biodiesel to oxidize and form deposits on engine nozzles, piston rings, and piston ring grooves (Sajjadi et al. 2016). This parameter is directly related to several important properties of biodiesel, such as oxidation stability and cold flow properties. Biodiesel with a high iodine value is more sensitive to oxidation and less stable.

The ASTM D6751-12 and NTC 5444:2006 standards limit its value to a maximum of 120, while the EN 14214, RANP 45/2014, and IS:15607 do not set the limit. However, EN 14214 limits the methyl content of linoleic acid in biodiesel to 12% and the content of FAME with four or more double bonds at a maximum of 1%, which also influences oxidation.

8.2.7 Higher Calorific Value

The higher heating value (HHV) of a fuel is defined as the amount of heat released in the complete combustion of the fuel when the water vapor from combustion is condensed. This property characterizes the energy content of fuels and therefore their efficiency generally increases with the number of carbon and hydrogen atoms, as well as with the increase in the proportion of these elements in relation to oxygen. Therefore, increasing the chain length of methyl esters favors HHV, and an increase in unsaturation causes HHV to decrease (Demirbas 1998). The EN 14214 standard requires an HHV of 35 while the ASTM D6751-12 standard establishes a range between 39 and 41 MJ/kg.

8.2.8 Acid Value

Acid Value (AV) is a measure of the level of free fatty acids present in the fuel expressed in terms of milligrams of potassium hydroxide (KOH) required to neutralize 1 g of fatty acid methyl esters that are not chain bound of glycerol. The content of free fatty acids (FFA) in the oil must be less than 1%, larger amounts require additional amounts of basic catalyst (NaOH or KOH) to neutralize the free fatty acids. In this case, a previous step of esterification is required to reduce the content of free fatty acids or to use enzymatic or acid catalysis (Demirbas 2009). AV is an important parameter to indicate the quality, age, and degree of purity of biodiesel during processing and storage, the limit specified for the acid number in biodiesel by ASTM D6751 and EN14214 standards is below 0.5 mg KOH/g biodiesel. In addition, free fatty acids in the presence of water cause soap and foam formation that can induce an increase in viscosity and make it difficult to separate glycerol (Demirbas 2009).

8.2.9 Other Parameters

The flashpoint is the temperature at which the fuel ignites when exposed to a flame or spark, the flashpoint of biodiesel is higher than that of petrodiesel, which makes it safer during storage and transportation, but It should be taken into account that the presence of alcohol lowers the flashpoint (Meher et al. 2006), so Standard EN 14214 limits the methanol content to a maximum of 0.2% by weight of biodiesel.

Regarding the purity of biodiesel, ideally it should have 100% fatty acid methyl esters (FAME) to provide good engine performance, but in practice, the low conversion of the oil together with the presence of secondary reactions and the difficulty of separating the products and by-products, they generate impurities in the biodiesel that affect the fluidity, raise the cloud point and can cause damage to the engine. The main impurities in biodiesel are alcohol, soaps, catalyst residues, mono, di and triglycerides, and glycerol. Low levels of total glycerin indicate a high degree of oil conversion, while high levels of free glycerin can cause problems with injector deposits and filter plugging (Silva et al. 2015). According to the parameters established in the international standards for the quality of biodiesel, the maximum values of total glycerin and free glycerin are 0.24% and 0.020% by weight of the final product, respectively. High levels of mono-, di-, and triglycerides affect the flashpoint, cause the clogging of filters, the formation of carbonaceous deposits in the injectors, and negatively affect the performance of biodiesel at low temperatures (Silva et al. 2015). Likewise, the maximum limit for linolenic acid is 12%, while the maximum content of fatty acids with more than three double bonds (PUFA) is 1%. To produce biodiesel with excellent properties, it is necessary to achieve a balance between CN and cold flow properties. Therefore, the oil must have long-chain TAG and a fatty acid profile with moderate levels of SFA that provide a good cetane number without affecting winter operation; the highest possible levels of MUFAs (such as C18:1) to provide good cold flow properties and the lowest possible levels of PUFAs to increase oxidation stability. In addition, a maximum of 1% free fatty acids and a maximum water content of 500 mg/kg.

8.2.10 Microalgae Biodiesel

In recent years, microalgae oil has been tested as a raw material to produce biodiesel showing similar and superior characteristics to those obtained with vegetable oils; these results depend largely on the species and the culture conditions (Santhakumaran et al. 2019). Since official methods for determining these properties require large amounts of biodiesel, are relatively expensive, and/or time consuming, scientists have developed empirical equations to predict some quality properties of biodiesel, such as density, kinematic viscosity, number of cetane (CN), higher heating value (HHV), oxidation stability, and cold flow properties, from the FAME profile determined by chromatography (Lapuerta et al. 2009, 2010; Park et al. 2008; Ramos et al. 2009). Table 8.2 shows the properties of biodiesel estimated for various species of microalgae.

Among the genus of microalgae identified as promising raw material for biodiesel production are some marine species such as *Nannochloropsis*, *Neochloris*, *Dunaliella*, and *Chlorella;* as well as freshwater such as *Chlorella*, *Scenedesmus*, *Acutudesmus*, *Chlorococcum*, *Ankistrodesmus*, and *Pavlova* among others. In general, the estimated properties of biodiesel produced from microalgae oils presented in Table 8.2 are within the limits established by international standards. A critical characteristic of biodiesel is the cetane number that indicates the ability of a fuel to ignite quickly, with some exceptions it is greater than 47 and some species exceed 65; High NCs are achieved by increasing the chain length and saturation of the fatty acids used in the oil.

Fuel density and viscosity affect fuel injection, engine performance, and emission characteristics. The density is in a wider range than that specified by international standards, it ranges between 0.803 and 1000 kg/m^3, while the kinematic viscosity ranges between 2 and 5.4 mm^2/s with lower values for *Botryococcus* and some species of *Chlorella*. Oxidation stability at 110 °C is the property that varies the most in microalgae biodiesel, it ranges between 4.51 and 32.46 hours, it is directly influenced by the length of the chain and the degree of saturation and is directly correlated with the number of cetane and inversely with the iodine number, which is a measure of the degree of unsaturation in the oil. Thus, for a consortium of *Chlorella* sp. with bacteria Moreno-Garcia et al. (2021) reported oxidative stability of 32.46 hours, an iodine value of 57.8 gI_2/100 g of biodiesel, and a CN of 58.33, complying with the ASTM 6751-02 and EN 14214 standards.

At a higher degree of unsaturation, biodiesel is more prone to oxidation, so its quality is affected over time in storage. However, unsaturation reduces the viscosity of biodiesel and improves cold flow properties, which are specific to each country depending on the season of the year, only the EN 14214:2012 standard establishes a value of −13 °C for the point cold filter plugging (CFPP). Since the culture conditions and the species determine the fatty acid profile, microalgae biodiesel exhibits great variability in cold flow properties ranging from 23.4 to −3.9 for CP, 18.5 to −15 for PP and from 60.75 to −16.20 °C for CFPP.

The selection of good quality oils is essential to obtain quality biodiesel, reduce the cost of production, and increase the yield in the transesterification process. The lipid productivity and fatty acid profile of microalgal oil vary quantitatively and qualitatively between species within and between

Table 8.2 Physical–chemical properties of microalgae biodiesel

Microalgae	Cetane number	Density kg/m^3 at 20 °C	Viscosity at 40 °C cSt	Oxidation stability (h)	Cloud point (°C)	Pour point (°C)	CFPP (°C)	Higher heating value (MJ/kg)	Iodine value (gI$_2$/ 100 g)	References
Acutudesmus obliquus	65.90	870	4.2	30	23.4	18.5	55.1	40.5		Pandit et al. (2017)
Ankistrodesmus braunii,	33.65	850	3.26				7.01	38.43	163.37	Fawzy et al. (2021)
Ankistrodesmus falcatus	23.74	850	3.26				0.38	38.43	207.89	Fawzy et al. (2021)
Botryococcus sp.	51.57	860	1.28	8.12	1.56		−7.33	38.98	93.24	Vishwakarma et al. (2019)
Calothrix	60.74	871	4.14	8.99			−3.61	39.55		Deshmukh et al. (2019)
Auxenochlorella protothecoides	55.26			7.29					97.64	Andeden et al. (2021)
Chlorella pyrenoidosa	50.15						4.62		103.07	Zhang et al. (2019)
Chlorella minutissima	53.35	920	5.4	9				39		Chandra and Amit (2019)
Chlorella sp.	51.12	860	1.24	8.47	4.17		−6.49	38.6	92.86	Vishwakarma et al. (2019)
Chlorella sp.	50.07	882	3.75	4.51			−8.07	39.52		Deshmukh et al. (2019)
Chlorella sp.	68.79	880	2.78	10.44	13.68	8.03	22.64	43.08	42.46	Arguelles and Martinez-Gos (2021)
Chlorella vulgaris	57.00	1000	4	7	14	9	17	38		Pandit et al. (2017)
Chlorella vulgaris	51.40	860	3.7			−15		38.7		Tayari et al. (2020)
Chlorella vulgaris	510	868.1	1.66			6		20.3	38.9	El-Sheekh et al. (2019)
Chlorella vulgaris	48.52			6.21			− 16.20		41.9	Chávez-Fuentes et al. (2018)
Chlorella zachariaii	42.80			5	−3.9	−11.1	−1.2	43.0		Santhakumaran et al. (2019)
Consortium of Chlorella sp. *and bacteria*	58.33	880		32.46	11.9		−5.58	39.6	57.8	Moreno-Garcia et al. (2021)
Chlorolobion sp.	65.17	890	2.79	8.93	14.47	8.88	3.68	43.42	60.08	Arguelles and Martinez-Gos (2021)
Chlorococcum oleofaciens	68.60						60.75		16.3	Rayati et al. (2020)
Chlorococcum	51.74	880	3.89	4.77			8.76	39.58		Deshmukh et al. (2019)
Chroococcus	51.54	881	3.93	4.82			8.43	39.63		Deshmukh et al. (2019)
Fasciculochloris boldii	63.3			12.8	−2.7	−9.2	−14.3	41.7	123.9	Santhakumaran et al. (2019)
Lyngbya putealis	59.27	873	4.09	8.57			− 0.894	39.55		Deshmukh et al. (2019)
Nostoc	53.64	878	4.06	5.3			5.96	39.65		Deshmukh et al. (2019)
Sphaerocystis antoni-kadavilaii	42.3			4.6	−0. 9	−7.8	−6.5	43.2	120.3	Santhakumaran et al. (2019)
Scenedesmus incrassatulus	61.12	850	3.27				−3.23	39.18	43.45	Fawzy et al. (2021)

(continued)

Table 8.2 (continued)

Microalgae	Cetane number	Density kg/m^3 at 20 °C	Viscosity at 40 °C cSt	Oxidation stability (h)	Cloud point (°C)	Pour point (°C)	CFPP (°C)	Higher heating value (MJ/kg)	Iodine value (gI$_2$/100 g)	References
Scenedesmus incrassatulus	62.00	803	3.78	19	9	3	−4	41		Arias-Peñaranda et al. (2013)
Scenedesmus abundans	52.15								94.06	Mandotra et al. (2014)

taxonomic groups, with their physiological state, culture conditions, harvest age, harvest process, and method of oil extraction. Microalgae adapt to environmental changes with diverse physiological responses, so when deciding if a microalga has potential as a feedstock for biodiesel production, it is essential to apply strategies that provide maximum lipid productivity and an adequate profile of fatty acids to obtain quality biodiesel. In this sense, research efforts have focused on the accumulation of lipids with high levels of saturated and monounsaturated fatty acids and low concentrations of polyunsaturated, by manipulating growth conditions.

8.3 Strategies to Improve Microalgae Biodiesel Properties

The synthesis and accumulation of lipids in cells occur when microalgae are subjected to stress conditions imposed by physicochemical or environmental stimuli. The main chemical factors are the availability of nutrients, the carbon source, the salinity, and the pH of the culture medium, while the physical ones are light intensity and temperature. In addition, the growth phase of the culture also affects the lipid content and fatty acid composition. A review of the effect of different culture strategies on lipid accumulation and fatty acid profile of various species of microalgae is presented in Table 8.3. It is observed that in general, the lipid content can vary with the species and the culture conditions. Values ranging from 20 to 80% of dry cell weight have been reported. The impact of each parameter is discussed in detail in the following sections.

8.3.1 Nutrient Deficient

Generally, the lipid content is increased in a stressed environment such as lack of nutrients (N, P, S, Fe). Nitrogen has been shown to be the main regulator of growth and lipid accumulation (Fernandes and Cordeiro 2022; Yu et al. 2019; Yuan et al. 2018). The source and concentration of nitrogen in the culture medium affect the growth and accumulation of lipids, investigations carried out to induce the synthesis of lipids on a small scale, agree that cultivation in nitrogen-deficient conditions limits the growth of microalgae, while there is great variation in lipid content and fatty acid profile. Therefore, they suggest a two-stage culture, cell growth, and lipid accumulation, which require different culture parameters; the first stage requires a medium rich in nutrients to favor biomass growth and the second stage under stress conditions to trigger lipid accumulation (Arias-Peñaranda et al. 2013; Gour et al. 2020; Mujtaba et al. 2012; Sun et al. 2014).

Optimization of lipid production in microalgae by nitrogen stress has been extensively studied. Mandotra et al. (2014) observed that lipid accumulation increased in *Scenedesmus abundans* from 43.93 to 67.59% of dry cell weight when the nitrogen source (KNO_3) was removed from the medium. However, the lipid productivity decreased from 124.71 to 60.16 mg/l.d when comparing a medium with 0.32 g/l KNO_3 with the control medium. This can be explained because nitrogen is an essential component for protein and nucleic acid synthesis and its absence causes cell division and cell growth to be impaired. The fatty acid profile of *Scenedesmus abundans* obtained on a large scale in a culture medium with 0.32 g/l of nitrogen contains 81% saturated and monounsaturated fatty acids and 18% polyunsaturated fatty acids, which give biodiesel a cetane number and iodine index (Table 8.2) within the specifications of the international standards presented in Table 8.1 (Mandotra et al. 2014). Similarly, Pruvost et al. (2011) showed that nitrogen starvation in the culture medium triggers the accumulation of lipids while affecting the sugar and protein contents, they managed to increase the total lipid content of *N. oleoabundans* from 23% in normal culture to 25–37% under nitrogen limiting conditions. In a recent study, Arguelles and Martinez-Gos (2021) evaluated the effect of nitrogen concentration ($NaNO_3$) on lipid accumulation and fatty acid profile of *Chlorolobion* sp. and *Chlorella* sp., reaching a maximum lipid production of 31.61 and 28.77% in the lowest nitrogen conditions (0.375 g/l $NaNO_3$) with a fatty acid profile rich in SFA and a lower proportion of MUFA and PUFA that give it to biodiesel high cetane number, good oxidative stability but poor cold flow properties as seen in Table 8.2.

Another study reported that the total lipid accumulation of *Picocystis salinarum* increased from 21.23% to 33.87% when changing from a culture medium rich in nitrogen (1 N) to one low in nitrogen (0.125 N). In the entire range of nitrogen concentration evaluated, the FAME profile has high proportions of SFA, mainly palmitic acid; however, it was observed that as nitrogen decreases in the medium, the SFA content increases (99%) while oleic acid decreases (Tarazona Delgado et al. 2021). In a recent study, Fernandes and Cordeiro (2022) evaluated the effect of nitrogen concentration (NO_3) in the culture medium on lipid accumulation and fatty acid profile of *Pavlova pinguis* and unlike most reported data, found that the increase in nitrogen concentration in the culture led to a higher lipid content (21%), especially polyunsaturated fatty acids (38%). While in low nitrogen culture conditions, a higher proportion of saturated and monounsaturated fatty acids was observed. In another study, a two-stage nitrogen starvation strategy, composed of (1) growth under nutrient-rich conditions followed by (2) cultivation under nitrogen starvation was used on *Scenedesmus incrassatulus*. The lipid content increased from 12 to 12.5% in the first stage (4 days) and up to 23.1 ± 0.6% in the deprivation stage of nitrogen (12 days). Under these conditions, the most abundant fatty acids (95%) were C16/C18, being 84% saturated and monounsaturated, with predominance of oleic acid (45%) (Table 8.3). This lipid profile gives *S. incrassatulus* a high potential as a raw material to produce biodiesel, as observed in the values of the main properties of biodiesel presented in Table 8.2 (Arias-Peñaranda et al. 2013).

Another study with the green microalga *Michonastes Afer*, the effect of the nitrogen source (nitrate, ammonium, urea) on the accumulation of lipids and their composition was evaluated, showing that sodium nitrate is the preferred source, the highest content of lipids (58.56%) was obtained with NaNO3 under nitrogen-deficient conditions (4.4 mM) and neutral pH, while the best productivity required some nitrogen (17.6 mM). The fatty acid profile is independent of the nitrogen source and highly dependent on concentration; under nitrogen-deficient conditions and neutral pH, a higher proportion of SFA and MUFA was obtained (Yuan et al. 2018).

Different studies observed a similar response under phosphorous stress. For example, Yu et al. (2019) evaluated the effect of phosphorus on lipid accumulation in *Isochrysis zhangjiangensis* in a medium with nitrogen sufficiency and deprivation. The highest lipid content was recorded in a medium with sufficient N and limited in P. The relative SFA content increased significantly with decreasing P concentration under conditions of N sufficiency. Lipids were characterized by high SFA content and MUFA (61.15%) with 28.19% oleic acid. Jaiswal et al. (2020) reported that phosphorus deprivation in the medium increased lipid accumulation (~53%) in *Scenedesmus* sp. Sonkar and Mallick (2018) examined the effect of phosphorus on lipid accumulation in five green microalgae (*Chlorella vulgaris, Chlorella minutissima, Chlamydomonas* sp., *Tetradesmus oblicua*, and *Tetradesmus lagerheimii*). Phosphate-sufficient, phosphate-free, two-stage, and sequential addition of phosphate strategies were used. The lipid content was similar in the treatment without phosphate and the treatment in two stages, being slightly higher in the absence of phosphate from the beginning. However, lipid productivity was better in two stages, that is when the microalgae were grown with sufficient nutrients and then transferred to a medium without phosphate. No influence of phosphate on the fatty acid profile was observed, in all microalgae, an SFA and MUFA content greater than 80% was observed, which means its suitability for biodiesel production.

Zhang et al. (2019) investigated the effect of N, P, and S deficiency on lipid accumulation in *Chlorella pyrenoidosa*. The results indicate that in all cases the lipid content increased, with a maximum content under conditions of nitrogen, phosphorus, and sulfur deficiency, of 48.90%, 38.85%, and 31.60%, respectively. The fatty acid profile was similar in all three cases, with a predominance of SFA and PUFA, which gave biodiesel high cetane numbers but poor cold flow properties. A study reported by Andeden et al. (2021) demonstrated that the lipid content of microalgae was significantly influenced by the pH and N limitation. They evaluated the effect of nitrogen deprivation associated with alkaline pH (pH 10) using a carbonate–bicarbonate buffer on the accumulation of triacylglycerols (TGA) in *Auxenochlorella protothecoides* KP7 for biodiesel production. They observed that this combination after 7 days of culture had a synergistic effect on TAG accumulation and fatty acid profile. The lipid content reached up to 26.5% of the dry cell weight and the productivity 17.1 mg/l.d; while the percentages of SFA, MUFA, and PUFA were 16.3, 58.4, and 25.1%, respectively. The properties of the biodiesel obtained from the oil *A. protothecoides*, meet the specifications of the standards presented in Table 8.1. This high value of CN can be specifically attributed to the abundance of saturated and monounsaturated methyl esters (>74%). A high CN causes better combustion, improving engine efficiency and reducing nitrogen oxides (NOx) in exhaust emissions (Ramos et al. 2009). Similarly, Yuan et al. (2018) reported that the lipid content and the fatty acid profile are highly dependent on pH, the highest proportion of saturated fatty acids in the biomass of *Michonastes Afer was* observed in cultures without pH control; when the pH was controlled under neutral conditions, the highest content of lipids with high contents of oleic acid was obtained.

Table 8.3 A review of the effect of different culture strategies on lipid accumulation and the lipid profile of various species of microalgae

Microalgae	Strategy	% lipid dry wt	Lipid Productivity mg/L.d	ΣSFA in lipids (%)	ΣMUFA in lipids (%)	ΣPUFA in lipids (%)	References
Scenedesmus abundans	0.32 g/l KNO_3	67.59	124.71	32.5	48.5	18.0	Mandotra et al. (2014)
Chlorolobion sp.	0.375 g/l $NaNO_3$	31.61	227.84	41.78	19.99	18.61	Arguelles and Martinez-Gos (2021)
Chlorella sp.,	0.375 g/l $NaNO_3$	28.77%	151.14	50.40	16.22	15.03	Arguelles and Martinez-Gos (2021)
Picocystis salinarum	Nitrogen 9.375 mg/l $NaNO_3$	33.87%		99.6	–	0.3	Tarazona Delgado et al. (2021)
Scenedesmus incrassatulus	Two-stage cultivation			32.13	51.59	6.32	Arias-Peñaranda et al. (2013)
Chlorella pyrenoidosa	Nutritional deficiency (nitrogen)	48.9	72.04	38.6	18.9	40.55	Zhang et al. (2019)
Auxenochlorella protothecoides	N starvation alkaline pH	26.5	17.1	16.3	58.4	25.1	Andeden et al. (2021)
Botryococcus sp.	BG-11 medium and 30 days	63.03		19.6	58.3	21.34	Vishwakarma et al. (2019)
Chlorella sp.	BG-11 medium and 30 days	50.27		24.9	53.8	20.1	Vishwakarma et al. (2019)
Fasciculochloris boldii	BBM medium		2.4	71.2	17.3	11.5	Santhakumaran et al. (2019)
Sphaerocystis antoni-kadavilaii	BBM medium		4.4	32.0	7.6	60.1	Santhakumaran et al. (2019)
Chlorella zachariaii	BBM medium		1.3	40.1	10.5	49.3	Santhakumaran et al. (2019)
Chlorella vulgaris	High light minimum N violet light.	83.87	7.81	66.93	25.56	7.51	Chávez-Fuentes et al. (2018)
Acutudesmus obliquus	0.4 M NaCl	43.4	23.8	92.8	5.8	4.3	Pandit et al. (2017)
Ankistrodesmus braunii	6th day 50 mM NaCl	34.4		42.97	2.09	54.95	Fawzy et al. (2021)
Ankistrodesmus falcatus	10th day 100 mM NaCl	53		29.85	2.77	67.37	Fawzy et al. (2021)
Scenedesmus incrassatulus	6th day 100 mM NaCl	37.7		68.76	18.31	12.94	Fawzy et al. (2021)
Chlorella vulgaris	BG11 0.4 M NaCl	49.5	18.9 mg/l.d	52.6	16.8	28.4	Pandit et al. (2017)
Chlorella minutissima	N, P, Fe, Zn, glucose, pH, T light intensity			39.11	20.07	37.68	Chandra and Amit (2019)
Chlorella vulgaris	Grown in outdoor pond.	32.3		3.19	44.34	51.94	Arutselvan et al. (2021)
Coelastrella sp.	Grown in lab	26.2		34.66	12.81	51.45	Arutselvan et al. (2021)
Scenedesmus dimorphus	Grown in outdoor pond	26.4		8.34	50.92	40.66	Arutselvan et al. (2021)
Chlorococcum sp.	Grown in lab	31.2		31.96	12.95	54.81	Arutselvan et al. (2021)
Consortium C. vulgaris + S. dimorphus.	Grown in outdoor pond	34.5		9.84	46.74	38.78	Arutselvan et al. (2021)
Chlorella vulgaris	Wastewater			26.9	23.4	44.8	Tayari et al. (2020)
Chlorococcum oleofaciens	400 μmol photons	59.18	198.32	82.3	9.15	8.55	Rayati et al. (2020)
C. vulgaris	Co-cultivation with bacterium	51.2	96.77	33.64	52.13	14.23	Wei et al. (2020)

(continued)

Table 8.3 (continued)

Microalgae	Strategy	% lipid dry wt	Lipid Productivity mg/L.d	ΣSFA in lipids (%)	ΣMUFA in lipids (%)	ΣPUFA in lipids (%)	References
Consortium of Chlorella sp. *and bacteria*	Wastewater 10% CO_2 40 μmol/m^{2}s			33.3	62.1	4.7	Moreno-Garcia et al. (2021)
Chlorella pyrenoidosa	Effluent anaerobic increase pH, temperature	30.2		20.1	25.6	52.2	Zhao et al. (2019)

8.3.2 Addition of Micronutrients

Mandal et al. (2019) evaluated the effect of iron (Fe), molybdenum (Mo), and zinc (Zn), on the lipid content and fatty acid composition of *Scenedesmus acutus, Chlorella* sp., *Graesiella emersonii,* and *Chlorophyta* sp. They showed that the addition of Fe and Zn favored the accumulation of lipids in *Chlorella sp., G. emersonii*, and *S. acutus*, however, the increase was not in the same proportion for all. *G. emersonii* recorded the largest increase (140.3%) in conditions supplemented with Zn and 105% in conditions supplemented with Fe. The addition of the three metals favored the accumulation of SFA and MUFA and the estimated properties for biodiesel are in accordance with the limits of international standards. In another study, Islami and Assareh (2019) investigated the effect of the addition of ferric iron to the culture medium on the lipid content of the green microalgae *Tetradesmus obliquus*. They found that the addition of 0.1 mmol/l of Fe^{+3} to the BG-11 medium allowed for a 2.22-fold increase in the lipid content. Similarly, it was evidenced that the addition of iron increased the content of SFA and decreased the MUFA and PUFA.

8.3.3 Temperature

Temperature has been reported to be an essential factor in lipid accumulation and strongly affects the fatty acid profile of microalgae. In many microalgae, a tendency to accumulate PUFA at low temperatures and SFA with increasing temperature has been observed. Xin et al. (2010) reported that *Scenedesmus* sp. can grow in a wide temperature range (10–30 °C), lipid accumulation was inversely correlated with temperature, reaching a maximum (35%) at 20 °C after 15 days of culture, with a higher percentage of PUFA. As the temperature increased (30 °C), the PUFA content decreased and the MUFA content increased. Similarly, Wei et al. (2015) reported that the increase in culture medium temperature led to a decrease in PUFA of *Tetraselmis subcordiformis* and *Nannochloropsis oculata* while the SFA increased.

8.3.4 Carbon Sources

Microalgae are photosynthetic microorganisms that grow in the presence of light and carbon dioxide, obtaining energy from photosynthesis (photoautotrophic mode). However, some species can also grow in heterotrophic (dark environments, with an additional source of organic carbon as an energy source) and/or mixotrophic (they can assimilate a variety of organic compounds in the dark and in the light) (Deng et al. 2018).

8.3.4.1 Inorganic Carbon (CO_2)

Phototrophic cultivation is the most common way of cultivating microalgae, due to the use of CO_2 as a carbon source. It has been shown that high CO_2 concentrations associated with other types of stress such as nitrogen deficiency favor the accumulation of lipids. Kumari et al. (2021) evaluated the interaction of nutrient bioavailability and CO_2 in *Chlorella vulgaris*. They found that the limitation of N and P associated with the 5% CO_2 supplementation favored a 32% increase in lipids with a high amount of saturated fatty acids, essential to provide biodiesel with a high cetane number and to ensure good burning properties ignition. Similar results were obtained by Chu et al. (2019) who reported a 2.2-fold increase in the lipid content of Chlorella PY-ZU1 through a two-stage culture, a first stage in the presence of 15% CO_2 until the nutrients were exhausted and a second stage with the addition of phosphorus.

Qi et al. (2019) evaluated the nocturnal lipid accumulation of a *C. vulgaris* strain under different stress strategies such as CO_2 level, mixing mode, aeration rate, NO_3^-–N concentration, temperature, and the initial concentration of the biomass. Among the stress conditions evaluated, N deficiency (NO_3–N) had the greatest effect on lipid accumulation; suggest a two-stage strategy, biomass production during the day and stress strategy at night under optimal conditions (aeration mode with 0.10 vvm, 15% CO_2, 5–10 mg/l NO_3–N, 30–35 ° C, approximately 2500 mg/l of biomass). The fatty acid profile, with a predominance of saturated fatty acids, was slightly affected by stress conditions, increasing the content of oleic acid and decreasing that of linolenic acid. The values reported for the properties of biodiesel confirm compliance with international quality standards. A study from Lakshmikandan et al. (2021) reported that for *Chlorella vulgaris* an increase in CO_2 concentration favors lipid accumulation and does not affect the fatty acid profile. Culture with 8% CO_2 promoted lipid accumulation (22.5%) with a composite profile of SFA and MUFA. The properties of biodiesel are within the limits established by international standards.

8.3.4.2 Organic Carbon

The role of glucose and light in the fatty acid profile of *Chlorella kessleri* was studied by Deng et al. (2018), in cultures with three different trophic conditions (photoautotrophic, heterotrophic, and mixotrophic). They found that in mixotrophic cultures, the lipid content decreased from 29.7 to 17.3% when the light intensity increased from 30 to 180 μE/m^2 s, with low light intensity being the most suitable for the accumulation of lipids. The fatty acid profile was influenced by light intensity and photoperiodicity, showing that under photoautotrophic conditions with a light intensity of 90 μE/m^2 s, fatty acids C16:0, C18:1, and C18:3 prevailed, which represent 28.6, 26.0, and 22.2%, respectively, while in the heterotrophic and mixotrophic cultures the PUFA content was >65%. In another study, Wang et al. (2019) investigated different mixtures of glucose and sodium acetate as carbon sources to enhance lipid accumulation in *Coccomyxa subellipsoidea*. Growth in a mixture of 20 g/l glucose and 12 g/l acetate was shown to increase lipid content (52.16%) and fatty acid profile with 18.12 SFA, 51.11%. of MUFA (48.69% oleic acid) and 20.65% PUFA allows the oil to be used as raw material to produce biodiesel.

8.3.5 Light Source and Intensity

Light intensity plays an important role in lipid accumulation under nitrogen deprivation conditions, it has been shown that a certain level of light energy must be supplied for maximal lipid synthesis to occur in an N-depleted condition. However, the correlation between light intensity and lipid accumulation is usually species specific (Chávez-Fuentes et al. 2018).

Chávez-Fuentes et al. (2018) evaluated the quality of biodiesel obtained from *Chlorella vulgaris* oil growing under stressful conditions of intensity and quality of light and nitrogen limitation in the culture medium. The results showed that the quality and quantity of light associated with other variables such as nitrogen limitation can be a useful strategy to increase lipid content. The highest content of lipids (70.92 and 83.87%) was presented under yellow and violet light sources at high light intensity (140 μE/m^2 s) and minimum content of nitrogen (30 mg/l). However, the wavelength did not affect the fatty acid profile, which remained above 80% of SFA and MUFA. Conditions that favor biodiesel properties such as oxidative stability, cetane number, and cold flow properties (Table 8.2). These results agree with Hadrich et al. (2018), who showed that nitrate deficiency and high light intensity favored the accumulation of lipids within the cells (18.83%) of *Chlorella* sp. with a fatty acid profile characterized by a large amount of SFA and MUFA (≈72%), mainly C16:0 and C18:1 and a low amount of PUFA (≈15%); a suitable proportion to produce quality biodiesel.

In another study, Sun et al. (2014) evaluated the effect of nitrogen deficiency, light intensity, and iron on the total lipid content and fatty acid profile of *Neochloris oleoabundans* in a two-stage culture. Demonstrating that N hunger and light intensity are the factors that most affect lipid accumulation. The maximum lipid productivity (51.58 mg/l/d) was found in culture under 2 days of N deprivation, 200 μmol/m^2s, and Fe^{3+} 0.037 mM, with lipid content of 33.36%. Saturated/monounsaturated fatty acids increased significantly with increasing light intensity and length of the N starvation period. Furthermore, they showed that two-stage culture composed of growth in nutrient-rich conditions followed by culture with starvation Nitrogen improved lipid accumulation and fatty acid profile. In a recent study, Rayati et al. (2020) evaluated the effect of light intensity on lipid accumulation in the freshwater microalga *Chlorococcum oleofaciens*. They found that the highest lipid content (59.18%) and lipid productivity (198.32 mg/l.d) were achieved with an intensity of 400 μmol photons/m^2 s. A culture illuminated with 50 μmol photons/m^2 s provided a profile of 50.08% SFA, 26.61% MUFA, and 23.31% PUFA, and as light intensity increases, SFA content increases to the detriment of the MUFA and PUFA, which give biodiesel high cetane numbers but low cold flow properties.

8.3.6 Salinity

Fawzy et al. (2021) examined the effect of nutrient stress (N, P, Fe) and salinity of *Ankistrodesmus braunii*, *Ankistrodesmus falcatus*, and *Scenedesmus incrassatulus*, on lipid accumulation and fatty acid profile. The results indicated that in all the microalgae the highest percentages of lipids were reached by saline stress and that, in addition, the harvest season and the species influence the accumulation of lipids. In the three microalgae evaluated, it was evidenced that the reduction of the nutrient content in the culture medium caused a gradual increase in the accumulation of lipids, obtaining the maximum on the sixth day of culture under different conditions. For example, for *A. braunii* and *S. incrasátulus*, maximum of 32.6% and 35.1%, respectively, were obtained with a 100% nitrogen deficiency, while for

A. falcatus it was 29.9% with a deficiency of 100% nitrogen. The maximum accumulation of lipids for *A. braunii* of 29.6% and *S. incrasatulus* of 19.9%, occurred with a phosphorus deficiency of 75%, while for *A. falcatus* it was 26.3% with a deficiency of 100% phosphorus. The highest lipid content for *A. braunii* of 22.1% was observed with 100% iron deficiency; while for *A. falcatus* and *S. incrassatulus* the maximum of 20.4% and 19.7%, respectively, occurred in cultures with 50% iron deficiency. Salinity stress favored the highest lipid accumulation in the three microalgae, showing that sensitivity to salt is different for each species. *A. braunii* with moderate salt stress (50 mM NaCl) reached its highest value of 34.4% at 6 days, while the other two reached higher accumulation with high salt stress (100 mM NaCl) *A. falcatus* 53% at 10 days and *S. incrassatulus* 37.7% at 6 days. The fatty acid profile of biomass grown under salt stress revealed that *A. braunii* and *A. falcatus* have low SFA and MUFA content and a high PUFA content (Table 8.3), so they are not suitable to produce biodiesel, as confirmed by the low NC values and high IV values (Table 8.2) that make it susceptible to oxidation during storage. Regarding *S. incrassatulus*, 87.07% of SFA and MUFA indicate its high potential as a source of oil to produce biodiesel, which is confirmed by the excellent properties of biodiesel presented in Table 8.2, which meet the specifications of international standards. These results are consistent with several studies that recommend different species of *Scenedesmus* for biodiesel production due to their high content of SFA and MUFAs under different growing conditions (Arias-Peñaranda et al. 2013: Mandotra et al. 2014; Samkhaniyani et al. 2017).

Gour et al. (2020) evaluated the effect of salt stress on lipid accumulation in *Scenedesmus quadricauda, Scenedesmus dimorphus, and Chlorella* sp., finding that salinity favors lipid accumulation in these freshwater species. The maximum lipid content (18.28, 30.70, and 32.19%) and lipid productivity (8.59, 13.81, and 10.27 mg/l.d) were achieved in cultures in BG-11 medium with 160 mM NaCl for *S. quadricauda, S. dimorphus*, and *Chlorella* sp., respectively. The accumulation of lipids in *S. dimorphus* reached 39.42% when applying saline stress (160 mM) after the growth phase (18 days of culture in BG-11 medium) compared to the control (without salinity stress, 20.4%). A similar trend was observed when Ummalyma et al. (2020) reported that freshwater microalgae can grow and accumulate lipids when they grow in seawater. Of 10 isolated strains, RAP-13, identified as *Chlorococcum* sp., was able to grow in a wide pH range (4–10) in a seawater medium, with maximum lipid values of 42% and a profile of fatty acids dominated by SFA and PUFA. Similarly, Suyono et al. (2015) used seawater, brackish water, and freshwater to study the effect of salinity on lipid accumulation in a consortium of microalgae. Showing that salinity favors the accumulation of lipids, the highest lipid content (13.58%) was found at 7 days when the microalgae were cultured in seawater.

Another study by Pugkaew et al. (2019) evaluated the effect of salinity changes relative to normal seawater level (30 ppt) on lipid accumulation and fatty acid composition of the marine microalga *Tetraselmis suecica*. Demonstrating that salinity changes did not significantly affect the lipid content or the fatty acid profile. Lipid content at 10 ppt salinity (5.3%) increased slightly compared to that obtained (4.3%) with seawater or higher salinity. The composition of fatty acids remained unchanged in the salinity range evaluated, and are from 36 to 38% SFA, 23 to 25% MUFA, and 36 to 40% PUFA.

The fact that saline stress favors the accumulation of lipids in the biomass of freshwater microalgae opens the possibility of using the nutrients of saltwater as a cheaper medium, with the additional advantage of less contamination by bacteria and fungi that generally cause decreased productivity in freshwater microalgae culture in open ponds.

8.3.7 Co-cultivation in Wastewater

Although it has been shown that the production of biodiesel from microalgae has multiple advantages, its cost is still very high. Therefore, in recent years, mixotrophic culture in wastewater has been proposed with a dual purpose. The remediation of the waters and the reduction of production costs of the microalgal biomass. Various studies have reported that a wide variety of microalgae can grow in wastewater. Chandra et al. (2021) evaluated the cultivation *of Chlorella minutissima, Scenedesmus abundans, Nostoc muscorum,* and *Spirulina* sp. in wastewater from the dairy industry. They demonstrated that the cultivation of polymicroalgae can be more favorable for wastewater remediation and the production of oils with the right fatty acid profile to produce biodiesel. Another study by Khalekuzzaman et al. (2021) evaluated the accumulation and lipid profile of a consortium of microalgae (*Chlorella vulgaris, Chlorella sorokiniana,* and *Scenedesmus simris002*) in transparent-colored photobioreactors (white, blue, and green) using the effluent of an anaerobic reactor as a source of nutrients (N, P). They showed that microalgae grown in white PBR had a high lipid content (38%) with a fatty acid profile composed of 26% SFA, 28% MUFA, and 46% PUFA. Additionally, the efficiency of turbidity, TSS, and COD removal greater than 92% was evidenced. In a recent study, Lee et al. (2021) compared the accumulation of lipids and the fatty acid profile of *Coelastrella* sp. by growing it in a photoautotrophic process in BG-11 medium and a heterotrophic process followed by a mixotrophic, process using raw piggery wastewater with acetate and high ammonia content. They showed that acetate in the wastewater alleviated ammonia stress and by replacing the BG-11 medium with pig house wastewater, lipid accumulation was promoted, the proportion of MUFA and SFA was increased, and the PUFA content was decreased. Additionally, 99% of ammonia, 92% of chemical oxygen demand (COD), and 100% of phosphorous were removed during the 4-day process. The biodiesel obtained showed good oxidative stability and high cetane number (Table 8.2) with values within the limits of international standards.

Enwereuzoh et al. (2020) compared the quality of biodiesel produced by *Tetradesmus obliquus, Heterochlorella luteoviridis*, and *Chlamydomonas reinhardtii* growing in standard modified tris-acetate-phosphate medium and in original and diluted wastewater from fish farms (FFW). For all strains, the fatty acid profile was similar in all treatments with SFA $>$ 22%, MUFA $>$40% (C18:1), and PUFA $<$12%, especially in the treatments with diluted wastewater. In all samples, the most important properties of biodiesel were within the range of international standards. *Chlamydomonas reinhardtii* presented the most suitable biodiesel properties, especially CFPP. This can be attributed to the inhibition of PUFA accumulation in *Chlamydomonas reinhardtii* cultured in both FFWs.

Jordaan et al. (2018) evaluated different types of effluents as culture media for heterotrophic growth and lipid accumulation in six mixed algal cultures. The results showed that lipid accumulation was mainly influenced by the microalgae consortia used. In general, the growth was low, the highest accumulation of lipids was observed in the acid mine drainage (84.39%) and the potato starch effluent (84.85%) for the consortium *Fragilaria crotonensis*, *Aulacoseira granulado*, *Cyclotella meneghiniana*, *Navicula* spp., and *Fragilaria ulna*. The fatty acid composition of the oil (effluent from canned fish and whey) showed high saturation, leading to good oxidative stability and acceptable cetane numbers, but poor cold flow properties.

Arutselvan et al. (2021) evaluated lipid accumulation in *Chlorella vulgaris, Coelastrella* sp., *Scenedesmus dimorphus*, *Chlorococcum* sp., and a *consortium (Chlorella vulgaris + Scenedesmus dimorphus)* under laboratory conditions and in open-air ponds in a medium supplemented with urea and phosphates. They showed that *S. dimorphus* and *C. vulgaris*, both individually and in consortium, accumulated more lipids under laboratory conditions. In laboratory culture, the fatty acid profile of all species was dominated by polyunsaturated fatty acids, while growing *S. dimorphus* and *C. vulgaris* in open-air ponds increased MUFA content and decreased PUFA content. It was evidenced that when cultivating *S. dimorphus* and *C. vulgaris* in a consortium there was a synergistic effect that increased the lipid content (Table 8.3). The fatty acid profile in the consortium presented a higher proportion of MUFA and PUFA, and in the open-air pond culture, an increase in MUFA and a decrease in PUFA were observed. These results suggest that the consortium of *S. dimorphus* and *C. vulgaris* has potential as feedstock for biodiesel due to its high oleic acid content and could be grown in a stressed medium (deprived of nutrients or salinity) to increase lipid accumulation and decrease the PUFA content.

In another study, the association of *Chlorella vulgaris* with bacteria was evaluated, showing that an initial alga: bacteria ratio of 40:1 under conditions of nitrogen deficiency favored a 2.2-fold increase in lipid productivity when compared to axenic culture, while the lipid content increased by 47% reaching values of 52%, with an increase in the MUFA content (Wei et al. 2020).

Recently the combination of various stress factors to achieve a synergistic effect on lipid accumulation has been investigated, as reported by Moreno-Garcia et al. (2021) who evaluated the influence of light intensity, CO_2 content, and carbon source on the lipid quality of a bacterial and microalgae consortium isolated from dairy wastewater, consisting mainly of *Chlorella* sp. They showed that low light intensities associated with high CO_2 and organic carbon content favor MUFA production and higher light irradiation improved SFA and PUFA ratios. The NH3-N present in the wastewater used as a nutrient source was related to a higher production of MUFA. The fatty acid profile (Table 8.3) obtained at low light intensity and CO_2 and organic carbon input (40 μmol m d^{-1}, 10% CO_2, Glycerol 3 g/l) suggests that the lipids extracted from algal biomass grown in a mixture of wastewater are suitable for biodiesel production, which is confirmed by the high cetane number and good cold flow properties, as well as high oxidation stability, which are within the standards international listed in Table 8.1. In another study, Chandra and Amit (2019) evaluated the effect of the culture medium (BBM, modified CHU-13, and BG-11) and different abiotic factors (N, P, Fe, Zn, pH, light, glucose, and photoperiod) on the accumulation of lipids in *Chlorella minutissima*. They reported that the modified CHU-13 medium with 300 mg/l nitrogen provided the highest lipid content (50.39 $\pm$ 0.018%). Of the abiotic factors studied, nitrogen, pH, and light intensity had the greatest influence on lipid accumulation. The properties of the biodiesel obtained from *C. minutissima* met the criteria of international standards. Similarly, El-Sheekh et al. (2019) showed that the accumulation of lipids in *Chlorella vulgaris* increased when several stress factors were combined, reaching 15.8% of lipids in an optimized medium with the substitution of 50% nitrate for urea and the addition of 10 g/l NaCl, 0.5 g/l of ferrous sulfate 1 g/l of sodium acetate.

Another study by Zhao et al. (2019) evaluated the combined effect of ammonia, temperature, and pH on lipid accumulation in *Chlorella pyrenoidosa* growing in the effluent of an anaerobic wastewater treatment plant. They showed that the increase in pH induced by free ammonia and temperature inhibited the growth of algae, but significantly promoted the accumulation of lipids. The highest lipid content (30.2%) was reached at 25 °C and pH 8.3–8.5 (25 °C, ammonium 280 mg/L), which was 1.6 times higher than that found under neutral conditions. In addition, increasing the pH also increased % PUFA (C18:2 and C18:3).

Wen et al. (2016) evaluated the effect of scaling the culture (200 ml, 10 L, 40 m^3) on the accumulation of lipids in the biomass of *Graesiella* sp. GBM-1. They reported that when going from a laboratory culture in a 200-ml column with artificial lighting to one of 40 m^3 with solar radiation, the lipid content decreased from 45.2 to 33.4%. They suggest that to balance low growth and lipid accumulation, the nitrate concentration should be between 1.18 and 3.53 mM.

8.4 Conclusion

The literature reviewed shows that the properties of biodiesel are influenced by the length of the carbon chain of the oil, its structure, and the presence of unsaturation. Saturated fatty acids such as palmitic and stearic promote high cetane number values, good oxidation stability, and high heat of combustion, while unsaturated fatty acids are attributed to good cold flow properties, low the melting point, and viscosity, but are responsible for the low oxidation stability.

Obtaining biodiesel with good ignition characteristics without deteriorating cold flow properties and oxidation stability requires that the oil used as raw material has an adequate balance between SFA and MUFA, as well as a very low content of PUFA.

In general, freshwater microalgae have a higher proportion of SFA and MUFA, while marine algae have a higher PUFA content. However, microalgae can adapt to various stress conditions by modifying the lipid content and the fatty acid profile, so by manipulating the culture conditions it is possible to achieve a certain fatty acid composition.

Different strategies were reviewed to increase lipid accumulation in microalgae biomass and achieve a profile with higher SFA and MUFA content. It is concluded that nitrogen deprivation stress and high light intensity are the factors that most increase lipid accumulation and promote the desired profile; however, since nitrogen is essential for cell growth, productivity decreases, so a two-stage culture that includes optimization of growth in a rich medium until the nutrients are depleted to naturally reach the deprivation stage where lipid production is triggered, is presented as a promising option. The increase in temperature, salt stress, and supplementation with CO_2 and metallic trace elements can also induce the accumulation of lipids with a high proportion of SFA and MUFA. The combination of these factors with nitrogen deprivation and high irradiation has been shown to have a synergistic effect and is specific to each strain of microalgae. Also, the use of microalgal consortiums and co-cultures with bacteria and fungi present in the effluents of some industrial processes rich in organic matter, nitrogen, and phosphorus is proposed, as cheap and abundant culture media with the additional benefit of elimination of water contaminants.

Additionally, there are other factors that influence the content of lipids and the profile of fatty acids, such as the choice of culture technology (photobioreactor), the methods of harvesting and extracting the oil, and the optimization of the variables of the transesterification process to obtain the biodiesel, which are outside the scope of this chapter.

References

Alptekin E, Canakci M (2008) Determination of the density and the viscosities of biodiesel–diesel fuel blends. Renew Energ 33:2623–2630. https://doi.org/10.1016/j.renene.2008.02.020

Andeden EE, Ozturk S, Aslim B (2021) Effect of alkaline pH and nitrogen starvation on the triacylglycerol (TAG) content, growth, biochemical composition, and fatty acid profile of *Auxenochlorella protothecoides* KP7. J Appl Phycol 33:211–225. https://doi.org/10.1007/s10811-020-02311-0

Arguelles ED, Martinez-Gos MR (2021) Lipid accumulation and profiling in microalgae *Chlorolobion* sp. (BIOTECH 4031) and *Chlorella* sp. (BIOTECH 4026) during nitrogen starvation for biodiesel production. J Appl Phycol 33:1–11. https://doi.org/10.1007/s10811-020-02126-z

Arias-Peñaranda MT, Cristiani-Urbina E, Montes-Horcasitas C, Esparza-García F, Torzillo G, Cañizares-Villanueva RO (2013) *Scenedesmus incrassatulus* CLHE-Si01: a potential source of renewable lipid for high quality biodiesel production. Bioresour Technol 140:158–164. https://doi.org/10.1016/j.biortech.2013.04.080

Arutselvan C, Narchonai G, Pugazhendhi A, LewisOscar F, Thajuddin N (2021) Evaluation of microalgal strains and microalgal consortium for higher lipid productivity and rich fatty acid profile towards sustainable biodiesel production. Bioresour Technol 339:125524. https://doi.org/10.1016/j.biortech.2021.125524

Çakmak A, Özcan H (2020) Biofuel additive production from glycerol and determination of its effect on some fuel properties. SN Appl Sci 2:1637. https://doi.org/10.1007/s42452-020-03308-7

Chandra R, Amit GUK (2019) Effects of various abiotic factors on biomass growth and lipid yield of *Chlorella minutissima* for sustainable biodiesel production. Environ Sci Pollut Res 26:3848–3861. https://doi.org/10.1007/s11356-018-3696-1

Chandra R, Pradhan S, Patel A, Ghosh U (2021) An approach for dairy wastewater remediation using mixture of microalgae and biodiesel production for sustainable transportation. J Environ Manag 297:113210. https://doi.org/10.1016/j.jenvman.2021.113210

Chávez-Fuentes P, Ruiz-Marin A, Canedo-López Y (2018) Biodiesel synthesis from *Chlorella vulgaris* under effect of nitrogen limitation, intensity and quality light: estimation on the based fatty acids profiles. Mol Biol Rep 45:1145–1154. https://doi.org/10.1007/s11033-018-4266-9

Chu F, Cheng J, Zhang X, Ye Q, Zhou J (2019) Enhancing lipid production in microalgae Chlorella PY-ZU1 with phosphorus excess and nitrogen starvation under 15% CO2 in a continuous two-step cultivation process. Chem Eng J 375:121912. https://doi.org/10.1016/j.cej.2019.121912

Demirbas A (1998) Fuel properties and calculation of higher heating values of vegetable oils. Fuel 77(9–10):1117–1120. https://doi.org/10.1016/S0016-2361(97)00289-5

Demirbas A (2009) Progress and recent trends in biodiesel fuels. Energ Convers Manage 50:14–34. https://doi.org/10.1016/j.enconman.2008.09.001

Deng X, Chen B, Xue C, Li D, Hu X, Gao K (2018) Biomass production and biochemical profiles of a freshwater microalga *Chlorella kessleri* in mixotrophic culture: effects of light intensity and photoperiodicity. Bioresour Technol 273:358–367. https://doi.org/10.1016/j.biortech.2018.11.032

Deshmukh S, Bala K, Kumar R (2019) Selection of microalgae species based on their lipid content, fatty acid profile and apparent fuel properties for biodiesel production. Environ Sci Pollut Res 26: 24462–24473. https://doi.org/10.1007/s11356-019-05692-z

El-Sheekh MM, Gheda SF, El-Sayed AEKB, Shady AMA, El-Sheikh ME, Schagerl M (2019) Outdoor cultivation of the green microalga *Chlorella vulgaris* under stress conditions as a feedstock for biofuel. Environ Sci Pollut Res 26:18520–18532. https://doi.org/10.1007/s11356-019-05108-y

EN14214 (2012) Automotive fuels–fatty acid methyl esters (FAME) for diesel engines-requirements and test methods. Berlin: Beuth-Verlag

Enwereuzoh U, Harding K, Low M (2020) Characterization of biodiesel produced from microalgae grown on fish farm wastewater. SN Appl. Sci. 2:970. https://doi.org/10.1007/s42452-020-2770-8

Fawzy MA, El-Otify AM, Adam MS, Moustafa SSA (2021) The impact of abiotic factors on the growth and lipid accumulation of some green microalgae for sustainable biodiesel production. Environ Sci Pollut Res 28:42547–42561. https://doi.org/10.1007/s11356-021-13781-1

Fernandes T, Cordeiro N (2022) High-value lipids accumulation by *Pavlova pinguis* as a response to nitrogen-induced changes. Biomass Bioenergy 158:106341. https://doi.org/10.1016/j.biombioe.2022.106341

Gour RS, Garlapati VK, Kant A (2020) Effect of salinity stress on lipid accumulation in *Scenedesmus* sp. and *Chlorella* sp. feasibility of stepwise culturing. Curr Microbiol 77:779–785. https://doi.org/10.1007/s00284-019-01860

Hadrich B, Akremi I, Dammak M, Barkallah M, Fendri I, Abdelkafi S (2018) Optimization of lipids' ultrasonic extraction and production from *Chlorella* sp. using response-surface methodology. Lipids Health Dis 17:87. https://doi.org/10.1186/s12944-018-0702-z

Huang GH, Chen F, Wei D, Zhang X, Chen G (2010) Biodiesel production by microalgal biotechnology. Appl Energy 87(1):38–46. https://doi.org/10.1016/j.apenergy.2009.06.016

Islami HR, Assareh R (2019) Effect of different iron concentrations on growth, lipid accumulation, and fatty acid profile for biodiesel production from *Tetradesmus obliquus*. J Appl Phycol 31:3421–3432. https://doi.org/10.1007/s10811-019-01843-4

Jaiswal KK, Banerjee I, Singh D, Sajwan P, Chhetri V (2020) Ecological stress stimulus to improve microalgae biofuel generation: a review. Octa J Biosci 8(1):48–54

Jordaan E, Roux-van der Merwe MP, Badenhorst J, Knothe G, Botha BM (2018) Evaluating the usability of 19 effluents for heterotrophic cultivation of microalgal consortia as biodiesel feedstock. J Appl Phycol 30:1533–1547. https://doi.org/10.1007/s10811-017-1341-x

Khalekuzzaman M, Kabir SB, Islam MB, Datta P, Alam MA, Xue J (2021) Enhancing microalgal productivity and quality by different colored photobioreactors for biodiesel production using anaerobic reactor effluent. Biomass Conv Bioref 11:767–779. https://doi.org/10.1007/s13399-020-00852-5

Knothe and Razon (2017) Biodiesel fuels. Prog Energy Combust Sci 58: 36–59. https://doi.org/10.1016/j.pecs.2016.08.001

Knothe G, Steidley KR (2005) Kinematic viscosity of biodiesel fuel components and related compounds. Influence of compound structure and comparison to petrodiesel fuel components. Fuel 84:1059–1065. https://doi.org/10.1016/j.fuel.2005.01.016

Kumari K, Samantaray S, Sahoo D, Tripathy BC (2021) Nitrogen, phosphorus and high CO2 modulate photosynthesis, biomass and lipid production in the green alga *Chlorella vulgaris*. Photosynth Res 148:17–32. https://doi.org/10.1007/s11120-021-00828-0

Lakshmikandan M, Murugesan AG, Wang S, Abomohra A, Jovita P, Kiruthiga S (2021) Sustainable biomass production under CO2 conditions and effective wet microalgae lipid extraction for biodiesel production. J Clean Prod 247:119398. https://doi.org/10.1016/j.jclepro.2019.119398

Lanjekar R, Deshmukh D (2016) A review of the effect of the composition of biodiesel on NOX emission, oxidative stability and cold flow properties. Renew Sust Energ Rev 54:1401–1411. https://doi.org/10.1016/j.rser.2015.10.034

Lapuerta M, Rodríguez-Fernández J, Font de Mora E (2009) Correlation for the estimation of the cetane number of biodiesel fuels and implications on the iodine number. Energy Policy 37:4337–4344

Lapuerta M, Rodríguez-Fernández J, Armas O (2010) Correlation for the estimation of the density of fatty acid esters fuels and its implications. A proposed Biodiesel Cetane. Chem Phys Lipids 163:720–727

Lee SA, Ko SR, Lee N, Lee JW, Le VV, Oh HM, Ahn CY (2021) Two-step microalgal (*Coelastrella* sp.) treatment of raw piggery wastewater resulting in higher lipid and triacylglycerol levels for possible production of higher-quality biodiesel. Bioresour Technol 332:125081. https://doi.org/10.1016/j.biortech.2021.125081

Lown AL, Peereboom L, Mueller SA, Anderson JE, Miller DJ, Lira CT (2014) Cold flow properties for blends of biofuels with diesel and jet fuels. Fuel 117:544–551. https://doi.org/10.1016/j.fuel.2013.09.067

Mandal MK, Saikia P, Chanu NK, Chaurasia N (2019) Modulación del contenido de lípidos y el perfil de lípidos mediante la suplementación de hierro, zinc y molibdeno en microalgas autóctonas. Environ Sci Pollut Res 26:20815–20828. https://doi.org/10.1007/s11356-019-05065-

Mandotra SK, Kumar P, Suseela MR, Ramteke PW (2014) Fresh water green microalga *Scenedesmus abundans*: a potential feedstock for high quality biodiesel production. Bioresour Technol 156:42–47. https://doi.org/10.1016/j.biortech.2013.12.127

McCormick R, Ratcliff M, Moens L, Lawrence R (2007) Several factors affecting the Biodiesel stability in accelerated standard tests. Fuel Process Technol 88:651–657. https://doi.org/10.1016/j.fuproc.2007.01.006

Meher LC, Vidya S, Naik SN (2006) Technical aspects of biodiesel production by transesterification—a review. Renew Sust Energ Rev 10:248–268. https://doi.org/10.1016/j.rser.2004.09.002

Mesa J (2020) Impactos de la producción de biodiésel en la cadena de valor de palma. Fedepalma

Moreno-Garcia L, Gariépy Y, Barnabé S, Raghavan V (2021) Factors affecting the fatty acid profile of wastewater-grown-algae oil as feedstock for biodiesel. Fuel 304:121367. https://doi.org/10.1016/j.fuel.2021.121367

Mujtaba G, Choi W, Lee CG, Lee K (2012) Lipid production by Chlorella vulgaris after a shift from nutrient-rich to nitrogen starvation conditions. Bioresour Technol 123:279–283. https://doi.org/10.1016/j.biortech.2012.07.057

Pandit PR, Fulekar MH, Karuna MSL (2017) Effect of salinity stress on growth, lipid productivity, fatty acid composition, and biodiesel properties in *Acutodesmus obliquus* and *Chlorella vulgaris*. Environ Sci Pollut Res 24:13437–13451. https://doi.org/10.1007/s11356-017-8875-y

Park JY, Kim DK, Lee JP, Park SC, Kim YJ, Lee JS (2008) Blending effects of biodiesels on oxidation stability and low temperature flow properties. Bioresour Technol 99:1196–1203. https://doi.org/10.1016/j.biortech.2007.02.017

Pruvost J, Van Vooren G, Le Gouic B, Couzinet-Mossion A, Legrand J (2011) Systematic investigation of biomass and lipid productivity by microalgae in photobioreactors for biodiesel application. Bioresour Technol 102(1):150–158. https://doi.org/10.1016/j.biortech.2010.06.153

Pugkaew W, Meetam M, Yokthongwattana K, Leeratsuwan N, Pokethitiyook P (2019) Effects of salinity changes on growth, photosynthetic activity, biochemical composition, and lipid productivity of marine microalga Tetraselmis suecica. J Appl Phycol 31:969–979. https://doi.org/10.1007/s10811-018-1619-7

Qi F, Pei H, Mu R, Ma G, WU D, Han Q (2019) Characterization and optimization of endogenous lipid accumulation in *Chlorella vulgaris* SDEC-3M ability to rapidly accumulate lipid for reversing nightly lipid loss. Biotechnol Biofuels 12: 151 doi:https://doi.org/10.1186/s13068-019-1493-9

Ramos MJ, Fernández CM, Casas A, Rodríguez L, Pérez A (2009) Influence of fatty acid composition of raw materials on biodiesel properties. Bioresour Technol 100(1):261–268. https://doi.org/10.1016/j.biortech.2008.06.039

Rao GL, Ramadhas AS, Nallusamy N, Sakthivel P (2010) Relationships among the physical properties of biodiesel and engine fuel system design requirement. Int J Energy Environ 1(5):919–992

Rayati M, Rajabi Islami H, Shamsaie Mehrgan M (2020) Light intensity improves growth, lipid productivity, and fatty acid profile of *Chlorococcum oleofaciens* (Chlorophyceae) for biodiesel production. Bioenergy Res 13:1235–1245. https://doi.org/10.1007/s12155-020-10144-5

Sajjadi B, Aziz A, Raman A, Arandiyan H (2016) A comprehensive review on properties of edible and non-edible vegetable oil-based biodiesel: composition, specifications and prediction models. Renew Sust Energ Rev 63:62–92. https://doi.org/10.1016/j.rser.2016.05.035

Samkhaniyani F, Najafpour GD, Ardestani F (2017) Evaluation of effective nutritional parameters for *Scenedesmus* sp. microalgae culturing in a photobioreactor for biodiesel production. Int J Environ Sci Technol 14:1037–1046. https://doi.org/10.1007/s13762-016-1212-z

Santhakumaran P, Kookal SK, Mathew L, Ray JG (2019) Bioprospecting of three rapid-growing freshwater green algae, promising biomass for biodiesel production. Bioenergy Res 12: 680–693. https://doi.org/10.1007/s12155-019-09990-9

Silva BF, Wendt EV, Castro JC, Oliveira AE, Carrim AJI, Vieira JDG, Sassi R, Sassi CFC, Silva ALV, Barboza GFO, Antoniosi Filho NR (2015) Analysis of some chemical elements in marine microalgae for biodiesel production and other uses. Algal Res 9:312–321. https://doi.org/10.1016/j.algal.2015.04.010

Sonkar S, Mallick N (2018) An alternative strategy for enhancing lipid accumulation in chlorophycean microalgae for biodiesel production. J Appl Phycol 30:2179–2192. https://doi.org/10.1007/s10811-018-1419-

Sun X, Cao Y, Xu H, Liu Y, Sun J, Qiao D, Cao Y (2014) Effect of nitrogen-starvation, light intensity and iron on triacylglyceride/carbohydrate production and fatty acid profile of *Neochloris oleoabundans* HK-129 by a two-stage process. Bioresour Technol 155:204–212. https://doi.org/10.1016/j.biortech.2013.12.109

Suyono EA, Haryadi W, Zusron M, Nuhamunada M, Rahayu S, Nugroho P (2015) The effect of salinity on growth, dry weight and lipid content of the mixed microalgae culture isolated from Glagah as biodiesel substrate. J Life Sci 9:229–233. https://doi.org/10.17265/1934-7391/2015.05.006

Tarazona Delgado R, Guarieiro M, Antunes PW, Cassini ST, Montoya Terreros H, de Oliveira FV (2021) Effect of nitrogen limitation on growth, biochemical composition, and cell ultrastructure of the microalga *Picocystis salinarum*. J Appl Phycol 33:2083–2092. https://doi.org/10.1007/s10811-021-02462-8

Tayari S, Abedi R, Abedi A (2020) Investigation on physicochemical properties of wastewater grown microalgae methyl ester and its effects on CI engine. Environ Climate Technol 24(1):72–87. https://doi.org/10.2478/rtuect-2020-0005

Ummalyma SB, Sukumaran RK, Pandey A (2020) Evaluation of freshwater microalgal isolates for growth and oil production in seawater medium. Waste Biomass Valor 11:223–230. https://doi.org/10.1007/s12649-018-0393-8

Vishwakarma R, Dhar DW, Saxena S (2019) Influence of nutrient formulations on growth, lipid yield, carbon partitioning and biodiesel quality potential of *Botryococcus* sp. and *Chlorella* sp. Environ Sci Pollut Res 26:7589–7600. https://doi.org/10.1007/s11356-019-04213-2

Wang Z, Luo F, Wang Z, Zhou R, Tang Y, Li Y (2019) The potential growth and lipid accumulation in *Coccomyxa subellipsoidea* triggered by glucose combining with sodium acetate. World J Microbiol Biotechnol 35:110. https://doi.org/10.1007/s11274-019-2682-1

Wei L, Huang X, Huang Z (2015) Temperature effects on lipid properties of microalgae *Tetraselmis subcordiformis* and *Nannochloropsis oculata* as biofuel resources. Chin J Ocean Limnol 33:99–106. https://doi.org/10.1007/s00343-015-3346-0

Wei Z, Wang H, Li X, Zhao Q, Yin Y, Xi L, Ge B, Qin S (2020) Enhanced biomass and lipid production by co-cultivation of *Chlorella vulgaris* with *Mesorhizobium sangaii* under nitrogen limitation. J Appl Phycol 32:233–242. https://doi.org/10.1007/s10811-019-01924-4

Wen X, Du K, Wang Z, Peng X, Luo L, Tao H, Xu Y, Zhang D, Geng Y, Li Y (2016) Effective cultivation of microalgae for biofuel production: a pilot-scale evaluation of a noveloleaginous microalga Graesiella sp. WBG-1. Biotechnol Biofuels 9:123. https://doi.org/10.1186/s13068-016-0541-y

Xin L, Hong-ying H, Jia Y (2010) Lipid accumulation and nutrient removal properties of a newly isolated freshwater microalga, *Scenedesmus* sp. LX1, growing in secondary effluent. N Biotechnol Special Issue on Biocatalysis and Agricultural Biotechnology 27(3): 59–63. https://doi.org/10.1016/j.nbt.2009.11.006

Yu SJ, Hu H, Zheng H, Wang YQ, Pan SB, Zeng RJ (2019) Effect of different phosphorus concentrations on biodiesel production from *Isochrysis zhangjiangensis* under nitrogen sufficiency or deprivation condition. Appl Microbiol Biotechnol 103:5051–5059. https://doi.org/10.1007/s00253-019-09814-y

Yuan C, Xu K, Sun J, Hu GR, Li FL (2018) Ammonium, nitrate, and urea play different roles for lipid accumulation in the nervonic acid—producing microalgae *Mychonastes afer* HSO-3-1. J Appl Phycol 30:793–801. https://doi.org/10.1007/s10811-017-1308-y

Zhang L, Wang N, Yang M, Ding K, Wang Y, Huo D, Hou C (2019) Lipid accumulation and biodiesel quality of *Chlorella pyrenoidosa* under oxidative stress induced by nutrient regimes. Renew Energy 143:1782–1790. https://doi.org/10.1016/j.renene.2019.05.081

Zhao XC, Tan XB, Yang LB, Liao JY, Li XY (2019) Cultivation of *Chlorella pyrenoidosa* in anaerobic wastewater: the coupled effects of ammonium, temperature and pH conditions on lipids compositions. Bioresour Technol 284:90–97

9 Dark Fermentation of Microalgae and Cyanobacteria for Hydrogen Production

Cigdem Demirkaya and Hector De la Hoz Siegler

Abstract

Hydrogen is an effective energy carrier, having high-energy density and releasing no pollutants upon combustion. Indirect biophotolysis is a two-stage process wherein photosynthesis and hydrogen evolution are effectively decoupled. In the first stage, microalgae or cyanobacteria use sunlight to fix carbon dioxide and fuel their growth. In the second stage, dark anaerobic fermentation of the accumulated biomass is used to generate hydrogen by the action of hydrogenases. This chapter presents an overview of the metabolic processes leading to the formation of hydrogen in dark anaerobic conditions, and the factors affecting hydrogen formation rate. Recent developments in dark fermentation for hydrogen production from microalgae and cyanobacteria and different strategies to improve hydrogen production efficiency are also presented.

Keywords

Hydrogen · Microalgae · Indirect biophotolysis · Cyanobacteria · Dark fermentation

9.1 Introduction

The high heating value of hydrogen, 141.65 MJ kg^{-1} (All and Basit 1993), and its clean combustion that results only in the formation of water vapor, make it an excellent energy carrier (Dawood et al. 2020; Goria et al. 2022). Thus, hydrogen has long been identified as a game-changing approach for abating carbon dioxide (CO_2) emissions and tackling climate change, as it can replace fossil fuels in different hard to decarbonize areas, including transportation, heating, and metallurgical processes.

C. Demirkaya · H. De la Hoz Siegler (✉)
Department of Chemical and Petroleum Engineering, University of Calgary, Calgary, AB, Canada
e-mail: h.siegler@ucalgary.ca

Hydrogen can be generated via different alternative methods, including coal gasification, steam methane reforming (SMR), water splitting, or biological pathways. Although hydrogen is a colorless gas, hydrogen is referred by different color denominations according to the method used in its production (Dawood et al. 2020; Hermesmann and Müller 2022). Currently, most of the world's hydrogen production is obtained from fossil fuels, primarily by steam reformation of natural gas, which accounts for about 60% of the global yearly production, coal gasification, and as a by-product of refinery operations (IEA 2021). The reforming reaction, where the atoms in the methane (CH_4) and water molecules are rearranged to form H_2 and CO, takes place at high temperatures (700 °C–1000 °C) and high pressure (Yu et al. 2021; Hermesmann and Müller 2022). These extreme reaction conditions translate into a highly energy-intensive process with significant CO_2 emissions. Hydrogen obtained by SMR is conventionally referred as gray hydrogen. When SMR is coupled to Carbon Capture and Storage (CCS), the hydrogen is designated as blue. This is done to indicate that this hydrogen has a lower carbon footprint than conventional gray hydrogen (Bauer et al. 2021). Although blue hydrogen has lower overall CO_2 emissions than grey hydrogen, it is still an energy-intensive process. Moreover, as energy is consumed by the carbon capture and compression equipment, blue hydrogen is more energy intensive than gray hydrogen.

Green hydrogen is produced by the electrolysis of water powered by electricity obtained from renewable sources (Dincer 2012; Hermesmann and Müller 2022). Although green hydrogen offers the lowest carbon and water footprint, it only accounted for a very small share of the global production at 30 kt of H_2 in 2020, or roughly 0.03% of the world's hydrogen supply (IEA 2021). Hydrogen can also be derived from biological sources, either from bioenergy through biomass gasification or by metabolic processes that directly generate hydrogen. Biologically derived hydrogen is still in its infancy, and therefore it has not yet received a color

A. de Jesús Martínez-Roldán (ed.), *Biotechnological Processes for Green Energy, and High Value Bioproducts by Microalgae, and Cyanobacteria Cultures*, Developments in Applied Phycology 13, https://doi.org/10.1007/978-3-031-43969-8_9

Table 9.1 Biological processes for hydrogen production

Mechanism	Reactions	Advantages	Disadvantages
Direct biophotolysis	$2H_2O \rightarrow O_2 + 4H^+ + 4e^-$	High-energy conversion efficiency	O_2 sensitivity; translucent H_2 impermeable material required
	$4H^+ + 4e^- \rightarrow H_2$		
	$2H_2O \rightarrow O_2 + 2H_2$		
Indirect biophotolysis	$H_2O + 6CO_2 \rightarrow C_6H_{12}O_6 + 6O_2$	Separation of H_2 production from O_2 evolution	Translucent H_2 impermeable material may be required
	$C_6H_{12}O_6 + 6H_2O \rightarrow 12H_2 + 6CO_2$		
	$12H_2O \rightarrow 6O_2 + 12H_2$		
	$N_2 + 8H^+ + 8e^- + 16\ ATP \rightarrow 2NH_3 + H_2 + 16\ ADP + 16Pi$		
Photofermentation	$C_xH_yO_z + (2x-z)\ H_2O \rightarrow (y/2 + 2x-2)\ H_2 + xCO_2$	No organic by-products	Translucent H_2 impermeable material required; Organic acids or waste organic material needed as substrate; produces CO_2
	$N_2 + 8H^+ + 8e^- + 16\ ATP \rightarrow 2NH_3 + H_2 + 16\ ADP + 16Pi$		
Dark fermentation	$C_6H_{12}O_6 + 2H_2O \rightarrow 2CH_3COOH + 4H_2 + 2CO_2$	High H_2 production rates	Organic carbon substrate required; many by-products; low H_2 yield; produces CO_2

Adapted from Brentner et al. (2010)

denomination. While some authors consider that biohydrogen is a type of green hydrogen, there is no general agreement on this classification.

Microalgae and cyanobacteria are attractive for their potential to fix CO_2 into biomass via photosynthesis and their ability to simultaneously produce biohydrogen (Kumari et al. 2017; Nagarajan et al. 2017; Jiménez-Llanos et al. 2020). Microalgae and cyanobacteria can produce hydrogen through four different mechanisms: direct biophotolysis, indirect biophotolysis, photofermentation, and dark fermentation, as well as combinations of these processes (Show et al. 2019).

Biophotolysis is the process of splitting water into hydrogen and oxygen by the action of a living organism using light as the energy source (Acar and Dincer 2018). Water splitting is a two-step process: water oxidation, and hydrogen reduction. In direct biophotolysis, these two steps occur simultaneously, driven by the energy of the absorbed photons. It offers the highest theoretical efficiency. Unfortunately, the hydrogen-reducing enzymes (i.e., hydrogenases) are quickly inhibited by oxygen. Therefore, there is a need to consume or remove the oxygen as soon as it is produced. Alternatively, the enzymes could be engineered to reduce or eliminate oxygen sensitivity.

Indirect biophotolysis involves the spatial and/or temporal decoupling of the two water splitting reactions. In this case, H_2 does not directly evolve from water but from the autofermentation of endogenously stored carbohydrates (Hallenbeck and Benemann 2002; Buitrón et al. 2017; Sivaramakrishnan et al. 2021). This process normally occurs in nitrogen-fixing cyanobacteria, where H_2 is a by-product of nitrogen fixation. In the absence of nitrogen, H_2 is the only reaction product.

Anaerobic fermentation can result in the formation of hydrogen gas. Purple bacteria use light to drive the conversion of organic carbon (usually low-energy organic acids) into biomass with CO_2 and hydrogen as by-products. The process of photofermentation requires the availability of an organic carbon source, but it can be used as a method of carbon recycling as it can start with low-value (in terms of free-energy content) molecules, such as acetate, butyrate, and lactate (Singh and Sarma 2022).

Dark fermentation takes place in the absence of light and results in the production of CO_2. In dark fermentation, the substrate is typically a carbohydrate that is degraded to produce hydrogen and other metabolites (Kamran 2021). It could potentially be used to convert high-energy organic carbon (e.g., cellulose and lignocellulose) into bioproducts and H_2, offering higher conversion efficiency than other fermentation technologies, such as ethanolic or lactic acid fermentation.

In Table 9.1, a summary of the four biological processes for hydrogen production and the key reactions involved is presented.

This chapter presents an overview of the metabolic processes that lead to the formation of hydrogen under dark anaerobic conditions, and the factors affecting hydrogen formation rate. Recent developments in dark fermentation for hydrogen production using cyanobacteria and microalgae, as well as strategies to improve production efficiency, are also presented.

9.2 Indirect Biophotolysis: A Process Overview

Indirect biophotolysis can be understood as the combination of photosynthesis and fermentation to produce H_2, see Fig. 9.1. During photosynthesis, cells capture energetic photons with different chromophores, mainly chlorophyll, and use the light energy to drive the reduction of CO_2 into carbohydrates and ultimately to produce biomass. Water is

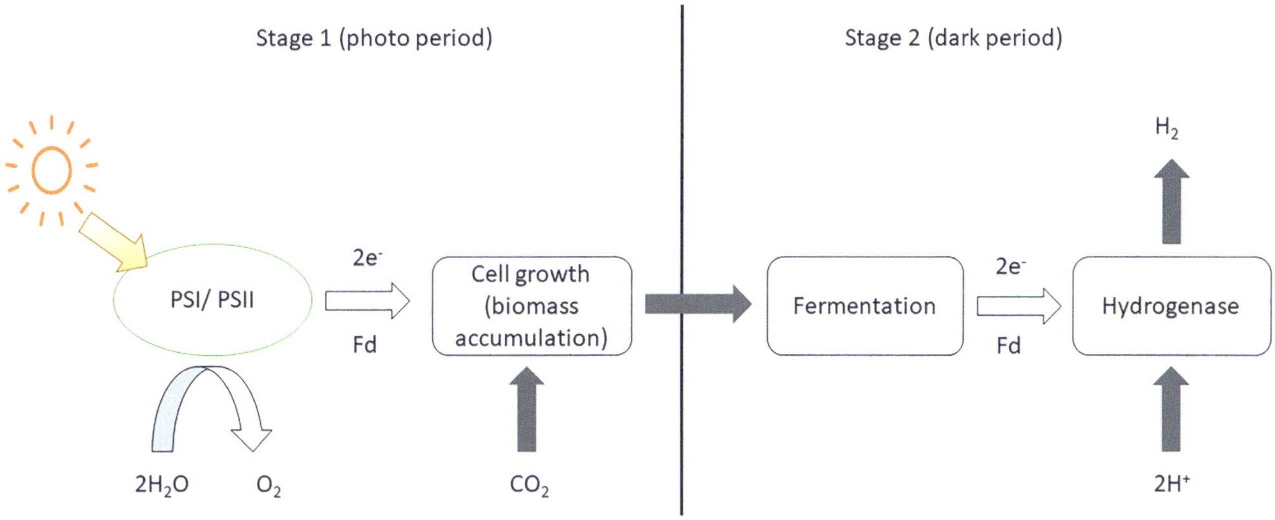

Fig. 9.1 Indirect biophotolysis processes of microalgae and cyanobacteria (adapted from Hallenbeck and Benemann 2002)

consumed by the photosynthesis reaction, as shown in Table 9.1. In the second stage, during dark conditions, reserve energy storage polymers like carbohydrates are consumed to fuel the cells' metabolism. Any oxygen present at the start of the dark phase is quickly consumed and if no additional oxygen is provided then an anoxic environment is formed. In the absence of oxygen, excess NADH and NADPH are created using fermentative pathways. The excess NAD(P)H is used to convert H^+ to H_2, thus restoring the redox balance in the cells.

The uncoupling between photosynthesis and fermentation can be achieved temporally by alternating periods of light and darkness in the same culture vessel, or spatially by transferring the biomass produced in the first stage from a photobioreactor or open pond system into an enclosed dark fermenter. Indirect biophotolysis can be realized with microorganisms that are naturally capable of performing both photosynthesis and dark fermentation, or by engineering these metabolic capabilities either by genetic manipulation or by building a consortium of phototrophs and fermentative heterotrophs.

H_2 production via indirect biophotolysis is affected by the culture conditions and the type of bioreactor used in each stage (Akhlaghi and Najafpour-Darzi 2020). In the photosynthetic stage, light intensity and photoperiod, temperature, pH, nutrient supply, and carbon source type and concentration are known to affect biomass productivity and composition (Gatamaneni et al. 2018). As hydrogen production in the second stage mostly relies on carbohydrate content, culture conditions in the photosynthetic stage need to be optimized to maximize carbohydrate content (Aikawa et al. 2012). Similarly, in the dark fermentation stage, culture conditions play a critical role activating, promoting, or suppressing different metabolic pathways, and ultimately controlling net hydrogen production rate and yield (Chandrasekhar et al. 2015). As the carbohydrate storages are consumed by the different cell metabolic processes, the activation of pathways that do not result in hydrogen formation reduces the yield of hydrogen (Razu et al. 2019).

Hydrogen production by fermentation has been shown to be a process operating close to thermodynamic equilibrium under biologically relevant conditions (Ananyev et al. 2012). Thus, operating conditions or setups that promote a quick removal of evolved hydrogen will result in a higher production rate and yield. Hydraulic retention time is of particular importance in the dark fermentation stage, as prolonged fermentation time may trigger a shift to unfavorable metabolic pathways (Chandrasekhar et al. 2015). In mixed cultures, a longer fermentation time might favor the activity of methanogenic organisms over that of acidogenic, hydrogen producing ones.

9.3 Key Enzymes in Hydrogen Production

The biological production of hydrogen is dependent upon the presence of a hydrogen-producing enzyme. The enzymes known to catalyze hydrogen evolution are all metalloenzymes, containing a transition metal at the catalytic active site (Hallenbeck and Benemann 2002). In microalgae and cyanobacteria, there are two major families of enzymes able to drive the production of hydrogen: nitrogenases and reversible or classical hydrogenases. A third family of enzymes, membrane-bound hydrogenases have not been found to produce any significant or measurable hydrogen (Das and Veziroglu 2001).

9.3.1 Nitrogenases

Many cyanobacteria, both filamentous and unicellular, can convert atmospheric nitrogen into ammonia by the action of nitrogenase, providing these cyanobacteria access to the largest nitrogen pool available on earth (Compaoré and Stal 2010). Nitrogenases are two-component protein complexes with a dinitrogenase (MoFe) protein and a dinitrogenase reductase (Fe) protein (Anwar et al. 2019). They use magnesium adenosine triphosphate (MgATP) and low-potential electrons to reduce a variety of substrates, including protons into H_2 (Hallenbeck and Benemann 2002). The expression of nitrogenase is regulated by nitrogen deficiency and oxygen presence, as the assembly and operation of nitrogenase is costly to the cells, and nitrogenases are highly sensitive to the presence of oxygen (Pratte and Thiel 2014). To protect these oxygen sensitive enzymes, nitrogenase activity is either only present inside the heterocysts, which are specialized cells in filamentous cyanobacteria that lack a functional photosystem, or are only activated in the dark, when no photosynthesis takes place and the intracellular environment becomes micro-aerobic (Compaoré and Stal 2010).

In N_2 fixation by nitrogenase, 8 mol of ferredoxin are oxidized consuming 16 mol of ATP and producing 2 mol of ammonia and 1 mol of hydrogen. If N_2 is not available, protons become the exclusive substrate and the action of nitrogenase will generate 1 mol of H_2 for every 4 mol of ATP and 2 mol of ferredoxin consumed (Huesemann et al. 2010). Due to this high-energy demand, the action of nitrogenase needs to be coupled with photosynthesis or the respiratory electron transport chain (Hatano et al. 2022). Glycolysis alone is not able to provide sufficient ATP for either N_2 fixation or H_2 production under dark anaerobic conditions (Hallenbeck and Benemann 2002).

9.3.2 Hydrogenases

Hydrogenases are electron transporters in many cellular reactions. They carry electrons from NADH, oxidize ferredoxin (Fd), and generate H_2 under anoxic conditions (Eq. 9.1) (Hallenbeck and Benemann 2002; Buitrón et al. 2017; Sivaramakrishnan et al. 2021).

$$\mathrm{NADPH} + 2\mathrm{H}^+ + 2\,\mathrm{Fd}^-_{\mathrm{reduced}} \rightarrow \mathrm{NADP}^+ + \mathrm{H}_2 + 2\,\mathrm{Fd}_{\mathrm{oxidized}} \quad (9.1)$$

Hydrogenases are widely found in microalgae, cyanobacteria, and fermentative heterotrophic bacteria, and are broadly classified into three main categories depending on the metals present in the active catalytic site: (i) [FeFe] hydrogenase, (ii) [NiFe] hydrogenase, and (iii) [Fe] only hydrogenase (Vignais and Billoud 2007; Lubitz et al. 2014). [FeFe] and [NiFe] hydrogenase have been reported in microalgae and cyanobacteria (Khetkorn et al. 2017). The hydrogen production rate of [FeFe] and [NiFe] hydrogenases is higher than nitrogenase, with turnover rates being 15 times higher in hydrogenases (Srirangan et al. 2011).

[FeFe] hydrogenases contain two irons with a dithionate cofactor (Wittkamp et al. 2018) and they are extremely sensitive to oxygen, getting easily inhibited in the presence of even trace amounts of oxygen. Consequently, synthesis of [FeFe] hydrogenases occurs only under anoxic conditions (Posewitz et al. 2009). [NiFe] hydrogenase contains two subunits, with a large subunit of [NiFe] as the active site and a small subunit containing Fe-S bonds (Volbeda et al. 1996; Peters et al. 2015). [NiFe] hydrogenase is more oxygen tolerant than [FeFe] hydrogenase and can work in the presence of oxygen (Goris et al. 2011). The more oxygen-tolerant [NiFe] hydrogenases have a modified [NiFe] active site with two cysteines, located at opposite ends, instead of glycine. The replacement of glycine in both locations increases oxygen tolerance. Oxygen tolerance decreases to just 1% oxygen concentration even if just one of the cysteines is replaced with glycine (Lukey et al. 2011).

9.3.2.1 Hydrogenase Activity in Microalgae

In green microalgae, hydrogenase enzymes play an important role in both indirect biophotolysis and fermentative hydrogen production pathways. *Chlamydomonas reinhardtii* is the most widely studied model microalgae, and consequently most of the knowledge we have regarding the activity and functioning of hydrogenases is based on this microalga (Kim et al. 2006; Tsygankov et al. 2006; Hemschemeier et al. 2008).

In most green microalgae, [FeFe] hydrogenases are the only hydrogen-producing enzymes. The biggest challenge for producing H_2 in microalgae is consequently the extremely high oxygen sensitivity of these hydrogenases, as H_2 yield is hindered if anoxic conditions are not continuously maintained. Therefore, [FeFe] hydrogenases are typically not active during photosynthetic activity (Bingham et al. 2012), and H_2 production in green microalgae typically does not occur under direct light. It has been shown, however, that hydrogen formation can take place under light if anoxic conditions are created by continuously removing oxygen (Mus et al. 2007). Induction of H_2 production in *C. reinhardtii* has also been achieved by creating sulfur deprivation, thus inhibiting the repair of PSII proteins that get damaged by light (Laurinavichene et al. 2004; Pollock et al. 2005; Kosourov et al. 2007). Inhibition of PSII results in reduced photosynthetic activity, and a lower oxygen evolution rate. As cells use oxygen for respiration to produce energy for cell maintenance, this quickly creates an anoxic environment that facilitates H_2 formation.

Table 9.2 Biohydrogen production by microalgae and cyanobacteria under various environmental conditions and responsible enzymes

Enzyme	Type	Microalgae/ cyanobacteria	Conditions	Hydrogen yield	References
Hydrogenase	[NiFe]	*Synechocystis* sp.	Medium A+, Temp 35 °C, nitrogen starvation, dark, anoxic	30 mol/10^{17} cells	McNeely et al. (2014)
	[NiFe]	*Synechocystis*	BG-11 medium, pH 8, Temp 28 °C, dark, anoxic	600 nmol mg of Chl a^{-1} min^{-1}	Gutekunst et al. (2018)
	[FeFe]	*Chlamydomonas reinhardtii*	TAP medium, pH 8, Temp 25 °C, two stage (light and dark), anoxic	6 mL L^{-1}	Saleem et al. (2012)
	[FeFe]	*Chlorella vulgaris*	BCC medium, pH 8, Temp 22 °C, dark, anoxic	0.35 mL h^{-1}	Alalayah et al. (2017)
	[NiFe]	*Synechococcus*	Medium A+, Temp 38 °C, dark, anoxic	15 nmol/10 cells mL^{-1}	Kumaraswamy et al. (2013)
	[Fe]	*Chlorella fusca*	Temp 20 °C, dark, anoxic	20 µmol mg of Chl a^{-1}	Winkler et al. (2002)
Nitrogenase		*Cyanothece* sp. Miami BG 043511 Anabena	Artificial wastewater, pH 8, Temp 30 ° C, two stage (light and dark), anoxic	16.4 µmol g^{-1}	Skizim et al. (2012)
		Anabena siamensis	BG-11 medium, Temp 21 °C, dark, anoxic	3.092–2.426 µmol mg of Chl a^{-1}	Andrić et al. (2017)
		Synechococcus sp.	BG-11 medium, Temp 25 °C, dark, anoxic	4.44 µmol mg of Chl a^{-1}	Kossalbayev et al. (2020)

Some studies have attempted to identify oxygen-tolerant [FeFe] hydrogenase or obtain oxygen-tolerant [FeFe] hydrogenases by genetic modification (Bhagat et al. 2020). Optimization of photoperiod can also help to minimize oxygen accumulation, thus preventing hydrogenase inhibition. Kosourov et al. (2018) achieved sustained efficient H_2 production by transferring the growing cultures of *C. reinhardtii* from continuous light to a train of strong light pulses superimposed on darkness or low background illumination. This method redirected the photosynthetic electron flow to the hydrogenase rather than toward CO_2 fixation and biomass formation, thus increasing the overall H_2 photoproduction yield.

The capability of fermentation under dark anoxia helps *C. reinhardtii* to survive by allowing it to produce ATP from reserves accumulated during photosynthesis (Catalanotti et al. 2013). During light periods, photosynthesis generates excess anabolic activity leading to the accumulation of polysaccharides and/or lipids. Then, during dark periods these reserve molecules are converted to pyruvate to fuel the cell's metabolism. Pyruvate acts as a substrate for several fermentative pathways that lead to the production of organic acids, acetyl-CoA, alcohols, CO_2, and H_2 (Catalanotti et al. 2013). Hydrogenases play an important role in the fermentative pathways that produce hydrogen (Das and Veziroglu 2001). The enzyme pyruvate ferredoxin oxidoreductase oxidizes pyruvate to acetyl-CoA and CO_2, generating reduced ferredoxin. Hydrogenases use the reduced ferredoxin to generate H_2 (Melis et al. 2000; Catalanotti et al. 2013). As there are several competing fermentative pathways that consume pyruvate, hydrogen yield is typically low under dark fermentation conditions.

9.3.2.2 Hydrogenase Activity in Cyanobacteria

Cyanobacteria have multiple enzymes involved in H_2 metabolism including [FeFe] and [NiFe] hydrogenases, as well as nitrogenases (Madamwar et al. 2000; Schütz et al. 2004). Several cyanobacteria strains have evolved different strategies to protect these oxygen sensitive enzymes. Furthermore, most cyanobacteria have [NiFe] hydrogenase or bidirectional [NiFe] hydrogenase, which are more tolerant to oxygen (Carrieri et al. 2011).

Cournac et al. (2004) showed that the bidirectional [NiFe] hydrogenase in *Synechocystis* sp. (PCC 6803) was insensitive to light but was reversibly inactivated by O_2. Reactivation of the enzyme was shown to occur quickly in the presence of NADH or NADPH. Interestingly, the analysis of this enzyme showed that the enzyme was constitutively expressed in the presence of oxygen during the light period but was activated under dark anoxic conditions, which is different from the *C. reinhardtii* hydrogenase enzyme that is produced and activated only under anoxic conditions.

Biohydrogen production in microalgae and cyanobacteria under various environmental conditions is given in Table 9.2.

9.3.2.3 Hydrogenase Activity in Heterotrophic Bacteria

Hydrogenases are widely found in both anaerobic and aerobic bacteria, including methanogenic, acetogenic, and nitrate- and sulfate-reducing bacteria. A multiplicity of hydrogenase forms has been found in bacteria, highlighting the central role that hydrogenases play in cell energetics (Calusinska et al. 2010; Kothari et al. 2012).

In micro-aerophilic environments, Clostridia are dominant in mixed communities capable of hydrogen production and are able to produce hydrogen via butyric and mixed acid fermentations. [FeFe] and [NiFe] hydrogenases have been identified in most *Clostridium* strains (Calusinska et al. 2010). During methanogenesis, several types of [NiFe]-hydrogenases are involved (Enzmann et al. 2018). [NiFe] hydrogenases in bacteria comprise a minimum of two subunits: the small and the large subunits. [NiFeSe]-hydrogenases are a subfamily of the [NiFe]-hydrogenase with a selenocysteine in the active site that is found in some *Desulfovibrio* species (Baffert et al. 2019). Hydrogen production through the activity of hydrogenases has also been reported in facultative anaerobes (e.g., *Escherichia coli*, *Enterobacter*, and *Citrobacter*), and some aerobic bacteria (e.g., *Alcaligenes* and *Bacillus*) (Kothari et al. 2012).

The diversity of hydrogenases in heterotrophic bacteria can be exploited to facilitate the conversion of microalgal biomass into hydrogen, even in microalgal strains that lack native hydrogenase activity. A combination of microalgae cultivation and dark fermentation has been shown to be a promising pathway to produce hydrogen (Ren et al. 2015; Wang and Yin 2018).

9.4 Factors Affecting Hydrogen Production

In a two-stage process for hydrogen production involving a photosynthesis stage and a dark fermentation stage, net hydrogen production is determined by factors affecting biomass production and the storing of reserve polysaccharides in the first stage as well as the activity of hydrogen-producing pathways in the second stage.

Growth rate and biomass productivity, as well as accumulation of reserve molecules in microalgal and cyanobacterial cultures are mainly affected by photoperiod, light intensity, nutrient availability, and temperature. High cell density and a high carbohydrate content helps to increase H_2 yield during dark fermentation, when the carbohydrate storages serve as feedstock. A higher initial carbohydrate content will result in increased total H_2 production. The activity of hydrogenases and nitrogenases is, as previously discussed, also affected by light, oxygen concentration, and nutrient deficiency. A proper understanding of these effects is needed to enable increased hydrogen production.

9.4.1 Photoperiod and Light Intensity

Light intensity and the duration of light/dark cycles, or photoperiod, play a major role in photosynthesis, cell growth rate, and cell composition (Laurinavichene et al. 2004; Vítová et al. 2011). To achieve higher cell density and increased carbohydrate content, photosynthetic efficiency must be increased. This can be achieved by the manipulation of light intensity and light/dark cycle (Iasimone et al. 2018). Increasing light intensity stimulates the production of dihydroxyacetone phosphate, a precursor for glucose and starch synthesis (Champigny 1985), thus affecting positively the accumulation of carbohydrates. Unfortunately, strong light intensity leads to high dissipation rated (>80%) of light energy as unused thermal energy through the activation of photoprotective mechanisms (such as non-photochemical quenching), diminishing overall photosynthetic efficiency. This high-energy dissipation is caused by a mismatch between the absorption and utilization of energy during dark reactions (Murata et al. 2007; Masojídek et al. 2010; Keren and Krieger-Liszkay 2011). Light management can increase photosynthetic efficiency by avoiding photoinhibition, while avoiding low photon fluxes that may lead to high chlorophyll content per reaction center and result in lower photochemical reaction rates (Strasser and Srivastava 1995).

Cheng et al. (2017) reported increased biomass production and carbohydrate accumulation when the light intensity was raised from 500 to 1800 μmol photon m^{-2} s^{-1}. In another study, a significant increase in carbohydrate composition was reported by the increase of light intensity under continuous illumination. The increase continued until the photo saturation of the photosynthetic system. When the light intensity increases above the saturation level, a decrease in photosynthesis and cell growth occurs due to photoinhibition (Ho et al. 2012). By increasing mean light intensity from 215 to 330 μmol photon $m^{-2s\,-\,1}$, Fernandes et al. (2010) observed an increase in starch accumulation in *Chlorella vulgaris* from 8.5 to 40% (dry mass basis).

9.4.2 Temperature

Temperature directly affects the microalgae and cyanobacteria growth rate and H_2 production capacity (Cui et al. 2021; Gonçalves et al. 2021). Temperature changes the specific reaction rates in all metabolic pathways, potentially altering fluxes distribution. Increased H_2 production in cyanobacteria and microalgae has been reported as temperature increases (Jabri et al. 2021). The optimal temperature for each organism varies, but it is typically in the range of 15–60 °C (Varshney et al. 2015; Patel et al. 2019). Temperatures between 20 and 30 °C have been used in many studies to evaluate and further increase H_2 production (Bernard and Rémond 2012; Ras et al. 2013). Temperatures outside this range have been found to result in lower H_2 production.

9.4.3 Nutrient Availability

Growth rate and cell composition are highly dependent on the concentration of essential nutrients. Culture medium formulation needs to be optimized to promote the activation and upregulation of metabolic pathways that favor the accumulation of desired metabolites. Accumulation of carbohydrates, lipids, and proteins can be achieved by selectively creating specific nutrient deficiencies (Paliwal et al. 2017) or through the addition of chemical elicitors to the medium (Uprety et al. 2021). For hydrogen production, it is desirable to maximize the polysaccharides content, as these are easier to ferment. Nitrogen, phosphorus, and sulfur deprivation have been shown to have a direct effect on photosynthetic electron transport chain.

9.4.3.1 Nitrogen

Nitrogen is required for cell growth and plays a key function in protein synthesis (Arumugam et al. 2013; Yaakob et al. 2021). Under nitrogen-deficient conditions, the photosynthetically fixed carbon flux is redirected from protein synthesis to the biosynthesis of reserve components (lipid or carbohydrates) (Huo et al. 2011; Bellou and Aggelis 2013; Debnath et al. 2021). As the synthesis of both triacylglycerol and starch use the same precursors, it has been hypothesis that the pathways leading to the accumulation of lipids and carbohydrates are competitive (Nagappan et al. 2020), with evidence pointing to an early induction of carbohydrate synthesis followed by late induction of neutral lipid biosynthesis under nitrogen deficiency (Debnath et al. 2021).

Moreover, as nitrogen deficiency have a negative effect on protein synthesis it leads to a gradual decline in growth and cell lysis (Eriksen et al. 2007); thus, it is not surprising that reports on the effect of nitrogen deficiency on carbohydrate accumulation are mixed. While nitrogen deprivation resulted in no change in carbohydrate content in *Nannochloropsis oceanica* and *Phaeodactylum tricornutum* (Valenzuela et al. 2012; Xiao et al. 2012), a significant increase in the rate of synthesis and total accumulation of carbohydrates have been reported in cultures of *Chlorella zofingiensis*, *Porphyridium purpureum*, *Chlorella vulgaris*, and *Tetraselmis suecica* (D'Souza and Kelly 2000; Zhu et al. 2014; Li et al. 2019).

Nitrogen depletion also regulates the expression and activity of hydrogen-producing enzymes. As nitrogen depletion reduces photosynthetic activity, it promotes the formation of an anoxic environment that activates the H_2 production mechanism (Azwar et al. 2014). In heterocyst cyanobacteria such as *Anabaena*, the formation of cysts is downregulated by the presence of nitrogen (Pratte and Thiel 2014), which also inhibits nitrogenase activity.

9.4.3.2 Phosphorus

Phosphorus is an essential nutrient needed for the synthesis of nucleotides, proteins, and phospholipids. Phosphorus deficiency in microalgae and cyanobacteria triggers a decrease in protein synthesis and the accumulation of lipids and carbohydrates, although this effect is less pronounced than in the case of nitrogen deficiency (Lakatos et al. 2019). Phosphorus starvation significantly enhanced carbohydrate accumulation in both green microalgae and cyanobacteria. The final carbohydrate content in phosphorus-deficient cultures has been reported as high as 55% (dry biomass basis) in *Chlorella vulgaris* (Brányiková et al. 2011), 63% in *Arthrospira platensis* (Markou et al. 2012), 44.5% in *Leptolyngbya limnetica*, and 40.4% in *Oscillatoria obscura* (Kushwaha et al. 2018).

Phosphorus deficiency also has been shown to promote hydrogenase activation in *Chlorella* sp. (Batyrova et al. 2015). Although the exact mechanism for activation in this case is still not clear, it can be hypothesized that as phosphorus plays a similar role as nitrogen in downregulating protein production it likely results in a reduction of the photosynthetic activity. This in turn will result in a transition to anaerobic conditions, activation of hydrogenase enzymes, and consequent H_2 production.

9.4.3.3 Sulfur

Sulfur is another essential nutrient with an important role in photosynthesis. Sulfur deprivation results in decreased chlorophyll content, inhibiting the repair of PSII proteins that get damaged by light and ultimately stopping photosynthetic activity (Laurinavichene et al. 2004; Eroglu and Melis 2011). In the absence of sulfur, O_2 evolution is initially reduced, until it completely stops. Starch overproduction has been reported under sulfur limitation conditions in *C. reinhardtii*, *C. vulgaris*, and *Tetraselmis* (Melis 2007; Brányiková et al. 2011; Yao et al. 2012). In the case of *C. reinhardtii*, a tenfold increase in starch content was reported under sulfur-deficient conditions leading to significant gains in hydrogen production during fermentation (Melis 2007). Recovery from sulfur starvation is, however, very fast. Thus, sulfur starvation is considered one of the most feasible nutrient control strategies to manage biomass composition in microalgal cultures (Lakatos et al. 2019).

By affecting photosynthetic activity, sulfur limitation can help also to upregulate the expression and activity of hydrogenases. Jo et al. (2006) adjusted the concentration of media components and observed improved H_2 generation by increasing nitrogen, phosphate, using *C. reinhardtii* under sulfur-deficient conditions. *Synechocystis* sp. strain PCC 6803 has been shown also to perform well under sulfur depletion (Cournac et al. 2002). However, in the case of *Aphanothece halophytica* sulfur deprivation reduced cell growth and hydrogenase activity (Taikhao et al. 2013).

9.5 Strategies to Improve Hydrogen Production

9.5.1 Bioreactor Design

Reactor design and operation should be optimized in terms of light intensity and orientation, photoperiod, CO_2 and O_2 mass transfer, mixing, and temperature profile, all of which have been shown to affect H_2 production (Nagarajan et al. 2017, 2021).

Photobioreactors (PBRs) are the cultivation systems where microalgae and cyanobacteria biomass are grown. They provide many advantages such as protection from contamination by unwanted microorganisms and a more controlled environment (Markou 2020). Their design is based on surface area and light distribution to efficiently deliver enough light to promote cell growth (Legrand et al. 2021; Ranganathan et al. 2022). Therefore, a large surface-to-volume ratio is beneficial for increased cell growth and biomass accumulation (Deprá et al. 2019). Control of photoperiod and dark conditions, and removal of oxygen or nitrogen, however, are needed to activate hydrogenase or nitrogenase enzyme (González-Camejo et al. 2019; Vélez-Landa et al. 2021). Closed PBR configurations, such as vertical or horizontal tubular, flat-plate, or helical-tubular reactors, can be useful and efficient in creating oxygen and/or nitrogen-free systems (Tamburic et al. 2011; Giannelli and Torzillo 2012; Torzillo and Chini Zittelli 2015).

Stirred tank reactors (STRs) are good for ensuring effective homogeneous mixing, high mass transfer rates, and better control of most of the process parameters (Brindhadevi et al. 2021). SRTs can be used for both microalgae growth and H_2 production under dark and anoxic conditions (Krishnan et al. 2019). STRs can be efficient in inducing and controlling anoxic and dark conditions to initiate fermentative pathways in microalgae or cyanobacteria.

A two-stage process that involves the integration of PBRs and STRs can be beneficial to increase H_2 production. In the first stage, microalgae and cyanobacteria biomass is grown under phototrophic conditions, and then it is transferred to the second reactor, where dark and anoxic conditions are applied to induce hydrogenase activity for H_2 production. Moreover, cells may be recycled back to the first stage for regrowth. Fedorov et al. (2005) demonstrated a two-stage H_2 production system with a maximum of 0.58 ml H_2 L^{-1} for more than 4000 h, growing *C. reinhardtii* in an aerobic environment and then transferring the biomass to a tank reactor operating under hypoxic conditions for H_2 production.

9.5.2 Genetic Modifications

Genetic engineering of microalgal or cyanobacterial strains can potentially enhance H_2 production under oxic conditions (Anwar et al. 2019). The target for genetic manipulations is to (1) decrease oxygen sensitivity of H_2 producing enzymes, (2) reduce the activity of uptake hydrogenases that consume but do not produce hydrogen, and (3) eliminate or downregulate competing pathways (Kruse and Hankamer 2010; Li et al. 2022).

It has been suggested that complete or partial inactivation of the PSII system with genetic modification can prevent oxygen inhibition and lead to increased H_2 production. This hypothesis is supported by the increased H_2 production under sulfur-deprived conditions. Lin et al. (2013) reported partial inactivation of PSII in *Chlorella* sp. by knocking down the PSBO subunit, which is linked to low photosynthetic activity. They reported that photobiological H_2 production increased by as much as tenfold compared to the wild type.

Uptake hydrogenases consume H_2, thus reducing net H_2 production. A 4.7-fold increase in H_2 production, compared to wild type under the same conditions, was obtained by knocking down the gene expressing uptake hydrogenase in the cyanobacterium *Anabaena* sp. PCC 7120, resulting in a hydrogen production rate of 101.33 μmol H_2 mg chl a^{-1} h^{-1} (Khetkorn et al. 2012; Shahkouhi and Motamedian 2020).

Another genetic engineering strategy for enhancing H_2 production is the elimination of competing metabolic pathways (Yacoby et al. 2011). Typically, hydrogen yield is less than one-third the maximum theoretical yield, as the substrate is used to form other fermentation products, such as acetate, acetone, etc. (Hallenbeck et al. 2012). Reserve molecules, like sugars, are broken down by either glycolysis or the oxidative pentose pathway (OPP). For hydrogen production, the OPP is the most efficient metabolic pathway, as it degrades the substrate all the way into H_2 and carbon dioxide (Stincone et al. 2015). Thus, it is beneficial to redirect the electrons to the OPP pathway rather than through glycolysis, which uses electrons to produce other products (Kim et al. 2011). The difference between the two pathways results in three times higher H_2 production from the OPP pathway. Therefore, it has been proposed that enhancing the flux through the OPP pathway may help enhance H_2 production (Skizim et al. 2012). McNeely et al. (2014) reported increased NADPH and H_2 production by partially redirecting the electrons to the OPP pathway in *Synechococcus* 7002. Khetkorn et al. (2012) investigated the inhibition of competing metabolic pathways to redirect the electron flow toward nitrogenase and bidirectional Hox-hydrogenase in *Anabaena siamensis* TISTR 8012. They reported a threefold increase in H_2 production with maximum hydrogen production rate of 22 μmol H_2 mg chl a^{-1} h^{-1}. Increased H_2 production was also reported after inhibition of the respiratory terminal oxidases (ΔctaI, ΔctaII, and Δcyd) in *Synechocystis* PCC 6803, which induced higher activity of bidirectional Hox-hydrogenase (Gutthann et al. 2007). Similarly, the interruption of the nitrate assimilation pathway in *Synechocystis* PCC 6803 resulted in higher H_2 production (Baebprasert et al. 2011).

9.5.3 Process Integration

Different process integrations with microalgae and cyanobacteria cultivation have been proposed to enhance H_2 production. One of the most promising strategies is co-cultivating microalgae and bacteria or using natural microbial communities. This strategy simultaneously results in both microalgal biomass production and oxygen generation, while the produced biomass and oxygen are consumed by bacteria creating an anoxic environment that enables hydrogen production. Demirkaya et al. (2022) demonstrated hydrogen production in a cyanobacterial community dominated by the cyanobacterium *Candidatus* "Phormidium alkaliphilum," simultaneously producing hydrogen, organic acids, and other valuable products under both hypoxic and anoxic conditions. This cyanobacterial community produced hydrogen with a maximum yield of 326.14 μmol g^{-1} ash-free dry weight. Hydrogen production occurred even with some oxygen initially present in the fermentation bioreactor. Fakhimi and Tavakoli (2019) co-cultured the green microalgae *C. reinhardtii* with either *Escherichia coli*, *Pseudomonas stutzeri*, or *Pseudomonas putida*, increasing hydrogen production up to 24%, 46%, and 32%, respectively. The increased hydrogen production was due to the bacteria rapidly consuming the oxygen produced by *C. reinhardtii*, creating in this way an oxygen-deficient environment that promoted hydrogen evolution. *C. reinhardtii* growth was also affected by the presence of the bacteria.

Another strategy is using microalgal and cyanobacterial biomass as feedstock for dark fermentation with heterotrophic bacteria. This process can also be combined with anaerobic digestion to utilize waste products to further obtain other energy products. Xia et al. (2014) proposed a three-stage process that integrates dark fermentation, photofermentation, and anaerobic digestion using *Chlorella pyrenoidosa* cultures and cassava starch to enhance hydrogen production. They reported a higher hydrogen production rate of 31.96 mL h^{-1} g^{-1} (Total volatile solids basis) compared with those from only microalgae or cassava starch alone. The three-stage hydrogen production system resulted in hydrogen yield of 664.2 mL g^{-1} and a methane yield of 126.0 mL g^{-1} (Total volatile solids basis).

9.6 Conclusion

Biological hydrogen production through indirect biophotolysis is a promising pathway for efficient and sustainable hydrogen production. Hydrogen yield and production rate can be enhanced by integration of microbial photosynthesis and dark fermentation, either by exploiting the natural hydrogen production ability of some microalgae and cyanobacteria or by co-culturing or sequential culturing with fermentative bacteria. With the current technological advances in genetic and metabolic engineering, the creation of oxygen-tolerant hydrogenases and the elimination of competitive metabolic pathways can further improve hydrogen production. The recovery and purification of other valuable products from dark fermentation, such organic acids or other energy products from anaerobic digestion, can improve the sustainability and overall cost-effectiveness of biological hydrogen production.

References

Acar C, Dincer I (2018) Hydrogen production. In: Dincer I (ed) Comprehensive energy systems. Elsevier, Amsterdam, pp 1–40

Aikawa S, Izumi Y, Matsuda F, Hasunuma T, Chang JS, Kondo A (2012) Synergistic enhancement of glycogen production in *Arthrospira platensis* by optimization of light intensity and nitrate supply. Bioresour Technol 108:211–215

Akhlaghi N, Najafpour-Darzi G (2020) A comprehensive review on biological hydrogen production. Int J Hydrogen Energ 45(43): 22492–22512

Alalayah WM, Al-Zahrani A, Edris G, Demirbas A (2017) Kinetics of biological hydrogen production from green microalgae *Chlorella vulgaris* using glucose as initial substrate. Energy Sources A Recov Util Environ Eff 39(12):1210–1215

All I, Basit MA (1993) Significance of hydrogen content in fuel combustion. Int J Hydrogen Energ 18(12):1009–1011

Ananyev GM, Skizim NJ, Dismukes GC (2012) Enhancing biological hydrogen production from cyanobacteria by removal of excreted products. J Biotechnol 162(1):97–104

Andrić I, Pina A, Ferrão P, Fournier J, Lacarrière B, Le Corre O, Taikhao S, Phunpruch S (2017) Increasing hydrogen production efficiency of N_2-fixing cyanobacterium *Anabaena siamensis* TISTR 8012 by cell immobilization. Energy Procedia 138:366–371

Anwar M, Lou S, Chen L, Li H, Hu Z (2019) Recent advancement and strategy on bio-hydrogen production from photosynthetic microalgae. Bioresour Technol 292:121972

Arumugam M, Agarwal A, Arya MC, Ahmed Z (2013) Influence of nitrogen sources on biomass productivity of microalgae *Scenedesmus bijugatus*. Bioresour Technol 131:246–249

Azwar MY, Hussain MA, Abdul-Wahab AK (2014) Development of biohydrogen production by photobiological, fermentation and electrochemical processes: a review. Renew Sust Energ Rev 31:158–173

Baebprasert W, Jantaro S, Khetkorn W, Lindblad P, Incharoensakdi A (2011) Increased H_2 production in the cyanobacterium *Synechocystis* sp. strain PCC 6803 by redirecting the electron supply via genetic engineering of the nitrate assimilation pathway. Metab Eng 13(5): 610–616

Baffert C, Kpebe A, Avilan L, Brugna M (2019) Hydrogenases and H_2 metabolism in sulfate-reducing bacteria of the *Desulfovibrio* genus. In: Poole RK (ed) Advances in microbial physiology, vol 74. Academic Press, London, pp 143–189

Batyrova K, Gavrisheva A, Ivanova E, Liu J, Tsygankov A (2015) Sustainable hydrogen photoproduction by phosphorus-deprived marine green microalgae *Chlorella* sp. Int J Mol Sci 16(2): 2705–2716

Bauer C, Treyer K, Antonini C, Bergerson J, Gazzani M, Gencer E, Gibbins J, Mazzotti M, McCoy ST, McKenna R, Pietzcker R, Ravikumar AP, Romano MC, Ueckerdt F, Vente J, van der Spek M (2021) On the climate impacts of blue hydrogen production. Sustain Energy Fuels 6(1):66–75

Bellou S, Aggelis G (2013) Biochemical activities in *Chlorella* sp. and *Nannochloropsis salina* during lipid and sugar synthesis in a lab-scale open pond simulating reactor. J Biotechnol 164:318–329

Bernard O, Rémond B (2012) Validation of a simple model accounting for light and temperature effect on microalgal growth. Bioresour Technol 123:520–527

Bhagat AK, Buium H, Shmul G, Alfonta L (2020) Genetically expanded reactive-oxygen-tolerant alcohol dehydrogenase II. ACS Catal 10(5):3094–3102

Bingham AS, Smith PR, Swartz JR (2012) Evolution of an [FeFe] hydrogenase with decreased oxygen sensitivity. Int J Hydrogen Energ 37(3):2965–2976

Brányiková I, Maršálková B, Doucha J, Brányik T, Bišová K, Zachleder V, Vítová M (2011) Microalgae – novel highly efficient starch producers. Biotechnol Bioeng 108:766–776

Brentner LB, Jordan PA, Zimmerman JB (2010) Challenges in developing biohydrogen as a sustainable energy source: implications for a research agenda. Environ Sci Technol 44(7):2243–2254

Brindhadevi K, Shanmuganathan R, Pugazhendhi A, Gunasekar P, Manigandan S (2021) Biohydrogen production using horizontal and vertical continuous stirred tank reactor – a numerical optimization. Int J Hydrogen Energ 46(20):11305–11312

Buitrón G, Carrillo-Reyes J, Morales M, Faraloni C, Torzillo G (2017) Biohydrogen production from microalgae. In: Gonzalez-Fernandez-C, Muñoz R (eds) Microalgae-based biofuels and bioproducts from feedstock cultivation to end-products. Elsevier, Duxford, pp 209–234

Calusinska M, Happe T, Joris B, Wilmotte A (2010) The surprising diversity of clostridial hydrogenases: a comparative genomic perspective. Microbiology 156:1575–1588

Carrieri D, Wawrousek K, Eckert C, Yu J, Maness PC (2011) The role of the bidirectional hydrogenase in cyanobacteria. Bioresour Technol 102(18):8368–8377

Catalanotti C, Yang W, Posewitz MC, Grossman AR (2013) Fermentation metabolism and its evolution in algae. Front Plant Sci 4:1–17

Champigny M (1985) Regulation of photosynthetic carbon assimilation at the cellular level: a review. Photosynth Res 6:273–286

Chandrasekhar K, Lee YJ, Lee DW (2015) Biohydrogen production: strategies to improve process efficiency through microbial routes. Int J Mol Sci 16(4):8266

Cheng D, Li D, Yuan Y, Zhou L, Li X, Wu T, Wang L, Zhao Q, Wei W, Sun Y (2017) Improving carbohydrate and starch accumulation in *Chlorella* sp. AE10 by a novel two-stage process with cell dilution. Biotechnol Biofuels 10(1):75

Compaoré J, Stal LJ (2010) Oxygen and the light–dark cycle of nitrogenase activity in two unicellular cyanobacteria. Environ Microbiol 12: 54–62

Cournac L, Mus F, Bernard L, Guedeney G, Vignais P, Peltier G (2002) Limiting steps of hydrogen production in *Chlamydomonas reinhardtii* and *Synechocystis* PCC 6803 as analysed by light-induced gas exchange transients. Int J Hydrogen Energ 27(11–12): 1229–1237

Cournac L, Guedeney G, Peltier G, Vignais PM (2004) Sustained photoevolution of molecular hydrogen in a mutant of *Synechocystis* sp. strain PCC 6803 deficient in the type I NADPH-dehydrogenase complex. J Bacteriol 186(6):1737

Cui X, Yang J, Cui M, Zhang W, Zhao J (2021) Comparative experiments of two novel tubular photobioreactors with an inner aerated tube for microalgal cultivation: enhanced mass transfer and improved biomass yield. Algal Res 58:102364

Das D, Veziroglu N (2001) Hydrogen production by biological processes: a survey of literature. Int J Hydrogen Energ 26(1):13–28

Dawood F, Anda M, Shafiullah GM (2020) Hydrogen production for energy: an overview. Int J Hydrogen Energ 45(7):3847–3869

Debnath C, Bandyopadhyay TK, Bhunia B, Mishra U, Narayanasamy D, Muthuraj M (2021) Microalgae: sustainable resource of carbohydrates in third-generation biofuel production. Renew Sust Energ Rev 150:111464

Demirkaya C, Vadlamani A, Tervahauta T, Strous M, De la Hoz SH (2022) Co-production of hydrogen and organic acids from alkaline cyanobacterial biomass via dark autofermentation. SSRN Electron J. https://doi.org/10.2139/ssrn.4082772

Deprá MC, Mérida LGR, de Menezes CR, Zepka LQ, Jacob-Lopes E (2019) A new hybrid photobioreactor design for microalgae culture. Chem Eng Res Des 144:1–10

Dincer I (2012) Green methods for hydrogen production. Int J Hydrogen Energ 37(2):1954–1971

D'Souza FM, Kelly GJ (2000) Effects of a diet of a nitrogen-limited alga (*Tetraselmis suecica*) on growth, survival and biochemical composition of tiger prawn (*Penaeus semisulcatus*) larvae. Aquaculture 181:311–329

Enzmann F, Mayer F, Rother M, Holtmann D (2018) Methanogens: Biochemical background and biotechnological applications. AMB Express 8:1

Eriksen NT, Riisgård FK, Gunther WS, Iversen JJL (2007) On-line estimation of O_2 production, CO_2 uptake, and growth kinetics of microalgal cultures in a gas-tight photobioreactor. J Appl Phycol 19: 161

Eroglu E, Melis A (2011) Photobiological hydrogen production: recent advances and state of the art. Bioresour Technol 102(18):8403–8413

Fakhimi N, Tavakoli O (2019) Improving hydrogen production using co-cultivation of bacteria with *Chlamydomonas reinhardtii* microalga. Mater Sci Energ Technol 2(1):1–7

Fedorov AS, Kosourov S, Ghirardi ML, Seibert M (2005) Continuous hydrogen photoproduction by *Chlamydomonas reinhardtii*. Appl Biochem Biotechnol 121(1):403–412

Fernandes BD, Dragone GM, Teixeira JA, Vicente AA (2010) Light regime characterization in an airlift photobioreactor for production of microalgae with high starch content. Appl Biochem Biotechnol 161:218–226

Gatamaneni BL, Orsat V, Lefsrud M (2018) Factors affecting growth of various microalgal species. Environ Eng Sci 35(10):1037–1048

Giannelli L, Torzillo G (2012) Hydrogen production with the microalga *Chlamydomonas reinhardtii* grown in a compact tubular photobioreactor immersed in a scattering light nanoparticle suspension. Int J Hydrogen Energ 37(22):16951–16961

Gonçalves AL, Almeida F, Rocha FA, Ferreira A (2021) Improving CO_2 mass transfer in microalgal cultures using an oscillatory flow reactor with smooth periodic constrictions. J Environ Chem Eng 9(6): 106505

González-Camejo J, Viruela A, Ruano MV, Barat R, Seco A, Ferrer J (2019) Effect of light intensity, light duration and photoperiods in the performance of an outdoor photobioreactor for urban wastewater treatment. Algal Res 40:101511

Goria K, Kothari R, Singh HM, Singh A, Tyagi VV (2022) Biohydrogen: potential applications, approaches, and hurdles to overcome. In: Sahay S (ed) Handbook of biofuels. Elsevier, London, pp 399–418

Goris T, Wait AF, Saggu M, Fritsch J, Heidary N, Stein M, Zebger I, Lendzian F, Armstrong FA, Friedrich B, Lenz O (2011) A unique iron-sulfur cluster is crucial for oxygen tolerance of a [NiFe]-hydrogenase. Nat Chem Biol 7(5):310–318

Gutekunst K, Hoffmann D, Westernströer U, Schulz R, Garbe-Schönberg D, Appel J (2018) In-vivo turnover frequency of the cyanobacterial NiFe-hydrogenase during photohydrogen production outperforms in-vitro systems. Sci. Reports 8(1):1–10

Gutthann F, Egert M, Marques A, Appel J (2007) Inhibition of respiration and nitrate assimilation enhances photohydrogen evolution under low oxygen concentrations in *Synechocystis* sp. PCC 6803. Biochim Biophys Acta Bioenerg 1767(2):161–169

Hallenbeck PC, Benemann JR (2002) Biological hydrogen production; fundamentals and limiting processes. Int J Hydrogen Energ 27(11–12):1185–1193

Hallenbeck PC, Abo-Hashesh M, Ghosh D (2012) Strategies for improving biological hydrogen production. Bioresour Technol 110: 1–9

Hatano J, Kusama S, Tanaka K, Kohara A, Miyake C, Nakanishi S, Shimakawa G (2022) NADPH production in dark stages is critical for cyanobacterial photocurrent generation: a study using mutants deficient in oxidative pentose phosphate pathway. Photosynth Res 1: 1–8

Hemschemeier A, Fouchard S, Cournac L, Peltier G, Happe T (2008) Hydrogen production by *Chlamydomonas reinhardtii*: an elaborate interplay of electron sources and sinks. Planta 227(2):397–407

Hermesmann M, Müller TE (2022) Green, turquoise, blue, or grey? Environmentally friendly hydrogen production in transforming energy systems. Prog Energy Combust Sci 90:100996

Ho SH, Chen CY, Chang JS (2012) Effect of light intensity and nitrogen starvation on CO_2 fixation and lipid/carbohydrate production of an indigenous microalga *Scenedesmus obliquus* CNW-N. Bioresour Technol 113:244–252

Huesemann MH, Hausmann TS, Carter BM, Gerschler JJ, Benemann JR (2010) Hydrogen generation through indirect biophotolysis in batch cultures of the nonheterocystous nitrogen-fixing cyanobacterium *Plectonema boryanum*. Appl Biochem Biotechnol 162:208–220

Huo Y-X, Cho KM, Rivera JGL, Monte E, Shen CR, Yan Y, Liao JC (2011) Conversion of proteins into biofuels by engineering nitrogen flux. Nat Biotechnol 29:346

Iasimone F, Panico A, De Felice V, Fantasma F, Iorizzi M, Pirozzi F (2018) Effect of light intensity and nutrients supply on microalgae cultivated in urban wastewater: Biomass production, lipids accumulation and settleability characteristics. J Environ Manag 223:1078–1085

IEA (2021) Global Energy Review 2021. International Energy Agency, Paris. https://www.iea.org/reports/global-energy-review-2021

Jabri H, Taleb A, Touchard R, Saadaoui I, Goetz V, Pruvost J (2021) Cultivating microalgae in desert conditions: evaluation of the effect of light-temperature summer conditions on the growth and metabolism of nannochloropsis QU130. Appl Sci 11(9):3799

Jiménez-Llanos J, Ramírez-Carmona M, Rendón-Castrillón L, Ocampo-López C (2020) Sustainable biohydrogen production by *Chlorella* sp. microalgae: a review. Int J Hydrog Energy 45(15):8310–8328

Jo JH, Lee DS, Park JM (2006) Modeling and optimization of photosynthetic hydrogen gas production by green Alga *Chlamydomonas reinhardtii* in sulfur-deprived circumstance. Biotechnol Prog 22(2): 431–437

Kamran M (2021) Bioenergy. In: Kamran M, Rayyan M (eds) Renewable energy conversion systems. Elsevier, London, pp 243–264

Keren N, Krieger-Liszkay A (2011) Photoinhibition: molecular mechanisms and physiological significance. Physiol Plant 142:1–5

Khetkorn W, Lindblad P, Incharoensakdi A (2012) Inactivation of uptake hydrogenase leads to enhanced and sustained hydrogen production with high nitrogenase activity under high light exposure in the cyanobacterium *Anabaena siamensis* TISTR 8012. J Biol Eng 6(1):1

Khetkorn W, Rastogi RP, Incharoensakdi A, Lindblad P, Madamwar D, Pandey A, Larroche C (2017) Microalgal hydrogen production – a review. Bioresour Technol 243:1194–1206

Kim MS, Baek JS, Yun YS, Jun Sim S, Park S, Kim SC (2006) Hydrogen production from *Chlamydomonas reinhardtii* biomass using a two-step conversion process: anaerobic conversion and photosynthetic fermentation. Int J Hydrog Energy 31(6):812–816

Kim YM, Cho HS, Jung GY, Park JM (2011) Engineering the pentose phosphate pathway to improve hydrogen yield in recombinant *Escherichia coli*. Biotechnol Bioeng 108(12):2941–2946

Kosourov S, Patrusheva E, Ghirardi ML, Seibert M, Tsygankov A (2007) A comparison of hydrogen photoproduction by sulfur-deprived *Chlamydomonas reinhardtii* under different growth conditions. J Biotechnol 128(4):776–787

Kosourov S, Jokel M, Aro EM, Allahverdiyeva Y (2018) A new approach for sustained and efficient H_2 photoproduction by *Chlamydomonas reinhardtii*. Energy Environ Sci 11(6):1431–1436

Kossalbayev BD, Tomo T, Zayadan BK, Sadvakasova AK, Bolatkhan K, Alwasel S, Allakhverdiev SI (2020) Determination of the potential of cyanobacterial strains for hydrogen production. Int J Hydrog Energy 45(4):2627–2639

Kothari R, Singh DP, Tyagi VV, Tyagi SK (2012) Fermentative hydrogen production – an alternative clean energy source. Renew Sust Energ Rev 16(4):2337–2346

Krishnan S, Md Din MF, Taib SM, Nasrullah M, Sakinah M, Wahid ZA, Kamyab H, Chelliapan S, Rezania S, Singh L (2019) Accelerated two-stage bioprocess for hydrogen and methane production from palm oil mill effluent using continuous stirred tank reactor and microbial electrolysis cell. J Clean Prod 229:84–93

Kruse O, Hankamer B (2010) Microalgal hydrogen production. Curr Opin Biotechnol 21(3):238–243

Kumaraswamy GK, Guerra T, Qian X, Zhang S, Bryant DA, Dismukes GC (2013) Reprogramming the glycolytic pathway for increased hydrogen production in cyanobacteria: metabolic engineering of NAD+-dependent GAPDH. Energy Environ Sci 6(12):3722–3731

Kumari S, Nasr M, Kumar S (2017) Technological advances in biohydrogen production from microalgae. In: Gupta SK, Malik A, Bux F (eds) Algal biofuels: recent advances and future prospects. Springer, Singapore, pp 347–360

Kushwaha D, Upadhyay SN, Mishra PK (2018) Growth of cyanobacteria: optimization for increased carbohydrate content. Appl Biochem Biotechnol 184:1247–1262

Lakatos GE, Ranglová K, Manoel JC, Grivalský T, Kopecký J, Masojídek J (2019) Bioethanol production from microalgae polysaccharides. Folia Microbiol 64:627–644

Laurinavichene T, Tolstygina I, Tsygankov A (2004) The effect of light intensity on hydrogen production by sulfur-deprived *Chlamydomonas reinhardtii*. J Biotechnol 114(1–2):143–151

Legrand J, Artu A, Pruvost J (2021) A review on photobioreactor design and modelling for microalgae production. React Chem Eng 6(7): 1134–1151

Li T, Xu J, Wu H, Jiang P, Chen Z, Xiang W (2019) Growth and biochemical composition of *Porphyridium purpureum* SCS-02 under different nitrogen concentrations. Mar Drugs 17:124

Li S, Li F, Zhu X, Liao Q, Chang JS, Ho SH (2022) Biohydrogen production from microalgae for environmental sustainability. Chemosphere 291:132717

Lin HD, Liu BH, Kuo TT, Tsai HC, Feng TY, Huang CC, Chien LF (2013) Knockdown of PsbO leads to induction of HydA and production of photobiological H_2 in the green alga *Chlorella* sp. DT. Bioresour Technol 143:154–162

Lubitz W, Ogata H, Rüdiger O, Reijerse E (2014) Hydrogenases. Chem Rev 114(8):4081–4148

Lukey MJ, Roessler MM, Parkin A, Evans RM, Davies RA, Lenz O, Friedrich B, Sargent F, Armstrong FA (2011) Oxygen-tolerant [NiFe]-hydrogenases: the individual and collective importance of supernumerary cysteines at the proximal Fe-S cluster. J Am Chem Soc 133(42):16881–16892

Madamwar D, Garg N, Shah V (2000) Cyanobacterial hydrogen production. World J Microbiol Biotechnol 16(8):757–767

Markou G (2020) Overview of microalgal cultivation, biomass processing and application. In: Konur O (ed) Handbook of algal science, technology and medicine. Elsevier, London, pp 343–352

Markou G, Chatzipavlidis I, Georgakakis D (2012) Carbohydrates production and bio-flocculation characteristics in cultures of *Arthrospira (Spirulina) platensis*: improvements through phosphorus limitation process. Bioenergy Res 5:915–925

Masojídek J, Vonshak A, Torzillo G (2010) Chlorophyll fluorescence applications in microalgal mass cultures. In: Suggett D, Prášil O, Borowitzka M (eds) Chlorophyll a fluorescence in aquatic sciences: methods and applications. Developments in applied phycology, vol 4. Springer, Dordrecht, pp 277–292

McNeely K, Kumaraswamy GK, Guerra T, Bennette N, Ananyev G, Dismukes GC (2014) Metabolic switching of central carbon metabolism in response to nitrate: application to autofermentative hydrogen production in cyanobacteria. J Biotechnol 182–183(1): 83–91

Melis A (2007) Photosynthetic H_2 metabolism in *Chlamydomonas reinhardtii* (unicellular green algae). Planta 226:1075–1086

Melis A, Zhang L, Forestier M, Ghirardi ML, Seibert M (2000) Sustained photobiological hydrogen gas production upon reversible inactivation of oxygen evolution in the green alga *Chlamydomonas reinhardtii*. Plant Physiol 122:127–136

Murata N, Takahashi S, Nishiyama Y, Allakhverdiev SI (2007) Photoinhibition of photosystem II under environmental stress. Biochim Biophys Acta Bioenergetics 1767:414–421

Mus F, Dubini A, Seibert M, Posewitz MC, Grossman AR (2007) Anaerobic acclimation in *Chlamydomonas reinhardtii*: anoxic gene expression, hydrogenase induction, and metabolic pathways. J Biol Chem 282(35):25475–25486

Nagappan S, Devendran S, Tsai PC, Jayaraman H, Alagarsamy V, Pugazhendhi A, Ponnusamy VK (2020) Metabolomics integrated with transcriptomics and proteomics: evaluation of systems reaction to nitrogen deficiency stress in microalgae. Process Biochem 91:1–14

Nagarajan D, Lee DJ, Kondo A, Chang JS (2017) Recent insights into biohydrogen production by microalgae – from biophotolysis to dark fermentation. Bioresour Technol 227:373–387

Nagarajan D, Dong CD, Chen CY, Lee DJ, Chang JS (2021) Biohydrogen production from microalgae—Major bottlenecks and future research perspectives. Biotechnol J 16(5):2000124

Paliwal C, Mitra M, Bhayani K, Bharadwaj SVV, Ghosh T, Dubey S, Mishra S (2017) Abiotic stresses as tools for metabolites in microalgae. Bioresour Technol 244:1216–1226

Patel A, Matsakas L, Rova U, Christakopoulos P (2019) A perspective on biotechnological applications of thermophilic microalgae and cyanobacteria. Bioresour Technol 278:424–434

Peters JW, Schut GJ, Boyd ES, Mulder DW, Shepard EM, Broderick JB, King PW, Adams MWW (2015) [FeFe]- and [NiFe]-hydrogenase diversity, mechanism, and maturation. Biochim Biophys Acta, Mol Cell Res 1853(6):1350–1369

Pollock SV, Pootakham W, Shibagaki N, Moseley JL, Grossman AR (2005) Insights into the acclimation of *Chlamydomonas reinhardtii* to sulfur deprivation. Photosynth Res 86(3):475–489

Posewitz MC, Dubini A, Meuser JE, Seibert M, Ghirardi ML (2009) Hydrogenases, hydrogen production, and anoxia. In: Harris EH, Stern DB, and Witman GB (eds) *Chlamydomonas* sourcebook. Elsevier, Oxford, 3(2) pp. 217–255

Pratte BS, Thiel T (2014) Regulation of nitrogenase gene expression by transcript stability in the cyanobacterium *Anabaena variabilis*. J Bacteriol 196(20):3609–3621

Ranganathan P, Pandey AK, Sirohi R, Tuan Hoang A, Kim SH (2022) Recent advances in computational fluid dynamics (CFD) modelling of photobioreactors: design and applications. Bioresour Technol 350:126920

Ras M, Steyer JP, Bernard O (2013) Temperature effect on microalgae: a crucial factor for outdoor production. Rev Environ Sci Biotechnol 12(2):153–164

Razu MH, Hossain F, Khan M (2019) Advancement of bio-hydrogen production from microalgae. In: Alam MA, Wang Z (eds) Microalgae biotechnology for development of biofuel and wastewater treatment. Springer, Singapore, pp 423–462

Ren HY, Liu BF, Kong F, Zhao L, Ren NQ (2015) Sequential generation of hydrogen and lipids from starch by combination of dark fermentation and microalgal cultivation. RSC Adv 5(94): 76779–76782

Saleem M, Chakrabarti MH, Abdul Raman AA, Hasan DB, Ashri Wan Daud WM, Mustafa A (2012) Hydrogen production by *Chlamydomonas reinhardtii* in a two-stage process with and without illumination at alkaline pH. Int J Hydrogen Energ 37(6):4930–4934

Schütz K, Happe T, Troshina O, Lindblad P, Leitão E, Oliveira P, Tamagnini P (2004) Cyanobacterial H_2 production – a comparative analysis. Planta 218(3):350–359

Shahkouhi AM, Motamedian E (2020) Reconstruction of a regulated two-cell metabolic model to study biohydrogen production in a diazotrophic cyanobacterium *Anabaena variabilis* ATCC 29413. PLoS One 15(1):1–18

Show KY, Yan Y, Zong C, Guo N, Chang JS, Lee DJ (2019) State of the art and challenges of biohydrogen from microalgae. Bioresour Technol 289:121747

Singh N, Sarma S (2022) Biological routes of hydrogen production: a critical assessment. In: Sahay S (ed) Handbook biofuels. Elsevier, London, pp 419–434

Sivaramakrishnan R, Shanmugam S, Sekar M, Mathimani T, Incharoensakdi A, Kim SH, Parthiban A, Edwin Geo V, Brindhadevi K, Pugazhendhi A (2021) Insights on biological hydrogen production routes and potential microorganisms for high hydrogen yield. Fuel 291:120136

Skizim NJ, Ananyev GM, Krishnan A, Dismukes GC (2012) Metabolic pathways for photobiological hydrogen production by nitrogenase- and hydrogenase-containing unicellular cyanobacteria *Cyanothece*. J Biol Chem 287(4):2777–2786

Srirangan K, Pyne ME, Perry Chou C (2011) Biochemical and genetic engineering strategies to enhance hydrogen production in photosynthetic algae and cyanobacteria. Bioresour Technol 102(18): 8589–8604

Stincone A, Prigione A, Cramer T, Wamelink MMC, Campbell K, Cheung E, Olin-Sandoval V, Grüning NM, Krüger A, Tauqeer Alam M, Keller MA, Breitenbach M, Brindle KM, Rabinowitz JD, Ralser M (2015) The return of metabolism: biochemistry and physiology of the pentose phosphate pathway. Biol Rev 90(3):927–963

Strasser RJ, Srivastava A (1995) Polyphasic chlorophyll a fluorescence transient in plants and cyanobacteria. Photochem Photobiol 61:32–42

Taikhao S, Junyapoon S, Incharoensakdi A, Phunpruch S (2013) Factors affecting biohydrogen production by unicellular halotolerant cyanobacterium *Aphanothece halophytica*. J Appl Phycol 25(2):575–585

Tamburic B, Zemichael FW, Crudge P, Maitland GC, Hellgardt K (2011) Design of a novel flat-plate photobioreactor system for green algal hydrogen production. Int J Hydrogen Energ 36(11): 6578–6591

Torzillo G, Chini Zittelli G (2015) Tubular photobioreactors. In: Prokop A, Bajpai RK, Zappi ME (eds) Algal biorefineries Vol 2: Product and refinery design. Springer, Heidelberg, pp 187–212

Tsygankov AA, Kosourov SN, Tolstygina IV, Ghirardi ML, Seibert M (2006) Hydrogen production by sulfur-deprived *Chlamydomonas reinhardtii* under photoautotrophic conditions. Int J Hydrogen Energ 31(11):1574–1584

Uprety BK, Morrison EN, Emery N, Farrow SC (2021) Customizing lipids from oleaginous microbes: leveraging exogenous and endogenous approaches. Trends Biotechnol 40(4):482–508

Valenzuela J, Mazurie A, Carlson RP, Gerlach R, Cooksey KE, Peyton BM, Fields MW (2012) Potential role of multiple carbon fixation pathways during lipid accumulation in *Phaeodactylum tricornutum*. Biotechnol Biofuels 5(1):40

Varshney P, Mikulic P, Vonshak A, Beardall J, Wangikar PP (2015) Extremophilic micro-algae and their potential contribution in biotechnology. Bioresour Technol 184:363–372

Vélez-Landa L, Hernández-De León HR, Pérez-Luna YDC, Velázquez-Trujillo S, Moreira-Acosta J, Berrones-Hernández R, Sánchez-Roque Y (2021) Influence of light intensity and photoperiod on the photoautotrophic growth and lipid content of the microalgae *Verrucodesmus verrucosus* in a photobioreactor. Sustainability 13(12):6606

Vignais PM, Billoud B (2007) Occurrence, classification, and biological function of hydrogenases: an overview. Chem Rev 107(10): 4206–4272

Vítová M, Bišová K, Umysová D, Hlavová M, Kawano S, Zachleder V, Čížková M (2011) *Chlamydomonas reinhardtii*: duration of its cell cycle and phases at growth rates affected by light intensity. Planta 233(1):75–86

Volbeda A, Garcin E, Piras C, De Lacey AL, Fernandez VM, Hatchikian EC, Frey M, Fontecilla-Camps JC (1996) Structure of the [NiFe] hydrogenase active site: evidence for biologically uncommon Fe ligands. J Am Chem Soc 118(51):12989–12996

Wang J, Yin Y (2018) Fermentative hydrogen production using pretreated microalgal biomass as feedstock. Microb Cell Factories 17(1):22

Winkler M, Heil B, Heil B, Happe T (2002) Isolation and molecular characterization of the [Fe]-hydrogenase from the unicellular green alga *Chlorella fusca*. Biochim Biophys Acta Gene Struct Expr 1576(3):330–334

Wittkamp F, Senger M, Stripp ST, Apfel UP (2018) [FeFe]-Hydrogenases: recent developments and future perspectives. Chem Commun 54(47):5934–5942

Xia A, Cheng J, Ding L, Lin R, Song W, Zhou J, Cen K (2014) Enhancement of energy production efficiency from mixed biomass of *Chlorella pyrenoidosa* and cassava starch through combined hydrogen fermentation and methanogenesis. Appl Energy 120:23–30

Xiao Y, Zhang J, Cui J, Feng Y, Cui Q (2012) Metabolic profiles of *Nannochloropsis oceanica* IMET1 under nitrogen-deficiency stress. Bioresour Technol 130:731–738

Yaakob MA, Mohamed RMSR, Al-Gheethi A, Ravishankar GA, Ambati RR (2021) Influence of nitrogen and phosphorus on microalgal growth, biomass, lipid, and fatty acid production: an overview. Cell 10(2):393

Yacoby I, Pochekailov S, Toporik H, Ghirardi ML, King PW, Zhang S (2011) Photosynthetic electron partitioning between [FeFe]-hydrogenase and ferredoxin:NADP+-oxidoreductase (FNR) enzymes in vitro. Proc Natl Acad Sci USA 108(23):9396–9401

Yao C, Ai J, Cao X, Xue S, Zhang W (2012) Enhancing starch production of a marine green microalga *Tetraselmis subcordiformis* through nutrient limitation. Bioresour Technol 118:438–444

Yu M, Wang K, Vredenburg H (2021) Insights into low-carbon hydrogen production methods: green, blue and aqua hydrogen. Int J Hydrogen Energ 46(41):21261–21273

Zhu S, Wang Y, Huang W, Xu J, Wang Z, Xu J, Yuan Z (2014) Enhanced accumulation of carbohydrate and starch in *Chlorella zofingiensis* induced by nitrogen starvation. Appl Biochem Biotechnol 174:2435–2445

Induction of Carotenoid Synthesis in Microalgae with Reference to Their Production Outdoors

10

Cecilia Faraloni and Giuseppe Torzillo

Abstract

Plants, algae, and photosynthetic bacteria all contain carotenoids, which are lipid-soluble natural compounds. They can act as both light-harvesting complex and photoprotectors. Due to their nature, they are able to neutralize the effect of the presence of singlet oxygen and free radicals, acting as quenchers; for this function, an important and crucial role as an antioxidant has been attributed to a large number of carotenoids. Their production has been studied in several microalgal species, which represent a natural source of these antioxidants. In particular, *Haematococcus*, *Chlamydomonas*, *Chlorella*, *Dunaliella*, diatoms such as *Phaeodactylum* and *Isochrysis,* and dinoflagellates are able to synthesize large amounts of carotenoids. Among the most powerful antioxidant carotenoids, the xanthophylls loroxanthin, neoxanthin, lutein, violaxanthin, antheraxanthin, zeaxanthin, and α-carotene and β-carotene are the ones most synthesized under photo-oxidative stress conditions. Under physiological stresses, such as high light exposure, nutrient limitation-starvation, excessive low-high temperatures, the photosynthetic activity decreases, and different metabolic pathways are activated. The study of the physiological response to different stresses helps to understand the mechanisms which regulate the accumulation of antioxidant compounds. This information can be useful for optimizing the growth conditions of microalgal strains, the high carotenoid producers, for increasing their productivity, in terms of both antioxidants and biomass, and for the scale-up of the process from laboratory to outdoor cultures.

C. Faraloni (✉)
Institute of BioEconomy, CNR, Florence, Italy
e-mail: cecilia.faraloni@ibe.cnr.it

G. Torzillo
Institute of BioEconomy, CNR, Florence, Italy

Universidad de Costa Rica, Universidad de Costa Rica, Costa Rica 2060
San José, Costa Rica

Keywords

Carotenoids · Xanthophylls · Antioxidants · Microalgae · Stress · Photobioreactor

10.1 Introduction

Microalgae convert H_2O and CO_2 into organic compounds, by means of light energy, evolving oxygen. Due to their nature, they are able to well-adapt and survive under a large variety of environmental stress, such as cold, heat, drought, high salinity, high light intensity, osmotic pressure, UV exposure, which promote free radicals intracellular accumulation and increase the risk of photo-oxidation.

Microalgae have therefore developed effective defensive mechanisms against free radicals (Pulz and Gross 2004), enabling them to produce high amounts of antioxidant and bioactive compounds. Their large biodiversity, efficient photosynthetic activity, and high growth rate, with respect to higher plants, in addition to simple nutritional requirements make them a valuable source of a large variety of primary and secondary carotenoids. Because of these peculiarities, the study of the physiology of microalgae has been widely intensified, with the aim to optimize microalgae the metabolic processes involved in the synthesis of added-value substances. In particular, many studies have been carried out to implement the productivity of microalgal biomass and metabolites, in outdoor culture, with the development of photobioreactors (PBRs), open or closed, with different geometry and volume, for the improvement of the light utilization efficiency for photosynthesis.

By virtue of their high nutritional value the microalgal biomass find several applications in aquaculture, as supplementary food in human diets and as natural colorant, in the cosmetic field. The numerous microalgal species represent therefore an important pool of biodiversity, encouraging the study of new compounds and new possibilities for their application. A suitable approach is the microalgal cultivation

A. de Jesús Martínez-Roldán (ed.), *Biotechnological Processes for Green Energy, and High Value Bioproducts by Microalgae, and Cyanobacteria Cultures*, Developments in Applied Phycology 13, https://doi.org/10.1007/978-3-031-43969-8_10

in PBRs and outdoors, to produce a large amount of added-value biomass.

10.2 Carotenoids in Microalgae and Their Role

A huge spectrum of colors ranging from yellow to orange, red, and purple, are characteristic of carotenoids. They are constituted by eight isoprene units with a 40-carbon skeleton, typically consisting of a polyene chain with nine conjugated double bonds and an end group at both ends.

The biosynthetic pathway of carotenoids is reported in Fig. 10.1.

Isopentenyl pyrophosphate (IPP, C5) and its isomer dimethylallyl diphosphate (DMAPP, C5) are converted at a ratio of 3:1 to geranylgeranyl pyrophosphate via the methylerythritol phosphate pathway (MEP) in the chloroplast (GGPP, C20). Phytoene synthase condenses two molecules of Geranylgeranyl diphosphate (GGPP) into (15Z)-phytoene (C40), which is then isomerized into lycopene via phytoene desaturases (PDS), ζ-carotene isomerase (Z-ISO), ζ-carotene desaturase (ZDS), and carotene isomerase (CrtISO) (Ren et al. 2021).

Carotenoids are classified into two categories, based on their chemical characteristics, as carotenes and xanthophylls. Xanthophylls, which include *β*-cryptoxanthin, lutein, zeaxanthin, astaxanthin, fucoxanthin, and peridinin, are carotenoids with oxygen atoms, and carotenes include, *β*-carotene, *β*,*ψ*-carotene (*γ*-carotene), and lycopene.

There are primary and secondary carotenoids. The primary carotenoids, like the xanthophylls and ß-carotene, are found in the chloroplast, where they play a role in harvesting light energy.

When cells are subjected to stress (low/high temperature, salt stress, nutrient starvation or deficiency, high light), the photosynthetic apparatus may be not able to efficiently use the absorbed light energy and the dissipation mechanism can be activated. A consistent amount of secondary carotenoids can also be accumulated to act as quencher of free radicals and reactive oxygen species.

The increase of carotenoid content is functional to the dissipation of the excess of energy, which occurs by a safe non-radiative pathway, which is the non-photochemical quenching (NPQ) of chlorophyll fluorescence (Fig. 10.2).

NPQ is a well-known and well-documented mechanism of quenching and it is characterized by at least three components: (i) qE, i.e., high-energy state quenching, associated with the formation of a ΔpH in the thylakoid lumen (Briantais et al. 1979) and the synthesis of zeaxanthin via de-epoxidation of violaxanthin (Yamamoto et al. 1962; Niyogi et al. 2004); (ii) qT, state transition quenching, a change in the relative antenna sizes of photosystem II (PSII) and photosystem I (PSI) (Allen 1992); and (iii) qI, photoinhibitory quenching, a slowly reversible component of NPQ which includes a stationary quenching, independent of pH gradient, caused by zeaxanthin binding to Lhcb proteins (Dall'Osto et al. 2005) which is mainly related to slow reversible PSII reaction centers damage (Aro et al. 1993).

Chlorophyll fluorescence quenching has been widely used to investigate non-photochemical reduction of the plastoquinone pool (PQ-pool) in both higher plants and algae (Baker 2008). The increase in the basal fluorescence level (F_0) measured under weak non-actinic light has been connected to the redox state of the PQ-pool. The investigation of non-photochemical reduction of the PQ-pool uses the measurement of chlorophyll induction kinetics as a supplemental approach. It is well acknowledged that the chlorophyll (Chl) *a* fluorescence rising kinetics show the gradual decrease of

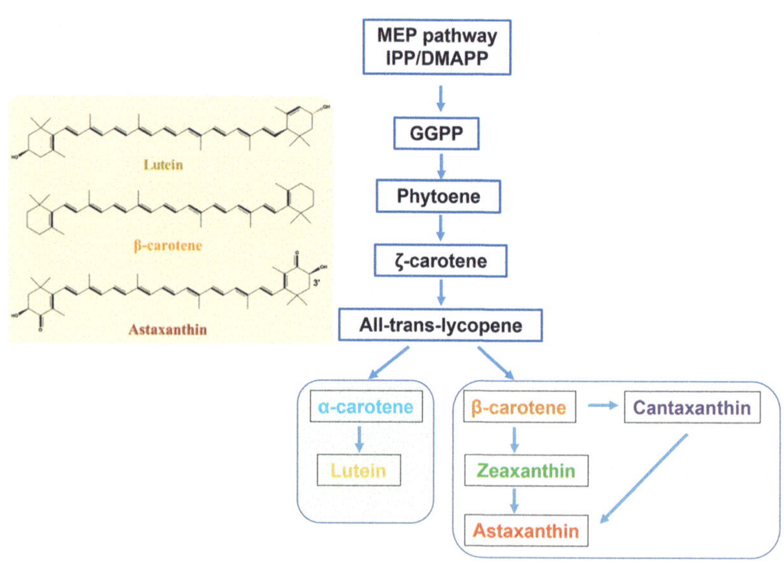

Fig. 10.1 Scheme of carotenoids synthesis pathways in microalgae

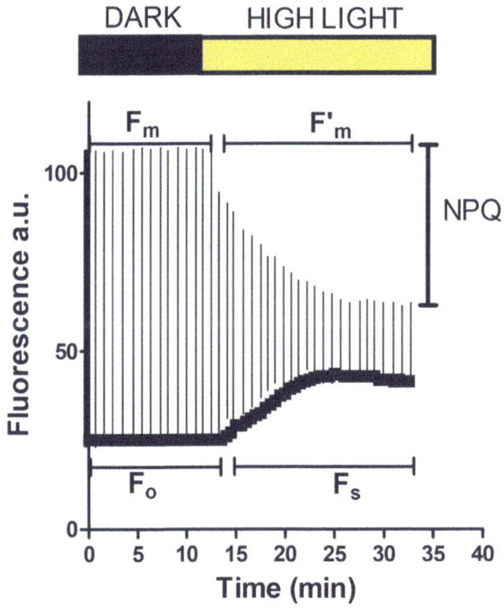

Fig. 10.2 Changes of NPQ measured by chlorophyll *a* fluorescence, during the shift from dark to high light conditions. The reduction of peak of chlorophyll *a* fluorescence means that part of the energy is not used for photochemistry but dissipated as heat

the photosynthetic electron transport chain, i.e., the conversion of Q_A to Q_A^- and the lowering of the PQ-pool. Therefore, fluorescence induction curves reflect the efficiency of the photosynthetic activity (Haldimann and Tsimilli-Michael 2002; Lazár 2006). By illumination of dark-adapted cells with actinic light it is possible to induce the chlorophyll *a* fluorescence polyphasic increase, from the minimum to the maximum fluorescence value (F_0 and F_m, respectively, O and P, respectively), as described by the curve reflecting the fluorescence increase, showed in the logarithmic form, called Kautsky curve (Fig. 10.3) (Neubauer and Schreiber 1987; Strasser et al. 1995). In this increase there are some

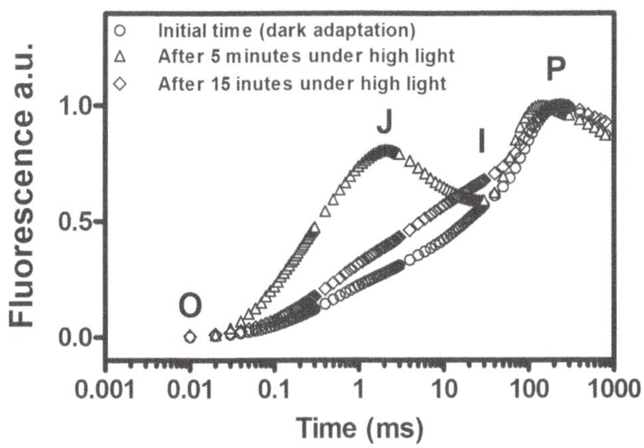

Fig. 10.3 Effect of high light intensity (1500 μmol photons m^{-2} s^{-1}) on the chlorophyll *a* fluorescence (OJIP) transient in *C. reinhardtii* cells after 15 min of dark adaptation (circles), 5 min of high light exposure (diamonds) and 15 min (triangles). Relative variable fluorescence is given by $V_t = (F_t - F_0)/(F_m - F_0)$

intermediate steps, the most relevant of which are evident at about 2 ms and 30 ms, indicated as J and I, respectively, according to the appearance of the curve, and the identification of the steps by Strasser et al. (Strasser et al. 1995). The changes to the shape of this curve are related to changes in the efficiency of the electron transport from PSII to PSI. In particular, the J step rise reflects the electron accumulation at PQ-pool level, hence revealing the amplitude of reduction of PQ (Haldimann and Tsimilli-Michael 2002; Hill et al. 2004).

As soon as photosynthetic microorganisms are alerted by the signals of a high reductive state level inside the cells specific mechanisms are activated to dissipate the accumulated electrons. These mechanisms involve the synthesis of antioxidant carotenoids (astaxanthin, lutein) and the xanthophyll cycle pigments: violaxanthin, antheraxanthin and zeaxanthin, and loroxanthin, fucoxanthin, mainly found in the marine microalgae *Phaeodactylum* and *Isochrysis* (Yamamoto 1979; Eskling et al. 1997).

Higher plants can also produce xanthophylls including violaxanthin, antheraxanthin, zeaxanthin, neoxanthin, and lutein, but microalgae, besides these, can produce xanthophylls such loroxanthin, astaxanthin, and canthaxanthin (Fig. 10.4). In diatoms and brown algae, fucoxanthin can also be found, in addition to another xanthophyll cycle, involving the de-epoxidation of diadinoxanthin (Ddx) to diatoxanthin (Dtx) (Fig. 10.4). By the induction of this cycle, the reduction of the singlet oxygen inside the cell is activated, and the cellular damage is avoided.

Among the microalgae, *Dunaliella salina, Haematococcus pluvialis, Botryococcus braunii, Nannochloropsis* sps., *Chlorococcum* sp., *Chlorella* sps., *Coelastrella striolata Scenedesmus* sp., and some diatoms are able to produce a large amount of lutein, canthaxanthin, β-carotene, astaxanthin, and fucoxanthin (Pulz and Gross 2004; Allen 2003; Wollman 2001). Some images of microalgal cells enriched with carotenoids are reported in Fig. 10.5. Important carotenoid market sectors include the feed, food, and beverage, pharmaceutical, cosmetic, aquaculture, and nutraceutical industries (Ambati et al. 2019). Global market size was estimated to be 1.5 billion USD, but high demand is expected to drive expansion, with a compound annual growth rate of 5.7% to 2 billion USD by 2022. β-carotene, lutein, fucoxanthin, and zeaxanthin are the most popular because of their range of applications. It has been reported (Gupta et al. 2021) that lutein makes up 225 million USD (15.6% of total) and β-carotene 246.2 million USD (17% of total). The market value of lutein is predicted to reach 357.7 million USD by 2022, whereas the cost of β-carotene is projected to range from 300 to 1500 USD/kg, contingent on carotenoid purity and demand.

The production of astaxanthin from *Haematococcus pluvialis* is profitable and already established on a global

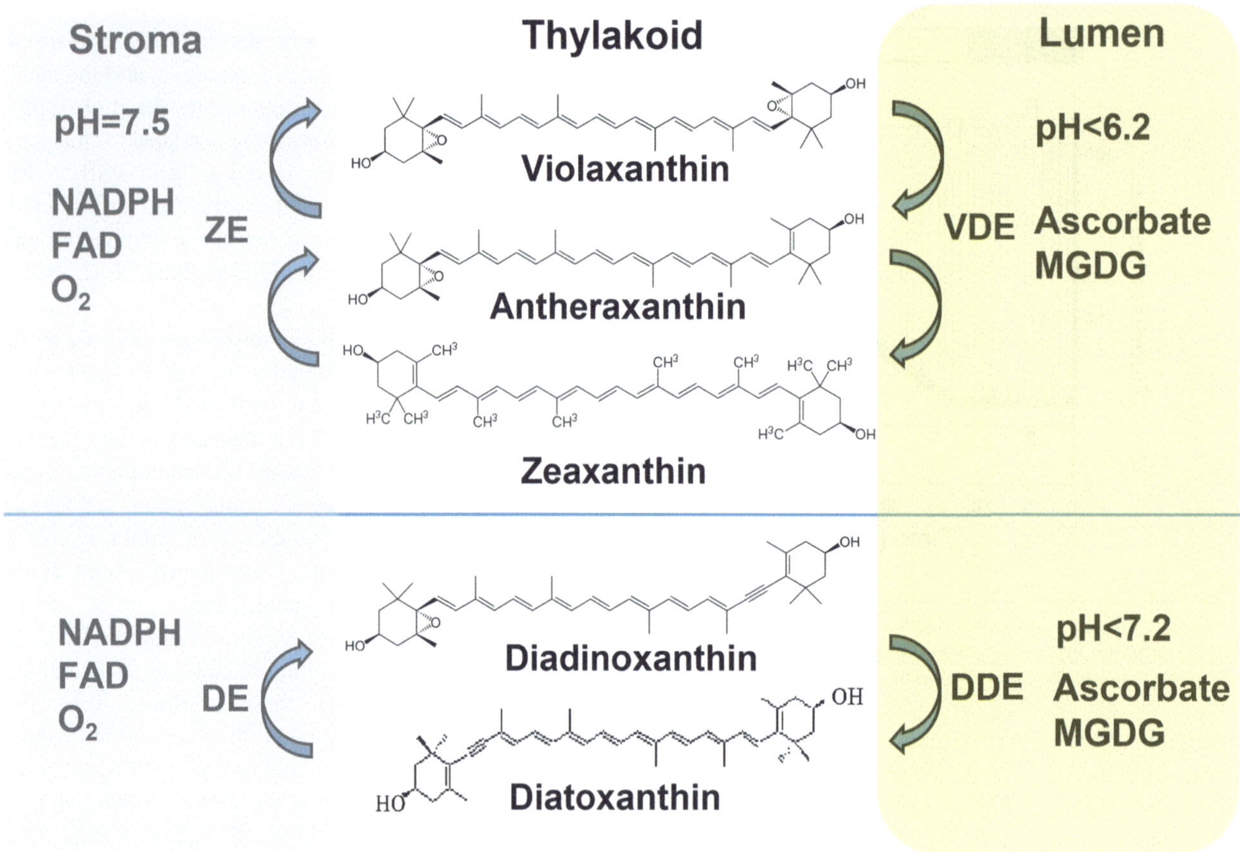

Fig. 10.4 Scheme of the induction of the xanthophyll cycle by means of violaxanthin de-epoxidase (VDE) and diadinoxanthin de-epoxidase (DDE), to synthesize zeaxanthin and diatoxanthin, respectively, and the epoxidation of zeaxanthin and diatoxanthin by means of zeaxanthin epoxidase (ZE) and diatoxanthin epoxidase (DE), respectively

scale. The production of astaxanthin in this microalga reaches as much as 5% per Dry Weigh (DW) (Butler et al. 2018) which is higher than the production of the bacterium *Paracoccus carotinifaciens* (2.2% DW). Beta-carotene production from *Dunaliella* is also increasing as well as the production of lutein, zeaxanthin, fucoxanthin. Marigold flowers are primary commercial source of lutein; however, microalgae biomass contains three to six times lutein per unit of dry matter.

The first step of astaxanthin production in *Haematococcus* occurs in the central part of the cell; with further synthesis its accumulation can be observed also in the external part and the cell becomes entirely red. Another important producer of antioxidant carotenoids is *Scenedesmus*, which can produce

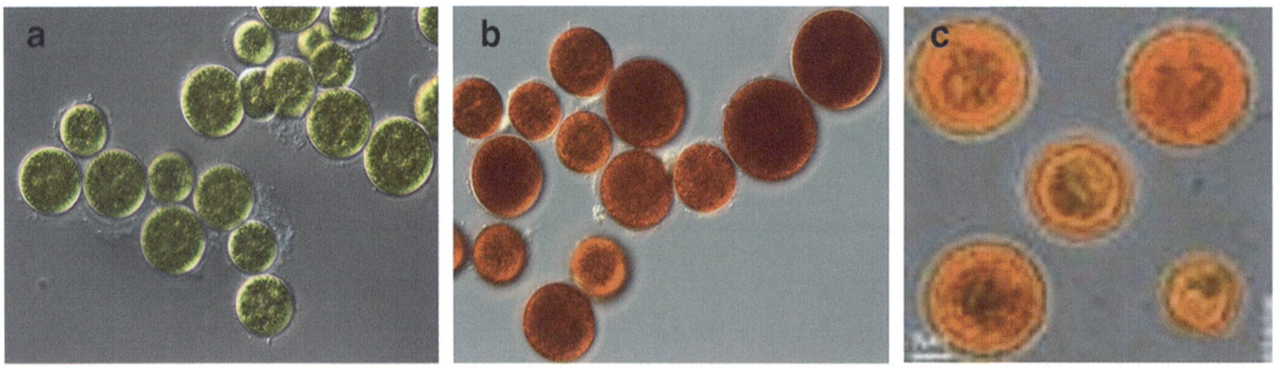

Fig. 10.5 Images at optical microscope of *H. pluvialis* cells in the green stage (**a**) and red stage (**b**) for the accumulation of astaxanthin. *Dunaliella* sp. and cells containing β-carotene (**c**)

large amounts of lutein, changing the color of the cells from green to yellow. *Chlorella vulgaris* is a green microalga well known for its function as supplementary food and for its high biomass content of protein and chlorophyll. Moreover, it has been reported that this microalga can produce high amounts of lutein, β-carotene, and zeaxanthin (1.18 mg g^{-1}, 0.31 mg g^{-1} and 0.24 mg g^{-1}, respectively), which displayed a good bioavailability (Serra et al. 2021). The unicellular green microalga *Dunaliella salina* can be considered another high carotenoid producer, mainly for β-carotene production. In this microalga, this pigment starts to accumulate within lipids bodies at one side of the cell, and then it accumulates all over the cell.

The green microalga *Asterarcys quadricellulare* has been isolated from fresh water and it is possible to increase its carotenoid production by optimizing nutrient media composition, pH, salinity, the exposure to different light intensity and/or quality, and duration (Singh et al. 2019). It is a new promising carotenoids producer. It is able to produce d 47.0, 28.7, 15.5 and 14.0 μg β-carotene, lutein, astaxanthin, and canthaxanthin mg^{-1} dry biomass, respectively. These findings are very interesting, as they underline the existence of a large biodiversity which makes microalgae a fruitful source of natural compounds.

10.3 Photoprotective Response in Microalgae

The optimal values of temperature, pH, salinity, light intensity, and nutritional composition of the medium are some of the factors that vary based on the strain, species, and origin. One of these that are crucial is light intensity. The physiological limitations of the photosynthetic system prevent it from using light irradiation over a particular degree of intensity.

Although the carotenoid synthesis is primarily induced by means of exposure to stress conditions like high light, lack or limitation of nutrients, extreme low or high temperatures, and salinity stress are all effective ways to induce the synthesis of antioxidant chemicals. However, these conditions are not ideal for increasing biomass production since they have a negative impact on growth.

When a stress is imposed, there are some parameters used to follow the behavior of the cells to monitor the photosynthetic activity during the induction of carotenoid synthesis. This procedure is helpful in managing all the production processes as it gives information on the level of cell damage and functionality.

Under stressful situations, the generation of singlet oxygen occurs, and when a free radical comes into contact with another molecule, requiring another electron to pair with, cellular damage takes place. A molecule is strongly attracted to the unpaired electron of a free radical, which causes the molecule to become a free radical in its own right. This occurrence is associated with the electron accumulation in the photosynthetic apparatus, in the electron transport chain, which is the prompting event alerting photosynthetic cells to activate defensive strategies, like large production of carotenoids. PQ-pool, one of the electron transport channels, has been extensively studied for its potential role in the control of photosynthetic activities. Because of this, the redox level of PQ-pool plays a crucial role in both the production of physiological responses to stress and the synthesis of carotenoids.

In fact, the over-reduction of PQ-pool, which is frequently linked to the induction of carotenoid synthesis, encourages the acidity of the thylakoid lumen, which in turn activates several carotenogenic enzymes. Low pH in the thylakoid lumen, in particular, promotes the de-epoxidation of violaxanthin to zeaxanthin via antheraxanthin (xanthophyll cycle) (Björkman 1987; Gilmore and Yamamoto 1993; Gilmore et al. 1995) (Fig. 10.4).

The synthesis of these carotenoids has double importance: (i) the de-epoxidation is a quenching reaction and (ii) xanthophylls have the ability to provide electrons (Mandal et al. 2009), and are effective in inhibiting the oxidation process at small concentrations.

10.4 Environmental Conditions Enhancing Synthesis of Antioxidant Compounds

All conditions affecting the growth and photosynthetic efficiency provoke the over-reduction of the photosynthetic chain and generation of free radicals. Some of the main factors inducing carotenoid synthesis are reported in Table 10.1. Some of the most antioxidant carotenoids produced by microalgae are reported in Table 10.2, where the productivity is also considered.

10.4.1 Light Intensity

The microalga *Chlamydomonas reinhardtii*, which is regarded as a model organism for physiological and biochemical investigations on photosynthesis, has been used extensively in the research on this phenomenon. In this microalga, zeaxanthin production was triggered by exposure to 800 mol photons m^{-2} s^{-1} within 10 min, while a partial induction of the cycle was seen at 300–350 mol photons m^{-2} s^{-1} (Niyogi et al. 1997).

With the induction of the synthesis of diatoxanthin, by the de-epoxidation of diadinoxanthin, the xanthophyll cycle is also induced in diatoms and dinoflagellates, such as *Phaeodactylum* and *Isochrysis*, respectively. In *P. tricornutum*, diadinoxanthin to diatoxanthin conversion is very rapid, as it may occur within 15 min, under sunlight

Table 10.1 Main environmental factors influencing the carotenoid synthesis in several microalgae species

Stress condition	Species	Effects on microalgal pigments	Reference
Light	*Haematococcus pluvialis*	An increase in light irradiance from 20 to 400 μmol photon m^{-2} s^{-1} increased 3.15-fold the amount of total carotenoids in microalgae	Gwak et al. (2014)
	Dunaliella salina	An increase in light irradiance enhanced carotenoid production after 2 days of stress The highest β-carotene production of 30 pg $cell^{-1}$ day^{-1} was obtained	Lamers et al. (2010) Xia et al. (2013)
	Chlamydomonas reinhardtii	Induction of the xanthophylls cycle and lutein synthesis	Faraloni et al. (2021); El-Mekkawi et al. (2019)
	Desmodesmus sp	A maximum lutein productivity of 3.6 mg L^{-1} day^{-1} was obtained	Lamers et al. (2010)
Nitrogen limitation/ starvation	*Chlorella vulgaris*	Low and excessive nitrogen stress increased and inhibited the synthesis of carotenoids, respectively	Abe et al. (2007)
	Chlorella zofingiensis	Carotenoid content was increased from 0.05 to 4.5 mg g^{-1}	Mulders et al. (2014a)
	Coelastrella striolata	Nitrogen limitation increased the carotenoid contents changing the color of the cells from green to red	Abe et al. (2007)
	Dunaliella salina	β-carotene productivity was increased from 1.28 to a maximum of 14.4 g L^{-1} day^{-1}	Lamers et al. (2012)
	Nannochloropsis gaditana	Induction of the xanthophylls cycle	Forjàn et al. (2022)
Sulfur limitation/ starvation	*Chlamydomonas reinhardtii*	Induction of the xanthophylls cycle and lutein synthesis	Faraloni and Torzillo (2010); Antal et al. (2006)
	Dunaliella salina	β-carotene production of 14.616 mg L^{-1}	Shaker et al. (2017)
Phosphorus limitation	*Nannochloropsis gaditana*	Induction of the xanthophylls cycle	Forjàn et al. 2022
Salinity stress	*Chlorella protothecoides*	The addition of 20 g NaCl L^{-1} increased the production of carotenoids	Campenni et al. (2012)
	H. pluvialis	The maximum astaxanthin content of 5.01 mg g^{-1} was obtained	Ding et al. (2019)
	Dunaliella salina	An increase of NaCl concentration from the optimum (2 M) increased the production of carotenoids	Hadi et al. (2008). Ben-Amoz, (1991)
	D. salina	30-fold increase in β-carotene production	Coesel et al. (2008)
High temperature	*Muriellopsis* sp.	A maximum lutein content of 0.54% DW. At 44 °C	Del Campo et al. (2000)
	Tetraselmis sp.	Lutein content increased to 3.17 mg g^{-1} DW and β-carotene content increased to 3.21 mg g^{-1} DW	Schülera et al. (2020)

Table 10.2 Microalgal carotenoid contents and productivity

Carotenoid	Content	Productivity/Yield	Microalgae	References
Lutein			*Chlorella sorokiniana* FZU60	Ma et al. (2020)
			Chlorella vulgaris UTEX 265	Gong and Bassi (2017)
			Chlorella vulgaris CS	McClure et al. (2019)
	7.4–42.0 mg g^{-1} DW	3.4–20.0 mg L^{-1} day^{-1}	*Scenedesmus* sp.	Přibyl et al. (2016)
			Chlorella sp. GY-H4	Wang et al. (2020)
			Chlorella sorokiniana MB-1-M12	Chen et al. (2019)
Astaxanthin			*Chlorella minutissima* MCC-27	Dineshkumar et al. (2016)
	50–70 mg g^{-1} DW	0.8–6.34 6.34 mg L^{-1} day^{-1}	*Haematococcus pluvialis*	Wan et al. (2014)
			Chromochloris zofingiensis	Liu et al. (2014)
β-Carotene	30–130 mg g^{-1} DW	–	*Dunaliella salina*	Rammuni et al. (2018)
			Nannochloropsis oceanica CCNM 1081	Mitra and Mishra (2019)
Zeaxanthin		0.72 mg L^{-1} day^{-1}	*Coelastrella* sp. M60	Pushpalatha et al. (2020)
	4.26–27 mg g^{-1} DW		*Chlorella ellipsoidea*	Koo et al. (2011)
			Mallomonas sp.	Petrushkina et al. (2017)
Fucoxanthin			*Isochrysis* zhanjiangensis	Li et al. (2019)
			Odontella aurita	Xia et al. (2013)
	16.4–24.0 mg g^{-1} DW	2.94–9.80 mg L^{-1} day^{-1}	*Isochrysis aff. Galbana*	Kim et al. (2012a)
			Tisochrysis lutea	Gao et al. (2020)
			Phaeodactylum tricornutum	Kim et al. (2012b)

exposure in outdoor cultures in tubular PBRs. The highest diatoxanthin amount could be detected when the light intensity was the highest (Torzillo et al. 2012). In particular, this microalga is well known for its capability of high production of fucoxanthin, a carotenoid mainly produced by marine microalgae. This pigment is associated with thylakoid membranes, and its role consists in driving the electrons to the electron transport chains, by transferring excitation energy to chlorophyll *a* (Jin et al. 2003; Mulders et al. 2014b). In *P. tricornutum*, fucoxanthin can reach 0.22–1.82% of the biomass, but in *Isochrysis* it can be found in much higher concentration, at conditions such as appropriate cell density, light intensity, and mixing.

Another powerful antioxidant carotenoid that is reported to grow in high light conditions is lutein. As a main carotenoid, this pigment is naturally present in the majority of photosynthetic cells, and under conditions of photo-oxidative stress, its production may rise. Under conditions of 1900 mol photons m^{-2} s^{-1} and 35 °C, the microalga *Scenedesmus almeriensis* in a tubular PBR outdoors was able to produce extremely high levels of lutein (around 5 mg m^{-2} d^{-1}) (Fernández-Sevilla et al. 2010). The simultaneous presence of high light and high warmth was the cause of this increase. In fact, microalgal strains typically thrive best at temperatures between 25 and 28 °C. Another main carotenoid that is naturally present in microalgal cells and may be over-synthesized under high light is β-carotene, which typically rises with high light exposure. One microalgae that is widely recognized for producing a lot of this pigment is *Dunaliella salina*. In a lab setting, with a light intensity of 200–1200 mol photons m^{-2} s^{-1}, at 30 °C, the generation rate of 13.5 mg L^{-1} d^{-1} has been reported (Kleinegris et al. 2011).

One of the most significant secondary carotenoids synthesized by microalgae is the red pigment astaxanthin. It is a very potent antioxidant that is mostly produced by *Haematococcus pluvialis*, especially in environments with intense light. It takes significantly longer to induce its production than the xanthophyll cycle since it requires at least a day of exposure to sunshine to see changes in the color of cells from green to red, whereas the xanthophyll cycle may be induced in a matter of minutes. However, *H. pluvialis* is a suitable option for carotenoid synthesis in outdoor cultivation since astaxanthin may be generated at extremely high concentrations, up to 5% of the biomass. The majority of *H. pluvialis* research investigations, conducted in outdoor cultures to use sunlight to stimulate astaxanthin synthesis, revealed that this microalga can grow and generate astaxanthin effectively at extremely high light intensity, primarily in the summer (Recht et al. 2014; Boussiba et al. 1992; Torzillo et al. 2003). The technology of astaxanthin with *Heamatococcus pluvialis* is well established and it is produced industrially in Israel and China (Shah et al. 2016; Lia et al. 2020).

10.4.2 Nutrient Limitation or Starvation

Along with high light irradiance, nutritional constraint is another significant stress that might happen when the environment is conducive to the synthesis of carotenoids.

Micronutrient limitation does not have as much of an impact on the activation of protective responses as macronutrient constraint, or famine. In fact, they play a direct role in many developmental processes, and their lack may affect the formation of reducing power and, as a result, the production of antioxidant carotenoids (Faraloni and Torzillo 2017).

10.4.3 Nitrogen

Nitrogen is one of the most important elements, as it is the constituent in many structural molecules of the cell, like proteins and enzymes, and is directly implicated in the growth process. For this reason, the evaluation of its concentration in the composition of the growth medium is very important.

The nitrogen starvation increases the synthesis of β-carotene in *Dunaliella*, as it was reported for this microalga that under high light exposure (Ben-Amoz 1991), nitrogen starvation and light intensity of 200 μmol photons m^{-2} s^{-1}, β-carotene can be found at 2.7% of the biomass (Lamers et al. 2012).

The increment of this pigment has been reported to be associated with the synthesis of total fatty acid, under both nitrogen starvation and high light stress (Rabbani et al. 1998). The production of this carotenoid and fatty acid biosynthetic pathway is strongly related, which may be explained by the fact that β-carotene is stored in lipid globules. This aspect is very interesting, and shows how the microalgal metabolic pathways are strongly associated to guarantee a useful and efficient response to the stresses.

The same relation can also be found for astaxanthin in *H. pluvialis*. In this microalga, the concomitant exposure to high light with nitrogen starvation induced both the lipid accumulation and the high astaxanthin synthesis, that was more than two times higher than the control (Liang et al. 2015).

However, when only nitrogen starvation was imposed, astaxanthin synthesis showed increased levels than in the control culture. Due to the cytoplasmatic buildup of the red pigment, astaxanthin production induces the change from the green stage to the red stage after 20 h, reaching 1.4% of dry weight.

In *Nannochloropsis gaditana*, the nitrogen limitation or deprivation induces the xanthophyll cycle, with the increment of violaxanthin and zeaxanthin. In this case the cells were previously grown on complete medium, in order to obtain a considerable amount of biomass, and then the cells were subjected to sulfur limitation or starvation. In this case the biomass resulted was enriched in the xanthophylls pigments (Forjàn et al. 2022). The volaxanthin increase was less evident, as this pigment is involved in the synthesis of zeaxanthin, via antheraxathin. However, its increment indicates a de-novo synthesis of violaxanthin, which is a signal of the strong induction of carotenoid synthesis.

10.4.4 Sulfur

Sulfur is fundamental to sustain protein synthesis and its deprivation or limitation causes the decrease of the protein content, affecting the photosynthetic activity and inducing the decline of the growth. This phenomenon is accentuated when microalgae are placed in closed PBRs, where the decline of photosynthetic oxygen evolution and the presence of oxygen respiration may create the occurrence of very low oxygen content or even anaerobic conditions, a very strong reductive environment. It has been extensively investigated in the microalga *C. reinhardtii*, which has been studied for the production of hydrogen under sulfur starvation. Under these conditions, a strong induction of the xanthophyll's cycle was observed, with a large increase of antheraxanthin and zeaxanthin and a de-novo synthesis of violaxanthin. Moreover, also the other xanthophyll lutein and β-carotene resulted increased (Faraloni and Torzillo 2010). In addition, anaerobiosis, but in complete medium, promoted the same response, although in a longer time (5 h, instead of some minutes). When aerobic conditions have been established, lutein and β-carotene lowered, as also antheraxanthin and zeaxanthin, revealing the occurrence of a recovery.

In *Nannochloropsis gaditana*, the sulfur limitation or deprivation induces the xanthophyll cycle, with the increment of both zeaxanthin and violaxanthin, as observed in the case of nitrogen (Forjàn et al. 2022).

10.4.5 Phosphorus

Phosphorus is very important for its involvement in the cellular composition of DNA, RNA, and phospholipids and the regulation of enzymatic activity, metabolic pathways, and production of ATP. It can be used in the form of polyphosphate or orthophosphate and its presence in the media composition guarantees an efficient photosynthetic activity (Belotti et al. 2014).

Phosphate limitation in combination with continuous high light exposure (2000 μmol photons m^{-2} s^{-1}) induced an increase in astaxanthin synthesis in *H. pluvialis,* in which cell contents reached 7% of biomass (Liyanaarachchi et al. 2020).

In *Nannochloropsis gaditana*, the phosphorus limitation or deprivation induces the xanthophyll cycle, with the increment of violaxanthin and zexanthin, as observed in the case of nitrogen (Forjàn et al. 2022).

10.4.6 Salt Stress

One of the most significant abiotic variables, influencing the growth and productivity of microalgae, is salt stress. Some microalgae have developed strategies to adapt to saline environment and under these unfavorable conditions they may regulate the synthesis of osmoregulatory compounds that can help the cells face the stress. In particular, osmotic stress can induce the increase of some polar lipids (glycolipids and phospholipids) (Benavente-Valdésa et al. 2016) incrementing the content in chloroplast and cellular membranes. As previously reported, lipid production is important for the location of some carotenoids and often their production is tightly associated with carotenoid synthesis.

Photosynthetic organisms experience a wide range of bioenergetic and biochemical changes under salt stress; in fact, an excess of the ions Na + and Cl lowers the environment's osmotic potential, which in turn reduces water intake. In microalgal cells, the impact of salt stress produces responses at different levels, changing membrane permeability, blocking ion homeostasis, increasing the catabolism of lipids and biopolymers, and changing the yield of the processes producing energy. As other stresses, the salt stress may have a strong impact on the growth rates in microalgal cultures, reducing the photosynthetic activity, to the same extent as high light exposure, and for this reason this stress can be considered a tool to promote carotenoid production. In *Chlorella zofingiensis*, the increase of NaCl concentration reduced the growth but promoted the increase of carotenoid content (Barghbani et al. 2012). In this microalga, even at low light intensity, large amounts of secondary carotenoids were accumulated, under salt stress, in particular cantaxanthin, but not astaxanthin, indicating that for astaxanthin the high light is a required factor (Li and Huang, 2009). The effect of salt stress in combination with light exposure is a crucial aspect, considering the impact of the stress on photosynthetic microorganisms. This has been evaluated in *C. vulgaris* by using different salt concentrations associated with different light-dark regimes. At the highest salt concentration of 0.43 M NaCl, the highest increase of astaxanthin content per cell was 4.8-fold increased, in concomitance with the increase in the lipid content, as observed for nitrogen limitation. In *C. vulgaris*, the salt stress, in a range of 0.34–0.051 M, caused a decline of the growth rate, which may be due to the imbalance in ion homeostasis, thereby reducing the photosynthetic activity, affecting light utilization and the metabolism of carbohydrates involved in osmotic regulation, and hence, generating the accumulation of reactive oxygen species (Benavente-Valdésa et al. 2016). This is considered a triggering factor for the induction of

Table 10.3 Advantages and disadvantages of photobioreactors, according with their characteristics

	Advantages	Disadvantages	References
Tubular	Large surface to volume ratio; controlled conditions (temperature, pH, evaporation, CO_2 consumption); high volumetric productivity; low contamination risk	High O_2 accumulation; high temperature; high cost of construction and operation	Huang et al. (2017); Yaakob et al. (2014)
Airlift column	Controlled conditions (temperature, pH, evaporation, CO_2 consumption); high volumetric productivity; low contamination risk; high mixing and mass transfer; low shear stress; low power intake	High cost of construction and operation; high cleaning cost	Huang et al. (2017); Yaakob et al. (2014)
Airlift flat panel	Controlled conditions (temperature, pH, evaporation, CO_2 consumption); high volumetric productivity; low contamination risk; low power intake and shear stress; good mixing and mass transfer; long lifespan; high illuminated surface/volume ratio	High cost of construction and operation	Huang et al. (2017); Yaakob et al. (2014)
Plastic bag	Controlled conditions (temperature, pH, evaporation, CO_2 consumption); high productivity; low contamination risk; low cost	Low mixing; short lifespan; easy to leakage	Huang et al. (2017); Yaakob et al. (2014)
Open ponds	Low cost of construction and operation; low energy input; easy cleaning and managing	Easy contamination and high pollution risk; low productivity; low mixing and light utilization; high cost for biomass harvesting	Huang et al. (2017); Yaakob et al. (2014)

carotenoid synthesis, as observed for nitrogen, sulfur, or phosphorus limitation. In addition, in this case the low biomass productivity can be associated with a high carotenoid yield.

10.5 Microalgae Cultivation in PBR to Produce Carotenoids

Cultivation of microalgae at a large scale may consider several aspects, for the achievement of a cost-effective production of microalgal biomass. The two distinct kinds of PBR are open and closed. Open PBR are ponds or their modifications; closed PBR (tubular, horizontal, vertical, flat panel, thin layer) may have different shapes, dimension, orientation, and inclination. The main characteristics, advantages and disadvantages are schematically reported in Table 10.3). The available commercial culture system units vary from 1000 liters for aquaculture feed to 3 × 104 liters for the generation of algal biomass, to 1 × 104 liters for tubular reactors, and up to 1 × 109 liters for the large ponds used for rearing Dunaliella in Australia (Borowitzka 2016). However, according to Borowitzka and Vonshak (2017, 345), a commercial facility must be able to generate at least 10 ton of high-value items annually. The biomass needed to manufacture 1 ton of product is shown in Fig. 10.6.

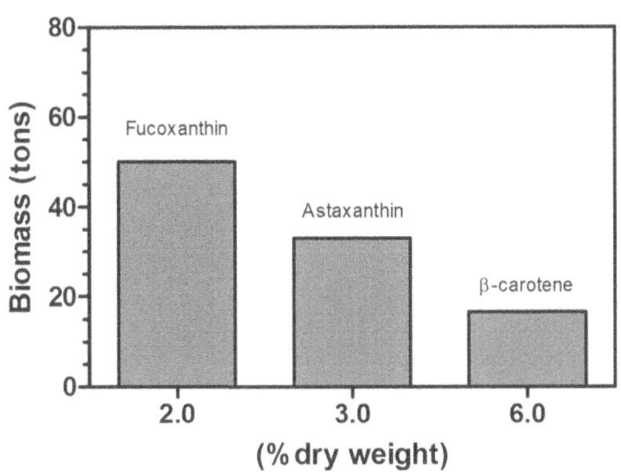

Fig. 10.6 Amount of biomass required to produce 1 ton of product per annum in *Dunaliella*, according to Borowitzka and Vonshak (2017)

Open PBRs can be economically advantageous for the cultivation of a contamination-resistant strain, such as extremophile strains, that are able to grow under extreme conditions, such as *Dunaliella salina*, extreme halophil, and also robust strains, such as *Chlorella* spp., with a high growth rate, which do not let other microorganisms take over. This kind of PBR is suitable under favorable climatic conditions, as it is difficult to have a full control of the main parameters important for optimization of the microalgal growth (mixing, temperature, pH, CO_2 enrichment and O_2 accumulation, and evaporation).

Closed PBRs offer the possibility to grow microalgae under controlled conditions, with a very low risk of contamination. They offer a high versatility and provide high productivity, with a high-quality biomass, thanks to the possibility to cultivate a wide range of strains, under precise conditions. Although they offer applications larger than the open PBRs, the closed PBRs have a considerable cost of construction and high operation costs, due to their complexity.

Considering the cultivation of microalgal strains devoted to a significant carotenoid production, the goal is to obtain a high productivity in terms of biomass cultivated under stress conditions which stimulated the carotenoid synthesis. This is a crucial aspect because the stress conditions which usually promote the increase of certain pigments (secondary ones) are not favorable for the growth.

For instance, the primary carotenoids, as lutein, are commonly present under favorable conditions for photosynthesis, so they can be present in biomass cultivated without particular stress, obtaining concomitantly high productivity, in terms of growth. By contrast, the synthesis of secondary carotenoids, such as β-carotene, astaxanthin, and zeaxanthin, is induced under stress conditions, which reduce the growth rate, as under these conditions the photosynthetic activity declines.

10.6 Concluding Remarks

A better understanding of the microalgal physiology of carotenoid outdoors is a prerequisite for increasing their production, as well as the selection of new species capable of producing carotenoids. Furthermore, it is important to select robust organisms capable of managing under stressful conditions in which the risk of pollution by other microalgal species and fungi can increase significantly. It is not a coincidence that species that are adapted to high salinity conditions, such as *Dunaliella,* are the most promising industrial species for the production of beta-carotene as they can more easily be grown in open ponds or lagunes. For other microalgae such as *Haematococcus* that grow in a not very selective medium, the use of closed systems is mandatory since in these systems it is easier to achieve the best growth conditions that favor the species being cultivated. Finally, closed systems are important for species such as *Nannochloropsis* and particularly for *Phaeodactylum* with optimal growth conditions between 20 and 25 °C which is easier to achieve with closed PBRs with an appropriate cooling system.

References

Abe K, Hattori H, Hirano M (2007) Accumulation and antioxidant activity of secondary carotenoids in the aerial microalga *Coelastrella Striolata* Var. Food Chem 100(2):656–661

Allen JF (1992) Protein phosphorylation in regulation of photosynthesis. Biochim Biophys Acta 1098:275–335

Allen JF (2003) State transitions- a question of balance. Science 299: 1530–1532

Ambati RR, Gogisetty D, Aswathanarayana RG, Ravi S, Bikkina PN, Bo L, Yuepeng S (2019) Industrial potential of carotenoid pigments from microalgae: current trends and future prospects. Crit Rev Food Sci Nutr 59:1880–1902

Antal T, Volgusheva AA, Kukarskih GP, Bulychev AA, Krendeleva TE, Rubin AB (2006) Effects of sulfur limitation on photosystem II functioning in *Chlamydomonas reinhardtii*as probed by chlorophyll a fluorescence. Physiol Plant 128:360–367

Aro EM, Virgin I, Anderson B (1993) Photoinhibition of photosystem II. Inactivation, protein damage and turnover. Biochim Byophys Acta 1143:113–134

Baker NR (2008) Chlorophyll fluorescence: a probe of photosynthesis in vivo. Annu Rev Plant Biol 59:89–113

Barghbani R, Rezaei K, Javanshir A (2012) Investigating the effects of several parameters on the growth of Chlorella vulgaris using Taguchi's experimental approach. Int J Biotechnol Wellness Ind 1: 128–133

Belotti G, de Caprariis B, De Filippis P, Scarsella M, Verdone N (2014) Effect of *Chlorella vulgaris* growing conditions on bio-oil production via fast pyrolysis. Biomass Bioenergy 61:187–195

Ben-Amoz A (1991) The biotechnology of cultivating *Dunaliella* for production of β-carotene rich algae. Bioresour Technol 38:233–235

Benavente-Valdésa JR, Aguilara C, Contreras-Esquivela JC, Méndez-Zavalab A, Montañezb J (2016) Strategies to enhance the production of photosynthetic pigments and lipids in chlorophycae species. Biotechnol Rep 10:117–125

Björkman O (1987) High irradiance stress in higher plants and interaction with other stress factors. Photosynth Res 4:11–18

Borowitzka MA (2016) Algal physiology and large-scale outdoor cultures of microalgae. In: Borowitzka MA, Beardall J, Raven JA (eds) The physiology of microalgae. Springer, Dordrecht, pp 601–652

Borowitzka MA, Vonshak A (2017) Scaling up microalgal cultures to commercial scale. Eur J Phycol 52(4):407–418

Boussiba S, Fan L, Vonshak A (1992) Enhancement and determination of astaxanthin accumulation in green alga *Haematococcus pluvialis*. Methods Enzymol 213:386–391

Briantais JM, Vermotte C, Picaud M, Krause GH (1979) A quantitative study of the slow decline of the chlorophyll a fluorescence in isolated chloroplasts. Biochim Biophys Acta 548:128–138

Butler T, McDougall G, Campbell R, Stanley M, Day J (2018) Media screening for obtaining *Haematococcus pluvialis* red motile macrozooids rich in astaxanthin and fatty acids. Biology 7:2

Campenni L, Nobre LB, Santos CA, Oliveira AC, Aires-Barros MR, Palavra AMF, Gouveia L (2012) Carotenoid and lipid production by the autotrophic microalga *Chlorella protothecoides* under nutritional, salinity, and luminosity stress conditions. Appl Microbiol Biotechnol 97(3):1383–1393

Chen JH, Chen CY, Hasunuma T, Kondo A, Chang CH, Ng IS, Chang JS (2019) Enhancing lutein production with mixotrophic cultivation of *Chlorella sorokiniana* MB-1-M12 using different bioprocess operation strategies. Bioresour Technol 278:17–25

Coesel SN, Baumgartner AC, Teles LM, Ramos AA, Henriques NM, Cancela et al (2008) Nutrient limitation is the main regulatory factor for carotenoid accumulation and for psy and pds steady state transcript levels in *Dunaliella salina* (*Chlorophyta*) exposed to high light and salt stress. Mar Biotechnol 10:602–611

Dall'Osto L, Caffari S, Bassi R (2005) A mechanism of nonphotochemical quenching energy dissipation, independent from PsbS, revealed by conformational change in the antenna protein CP26. Plant Cell 17:1217–1232

Del Campo JA, Moreno J, Rodriguez H, Vargas MA, Rivas J, Guerrero MG (2000) Carotenoid content of chlorophycean microalgae: factors determining lutein accumulation in *Muriellopsis* sp (*Chlorophyta*). J Biotechnol 76:51–59

Dineshkumar R, Subramanian G, Dash SK, Sen R (2016) Development of an optimal light-feeding strategy coupled with semicontinuous reactor operation for simultaneous improvement of microalgal photosynthetic efficiency, lutein production and CO_2 sequestration. Biochem Eng J 113:47–56

Ding W, Cui J, Zhao YT, Han BY, Li T, Zhao P et al (2019) Enhancing *Haematococcus pluvialis* biomass and γ-aminobutyric acid accumulation by two-step cultivation and salt supplementation. Bioresour Technol 2019:285. https://doi.org/10.1016/j.biortech.2019.121334

El-Mekkawi SA, Hussein HS, El-Enin SAA, El-Ibiari NN (2019) Assessment of stress conditions for carotenoids accumulation in *Chlamydomonas reinhardtii* as added-value algal products. Bull Natl Res Cent 43:130

Eskling M, Arvidsson PO, Akerlund HE (1997) The xanthophyll cycle, its regulation and components. Physiol Plant 100:806–816

Faraloni C, Torzillo G (2010) Phenotypic characterization and hydrogen production in *Chlamydomonas reinhardtii* Q_B binding D1 protein mutants under sulfur starvation: changes in chlorophyll fluorescence and pigment composition. J Phycol 46(4):788–799

Faraloni C, Torzillo G (2017) Synthesis of antioxidant carotenoids in microalgae in response to physiological stress. Intech Open. https://doi.org/10.5772/67843

Faraloni C, Di Lorenzo T, Bonetti A (2021) Impact of light stress on the synthesis of both antioxidants polyphenols and carotenoids, as fast Photoprotective response in *Chlamydomonas reinhardtii*: new prospective for biotechnological potential of this microalga. Symmetry 213(11):2220

Fernández-Sevilla JM, Acién-Fernández FG, Molina-Grima E (2010) Biotechnological production of lutein and its applications. Appl Microb Biotechnol 86:27–40

Forján E, Garbayo I, Casal C, Vílchez C (2022) Enhancement of carotenoid production in Nannochloropsis by phosphate and Sulphur limitation. Communicating current research and educational topics and trends in applied microbiology A. Méndez-Vilas (Ed.)

Gao F, Cabanelas IIT, Wijffels RH, Barbos MJ (2020) Process optimization of fucoxanthin production with *Tisochrysis lutea*. Bioresour Technol 315:123894

Gilmore AM, Hazlett TL, Govindjee (1995) Xanthophyll cycle-dependent quenching of photosystem II chlorophyll *a* fluorescence-formation of a quenching complex with a short fluorescence lifetime. Proc Natl Acad Sci USA 92:2273–2277

Gilmore AM, Yamamoto H (1993) Linear models relating xanthophylls and acidity to non-photochemical fluorescence quenching. Evidence that antheraxanthin explains zeaxanthin-independent quenching. Photosynth Res 35:67–68

Gong M, Bassi A (2017) Investigation of *Chlorella vulgaris* UTEX 265 cultivation under light and low temperature stressed conditions for lutein production in flasks and the coiled tree photo-bioreactor (CTPBR). Appl Biochem Biotechnol 183:652–671

Gupta AK, Seth K, Maheshwari K, Baroliya PK, Meena M, Kumar A, Vinayak V, Harish (2021) Biosynthesis and extraction of high-value carotenoid from algae. Front Biosci (Landmark Ed) 6:171–190

Gwak Y, Hwang YS, Wang B, Kim M, Jeong J, Lee CG et al (2014) Comparative analyses of lipidomes and transcriptomes reveal a concerted action of multiple defensive systems against photooxidative stress in *Haematococcus pluvialis*. J Exp Bot 65:4317–4334

Hadi MR, Shariati M, Afsharzadeh S (2008) Microalgal biotechnology: carotenoid and glycerol production by the green algae *Dunaliella* isolated from the gave-Khooni salt marsh, Iran. Biotechnol Bioprocess Eng 13(5):540–544

Haldimann P, Tsimilli-Michael M (2002) Mercury inhibits the non-photochemical reduction of plastoquinone by exogenous NADPH and NADH: evidence from measurements of the polyphasic chlorophyll *a* rise in spinach chloroplasts. Photosynth Res 74:37–50

Hill R, Larkum AWD, Frankart C, Kühl M, Ralph PMJ (2004) Loss of functional photosystem II reaction centres in zooxanthellae of corals exposed to bleaching conditions: using fluorescence rise kinetics. Photosynth Res 82:59–72

Huang Q, Jiang F, Wang L, Yang C (2017) Design of Photobioreactors for mass cultivation of photosynthetic organisms. Engineering 3: 318–329

Jin E, Polle JEW, Lee HK, Hyun SM, Chang M (2003) Xanthophylls in microalgae from biosynthesis to biotechnological mass production and application. J Microbiol Biotechnol 13:165–174

Kim SM, Kang SW, Kwon ON, Chung D, Pan CH (2012a) Fucoxanthin as a major carotenoid in Isochrysis aff. Galbana: characterization of extraction for commercial application. J Korean Soc Appl Biol Chem 55:477–483

Kim SM, Jung YJ, Kwon ON, Cha KH, Um BH, Chung D, Pan CH (2012b) A potential commercial source of Fucoxanthin extracted from the microalga *Phaeodactylum tricornutum*. Appl Biochem Biotechnol 166:1843–1855

Kleinegris DMM, Janssen M, Brandeburg WA, Wijffels RH (2011) Continuous production of carotenoids from *Dunaliella salina*. Enzyme Microb Biotechnol 85:289–295

Koo SY, Cha KH, Song DG, Chung D, Pan CH (2011) Optimization of pressurized liquid extraction of zeaxanthin from *Chlorella ellipsoidea*. Environ Boil Fishes 24:725–730

Lamers PP, van de Laak CCW, Kaasenbrood PS, Lorier J, Janssen M, De Vos RCH et al (2010) Carotenoid and fatty acid metabolism in light-stressed *Dunaliella salina*. Biotechnol Bioeng 106:638–648

Lamers PP, Janssen M, De Vos RCH, Bino RJ, Wijffels RH (2012) Carotenoid and fatty acid metabolism in nitrogen-starved *Dunaliella salina*, a unicellular green microalga. J Biotechnol 162:21–27

Lazár D (2006) The polyphasic chlorophyll a rise measured under high intensity of exciting light. Funct Plan Biol 33:9–30

Li Y, Huang J (2009) High-light and sodium chloride stress differentially regulate the biosynthesis of astaxanthin in Chlorella zofingiensis. J Phycol 45(3):635–641

Li Y, Sun H, Wu T, Fu Y, He Y, Ma X, Chen F (2019) Storage carbon metabolism of Isochrysis zhangjiangensis under different light intensities and its application for co-production of fucoxanthin and stearidonic acid. Bioresour Technol 282:94–102

Lia X, Wanga X, Duana C, Yia S, Gaob Z, Xiaoc C, Agathosd SN, Wange G, Lia J (2020) Biotechnological production of astaxanthin from the microalga Haematococcus pluvialis. Biotechnol Adv 43: 107602

Liang C, Zhai Y, Xu D, Ye N, Zhang X, Wang Y, Zhang W, Yu J (2015) Correlation between lipid and carotenoid synthesis and photosynthetic capacity in *Haematococcus pluvialis*. Gracias Y Aceites 66(2): e077

Liu J, Sun Z, Gerken H, Liu Z, Jiang Y, Chen F (2014) *Chlorella zofingiensis* as an alternative microalgal producer of Astaxanthin: biology and industrial potential. Mar Drugs 12:3487–3515

Liyanaarachchi VC, Nishshanka GKSH, Premaratne RGMM, Ariyadasa TU, Nimarshana PHV, Malik A (2020) Astaxanthin accumulation in the green microalga Haematococcus pluvialis: effect of initial phosphate concentration and stepwise/continuous light stress. Biotechnol Rep 28:e00538

Ma R, Zhang Z, Ho SH, Ruan C, Li J, Xie Y, Shi X, Liu L, Chen J (2020) Two-stage bioprocess for hyper-production of lutein from microalga *Chlorella sorokiniana* FZU60: effects of temperature, light intensity, and operation strategies. Algal Res 52:102119

Mandal S, Yadav S, Yadav S, Nema RK (2009) Antioxidants a review. J Chem Pharmac Res 1(102):104

McClure DD, Nightingale JK, Luiz A, Black S, Zhu J, Kavanagh JM (2019) Pilot-scale production of lutein using *Chlorella vulgaris*. Algal Res 44:101707

Mitra M, Mishra S (2019) A comparative analysis of different extraction solvent systems on the extractability of eicosapentaenoic acid from the marine eustigmatophyte *Nannochloropsis oceanica*. Algal Res 38:101387

Mulders KJM, Janssen JH, Martens DE, Wijffels RH, Lamers PP (2014a) Effect of biomass concentration on secondary carotenoids and triacylglycerol (TAG) accumulation in nitrogen-depleted *Chlorella zofingiensis*. Algal Res 6:8–16

Mulders KJM, Lamers PP, Martens DE, Wijffels RH (2014b) Phototrophic pigment production with microalgae: biological constraints and opportunities. J Phycol 50:229–242

Neubauer C, Schreiber U (1987) The polyphasic rise of chlorophyll fluorescence upon onset of strong continuous illumination: 1. Saturation characteristics and partial control by photosystem II acceptor side. Z Naturforsch 42:1246–1254

Niyogi KK, Björkman O, Grossman A (1997) The role of specific xanthophylls in photoprotection. Proc Natl Acad Sci 94:14162–14167

Niyogi KK, Li XP, Rosemberg V, Jung HS (2004) Is the PsbS the side of non-photochemical quenching in photosynthesis? J Exp Bot 56: 375–382

Petrushkina M, Gusev E, Sorokin B, Zotko N, Mamaeva A, Filimonova A, Kulikovskiy M, Maltsev Y, Yampolsky I, Guglya E et al (2017) Fucoxanthin production by heterokont microalgae. Algal Res 24:387–393

Přibyl P, Pilný J, Cepák V, Kaštánek P (2016) The role of light and nitrogen in growth and carotenoid accumulation in *Scenedesmus* sp. Algal Res 16:69–75

Pulz O, Gross W (2004) Valuable products from biotechnology of microalgae. Appl Microbiol Biotechnol 65(6):635–648

Pushpalatha S, Sangeetha R, Ariraman S, Ashokkumar B, Varalakshmi P (2020) Photocatalyst (TiO2) as an enhancer: an attempt to enhance the production of carotenoids and lipids with the combined oxidative stresses in *Coelastrella* sp. MClean Technol Environ Policy 23:41–53

Rabbani S, Beyer P, von Lintig J, Hugueney P, Kleinig H (1998) Induced β-carotene synthesis driven by triacylglycerol deposition in the unicellular alga *Dunaliella bardawil*. Plant Physiol 116(4): 1239–1248

Rammuni M, Ariyadasa TU, Nimarshana P, Attalage R (2018) Comparative assessment on the extraction of carotenoids from microalgal sources: Astaxanthin from *H. Pluvialis* and β-carotene from *D. Salina*. Food Chem 277:128–134

Recht L, Töpfer N, Batushansky A, Sikron N, Zarka A, Gibon Y, Nikolosky Z, Fait A, Boussiba S (2014) Metabolite profiling and integrative modeling reveal metabolic constraints for carbon portioning under nitrogen starvation in the green algae *Haematococcus pluvialis*. JBC 289(44):30387–30403

Ren Y, Sun H, Deng J, Huang J, Chen F (2021) Carotenoid production from microalgae: biosynthesis, salinity responses and novel biotechnologies. Mar Drugs 19:713

Schülera LM, Santosa T, Pereiraa H, Duartea P, Gangadhara KN, Florindob C, Schulzec PSC, Barreiraa L, Varela JCS (2020) Improved production of lutein and β-carotene by thermal and light intensity upshifts in the marine microalga *Tetraselmis* sp. CTP4. Algal Res 45:101732

Serra AT, Silva SD, Pleno de Gouveia L, Alexandre AMRC, Pereira CV, Pereira AB, Partidário C, Silva NE, Bohn T, Gonçalves VSS et al (2021) A single dose of marine Chlorella vulgaris increases plasma concentrations of lutein, β-carotene and zeaxanthin in healthy male volunteers. Antioxidants 10:1164

Shah MMR, Liang Y, Cheng JJ, Daroch M (2016) Astaxanthin-producing green microalga Haematococcus pluvialis: from single cell to high value commercial products. Front Plant Sci 7:531

Shaker S, Morowvat MH, Ghasemi Y (2017) Effects of sulfur, iron and manganese starvation on growth, β-carotene production and lipid profile of *Dunaliella salina*. J Young Pharm 9(1):43–46

Singh DP, Khattar JS, Rajput A, Chaudhary R, Singh R (2019) High production of carotenoids by the green microalga *Asterarcys quadricellulare* PUMCC 5.1.1 under optimized culture conditions. PLoS One 14(9):e0221930

Strasser R, Srivastava A, Govindge G (1995) Polyphasic chlorophyll a fluorescence transient in plants and cyanobacteria. Photochem Photobiol 61:32–42

Torzillo G, Goksan T, Faraloni C, Kopecky J, Masojidek J (2003) Interplay between photochemical activities and pigment composition in an outdoor culture of *Haematococcus pluvialis* during the shift from the green to red stage. J Appl Phycol 15:127–136

Torzillo G, Faraloni C, Silva AM, Kopecky J, Pilný J, Masojídek J (2012) Photoacclimation of *Phaeodactylum tricornutum* (Bacillariophyceae) cultures grown in outdoors photobioreactors and open ponds. Eur J Phycol 47(2):169–181

Wan M, Hou D, Li Y, Fan J, Huang J, Liang S, Wang W, Pan R, Wang J, Li S (2014) The effective photoinduction of *Haematococcus pluvialis* for accumulating astaxanthin with attached cultivation. Bioresour Technol 163:26–32

Wang X, Zhang MM, Sun Z, Liu SF, Qin ZH, Mou J, Zhou ZG, Lin CSK (2020) Sustainable lipid and lutein production from *chlorella* mixotrophic fermentation by food waste hydrolysate. J Hazard Mater 400:123258

Wollman FA (2001) State transitions reveal the dynamics and flexibility of the photosynthetic apparatus. EMBO J 20:3623–3630

Xia S, Wang K, Wan L, Li A, Hu Q, Zhang C (2013) Production, characterization, and antioxidant activity of fucoxanthin from the marine diatom *Odontella aurita*. Mar Drugs 11:2667–2681

Yaakob Z, Kamarudin KF, Rajikumar R, Takriff MS, Badar SN (2014) The current methods for the biomass production of the microalgae from wastewaters: an overview. World Appl Sci J 31(10): 1744–1758

Yamamoto HY (1979) Biochemistry of the violaxanthin cycle in higher plants. Pure Appl Chem 51:6–648

Yamamoto HY, Nakayama TOM, Chichester CO (1962) Studies on the light and dark interconversion of leaf xanthophylls. Arch Biochem Biophys 97:168–173

11 Effect of Dietary Supplementation with Microalgae Biomass on Gastrointestinal Tract Health

Yurixhi Favela-Galindo, Alfredo de Jesús Martínez-Roldán, and Silvia Marina González-Herrera

Abstract

The microalgal biomass has high amounts of different molecules, and many of them have a beneficial effect on human and animal health. These molecules include short-chain fatty acids, antioxidants, carotenoids, amino acids, and polysaccharides. The presence of these molecules causes the consumption of microalgal biomass to improve the function of different organs and systems in the body. Microalgae have a high protein content, including almost the essential amino acids; the amount of polyunsaturated fatty acids can reach at least 25% of the total lipids, and some contain omega-3 and -6 fatty acids. Biomass also includes vitamins, such as A, members of the B complex, nicotinic acid, biotin, and folic acids. This allows many health benefits to be obtained by ingestion, such as weight loss, serum glucose, enhanced inflammatory biomarkers, and liver function. In particular, for the gastrointestinal tract, supplementation with microalgal biomass increases the survival of lactobacilli, such as *Lactobacillus brevis*, *Lactobacillus plantarum*, and *Bifidobacterium longum*. In this chapter, we analyze the information published about the beneficial effects on the gastrointestinal tract by dietary supplementation with microalgae biomass in humans and lab animals, including dosages, effects, and molecules involved.

Keywords

Gastric health · Gut flora · Biomass · Prebiotic

11.1 Introduction

Microalgae are considered the oldest group of photosynthetic microorganisms because cyanobacteria appeared about 3.5 billion years ago. This group is very diverse and some estimations suggest the existence of approximately 200,000 different species (Koyande et al. 2019). This microorganism has been employed as feed for centuries, and recently, microalgae and cyanobacteria have been produced with the objective not only to produce feed but also to obtain high-value products. Many of these high-value products have interesting biological activities, some of which are pigments, amino acids, antivirals, antibiotics, immunostimulants, antioxidants, and vitamins (Flórez-Miranda et al. 2015; Koyande et al. 2019; Sawant and Mane 2018). Despite the important potential of many of these molecules, today, there are many challenges in the biomass production process, the biomass harvest, the extraction of the compounds, and the purification processes, to ensure the purity and quality required to be used in humans (Martínez-Roldán and Ibarra-Berumen 2019); no matter the challenges, there is no doubt about the many benefits to health caused by the consumption of microalgal and cyanobacteria biomass or even specific fractions of the biomass (Koyande et al. 2019).

There are many reports about effects in many biological processes caused by the intake of biomass of microalgae and cyanobacteria, including the diminution of the risk of colitis, the improvement of gastric ulcers, and the reduction of gastric constipation (Camacho et al. 2019; Kiran and Venkata Mohan 2021). Some studies highlight the anti-inflammatory effects and improvements in the integrity of the gastrointestinal system or even in the role the biomass plays in the prevention of gastric cancer (Barkia et al. 2019; Camacho et al. 2019; Kiran and Venkata Mohan 2021; Lupette and

Y. Favela-Galindo · S. M. González-Herrera (✉)
Maestría en Ciencia y Tecnología en Alimentos Funcionales. Tecnológico Nacional de México/ IT de Durango, Durango, Mexico
e-mail: sgonzalez@itdurango.edu.mx

A. de Jesús Martínez-Roldán
Investigador Nacional Nivel I, Catedrático CONAHCYT - TecNM/ITD, Departamento de Ingenierías Química y Bioquímica, División de Estudios de Posgrado, Maestría en Sistemas Ambientales, Durango, Mexico

Maestría en Sistemas Ambientales. Tecnológico Nacional de México/ IT de Durango, Durango, Mexico
e-mail: adjmartinezro@conacyt.mx

A. de Jesús Martínez-Roldán (ed.), *Biotechnological Processes for Green Energy, and High Value Bioproducts by Microalgae, and Cyanobacteria Cultures*, Developments in Applied Phycology 13, https://doi.org/10.1007/978-3-031-43969-8_11

Benning 2020; Nazih and Bard 2018). Some studies include in vitro assays to evaluate the effect of the addition of microalgal biomass (complete or hydrolysated) to cultures of gastric bacteria, including genera, such as *Lactobacillus*, *Enterococcus*, and *Bifidobacterium*. These studies proved a prebiotic effect of the microalgal biomass because its addition stimulates the beneficial action, growth, and diversity of this type of bacteria (Beheshtipour et al. 2012; De Jesus Raposo et al. 2016; de Medeiros et al. 2021; Patel et al. 2021). Recent studies highlight de prebiotic effect of strains of *Spirulina* (Barkia et al. 2019; Camacho et al. 2019; de Medeiros et al. 2021; Kiran and Venkata Mohan 2021; Nazih and Bard 2018), *Chlorella* (Barkia et al. 2019; Camacho et al. 2019; de Medeiros et al. 2021; Kiran and Venkata Mohan 2021; Patel et al. 2021), and other less common genera, such as *Phaeodactylum*, *Porphyridium*, *Haematococcus*, *Isochrysis*, *Schizochytrium*, *Nannochloropsis*, *Nostoc*, and *Nannochloropsis* (Camacho et al. 2019; Kiran and Venkata Mohan 2021; Lupette and Benning 2020). In the majority of these studies, the beneficial effect is related to the amount of polysaccharides, fatty acids (saturated or unsaturated), amino acids, antioxidants, and other molecules in the biomass (Kiran and Venkata Mohan 2021).

11.2 Bioactive Compounds

Today, approximately 75% of the microalgal biomass produced is commercialized as tablets, pills, capsules, or powder (Galasso et al. 2019). Nevertheless, in biomass, there are many compounds with the potential to be used and commercialized as nutraceuticals that have higher market prices than pills or powders (Singh et al. 2005). Some of these high-value products include antioxidants, such as carotenes, ascorbic acid, glutathione, chlorophylls, polysaccharides, sulfated polysaccharides, fatty acids, and unsaturated fatty acids (eicosapentaenoic, arachidonic, linoleic, oleic, docosahexaenoic, etc.), phycobilins, vitamins, sterols (such as phytosterol), and others (Cañizares-Villanueva et al. 2022; Kiran and Venkata Mohan 2021). All these compounds have proven beneficial effects on health not only in animals but also in humans, such as preventing infectious processes; reducing the levels of reactive oxygen species; diminishing inflammatory processes; anticancer, antiviral, and antimicrobial activities; and others (Cañizares-Villanueva et al. 2022; Patel et al. 2021; Singh et al. 2005). Table 11.1 lists some microalgal and cyanobacteria species, highlighting the nutraceutical compound and the proposed application for it.

Because of the presence of polysaccharides, amino acids, carbohydrates, and fatty acids, the biomass of microalgae and cyanobacteria is proposed to be a coadjutant in the medical treatment of the digestive system, such as colitis, ulcers, gastritis, and several other ailments such as cancer.

11.3 Microalgae as an Adjuvant in Gastric Diseases

The employment of astaxanthin is important in the treatment of gastric diseases with microalgal biomass or a fraction thereof. This pigment was employed in mice to diminish the colonization by *Helicobacter pylori* in the gastric tract and reduce the abdominal inflammation and bacterial amount; this study was carried out by supplementation of the diet with astaxanthin-rich *Haematococcus* biomass (Kang and Kim 2017). Another study employed extracts of *Chlorococcum* sp. rich in the same pigment and the consumption caused a reduction in cytokine production in *H. pylori*-infected BALB/c mice. The authors propose the enrichment of diets with *Chlorococcum* sp. biomass as an effective therapy against cases of infection with *H. pylori*, but they highlight more research on the topic to understand the mechanism and the molecules involved in the process (Liu and Lee 2003). Similar benefits have been observed when guinea pigs and humans with chronic infections of *H. pylori* consume biomass rich in beta-carotene. In the majority of cases, the effect is attributed to the capability of beta carotene to reduce the amount of reactive oxygen species produced by the infection of *H. pylori* (Kang and Kim 2017).

In the case of colon diseases, there are reports of beneficial effects of the consumption of microalgal biomass. Ávila-Román et al. (2016) supplied *Chlamydomonas debaryana* biomass to recurrent 2,4,6-trinitrobenzenesulfonic acid (TNBS)–induced colitis mice; the authors observed inhibition in the weight loss caused by the disease, improvements in the status and health of the mouse colon, and an important diminution in the amount of proinflammatory cytokines (Ávila-Román et al. 2016). The studies carried out in humans have demonstrated diminution in gastric inflammation, lower generation of flatulence, and enhancements in intestinal movements by the consumption of 1 to 3 g of *Chlamydomonas reinhardtii* biomass by 30 days (Fields et al. 2020).

11.4 Microalgae as Prebiotics

The microalgal biomass has been proposed to be an effective prebiotic because of the presence of fibers and polysaccharides in the biomass. The prebiotic concept was defined as a substance that favored the growth of microorganisms and was recently modified and is defined as a nondigestible component in food with the capability to stimulate the growth and proliferation of bacteria considered probiotics (Gupta and Garg 2009). Probiotic microorganisms have proven to play an important role in not only intestinal wellness but also nutrient absorption and the prevention of infections. The group called probiotic bacteria includes

Table 11.1 Molecules with nutraceutical potential have been reported for different species of microalgae and cyanobacteria (Kiran and Venkata Mohan 2021; Perković et al. 2022; Talero et al. 2015)

Specie	Nutraceutical	Molecule	Application
Chlorella vulgaris	Lutein	Pigment	Anticancer
Spirulina platensis	B12 vitamin	Vitamin	
	Phycocyanin	Protein-like pigment	Anticancer
Haematococcus pluvialis	Astaxanthin	Pigment	Antioxidant
Dunaliella salina	Betacarotene	Pigment	Cardioprotective agents
Dunaliella tertiolecta	Violaxanthin	Pigment	Anticancer
Phaeodactylum tricornutum	Fucoxanthin	Pigment	Anticancer
Porphyridium cruentum	Sulfated polysaccharides	Polysaccharide	Antiviral
Nannochloropsis oculata	Enriched EPA oil	Oil	Cardioprotective agents
Schizochytrium sp	Docosahexaenoic acid	Polyunsaturated fatty acid	Rheumatoid arthritis treatment
Crypthecodinium cohnii	Docosahexaenoic acid	Polyunsaturated fatty acid	Rheumatoid arthritis treatment

strains of the genera *Bifidobacterium*, *Lactobacillus*, *Escherichia*, *Streptococcus*, *Enterococcus*, and *Bacillus* and even some fungal strains, such as *Saccharomyces* (Gupta and Garg 2009; Holzapfel and Schillinger 2002).

The prebiotic compounds include carbohydrates, such as inulin or fructooligosaccharides (Holzapfel and Schillinger 2002); many other compounds have been recently proposed as prebiotics, and some are microalgae that have an important beneficial effect on probiotic bacteria. However, the diversity of microalgal strains makes the study of specific strains and compounds derived from microalgal biomass necessary to understand the mechanism that causes the stimulation of the prebiotic bacteria (Perković et al. 2022). There are few studies employing specific strains of probiotic bacteria, but the number of studies proving the beneficial effect of the consumption of microalgal biomass on gastric health in both animals and humans allows us to propose microalgae as an effective prebiotic. Nevertheless, it is necessary to understand the components of the biomass that cause the effect and the chemical mechanisms that allow the stimulation of probiotic bacteria (de Medeiros et al. 2021; Fields et al. 2020).

Some studies have focused on evaluating the effect of specific fractions on intestinal health. For example, Taylor et al. (2020) evaluated the effect of supplying B-glucan and paramylon (polysaccharides) from microalgal biomass to mice susceptible to colitis and found an improvement in the immune response and a recovery in the health in the gastric tract caused by the consumption of polysaccharides (Taylor et al. 2020).

11.5 In vitro Studies of the Beneficial Effect on Gastric Bacteria

There are reports of the beneficial effect of the addition of powdered microalgal biomass on probiotic bacteria. The majority includes biomass of genera, such as *Chlorella*, *Spirulina*, and *Dunaliella*, and these microalgae have proven beneficial effects on human health because of their higher amounts of antioxidants, amino acids, and carbohydrates. Additionally, their biomass production processes are very developed, and the production cost allows the commercialization of biomass at low prices (Martínez-Roldán and Cañizares–Villanueva 2019). Specifically, the effect on the probiotic bacteria caused by the addition of microalgal biomass includes increases in the growth of acid lactic bacteria, the reduction of the lag phase in the cultures of prebiotic microorganisms, and increases in the viability and the host immunostimulant effect. Another author proposed the possibility of producing lactic products enriched with microalgal biomass with the objective of improving the viability of the bacteria and causing extra benefits to the host (Patel et al. 2021).

De Medeiros et al. (2021) studied the in vitro effect of adding microalgal biomass to human colonic microbiota; the study included biomass of *Chlorella vulgaris*, *Desmodesmus maximus*, *Chlorococcum* sp., and *S. platensis,* and the effect was compared with cultures supplemented with fructooligosaccharides. They observed an increase in the abundance of *Lactobacillus–Enterococcus* and *Bifidobacterium* and a decrease in the abundance of Prevotellaceae–Bacteroidaceae, *Clostridium histolyticum*, and *Eubacterium rectale–Clostridium coccoides*. The effect was attributed to the presence of nystose, raffinose, and phenolic compounds, as well as insoluble fibers and kestose.

In addition, there are studies concerning the viability of the lactic bacteria in fermented products. In a research employing yogurt containing *Lactobacillus acidophilus*, *Bifidobacterium lactis*, *L. delbrueckii* ssp. *bulgaricus*, and *Stresptococcus themophilus*, the effect of the addition of biomass of *C. vulgaris* and *S. plantensis* was evaluated. The employment of *C. vulgaris* biomass increases the viability of the lactobacillus and the bifidobacteria during the fermentation process and in the storage period. For the products containing *S. platensis*, the pH decreases very slowly and the final acidity of the final product was improved. These benefits were observed with the addition of only 0.5% to 1% of biomass (Beheshtipour et al. 2012). Other authors propose the development of fermented foods added with *Chlorella* biomass and bifidobacteria because the microalgal biomass potentiates the beneficial effect of the bifidobacteria in the gastrointestinal tract (Hyrslova et al. 2021).

Additionally, there is a study demonstrating a synergistic effect between *Chlorella* biomass and the beneficial effects of *B. longum* and *L. plantarum* against rotavirus, the major diarrhea-causing pathogen worldwide. Nevertheless, it is important to highlight not only the antiviral effect of *Chlorella* itself but also the fact that the use of prebiotic bacteria increases the effect (Cantú-Bernal et al. 2020).

11.6 Co-culture of Microalgae-Benefic Bacteria

Another strategy proposed to employ microalgae as prebiotics is the co-culture of microalgae and probiotic bacteria. The co-culture strategy has been proven to be an efficient strategy for the development of biotechnologies and is promising for producing products with pre- and probiotic activities or even both (Perković et al. 2022). The co-culture strategy has many operative advantages, such as the diminution of contamination risks, the reduction of operative costs, or the elimination of downstream phases for harvesting or purification (Kapoore et al. 2021). Some studies of microalgae and bacteria co-culture demonstrated that the microalgal cell could be used as a holder to immobilize the bacterial cells. In a report employing a co-culture of *Bacillus casei* and *Chlorella*, with the bacteria attached to the cell wall of *Chlorella*, the microalgae acted as a kind of biocarrier. The adhesion of the bacteria to *Chlorella* causes improvements in the transit across the digestive system of the host, and this type of encapsulation increases the number of viable cells that arrive at the intestine, increasing the probiotic potential of the product (Emad and Elbialy 2019).

11.7 Conclusion

The supplementation of both animal and human diets with microalgal biomass benefits not only the digestive health but also the immune system, as well as the diminution in free radicals. The consumption of probiotic bacteria has important health benefits, and the use of both microalgae and probiotic bacteria can improve the benefits by a synergistic effect. Therefore, it is necessary to study mixtures of different microalgal biomasses (with different compositions) and probiotic bacteria in vitro and in vivo, to understand the molecules that cause the beneficial effects. In addition, it is important to study the possibility of developing co-cultures of probiotic bacteria and microalgae, because, in a co-culture, the operational cost is reduced; but this technology needs to be improved as the optimal culturing conditions of both microorganisms are very different, which impedes maximal biomass production.

Acknowledgments Favela–Galindo Y. thanks the CONAHCYT for the scholarship for the accomplishment of her Master studies, and all the authors thank the Tecnológico Nacional de México/IT de Durango for the support provided for this investigation.

References

Ávila-Román J, Talero E, Rodríguez-Luna A, García-Mauriño S, Motilva V (2016) Anti-inflammatory effects of an oxylipin-containing lyophilised biomass from a microalga in a murine recurrent colitis model. Br J Nutr 116:2044–2052

Barkia I, Saari N, Manning SR (2019) Microalgae for high-value products towards human health and nutrition. Mar Drugs 17:304

Beheshtipour H, Mortazavian AM, Haratian P, Khosravi-Darani K (2012) Effects of Chlorella vulgaris and Arthrospira platensis addition on viability of probiotic bacteria in yogurt and its biochemical properties. Eur Food Res Technol 235:719–728

Camacho F, Macedo A, Malcata F (2019) Potential industrial applications and commercialization of microalgae in the functional food and feed industries: a short review. Mar Drugs 17:312

Cañizares-Villanueva RO, Rojo-Gómez E, Arroyo-Sáchez BI, Martínez-Roldán AJ (2022) Chlorella as a source of antioxidant compounds. In: Marutholi M (ed) Chlorella and its health benefits, 1st edn. Nova Science Publishers

Cantú-Bernal S, Domínguez-Gámez M, Medina-Peraza I, Aros-Uzarraga E, Ontiveros N, Flores-Mendoza L, Gomez-Flores R, Tamez-Guerra P, González-Ochoa G (2020) Enhanced viability and anti-rotavirus effect of Bifidobacterium longum and lactobacillus plantarum in combination with Chlorella sorokiniana in a dairy product. Front Microbiol 11:875

De Jesus RMF, Bernardo De Morais AMM, De Morais Santos Costa RM (2016) Emergent sources of prebiotics: seaweeds and microalgae. Mar Drugs 14:27

de Medeiros VPB, de Souza EL, Albuquerque TMR, da Costa Sassi CF, dos Santos Lima M, Sivieri K, Colombo Pimentel T, Magnani M (2021) Freshwater microalgae biomasses exert a prebiotic effect on human colonic microbiota. Algal Res 60:102547

Emad M, Elbialy Z (2019) The algal biomass as a mechanical carrier for the lactobacillus bacteria and its uses in the food supplementation. Alex J Vet Sci 63:127

Fields FJ, Lejzerowicz F, Schroeder D, Ngoi SM, Tran M, McDonal D, Jiang L, Chan JT, Knight R, Mayfield S (2020) Effects of the microalgae *Chlamydomonas* on gastrointestinal health. J Funct Foods 65:103738

Flórez-Miranda L, Martínez-Roldán AJ, Cañizares-Villanueva RO (2015) Microalgae in livestock nutrition. In: Torres-Bustillos L (ed) Microalgae and other phototrophic bacteria: culture, processing, recovery and new products. Nova Science Publishers, USA, pp 87–110

Galasso C, Gentile A, Orefice I, Ianora A, Bruno A, Noonan DM, Sansone C, Albini A, Brunet C (2019) Microalgal derivatives as potential Nutraceutical and food supplements for human health: a focus on cancer prevention and interception. Nutrients 11:1226

Gupta V, Garg R (2009) Probiotics. Indian J Med Microbiol 27:202–209

Holzapfel WH, Schillinger U (2002) Introduction to pre- and probiotics. Food Res Int 35:109–116

Hyrslova I, Krausova G, Smolova J, Stankova B, Branyik T, Malinska H, Huttl M, Kana A, Doskocil I, Curda L (2021) Prebiotic and immunomodulatory properties of the microalga *Chlorella vulgaris* and its synergistic triglyceride-lowering effect with bifidobacteria. Fermentation 7:125

Kang H, Kim H (2017) Astaxanthin and β-carotene in *helicobacter pylori*-induced gastric inflammation: a mini-review on action mechanisms. J Cancer Prev 22:57

Kapoore RV, Wood EE, Llewellyn CA (2021) Algae biostimulants: a critical look at microalgal biostimulants for sustainable agricultural practices. Biotechnol Adv 49:107754

Kiran BR, Venkata Mohan S (2021) Microalgal cell biofactory—therapeutic nutraceutical and functional food applications. Plants 10:836

Koyande AK, Chew KW, Rambabu K, Tao Y, Chu DT, Show PL (2019) Microalgae: a potential alternative to health supplementation for humans. Food Sci Human Wellness 8:16–24

Liu BH, Lee YK (2003) Effect of total secondary carotenoids extracts from *Chlorococcum* sp. on *helicobacter pylori*-infected BALB/c mice. Int Immunopharmacol 3:979–986

Lupette J, Benning C (2020) Human health benefits of very-long-chain polyunsaturated fatty acids from microalgae. Biochimie 178:15–25

Martínez-Roldán AJ, Cañizares–Villanueva RO (2019) Biodiesel production from microalgae as a biorefinery concept. In: Rios Pinto LF, Marques Takase E, Silva Santana H (eds) A closer look at biodiesel production. Nova Science Publishers

Martínez-Roldán AJ, Ibarra-Berumen J (2019) Employment of wastewater to produce microalgal biomass as a biorefinery concept. In: Alam M, Wang Z (eds) Microalgae biotechnology for development of biofuel and wastewater treatment. Springer Singapore, Singapore, pp 487–504

Nazih H, Bard JM (2018) Microalgae in human health: interest as a functional food. In: Levine IA, Fleurence J (eds) Microalgae in health and disease prevention. Elsevier, pp 211–226

Patel AK, Singhania RR, Awasthi MK, Varjani S, Bhatia SK, Tsai ML, Hsieh SL, Che CW, Dong CD (2021) Emerging prospects of macro- and microalgae as prebiotic. Microb Cell Factories 20:1–16

Perković L, Djedović E, Vujović T, Baković M, Paradžik T, Čož-Rakovac R (2022) Biotechnological enhancement of probiotics through co-cultivation with algae: future or a trend? Mar Drugs 15: 142

Sawant SS, Mane VK (2018) Nutritional profile, antioxidant, antimicrobial potential, and bioactives profile of *Chlorella emersonii* kj725233. Asian J Pharm Clin Res 11:220–225

Singh S, Kate BN, Banecjee UC (2005) Bioactive compounds from cyanobacteria and microalgae: an overview. Crit Rev Biotechnol 25: 73–95

Talero E, García-Mauriño S, Ávila-Román J, Rodríguez-Luna A, Alcaide A, Motilva V (2015) Bioactive compounds isolated from microalgae in chronic inflammation and cancer. Mar Drugs 13: 6152–6209

Taylor HB, Gudi R, Brown R, Vasu C (2020) Dynamics of structural and functional changes in gut microbiota during treatment with a microalgal β-Glucan, Paramylon and the impact on gut inflammation. Nutrients 12:2193

Sustainable Production of Diatom-Based Omega-3 Fatty Acids 12

Aishwarya Mogal, Sarvjeet Kukreja, and Shristy Gautam

Abstract

Diatoms constitute a prominent group of phytoplankton characterized by their siliceous cell wall. They account for 20–25% of oxygen release and carbon dioxide fixation in the world. Diatoms yield bioactive compounds, one of which is lipids, and store fatty acids to produce different valuable products like omega-3 fatty acids, such as EPA and DHA, which are polyunsaturated fatty acids (PUFAs). Omega-3 fatty acids (EPA and DHA) have many health and medical benefits, including prevention and treatment of various diseases, and therefore taken on a regular basis. Due to this, there is an increased consumption of omega-3 fatty acids. Many studies found EPA and DHA content in various fatty fish which led to pressure on fish stock. An alternative source for producing omega-3 fatty acids is diatoms, the primary producer microalgae. This chapter discusses the importance of omega-3 fatty acids (EPA and DHA), lipid biosynthesis and factors influences lipid accumulation.

Keywords

Eicosapentaenoic acid · EPA · Docosahexaenoic acid · DHA · Diatoms · Omega-3 fatty acids

A. Mogal
Department of Molecular Biology and Genetic Engineering, School of Bioengineering and Biosciences, Lovely Professional University, Phagwara, India

S. Kukreja
Department of Agronomy, Lovely Professional University, Phagwara, India

S. Gautam (✉)
Department of Molecular Biology and Genetic Engineering, School of Bioengineering and Biosciences, Lovely Professional University, Phagwara, India

Biosciences and Biotechnologies Institute of Aix-Marseille (BIAM), CEA Cadarache, Saint-Paul-lez-Durance, France

12.1 Introduction

Algae has served as a source of high-value materials used in a wide range of industries, including pharmaceuticals, nutraceuticals, supplements, biotechnology, energy, and environmental applications. Diatoms are unicellular eukaryotes that live within silica-based (SiO2) cell walls (Battarbee et al. 2002) and are among the most common types of photosynthetic algae. The ability of diatoms to synthesize a variety of high-value materials has received more attention. Fatty acids (FAs) in diatoms are mostly in the C14:0 to C22:6 range. The most prevalent and significant FAs present in relatively large amounts in several diatom species include myristic acid (C14:0), palmitic acid (C16: 0), palmitoleic acid (C16:1n-7), eicosapentaenoic acid (EPA) (C20:5), and docosahexaenoic acid (DHA) (C22:6). Diatoms are autotrophic microalgae, can fix 20% carbon dioxide (CO_2), and accounts for 40% total oceanic productivity. Diatoms produce and accumulate omega-3 (ω-3) FAs, a type of polyunsaturated fatty acids (PUFAs), by lipid biosynthesis, in the form of triacylglycerides (TAGs). Omega-3 FAs have 20 or more carbon atoms and have the first double bond between the third and fourth carbon atoms at the methyl/omega end. Omega-3 FAs have been shown to reduce obesity, metabolic syndrome, arteriosclerosis, diabetes, autoimmune disorders such as inflammatory bowel disease, cancer, rheumatoid arthritis, asthma, microbial, the flu, and malignancies. In cardiovascular system safety, treatment of psychiatric disorders, and in some pregnancy-related problems, EPA and DHA are significant. DHA plays a unique role in the development of the brain and eyes. Due to this property, ω-3 FAs are considered as nutraceutical products with nutritious and medicinal value to prevent or cure diseases. Omega-3 FAs (EPA/DHA) cannot be synthesized in the human body to meet daily requirements. To meet their everyday needs, humans have to rely on sources like seafood, fish, oysters, and plant sources like flax seeds, chia, and walnuts. The European Food Safety

A. de Jesús Martínez-Roldán (ed.), *Biotechnological Processes for Green Energy, and High Value Bioproducts by Microalgae, and Cyanobacteria Cultures*, Developments in Applied Phycology 13, https://doi.org/10.1007/978-3-031-43969-8_12

Authority (EFSA) Panel on Dietetic Products, Nutritious, and Allergies (NDA, under the European Commission) recommends a daily dose of 250 mg EPA/DHA (Patel et al. 2021), which cannot be obtained from seafood, as it would be a burden on them (EFSA Dietetic Products and Allergies 2014). Other sources for ω-3 FAs are plants, which only synthesize linolenic acid (LA), γ-linolenic acid (GLA), and α-linolenic acid (ALA) and cannot synthesize EPA and DHA. To overcome all the hurdles, diatoms are considered an alternative source of ω-3 FAs. This chapter reviews ω-3 FAs production in diatoms and their use as a nutraceutical.

Nutraceuticals are food components that have nutritious and medicinal value to prevent or cure diseases. Omega-3 and -6, among PUFAs, are considered a family of nutraceuticals having properties of health benefits (Hamilton et al. 2020; Patel et al. 2021). Omega-3 and -6 FAs are primarily present in diatoms, autotrophic organisms found in broad nutrient and environment stress like hot springs, sea ice, wastewater, and freshwater, fixing 20% of atmospheric carbon, thereby decreasing CO_2 and nitrogen oxide content released from different sources (Marella et al. 2020), and achieving 40% of total oceanic productivity (McNair et al. 2018).

Diatoms are the primary group of ω-3 FAs, which produce and accumulate as TAGs (Vaezi 2015). Hence, they receive special attention for their benefits in human health for their anti-inflammatory, antioxidant, antibacterial, and antiviral properties (Ścieszka and Klewicka 2019). Diatoms are a major group of microalgae inhabiting water bodies (both marine and freshwater) where sufficient sunlight is received. Trillions of diatoms have been reported to be found on earth; each species of diatom possesses unique frustules (cell wall) pattern and shape, which is the characteristic feature of a diatom species. They are found to generate 20% of global oxygen produced annually and contribute to half of the organic matter present on earth (Field et al. 1998). Diatoms have the form of bands, fans, zigzags, and stars, either alone or in colonies. Individual cells vary between 2 and 200 microns in size. In the presence of sunlight and sufficient nutrients, living diatoms duplicate themselves every 24 h by asexual multiple fission. The cell wall around diatoms made up of silica is the distinctive feature of diatom anatomy. These are described as "jewels of the sea" and "living opals." Frustules, a defining characteristic of diatoms, are made of siliceous cell wall, and fit together very tightly so that any organic molecules across the wall have to pass through pores/slits (Li et al. 2014). Diatoms inhabit wet or moist place and live freely (planktonic) or with the attachment of substratum (benthic) having a specialized system secreting mucilage for the attachment or gliding along a surface, forming layers of cells of microorganisms which are tightly attached to each other known as biofilms (Martins et al. 2013). Since the nineteenth century, diatoms have been studied in India. A number of Indian and other researchers studied the Indian subcontinent's marine, freshwater, and fossil diatoms. Around 7000 diatomic taxa inhabit the Indian subcontinent's marine, brackish, fresh, and fossil areas (Patel et al. 2019).

12.2 Microalgae as the Initial Producers of Omega-3 Fatty Acid

The most important primary producers of marine ecosystems are microalgae (diatoms) which transform sunlight and CO_2 into biomass, including lipids, proteins, carbohydrates, and fats. Microalgae that accumulate at different trophic levels (Postma et al. 2016) also produce high-value FAs such as ω-3 FAs. Lipid content undergoes changes from the lowest level of trophy such as zooplanktons, crustaceans, mollusks, and some fish, which in turn affects the accumulation of lipids in higher species and humans. Omega-3 and -6, long-chain PUFAs, with 20 or more carbon atoms (Vaezi 2015), can be differentiated based on the double bond present between carbon atoms at methyl or omega end. If the first double bond is between the third and fourth carbon atoms, it is ω-3, and if it is between the sixth and seventh carbon atoms, it is ω-6 (Patel et al. 2021). ALA, EPA, DHA are the types of ω-3 while LA, GLA, and arachidonic acid (ARA) are the types of ω-6 having human benefits (Alam et al. 2020). Among ω-3 FAs, EPA and DHA play a vital nutritious role. Consumption of EPA/DHA daily can reduce a person's risk from suffering metabolic syndrome, cardiovascular diseases, arrhythmias, type 2 diabetes, inflammation, hypertension, obesity, and anthero-thrombosis due to anti-inflammatory, antithrombotic, and anti-arrhythmic properties (Vaezi 2015). Those who suffer from autoimmune disorders such as asthma, inflammatory bowel disease, and rheumatoid arthritis have high cytokine levels. To reduce the cytokine levels in the patients, they are treated with EPA and DHA which have potent immunomodulatory activities, either through transcription factor, gene expression, and intracellular signaling pathways or through the amount of eicosanoids (Vaezi 2015).

DHA plays a vital role in treating cancer as it confers apoptosis and neurological disorders (Vaezi 2015). International authorities like the EFSA recognize DHA, as a crucial ω-3 fatty acid, as it regulates development and function of the brain and vision if taken regularly during pregnancy, lactation, and throughout a child's infancy (Castejón and Señoráns 2020). Both ω-3 and ω-6 are essential bioactive compounds for health, but humans cannot synthesize them on their own to meet daily requirements (Alam et al. 2020). To meet their daily need of ω-3 FAs, seafood like fish (tuna, mackerel, sardines, and salmon), shrimps, and oysters (Alam et al. 2020) are used as a source (Oliver et al. 2020). The Long-chain polyunsaturated fatty acid (LC-PUFAs) in fish are generally present as free FAs containing approximately 180 mg EPA and 120 mg DHA in 1000 mg fish oil. This suggests that each TAG molecule present consists of one LC-omega-3 FAs (Schuchardt and Hahn 2013).

Considering the growing population, known health benefits, and daily recommendations intake, there is a massive burden on fish stock resulting in the depletion of reserves (Hamilton et al. 2015). Also, the preference for a vegan diet and vegetarian food and the low intake of EPA and DHA lead to the deficiency of ω-3 FAs (Dams et al. 2020). Large fishes do not produce PUFAs; they obtain it by bioaccumulation, preying upon small fishes that feed on zooplankton, phytoplankton, and microalgae (Alam et al. 2020). Another major problem is the polluted marine environment with heavy metals and polychlorinated biphenyl compounds contaminating fishes (Patel et al. 2021). So, fish cannot be relied on for ω-3 FAs. . Some seeds are good sources of ω-3 Fas, namely flaxseed, chia, and walnuts (Patel et al. 2021). The parent ALA is abundantly present in vegetable oils, able to convert it into EPA and DHA, but not to an extent to provide any medicinal, curable, and therapeutic benefits associated with it and it cannot meet daily requirements (Kaur et al. 2014).

To overcome all the hurdles, microalgae are considered to be an alternative to fish oils and plant oils for a reliable, long-term, and sustainable source of ω-3 FAs (Hamilton et al. 2015; Hamilton et al. 2020). Algae, the most common organisms on Earth, are found in nearly every climatic and livable condition, from terrestrial (soil) to aqueous environments (freshwater, marine water, brackish water) (Adarme-Vega et al. 2012; TV 2021). The estimated species, according to Guiry (2012), is 35 crores, and according to the Internet database Algaebase (http://www.algaebase.org), it is over 150,000 species (Ścieszka and Klewicka 2019). They are highly variable in their shapes and sizes and range from simple unicellular to complex multicellular eukaryotic organisms (RATH 2012) having all the organelles like cell wall, developed nucleus, chloroplast having pigments, mitochondria, and chlorophyll. Microalgae are classified based on cell wall components, cytological and morphological aspects, types of reserve metabolites, and pigments, e.g., Chlorophyta consists of chlorophylls and Phaeophyta group has the presence of fucoxanthin. In contrast, Rhodophyta has phycoerythrin and phycocyanin embedded in them (Singh and Sharma 2012). They can double their population in a concise period due to their faster growth rate and hence have a high biomass production rate, therefore used for a long time (Sharma and Sharma 2017). Algae constitute a large group of microorganisms (Patel et al. 2021) capable of converting atmospheric CO_2 and other nutrients into valuable bioactive secondary metabolites such as proteins, vitamins like tocopherols, carbohydrates like polysaccharides, lipids like PUFAs which are ω-3 FAs, monounsaturated FAs (MUFAs), sterols, and pigments like chlorophylls, carotenoids, and phycobilins (Pasquet et al. 2011; Ścieszka and Klewicka 2019).

Microalgae are induced with certain stress conditions, influencing their metabolism to grant high lipid production and accumulation. For example, with nitrogen starvation and optimal growth conditions, cells begin to accumulate lipids, resulting in an increase in lipid content, increasing the content of ω-3 FAs. Production of both lipids and ω-3 FAs can increase by varying optimal growth conditions (Adarme-Vega et al. 2012). EPA content yield in green algae *Nannochloropsis oculata* is 19%, whereas, in red algae *Rhodella violacea*, it accounts for 46%, and in *R. maculate*, it accounts for 40% of total lipids when subjected to 14 °C. In *Isochrysis galbana*, DHA content constituted 15% of total lipids when subjected to 8 °C (Aussant et al. 2018; Patel et al. 2021).

Research has been carried out on microalgae to produce lipids and their DHA and EPA content. Still, oleaginous diatoms family of microalgae group has not been explored (Patel et al. 2021) much for lipid production and its use in nutraceuticals. Diatoms are photoautotrophic but can change their metabolism to heterotrophic conditions when the sun's rays do not reach the bottom of water bodies, as it is the only way to survive. Because of this characteristic, diatoms can be cultivated worldwide in every climatic condition (Winwood 2013; Oliver et al. 2020). Some macronutrients are required in heterotrophic conditions; ammonium or yeast extract acts as a nitrogen source, phosphate acts as a phosphorous source, while glucose acts as a carbon source (Oliver et al. 2020). The production rate of lipids in the heterotrophic condition is 70% of their biomass. Diatoms also survive in mixotrophic conditions where the sole carbon source is provided externally (Patel et al. 2021). One of the well-known diatoms, *Phaeodactylum tricornutum*, is used for the production of lipids; when *P. tricornutum* was shifted from its optimal growth temperature of 25 °C to 10 °C, an increase in EPA content was reported from 13.1% to 24.2% (Patel et al. 2021).

12.3 Omega-3 Fatty Acid Production in Diatoms

Diatoms are developing a wide variety of compounds that help them adapt and survive under different conditions. During the growth stages of diatoms, FAs accumulate in the cells serving as energy source during unfavorable growth conditions or cell division. The concentration of FAs in the marine form varies from 10% to 50% (w/w), providing a high lipid content between 30 and 70% dry weight. Omega-3 FAs have high-energy content, thus stored in cells for cellular function. Numerous strains of microalgae such as

Phaeodactylum, Nannochlopsis, Thraustochythrium, and *Schizochythrium* have been studied extensively for ω-3 FA content. High-value FAs such as ω-3 FAs from microalgae, diatom, can be commercially manufactured through cell growth in media that are optimized for greater productivity; the media should contain the optimum amount of diverse nutritional sources and the pH and temperature kept to optimum. This optimization is required for higher biomass production. By controlling the concentration of various constituents of the media, a high yield of biomass can be obtained. Various advanced biotechnological techniques and metabolic engineering techniques can be used to selectively breed and obtain high-yielding DHA-producing diatom species. Therefore, such procedures would help in the commercial production of DHA thus addressing the global needs (Fig. 12.1).

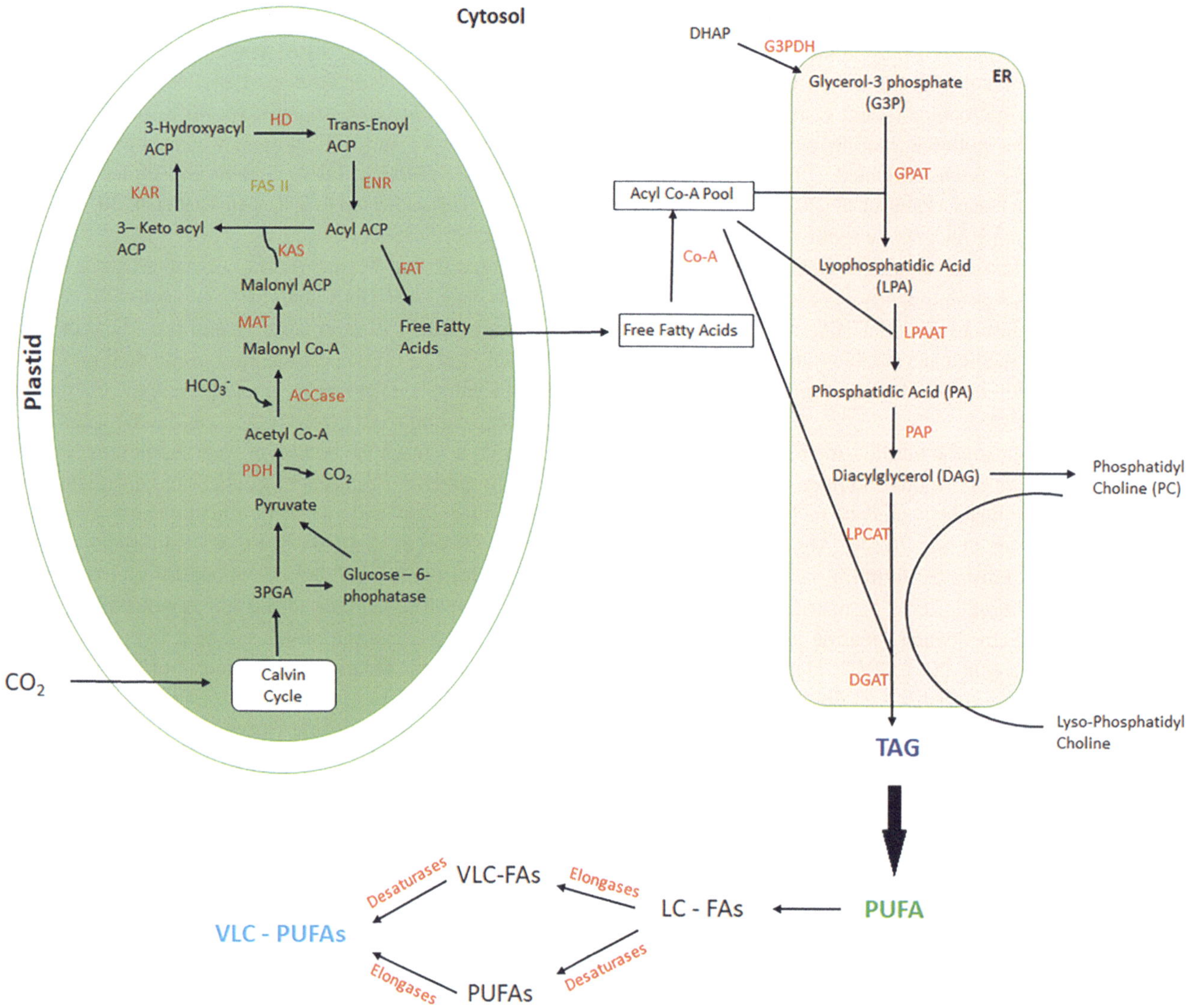

Fig. 12.1 Schematic overview of lipid biosynthesis in microalgae. PDH, pyruvate dehydrogenase complex; ACCase, acetyl CoA carboxylase; MAT, malonyl CoA: ACP transacylase; KAS, 3-ketoacyl ACP synthase; KAR, 3-ketoacyl ACP reductase; HD, 3-hydroxy acyl ACP dehydratase; ENR, enoyl ACP reductase; FAT, fatty acyl ACP thioesterase; Co-A, coenzyme A; DHAP, dihydroxyacetone phosphate; G3PDH, glycerol-3-phosphate dehydrogenase; GPAT, glycerol-3-phosphate acyltransferase; LPAAT, lysophosphatidic acid acyltransferase; PAP, phosphatidic acid phosphatase; LPCAT, lysophosphatidylcholine acyltransferase; DGAT, diacylglycerol acyltransferase; TAG, triacylglycerol; PUFA, polyunsaturated FAs; LC-FAs, long-chain FAs; VLC-FAs, very-long-chain FAs; VLC-PUFAs, very long-chain polyunsaturated FAs; ACP, acyl carrier protein. [Modified from Vaezi (2015), Sayanova et al. (2017), Zulu et al. (2018)]

12.4 Factors Affecting Lipid Accumulation

Many factors influence lipogenesis and its accumulation in diatoms, including carbon intake, nutrient stress, and environmental conditions. When conditions are unfavorable for diatom growth, they account for 80% of total lipid accumulation. Diatoms are autotrophic; hence, they use light energy and fix CO_2 from the atmosphere, which they then use for various metabolic processes. As carbon in diatoms is used to build storage compounds, there will be more lipid buildup if carbon consumption is elevated (Kaur et al. 2021). Many studies have been conducted on diatoms like *Thalassiosira pseudonana, P. tricornutum, T. weissflogii* for carbon uptake and its use in lipid metabolism and assumed that diatoms use Calvin–Benson carbon fixing cycle to fix carbon. *P. tricornutum,* under carbon deprivation, accumulates PUFAs, like EPA and DHA, with a high degree of unsaturation (Obata et al. 2013). Diatoms can be cultured heterotrophically by providing an extra organic carbon source in dark conditions, or mixotrophically by adding carbon sources such as glycerol, glucose, acetate, and starch (Zulu et al. 2018) and ammonia as a nitrogen source which enhances the growth of diatoms (Sayanova et al. 2017). The metabolism of lipids in diatoms is also enhanced by nutrient stress such as nitrogen, phosphorus, and silica. Many studies have been conducted on the diatoms *T. pseudonana, C. muelleri, T. weissflogii,* and *P. tricornutum* in which nitrogen limitation was employed as a nutrient stress and it has been observed that growth rate has reduced with increase in lipid content (d'Ippolito et al. 2015; Kaur et al. 2021).

Silicon is an important nutrient for diatom growth and the main component of cell wall; however, in a silicon-deficient environment, lipid accumulation rises. In one of the studies, it was shown that lipid content increased from 28% to 61% dry weight under silicon starvation (Levitan et al. 2014). Under silicon deficiency, the diatoms *T. pseudonana* and *Chaetoceros gracilis* generated a high amount of lipid (Zulu et al. 2018). Silicon deficiency, like nitrogen deficiency, has no physiological consequences on the diatoms (Levitan et al. 2014).

Phosphorous constraint is the third nutrient stress that diatoms can be subjected to in order to increase their lipid content. Phosphorous plays a crucial role in respiration and photosynthesis and is a major aspect of nucleic acid and phospholipids. Studies have shown increased levels of TAG in *P. tricornutum* (Abida et al. 2015) and *Chaetoceros* sp. (Obata et al. 2013). Other studies reveal a decline in EPA, suggesting that it is being restricted to save energy (Yang et al. 2014), but 6% increase in EPA production in the TAG pool was observed by Abida and colleagues (Abida et al. 2015).

The temperature has a significant impact on diatom growth. According to many studies conducted, a drop in temperature resulted in a higher yield of EPA and DHA. When the temperature was lowered from 25°°C to 10°°C, an increase in EPA content was found in *P. tricornutum* (Mimouni et al. 2018). Diatoms like *C. calcitrans, C. gracilis, T. pseudonana,* and *C. simplex* when subjected to 10 °C showed high PUFA content (Thompson et al. 1992). In *Cylindrotheca closterium,* at low temperature stationary phase, higher lipid content was seen (Almeyda et al. 2020). Similar results were obtained when *Odontella aurita* was stressed at low temperatures (Mimouni et al. 2018).

Another way to boost lipid synthesis in diatoms is light intensity. When the diatom *P. tricornutum* was exposed to 60 mol photon $m^{-2} s^{-1}$ of low-intensity light, a high yield of TAG was achieved (Remmers et al. 2017). On the contrary, when the marine diatom *Skeletonema costatum* was subjected to elevated light intensity, high amount of EPA was recovered (Sayanova et al. 2017). Prolonged exposure to light induces the synthesis of lipids, but it also causes the generation of reactive oxygen species, which harms chlorophyll (Goh et al. 2019). Salinity induces fluidity and permeability of the cell membrane of diatoms creating high osmotic pressure due to excess intake of salts (Mimouni et al. 2018). Marine diatom *N. laevis* shows enhancement in EPA concentration when subjected to 20 g/L NaCl (Chen et al. 2008).

12.5 Extraction and Purification of Omega-3 Fatty Acids

Diatoms are cultivated by means of the use of photobioreactors, open ponds, breeds, or hybrids to improve cell density. It uses flocculation or centrifugation to separate the algal cells from the culture media filtration, followed by drying to improve extraction (Martins et al. 2013). After that, lipid extraction is usually done using an organic solvent that is non-water miscible. The method, Bligh and Dyer, (Peng et al. 2014) uses methanol/chloroform solvent mixture for cell disruption. Bligh and Dyer are often the basis of the small extraction protocol. Nevertheless, the removal of larger measurements is usually done using hexane as a solvent. Saturated fatty acids will be precipitated more when the temperature of oil gets reduced (Patel et al. 2019). Processing often relies on the physique characteristics of the microalgae strain (e.g., properties of cell wall and cellular size) in order to improve the shell life, quality, and quantity of harvest and extraction techniques using the final product. Fresh products or dry pellets in aquaculture are used to maintain the nutritional content. The very first step in this process is to dewater microalgal biomass first by filtration, dissolved flotation, flocculation, sediment, and subsequently drying up into pellets, or administering directly to animals. Further extraction and purification processes are needed when produced in industries such as pharmaceutical industry.

Winterization, supercritical extraction, and partial (molecular) distillation are the recent techniques used to extract and purify PUFA from microalgae (Patel et al. 2019).

12.6 Biorefinery Approach for the Production of Omega-3 Fatty Acid

Diatom's ability to produce a wide range of products, including proteins, carbohydrates, and oils, has prompted the improvement of a biorefinery technique for large-scale production of these key metabolites from diatom (Peng et al. 2014). Similarly, in petrochemical industries where the crude oil produced is used for a range of other applications and production, the microalgal oils can also be engineered to produce bioproducts in the similar way. Lipids produced from diatom can be fractionally separated to produce high-valued lipids and FAs. The remaining fraction of the lipids can be used for the production of biofuels mainly in the processing of biodiesel. Biodiesel produced from microalgae will be able to address the global energy crises due to shortage of fossil fuels. After the processing of the lipids, the biomass that remains can be used as animal food supplement, which is a sustainable production that reduces the dependence on plant for animal fodder (Khan et al. 2018).

12.7 Challenges

Diatoms, a type of microalgae, are a third-generation feedstock that does not compete with food crops for nutrients, water intake, and arable land, but is more environmental friendly. However, using diatoms for long-term ω-3 fatty acid synthesis poses several difficulties. The cultivation and harvesting of appropriate diatom species on industrial scale is a critical stage. Toxic heavy metal pollution, water composition, and nutrients can all be found in diatoms cultured from wastewater reservoirs or treatment plants. The next task is to modify and analyze growth parameters such as temperature, salinity, light intensity, pH, and nutrient stress such as nitrogen, phosphorus, silicon, and carbon intake in order to increase ω-3 FA synthesis. It is important to develop a more sustainable method to the production of ω-3 FAs that includes low-cost equipment, nontoxic solvents, and minimal product loss.

12.8 Conclusion

Diatoms are important not only because they are diverse, ecological, have high biomass production rates, no competition with food crops for water, nutrients, and have a high ability to fix atmospheric CO_2 but also because they are widely applied in environmental monitoring, nanotech, medicine, farming, and food industries. They are a potent source for the production of various high-value products such as ω-3 FAs (DHA and EPA). DHAs are being produced from fishes, but due to global decline of fish stock it can no longer provide a sustainable source of ω-3 FAs (Peng et al. 2014); also fishes are not the primary producers of ω-3 FAs. Using microalgae such as diatoms proves to be a sustainable approach toward solving this problem as they have a higher ability to produce ω-3 FAs than any other sources. DHAs have several health benefits and people have been using DHA as food supplements for a long. Three distinct problems, namely human health, transport energy, and food safety, can be resolved if this development process is done at a large scale. In the last few years, algal biotechnology has developed and prospered into the global industry. Research and approaches are ongoing to achieve economically viable production of microalgal lipids. In addition to helping people develop a better understanding of their distributions, the discovery of DHA and EPA compounds from unexplored regions often contributes to various commercial purposes and is essential for humans to keep their bodies healthy and disease-free and without any limitations.

References

Abida H, Dolch L-J, Meï C, Villanova V, Conte M, Block MA, Finazzi G, Bastien O, Tirichine L, Bowler C (2015) Membrane glycerolipid remodeling triggered by nitrogen and phosphorus starvation in Phaeodactylum tricornutum. Plant Physiol 167(1):118–136

Adarme-Vega TC, Lim DK, Timmins M, Vernen F, Li Y, Schenk PM (2012) Microalgal biofactories: a promising approach towards sustainable omega-3 fatty acid production. Microb Cell Factories 11(1):1–10

Alam MA, Xu J-L, Wang Z (2020) Microalgae biotechnology for food, health and high value products. Springer

Almeyda MD, Scodelaro Bilbao PG, Popovich CA, Constenla D, Leonardi PI (2020) Enhancement of polyunsaturated fatty acid production under low-temperature stress in Cylindrotheca closterium. J Appl Phycol 32(2):989–1001

Aussant J, Guihéneuf F, Stengel DB (2018) Impact of temperature on fatty acid composition and nutritional value in eight species of microalgae. Appl Microbiol Biotechnol 102(12):5279–5297

Battarbee RW, Jones VJ, Flower RJ, Cameron NG, Bennion H, Carvalho L, Juggins S (2002) Diatoms. In: Tracking environmental change using lake sediments. Springer, pp 155–202

Castejón N, Señoráns FJ (2020) Enzymatic modification to produce health-promoting lipids from fish oil, algae and other new omega-3 sources: a review. New Biotechnol 57:45–54

Chen GQ, Jiang Y, Chen F (2008) Salt-induced alterations in lipid composition of diatom Nitzschia Laevis (Bacillariophyceae) under heterotrophic culture condition 1. J Phycol 44(5):1309–1314

d'Ippolito G, Sardo A, Paris D, Vella FM, Adelfi MG, Botte P, Gallo C, Fontana A (2015) Potential of lipid metabolism in marine diatoms for biofuel production. Biotechnol Biofuels 8(1):1–10

Dams S, Holasek S, Tsiountsioura M, Edelsbrunner M, Dietz P, Koefeler H, Malliga D-E, Gürbüz A, Meier-Allard N, Poncza B (2020) Effects of a plant-based fatty acid supplement and a powdered fruit, vegetable and berry juice concentrate on omega-3-indices and serum micronutrient concentrations in healthy subjects. Int J Food Sci Nutr 71(6):769–780

EFSA Dietetic Products N, Allergies (2014) Scientific opinion on health benefits of seafood (fish and shellfish) consumption in relation to health risks associated with exposure to methylmercury. EFSA J 12(7):3761

Field CB, Behrenfeld MJ, Randerson JT, Falkowski P (1998) Primary production of the biosphere: integrating terrestrial and oceanic components. Science 281(5374):237–240

Goh BHH, Ong HC, Cheah MY, Chen W-H, Yu KL, Mahlia TMI (2019) Sustainability of direct biodiesel synthesis from microalgae biomass: a critical review. Renew Sust Energ Rev 107:59–74

Guiry MD (2012) How many species of algae are there? J Phycol 48 (5):1057–1063. https://doi.org/10.1111/j.1529-8817.2012.01222.x

Hamilton ML, Warwick J, Terry A, Allen MJ, Napier JA, Sayanova O (2015) Towards the industrial production of omega-3 long chain polyunsaturated fatty acids from a genetically modified diatom Phaeodactylum tricornutum. PLoS One 10(12):e0144054

Hamilton HA, Newton R, Auchterlonie NA, Müller DB (2020) Systems approach to quantify the global omega-3 fatty acid cycle. Nature Food 1(1):59–62

Kaur N, Chugh V, Gupta AK (2014) Essential fatty acids as functional components of foods-a review. J Food Sci Technol 51(10): 2289–2303

Kaur R, Kaur L, Gautam S (2021) Role of lipid droplets in diatoms for biofuel production. J App Biol Biotech 9(5):1–9

Khan MI, Shin JH, Kim JD (2018) The promising future of microalgae: current status, challenges, and optimization of a sustainable and renewable industry for biofuels, feed, and other products. Microb Cell Factories 17(1):1–21

Levitan O, Dinamarca J, Hochman G, Falkowski PG (2014) Diatoms: a fossil fuel of the future. Trends Biotechnol 32(3):117–124

Li H-Y, Lu Y, Zheng J-W, Yang W-D, Liu J-S (2014) Biochemical and genetic engineering of diatoms for polyunsaturated fatty acid biosynthesis. Mar Drugs 12(1):153–166

Marella TK, López-Pacheco IY, Parra-Saldívar R, Dixit S, Tiwari A (2020) Wealth from waste: diatoms as tools for phycoremediation of wastewater and for obtaining value from the biomass. Sci Total Environ 724:137960

Martins DA, Custódio L, Barreira L, Pereira H, Ben-Hamadou R, Varela J, Abu-Salah KM (2013) Alternative sources of n-3 long-chain polyunsaturated fatty acids in marine microalgae. Mar Drugs 11(7):2259–2281

McNair HM, Brzezinski MA, Krause JW (2018) Diatom populations in an upwelling environment decrease silica content to avoid growth limitation. Environ Microbiol 20(11):4184–4193

Mimouni V, Couzinet-Mossion A, Ulmann L, Wielgosz-Collin G (2018) Lipids from microalgae. In: Microalgae in health and disease prevention. Elsevier, pp 109–131

Obata T, Fernie AR, Nunes-Nesi A (2013) The central carbon and energy metabolism of marine diatoms. Meta 3(2):325–346

Oliver L, Dietrich T, Marañón I, Villarán MC, Barrio RJ (2020) Producing omega-3 polyunsaturated fatty acids: a review of sustainable sources and future trends for the EPA and DHA market. Resources 9(12):148

Pasquet V, Chérouvrier J-R, Farhat F, Thiéry V, Piot J-M, Bérard J-B, Kaas R, Serive B, Patrice T, Cadoret J-P (2011) Study on the microalgal pigments extraction process: performance of microwave assisted extraction. Process Biochem 46(1):59–67

Patel A, Matsakas L, Hrůzová K, Rova U, Christakopoulos P (2019) Biosynthesis of nutraceutical fatty acids by the oleaginous marine microalgae Phaeodactylum tricornutum utilizing hydrolysates from organosolv-pretreated birch and spruce biomass. Mar Drugs 17(2): 119

Patel A, Karageorgou D, Katapodis P, Sharma A, Rova U, Christakopoulos P, Matsakas L (2021) Bioprospecting of thraustochytrids for omega-3 fatty acids: a sustainable approach to reduce dependency on animal sources. Trends Food Sci Technol 115:433–444

Peng K-T, Zheng C-N, Xue J, Chen X-Y, Yang W-D, Liu J-S, Bai W, Li H-Y (2014) Delta 5 fatty acid desaturase upregulates the synthesis of polyunsaturated fatty acids in the marine diatom Phaeodactylum tricornutum. J Agric Food Chem 62(35):8773–8776

Postma P, Barbosa M, Wijffels R, Eppink M, Olivieri G (2016) Microalgal biorefinery for bulk and high-value products: product extraction within cell disintegration. In: Handbook of electroporation, pp 1–20

Rath B (2012) Microalgal bioremediation: current practices and perspectives. J Biochem Technol 3(3):299–304

Remmers IM, Martens DE, Wijffels RH, Lamers PP (2017) Dynamics of triacylglycerol and EPA production in Phaeodactylum tricornutum under nitrogen starvation at different light intensities. PLoS One 12(4):e0175630

Sayanova O, Mimouni V, Ulmann L, Morant-Manceau A, Pasquet V, Schoefs B, Napier JA (2017) Modulation of lipid biosynthesis by stress in diatoms. Philos Trans R Soc B: Biol Sci 372(1728): 20160407

Schuchardt JP, Hahn A (2013) Bioavailability of long-chain omega-3 fatty acids. Prostaglandins Leukot Essent Fat Acids 89(1):1–8

Ścieszka S, Klewicka E (2019) Algae in food: a general review. Crit Rev Food Sci Nutr 59(21):3538–3547

Sharma P, Sharma N (2017) Industrial and biotechnological applications of algae: a review. J Adv Plant Biol 1(1):01

Singh R, Sharma S (2012) Development of suitable photobioreactor for algae production–a review. Renew Sust Energ Rev 16(4):2347–2353

Thompson PA, Mx G, Harrison PJ, Whyte JN (1992) Effects of variation in temperature. II. On the fatty acid composition of eight species of marine phytoplankton 1. J Phycol 28(4):488–497

TV J (2021) The in herentpotential of algae for forthcoming future: a comprehensive review. Ann Romanian Soc Cell Biol 324:12452–12462

Vaezi R (2015) Enhancing the production of omega-3 polyunsaturated fatty acids in marine diatoms. University of East Anglia

Winwood R (2013) Algal oil as a source of omega-3 fatty acids. In: Food enrichment with Omega-3 fatty acids. Elsevier, pp 389–404

Yang ZK, Zheng JW, Niu YF, Yang WD, Liu JS, Li HY (2014) Systems-level analysis of the metabolic responses of the diatom P haeodactylum tricornutum to phosphorus stress. Environ Microbiol 16(6):1793–1807

Zulu NN, Zienkiewicz K, Vollheyde K, Feussner I (2018) Current trends to comprehend lipid metabolism in diatoms. Prog Lipid Res 70:1–16

Metabolites from Microalgal Cultures as Potential Sources for the Pharmaceutical Industry 13

María Luján Flores, Mariana Jiménez-Veuthey, and Osvaldo León Córdoba

Abstract

Microalgae are industrially valuable due to their metabolic potential and ability to biosynthesize different bioactive components such as carbohydrates, proteins, phenols, fatty acids, pigments, sterols, and terpenes, which have potential use in the pharmaceutical, cosmetic, agronomic, food, and other industries. This explains the growing interest in the search for new compounds from microalgae obtained both through traditional cultures and, more recently, from particular nutrient sources such as industrial effluents. In the latter, microalgae can be used not only as bioremediation agents but also as a source of nutrients and bioactive components. This chapter reviews the most recent aspects related to obtaining microalgal biomass with high performance and quality for its application in health-related industries. In this sense, this chapter analyzes the operating conditions of the culture and the abiotic factors that most influence the metabolite concentration in microalgal biomass. Likewise, the authors describe the methodologies for metabolite extraction, fractionation, and analysis and present examples that have demonstrated their potential pharmaceutical application, such as strains of *Euglena* and *Scenedesmus*.

Keywords

Bioactive metabolites · Culture · *Euglena* · Metabolism · Microalgal biomass · Pharmaceutical importance · Pharmacognostical protocol · *Scenedesmus*

M. Luján Flores (✉)
Farmacognosia, Grupo de Investigación Química, Bioactividad y Metabolismo de Recursos Naturales Patagónicos (GQBMRNP) y Area de Análisis Instrumental (AAI) – Centro Regional de Investigación y Desarrollo Científico Tecnológico (CRIDECIT), Facultad de Ciencias Naturales y Ciencias de la Salud (FCNyCS), Universidad Nacional de la Patagonia San Juan Bosco (UNPSJB), Comodoro Rivadavia, Argentina

M. Jiménez-Veuthey
Laboratorio de Análisis de Metales en Alimentos y Otros Sustratos (LAMAS), Facultad de Ciencias de la Alimentación, Universidad Nacional de Entre Ríos (UNER), Instituto de Ciencia y Tecnología de los Alimentos de Entre Ríos (ICTAER), Concordia, Argentina

CONICET, Consejo Nacional de Investigaciones Científicas y Técnicas, Buenos Aires, Argentina
e-mail: mariana.jimenezveuthey@uner.edu.ar

O. León Córdoba
Química Biológica II, Grupo de Investigación Química Bioactividad y Metabolismo de Recursos Naturales Patagónicos (GQBMRNP) y Area de Análisis Instrumental (AAI), Laboratorio de Cromatografía y Análisis Instrumental (LACROMI) – Centro Regional de Investigación y Desarrollo Científico Tecnológico (CRIDECIT), Facultad de Ciencias Naturales y Ciencias de la Salud (FCNyCS), Universidad Nacional de la Patagonia San Juan Bosco (UNPSJB), Comodoro Rivadavia, Argentina

13.1 Introduction

Algae make up a heterogeneous group of photosynthetic organisms that inhabit different ecosystems. They are found not only in marine and freshwater environments, but can also colonize deserts, volcanoes, acidic environments, and even frozen environments. They have a high tolerance to variations in temperature, salinity, pH, light intensity, humidity, and nutrients, and can grow either alone or in symbiosis with other organisms (Barsanti et al. 2008; Khan et al. 2018).

According to their size, they are divided into macroalgae and microalgae. The former are multicellular, present different sizes visible to the human eye, and inhabit mainly marine environments, whereas the latter are simple, microscopic, photoautotrophic organisms, which can be cyanobacteria-like prokaryotes (Chloroxybacteria) or green algae-like eukaryotes (Chlorophyta). Prokaryotes are simple and small single-celled organisms, whereas eukaryotes are larger, more complex, and can be either unicellular or multicellular (Levasseur et al. 2020).

Microalgae have been traditionally classified according to their photosynthetic pigments. However, the present systems of classification take into account other criteria, including their cytological and morphological characters, cell wall constituents, and chemical nature of their storage products. Some of the methods commonly used to identify and classify

A. de Jesús Martínez-Roldán (ed.), *Biotechnological Processes for Green Energy, and High Value Bioproducts by Microalgae, and Cyanobacteria Cultures*, Developments in Applied Phycology 13, https://doi.org/10.1007/978-3-031-43969-8_13

algal species include morphological microscopic observations, molecular-based classification using specific short gene sequences and, more recently, semi-automated or fully automated classification using a flow cytometer combined with computational methods (Sosik and Olson 2007; Drews et al. 2013; Correa et al. 2021).

Microalgae have the ability to biosynthesize different bioactive components such as carbohydrates, proteins, phenols, fatty acids, pigments, sterols, and terpenes, which are essential and nonessential compounds produced as defense systems against adverse environmental conditions and predators (Gómez Luna 2007; Rizwan et al. 2018; Mishraa et al. 2021). The interest in these compounds is that they have shown potential use in the pharmaceutical, cosmetic, agronomic, food, and other industries and also have applications in wastewater treatment and atmospheric CO_2 mitigation.

The industrial cultivation of microalgae to produce biofuels and bioproducts has increased dramatically over the last few decades. Microalgae are produced in quantity and sold directly as food and nutrient supplements, while their processed products or extracts are used in biopharmaceuticals and cosmetics (Levasseur et al. 2020).

To obtain microalgal biomass for industrial applications, it must be taken into account that each step of the culture process can condition the quality of the products (Rizwan et al. 2018). The first step is to select the microalgal species considering the probability of obtaining sustainable biomass and to select the environmental conditions. The latter is important because favorable conditions will favor some species while stress conditions may enhance the production of metabolites of interest by other species (Rizwan et al. 2018). Next, it is important to select the most suitable system for the cultivation of the selected species taking into account the availability of nutrients, the favorable environmental factors for that species, the destination of the biomass produced, and economic aspects. The system can be either continuous or discontinuous, and either open or closed. In general, open systems are cheaper, with greater durability and good biomass productivity, whereas closed systems are more expensive, because they imply the use of closed photobioreactors that allow achieving a controlled environment. Productivity depends on various parameters that must be operationally in balance, including the microalgal strain, the amount of nutrients, the quantity and quality of the light, the temperature, the pH, and the culture substrate. Photosynthesis requires a carbon source and light supply, whereas growth requires the presence of nitrogen and phosphorus. All these factors and operational parameters condition the metabolism of the species, as also occurs with macroalgae and plants (Rizwan et al. 2018).

Once the optimum growth has been reached, the biomass obtained is collected, which implies removing a large amount of water. This can be done by various methods, including filtration, ultrafiltration, centrifugation, sedimentation, either alone or in combination, e.g., with flocculation. At this stage, it is crucial to take into account the destination of the biomass produced; in the case the biomass is destined for the pharmaceutical industry, which is the main focus of this chapter, it is necessary to prevent both the degradation of the metabolites of interest and the toxicity due to the presence of artifacts generated during handling, since this can condition the efficacy of the products obtained (Richmond 2004). In this case, as well as in the case the biomass is destined for the food industry, it is recommended to use continuously operating refrigerated centrifuges, since they allow large volumes of biomass to be processed in a short time and at certain temperatures.

It is also important that the collection method allows achieving an optimum biomass density (Grima et al. 2003; Richmond 2004). For this, it must be considered that the humidity content is acceptable, for which drying or dehydration may be necessary. The latter must be done mechanically, e.g., using lyophilization, spray-drying, or tunnel drying. Although natural drying in the sun is cheaper, this drying method is not recommended because it causes alterations in the chemical structure of photo-unstable metabolites, with a consequent effect on the quality of the final product.

The biomass obtained in optimal conditions for its subsequent use in the pharmaceutical industry must be then treated for the subsequent extraction of the products of interest. For this, it is important to induce cell disruption so that the metabolites in the microalgal cells can be released. This can be done by various mechanical methods, which must be chosen taking into account the composition of the cell wall of the microalgae and the nature of the products to be obtained. The methods include the use of manual mortars, mills, cell homogenizers, ultrasound alone or combined with material freezing, and vacuum. While there are nonmechanical methods, e.g., enzymatic reactions, chemical reactions at different pH or organic solvents, these cannot be used universally because they alter the chemical composition of the biomass. In the cases in which the material is not sifted during the process, once cell disruption has been achieved, sifting should be carried out to achieve a homogeneous particle size to optimize extraction (Rizwan et al. 2018).

After obtaining the material in adequate conditions to produce the products of interest, the next step is the extraction. The choice of the extraction system or systems to be used will depend on the objective, that is, on the products to which it is directed. The extraction may be general or specific for certain chemical groups; it can also be done with a single solvent or sequentially and exhaustively using solvents of increasing polarity. In the latter case, a usual sequence is *n*-hexane, followed by dichloromethane, ethyl acetate, methanol, and water. In addition, each solvent may be either unitary, such as dichloromethane or water, or a mixture of two or more, such as 96° ethanol, which constitutes a hydroalcoholic mixture. On the other hand, the extraction systems can be simple discontinuous or semi-continuous, or, on a larger scale, continuous. In addition, they may or may not include agitation. In the case they include agitation, it may be mechanical, magnetic, or orbital, and the process is called turboextraction. Regarding the temperature, the extraction can be at room or hot temperature; in the latter case, the temperature should not reach the boiling point of the solvent. It is also possible to work at temperatures lower than the ambient temperature, e.g., in the range between 8 and 15 °C. The high-cost equipment is used at an industrial scale, especially for supercritical fluids where CO_2 is the solvent of choice for the pharmaceutical, food, and cosmetic industries (Rizwan et al. 2018).

Once the extraction is complete, the extracts are recovered by refrigerated centrifugation and vacuum filtration. The extracts are concentrated under reduced pressure and dried in vacuum ovens, or by spraying. In the case of aqueous extracts, they are dried by lyophilization. Subsequently, they are subjected to fractionation and purification of the metabolites of interest. These topics will be explored in greater depth in the next sections of the chapter (Rizwan et al. 2018).

Regarding the application of microalgae in health, these organisms represent a novel source of primary and secondary metabolites such as carbohydrates, sulfated polysaccharides, peptides, proteins, phenol derivatives, terpenes, steroids, fatty acids, and pigments (Rizwan et al. 2018; Mishraa et al. 2021). These bioactive compounds exhibit a wide variety of physicochemical and pharmacological properties, like gelling and therapeutic properties, can be used as drug carriers, and present important antioxidant, antimicrobial, anticancer, or antiviral properties (Khan et al. 2018). They are thus of biotechnological, pharmaceutical, and cosmetic interests (Khan et al. 2018).

This chapter reviews the bioactive metabolites produced by microalgae and their wide application in the pharmaceutical industry.

13.2 Microalgal Metabolism and Cultivation

13.2.1 Microalgal Metabolism

As mentioned, microalgae are a diverse group of eukaryotic and prokaryotic photosynthetic microscopic algae with great ecological and economic impact. They can exist individually, in chains, or in groups, and grow in different environments such as freshwater and marine water. They can grow in harsh conditions and are able to withstand anaerobiosis, cold, heat, photooxidaton, and salinity (Patras et al. 2018; Marrez et al. 2019). The versatility of the metabolism of microalgae allows them to react quickly to changes in the environment. This property is important and of interest for its use to obtain metabolites for industrial applications.

As mentioned, primary metabolites produced by microalgae include carbohydrates, proteins, lipids, and fatty acids. Primary metabolism in turn provides intermediates for the synthesis of more complex metabolites. Carbohydrates are complex and, depending on the species and environmental factors, can include neutral sugars, amino sugars, uronic, and sulfonic acids, many of them with biological activity. Proteins are of interest not only as a source of essential amino acids but also for their biological activity and potential use in the industry (Patras et al. 2018; Marrez et al. 2019). Lipids are of great importance both to develop biofuels and to obtain essential fatty acids. Polyunsaturated fatty acids (PUFAs) are of great commercial interest due to their multiple applications. In particular, the PUFAs eicosapentaenoic acid (EPA, 20:5n-3) and docosahexaenoic acid (DHA, 22:6n-3) are produced by marine microalgae. Although EPA has been found in several classes of marine microalgae, only several species have shown high production capacity.

Secondary metabolites show a more limited distribution, typical of the species under study (Becker 1994; Chisti 2007; Kamalanathan and Quigg 2019). These metabolites are not vital to basic cellular function, but give microalgae advantages for their development against other organisms (Fig. 13.1). They are compounds with an immense diversity of biological and structural activities, often specific for certain species or strains. Thus, the study of these metabolites has been fundamental in the discovery of drugs with potential pharmaceutical value (Davies 2007; Maschek and Baker 2007; Li and Vederas 2009; Sasso et al. 2012).

Secondary metabolites synthesized by microalgae include various types of terpenoids. These metabolites are synthesized by effective biosynthesis pathways, including the eukaryotic mevalonic acid (MVA) pathway and the prokaryotic methyl-D-erythritol 4-phosphate (MEP) pathway. The former takes place in the cytosol and uses three molecules of acetyl-CoA for the production of an isopentenyl diphosphate (IPP) or dimethylallyl diphosphate (DMAPP) molecule, whereas the latter takes place in the chloroplast and uses a GAP molecule and a pyruvate molecule for the production of an IPP or DMAPP molecule (Clomburg et al. 2019). After long-term evolution, some eukaryotic microalgae such as *Chlamydomonas* and *Nannochloropsis* sp. have lost the MVA ancestral pathway and have only the MEP pathway (Davies et al. 2015).

Secondary metabolites can also be synthesized by the shikimic acid pathway (whose feedstock molecules are phosphoenol pyruvate (PEP) and D-erytrhose-4-phosphate). This pathway allows microalgae to synthesize a large number of precursors of aromatic compounds of diverse skeletal

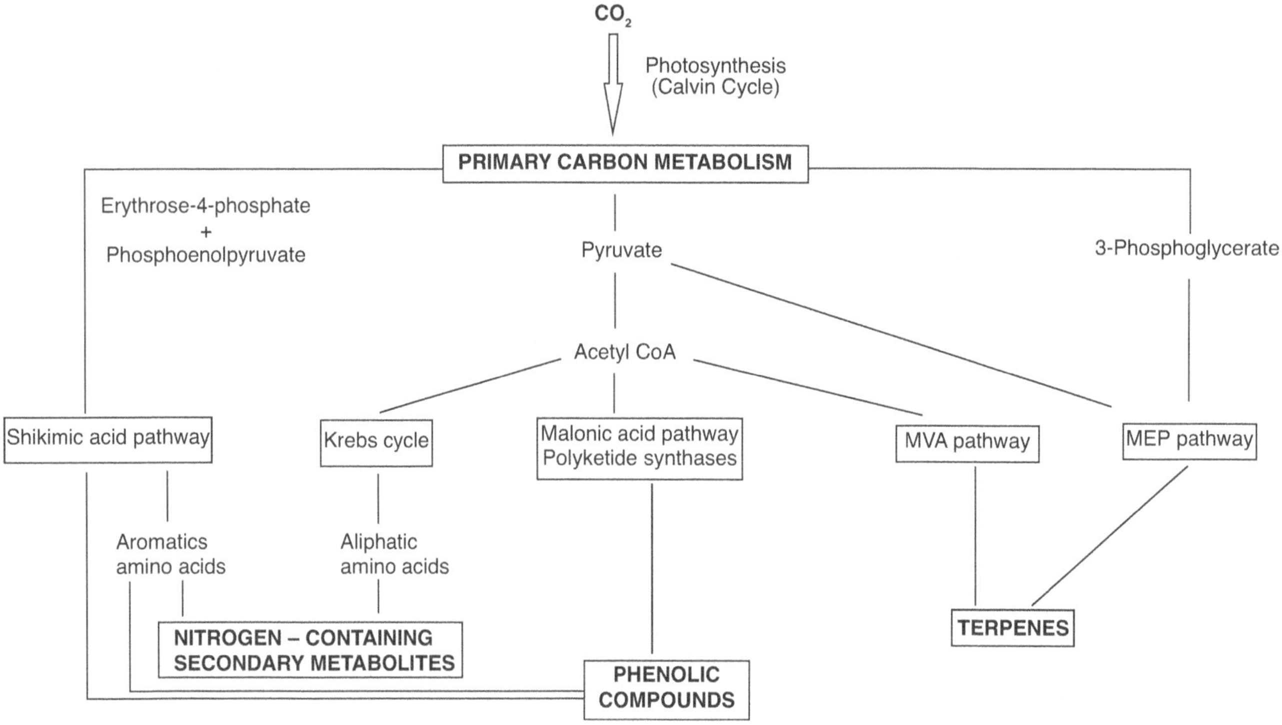

Fig. 13.1 Metabolic pathways

patterns and substitutions. Chorismic acid, an important branching point product in this pathway, also serves as an important precursor. These precursors are used by natural products with C_6-C_1, C_6-C_3, or C_6-C_3-C_3-C_6 units as their complete skeletal carbon content or with the C_3-C_6 of the C_6-C_3-C_6 system as their partial skeletal carbon content (Talapatra and Talapatra 2015).

Because of their antioxidant activities, aromatic compounds such as phenolic compounds exhibit several health-related properties, including anticancer, antiviral, and anti-inflammatory activities, effects on capillary fragility, and ability to inhibit human platelet aggregation (Dillard and German 2000; Dai and Mumper 2010; Lopez et al. 2015).

Phenolic compounds can also be synthesized by the polyketide pathway. The initial carbon chain of a polyketide is formed by intermolecular Claisen condensation of malonyl coenzyme A, the latter of which is formed from acetyl coenzyme A by carboxylation. The enzyme complexes that synthesize these compounds are called polyketide synthase precursors (Talapatra and Talapatra 2015).

In microalgae, phenolic compounds are more studied in dinoflagellate species due to the importance of the toxins they produce. Dinoflagellate polyketides include polyether compounds and macrolides (macrolactones), such as the amphidinolide toxins from *Amphidinium* species (Kobayashi and Kubota 2007; Sasso et al. 2012). These compounds are studied because they exhibit medicinally important antibiotic, anticancer, antifungal, antiparasitic, and immunosuppressive activities (Talapatra and Talapatra 2015).

The enormous variety of secondary metabolites that can be synthesized by microalgae makes it necessary to study them not only for their well-known uses but also as possible producers of secondary metabolites of interest, especially in the pharmaceutical industry.

13.2.2 Microalgal Cultivation

Microalgae can be grown for wastewater treatment or to obtain primary and/or secondary metabolites. As mentioned, they can produce a wide range of bioproducts, including polysaccharides, lipids, pigments, proteins, vitamins, and bioactive compounds (Khan et al. 2018), and are a rich source of carbon compounds that have different applications as biofuels, health supplements, pharmaceuticals, and cosmetics (Das et al. 2011).

Microalgae are very versatile organisms, a fact that allows them to develop in different conditions. They present four modes of metabolic growth:

1. *Autotrophy*: in which light is the only energy source and inorganic carbon is the only carbon source
2. *Heterotrophy*: in which energy and carbon are derived exclusively from an organic carbon source, such as

glucose, glycerol, and acetate, and growth can proceed without light supply
3. *Mixotrophy*: in which energy is obtained from light and organic carbon sources, and carbon is obtained from organic and inorganic carbon sources
4. *Photoheterotrophy*: in which carbon is obtained from organic compounds but strictly with a light supply (Rizwan et al. 2018; Kamalanathan and Quigg 2019)

As examples, it can be mentioned that *Arthrospira platensis*, *Chlorella vulgaris*, and *Haematococcus pluvialis* can grow in photoautotrophic, heterotrophic, and mixotrophic conditions, while *Scenedesmus acutus* and *Selenastrum capricornutum* can grow either photoautotrophically, heterotrophically, or photoheterotrophically (Chojnacka and Marquez-Rocha. 2004; Rizwan et al. 2018).

The cultivation of microalgae depends first on the species being cultivated and secondly on its objective. General factors to consider are light intensity, temperature, the availability of nutrients (mainly nitrogen, phosphorus, and potassium), pH, salinity, and the amount of inorganic carbon, oxygen, and CO_2 (Mata et al. 2010). In addition, other factors to take into account to determine the success of the culture include the methods used for stirring and mixing, the bioreactor dimensions, the harvest frequency, and the dilution rate (Khan et al. 2018).

13.2.2.1 Light Intensity

Light intensity is one of the main limiting factors in microalgal culture (Contreras-Flores et al. 2003). The duration and intensity of light directly affects the photosynthesis of microalgae and influences the biochemical composition of microalgae and biomass yield (Hernández-Pérez and Labbé 2014). In the absence of nutrient limitation, photosynthesis increases with increasing light intensity, reaching the maximum species-specific growth rate at the point of light saturation (Park et al. 2011). Past this point, the photoinhibition point is reached, with detrimental results for microalgae, including loss of photosynthetic efficiency and crop productivity (Becker 2004; Park et al. 2011; Hernández-Pérez and Labbé 2014; Khan et al. 2018). Thus, in each case, the optimal light intensity for each species must be experimentally determined to maximize photosynthesis, have a minimum rate of photorespiration, and avoid photoinhibition. According to several studies, the most suitable for algal growth is 16 h light/8 h dark. To avoid photoinhibition, also called the self-shadowing effect, in which algae in lower layers are shaded by those in upper layers, it is also necessary to guarantee proper penetration and uniform light distribution (Park et al. 2011; Hernández-Pérez and Labbé 2014; Khan et al. 2018).

13.2.2.2 Temperature

Algal production increases proportionally with temperature until it reaches the optimum temperature for each species. Above this, respiration increases and photorespiration reduces overall productivity. The optimum temperature varies among species, but is generally between 20 and 30 °C (Singh and Singh 2015). An increase or decrease in temperature beyond the optimum point slows down or even stops the growth and activity of algae (Bechet et al. 2017). Some thermophilic algae such as *Anacystis nidulans* can withstand temperatures of up to 40 °C and some algae can grow in hot springs at temperatures near 80 °C (Khan et al. 2018).

Temperature can also be used as a stress treatment to induce the production of valuable metabolites. Although the optimal temperature for various species of microalgae ranges between 27 and 31 °C (Kitaya et al. 2005; Khan et al. 2018), some algae, such as *C. vulgaris*, produce more carbohydrates and lipids if grown at 25 °C than at 30 °C (Converti et al. 2009).

13.2.2.3 Nutrients

Although the basic nutrient requirements for all microalgal species are the same, different species may have different nutritional needs. The main macronutrients necessary for microalgae are carbon, nitrogen, and phosphorus, since they form their backbone ($CH_{1.7}$, $O_{0.4}$, $N_{0.15}$, $P_{0.0094}$) (Khan et al. 2018). Some species of marine microalgae also require silicon as a macronutrient (Hernández-Pérez and Labbé 2014).

After carbon, nitrogen is the most important nutrient for microalgae, and is incorporated as nitrate (NO_3^-), ammonium (NH_4^+), or urea (Grobbelaar 2004; Hernández-Pérez and Labbé 2014; Khan et al. 2018). Nitrogen is also a critical factor in regulating the lipid content of microalgae (Park et al. 2011). Typically, microalgae have a lipid content of approximately 20%, but when nitrogen becomes the limiting factor for growth, the accumulation of lipid levels increases by more than 40%. This may reduce algal production, suggesting that the two conditions, high lipid content and high productivity, may be mutually exclusive (Ho et al. 2013). The other main nutrient is phosphorus, which is incorporated as HPO_4^{2-} or $H_2PO_4^-$. It is essential in many cellular processes, and its deficit in the culture medium is a limitation to growth.

The microalgal culture can be added with bacteria, which degrade nutrients into forms such as ammonia or nitrate, which can be easily assimilated by microalgae. This mixture may not only provide a uniform distribution of nutrients, air, and CO_2 to the culture but also improve the penetration and uniform distribution of light, preventing biomass aggregation (Khan et al. 2018).

13.2.2.4 pH, Salinity, and CO_2

The pH of the microalgal culture is influenced by several factors such as the algal productivity and respiration, the alkalinity and ionic composition of the culture medium, the autotrophic and heterotrophic microbial activity, and the efficiency of the CO_2 addition system (Park et al. 2011; Hernández-Pérez and Labbé 2014).

Different microalgal species have optimal growth at different pH values. However, most grow well at pH values ranging from 6 to 8 (Khan et al. 2018), and only a few can withstand a pH range as wide as that tolerated by *C. vulgaris* (Lam and Lee 2012), which can grow over a wide pH range.

If the pH of the culture is increased, the salinity of the culture will also increase, which is very damaging to algal cells. pH can be controlled with the addition of acid or base or with an automated CO_2 injection system, which also allows supplying the CO_2 necessary for high-productivity crops. CO_2 needs can be considerable, for which the associated costs for a commercial crop can also be high. Hence, there is an importance to look for an inexpensive way to mitigate these costs (Hernández-Pérez and Labbé 2014).

13.2.2.5 Stress Conditions

Microalgae usually produce secondary metabolites when grown under stressful conditions, such as strong light, high salinity, high temperature, deprivation of nitrogen or other nutrients, short-term exposure to UV radiation, or a combination of these factors. In addition, in the mixotrophic growth mode under these conditions, certain molecules accumulate. Thus, this growth mode is an efficient way to induce the accumulation of secondary metabolites (Márquez-Rocha et al. 2019). In particular, stress conditions can influence the production of polyphenols (Onofrejová et al. 2010). In this regard, Abd El-Baky et al. (2009) studied the influence of the growth conditions on the levels of polyphenols produced by the microalga *Spirulina maxima* and found that they were strongly affected by the concentrations of nitrogen ($NaNO_3$) and L-phenylalanine in the culture medium. Worms et al. (2006) reported that algae can also regulate their internal cellular environment by making toxic metal ions less bioavailable through intracellular trace metal binding or sequestration by metal chelators. Several classes of metal chelators are involved in this intracellular complexation and cell protection (López et al. 2015).

13.2.2.6 Genetic Engineering

The development of genetic engineering has made it possible to improve the production of secondary metabolites in microalgae. The strategies used can regulate the metabolic flux or directly build new biosynthetic pathways for terpenoid production by rewriting the biosynthetic pathway (Schmitz et al. 2017). For example, Xu et al. (2016) found that an effective strategy to improve terpene production in microalgae is the overexpression of the enzymes hydroxyl methylglutaryl CoA reductase (HMGR) and 1-deoxy-D-xylulose 5-phosphate synthase, which are the limiting enzymes of the MVA pathway and the MEP pathway, respectively. In *Phaeodactylum tricornutum*, the overexpression of HMGR has been found to increase the accumulation of squalene, cycloartenol, and obtusifoliol by ten-fold, three-fold, and 2.5-fold, respectively (Jaramillo-Madrid et al. 2020). On the other hand, when the compound of interest is a metabolic intermediate rather than the final product, the strategy used is enzyme inhibition. In this case, an enzyme that catalyzes the subsequent step to the intermediate of interest can be inhibited. An example of this is the blocking of the synthesis of the enzyme squalene epoxidase in *Chlamydomonas reinhardtii* (Kajikawa et al. 2015; Huang et al. 2021).

13.3 Obtaining Products from Microalgae

13.3.1 Introduction

As mentioned, microalgae, in addition to being recognized for their great bioremediation capacity due to the high capture and biotransformation of atmospheric CO_2, constitute an important raw material to obtain numerous high-value bioactive metabolites (Molina Grima et al. 2003; Ventura et al. 2017). However, intrinsic factors such as temperature, salinity, light, and nutrient availability affect the chemical composition of the biomasses obtained (Balasubramaniam et al. 2021).

The first unialgal cultures were produced in 1890 with the microalga *C. vulgaris*. Much later, after 1948, research efforts in the United States, Japan, and Germany were focused on new applications of algal cultures, mainly for their use in the food industry (Burlew 1953; Dixon and Wilken 2018). This promoted the industrial-scale production of *C. vulgaris* in Japan in the 1960s, both for food and medicinal purposes (Borowitzka 2013). Subsequently, other species were further studied in different countries (Milledge 2010). Between 1980 and 1990, the United States promoted research and programs related to the production of biofuels from microalgae. Then, the interest in other derivatives, such as bioactive compounds for their application in health, was expanded. The most used genera today are *Dunaliella*, *Chlorella*, *Haematococcus*, *Nannochloropsis*, *Chlamydomonas*, and *Spirulina*, mainly because of the production of pigments, unsaturated fatty acids, products for human and animal health, food, and raw materials for cosmetics and biofuels (Dixon and Wilken 2018). Species such as *Spirulina platensis*, *C. vulgaris*, and *Dunaliella salina* are considered of high value in food, being used as functional foods and supplements.

In recent years, more and more microalgal species are investigated for their nutritional and pharmacological properties, as well as for the knowledge of the metabolites they biosynthesize, which contribute to their properties. Among the latter, carbohydrates, proteins, unsaturated fatty acids, phenols, steroids, triterpenes, and pigments stand out. In addition, through modifications in the cultivation conditions, the metabolism can be directed toward certain products, thus increasing the value of the raw material. However, despite the high industrial value of these products, it is necessary to go even further in marketing considerations, taking into account the analysis of costs based on the objectives for each stage of production. On the other hand, the use of large volumes of organic solvents in extractive processes entails a high environmental impact, which increases the cost of the processes since the disposal of waste is high-cost. Also, the use of more sophisticated and thus more expensive equipment, as well as specialized human resources, reduces profitability (Ventura et al. 2017). This leads to an increasing search to optimize methodologies with less environmental impact, which allows a higher yield in biomass and products destined for different purposes in a way as that maximizes the use of production (Dixon and Wilken 2018).

In this context, this chapter will next describe the steps included in the process used to obtain products from microalgae, highlighting the most used procedures.

13.3.2 Obtaining the Microalgal Biomass

As indicated in the Introduction to this chapter, the production of microalgae-derived metabolites requires the proper selection of the species taking into account the objectives to which the production is directed, the selection of the appropriate culture methodology for the algae (Borowitzka 1999; Pulz et al. 2001; Molina Grima et al. 2003), the collection or recovery of the biomass, and further downstream processing to purify the metabolite from the biomass. This last part includes the extraction, fractionation, and purification steps (Molina Grima et al. 2003). Finally, the analyses carried out on the extracts, fractions, and/or metabolites will be crucial to achieve the optimum quality for the intended application. In the case of the pharmaceutical industry, the qualitative and quantitative chemical constitution of the extract and/or fraction to be used and the purity of the metabolite, among other parameters, are essential.

Figure 13.2 shows a pharmacognostical protocol to obtain products of pharmaceutical importance from microalgae.

13.3.2.1 Selection of the Species and Culture Methodology

Selecting the species supposes prior knowledge of the optimal conditions that will allow the generation of the largest amount of biomass enriched in the metabolites of interest. Although it has been described that, in general, the production of microalgae is timeless, when culture media such as industrial effluents are used, variations in metabolic production may occur depending on the season of the year in which they are obtained. Therefore, preliminary information of the species in relation to the biomass yield, the type of metabolites that it biosynthesizes, and the amount of metabolites that can be obtained under certain conditions is crucial.

Once the culture conditions have been optimized, as described in other sections of this chapter, either in open-air systems where the lighting comes from natural light (ponds, canals), closed systems (fundamentally in photobioreactors located either outdoors -to use sunlight- or indoors), or in hybrid systems which try to prevent contamination and reduce costs, the next step will be to recover the biomass produced (Molina Grima et al. 2003; Hernández-Pérez and Labbé 2014; Narala et al. 2016; Khan et al. 2018; Tan et al. 2018).

13.3.2.2 Collection of Microalgae: Recovery of Biomass

The collection or harvesting of biomass requires the separation of the liquid medium in which the biomass was obtained, which generally corresponds to large volumes. This step is considered one of the most complex and costly in production and depends largely on the culture systems used (Rawat et al. 2011; Hernández-Pérez and Labbé 2014). The liquid can be extracted by different methods, mainly centrifugation, filtration, and sedimentation by gravity, which may be preceded by flocculation. The small size of microalgae and the amount of culture liquid make their separation difficult and generate a cost that can reach 20 or 30% of the total cost of biomass production (Zhu 2015; Dixon and Wilken 2018).

When selecting the collection method, the acceptable level of humidity that the biomass can conserve should also be considered. The product obtained by gravity sedimentation is generally more diluted than that recovered by centrifugation or filtration, so it may require thermal drying to optimize its conservation; this step increases the cost of production (Molina Grima et al. 2003; Dixon and Wilken 2018).

The main biomass recovery methods are described below.

Fig. 13.2 General pharmacognostic protocol diagram

1- Collection, classification, histochemistry, conservation, drying, grinding and sieving

2- Preliminary extraction

3- Exhaustive extraction

4- Fractionation of the products obtained

5- Analysis of the isolated products and structural determination

6- Evaluation of the possible application in Health and related industries

Phycochemical screening

Biosynthetic pathway research

Determination of primary and specific bioactivity

13.3.2.2.1 Centrifugation

Most microalgae can be collected by centrifugation. The centrifuges that can be selected and used in biomass recovery can be consulted in Chisti and Moo-Young (1991). This method is fast and efficient, so although it consumes a lot of energy, it is widely used, e.g., to produce concentrates for hatcheries in aquaculture. The cost can be high because it requires both large amounts of energy and specialized equipment (Chen et al. 2011; Hernández-Pérez and Labbé 2014). Thus, centrifugation is feasible when the value of the resulting products is high (Ho et al. 2011).

Heasman et al. (2000) evaluated the degree of cell recovery and the effects on cell viability under different centrifugation conditions. They observed that, when working at 13,000 g, the cell collection efficiency was greater than 95%, whereas, when working at 6000 and 1300 g, the efficiency decreased to 60 and 40%, respectively. Regarding cell viability, these authors observed that this depended significantly on the microalgal species and the centrifugation method used.

13.3.2.2.2 Filtration

This method can be a relatively slow process; for this reason, it is carried out under pressure or by vacuum. It is used, e.g., for larger microalgae such as *Spirulina platensis* or *Coelastrum proboscideum* (Mohn 1980), but not recommended for smaller species such as those of the genera *Scenedesmus* or *Chlorella* (Harun et al. 2010; Chen et al. 2011; Ho et al. 2011; Rawat et al. 2011; Hernández-Pérez and Labbé 2014).

Filtration implies the use of different types of filters which must be inert and innocuous in order not to alter or contaminate the biomass. As an example, diatomaceous earth can be used for biomass recovery of *Dunaliella* species grown in ponds.

Microfiltration and ultrafiltration through membranes, the former of which is suitable for fragile cells, are possible alternatives to conventional filtration (Petrusevski et al. 1995). Although microfiltration membranes have been shown to provide higher initial fluxes than ultrafiltration membranes, they tend to clog more easily but can be used for a long term under conditions of low pressure and relatively low flow rate. In general, if the volume to be filtered is small, microfiltration can be more profitable than centrifugation; for larger scales, centrifugation is a cheaper method to recover biomass (Mackay and Salusbury 1988), because, in general, filtration is expensive due to the requirements of energy, filters or membranes, and continuous pumping of the biomass (Rawat et al. 2011; Hernández-Pérez and Labbé 2014).

13.3.2.2.3 Sedimentation

Gravity sedimentation may be the method of choice for biomass recovery from wastewater-based processes (Venkataraman 1978). It is a common method as it is favorable for various types of microalgae and is energy efficient (Chen et al. 2011; Rawat et al. 2011; Abdel-Raouf et al. 2012; Hernández-Pérez and Labbé 2014). Sedimentation or decantation tanks can also be used. Flotation or flocculation can also be used previously (Shelef 1978).

13.3.2.2.4 Flotation

This method is described as promising for the collection of small unicellular algae and in laboratory-scale trials (Chen et al. 2011; Rawat et al. 2011; Hernández-Pérez and Labbé 2014). It consists of the addition of gases, such as ozone or air, which interact with the charge of the cell walls, allowing the microalgae to float and separate. However, ozone increases costs, which must be considered in the production design (Rawat et al. 2011).

13.3.2.2.5 Flocculation

Conventional flocculation works by the charge dispersion mechanism (Abdel-Raouf et al. 2012; Hernández-Pérez and Labbé 2014). It consists of flocculation by chemical techniques, electroflocculation, or bioflocculation (Chen et al. 2015; Dixon and Wilken 2018).

The addition of cationic flocculants neutralizes or reduces the negative surface charge of the microalgae in suspension, favoring aggregation; when combined with filtration, sedimentation, or flotation, they increase the efficiency of those processes (Elmaleh et al. 1991; Rawat et al. 2011).

Flocculants must be inexpensive, inert, nontoxic, and effective at low concentrations. Polyvalent metal salts such as ferric chloride, aluminum sulfate, and ferric sulfate are in common use. The coagulation efficiency of metal ions increases with increasing ionic charge. For example, aluminum sulfate is widely described to flocculate microalgal biomass in wastewater treatment, whereas alum is an effective flocculant for *Scenedesmus* and *Chlorella* (Golueke and Oswald 1965), but cannot be used when the biomass will be used in health, food, or aquaculture.

An alternative to the use of metal salts is cationic polymers. These are effective over a wider pH range, although effectiveness depends on many factors such as the polymer molecular mass, the molecular charge density, the concentration of flocculant used, the biomass concentration, the ionic strength, the molecular weight, and the pH of the culture medium. In general, a significant molecular mass, a high charge density, and a high concentration of polyelectrolytes improve the flocculant capacity.

Chitosan, a D-glucosamine polysaccharide, has been shown to be effective for the flocculation of various species of microalgae; however, its flocculant capacity decreases with increasing salinity (Morales et al. 1985; Lubián 1989). It has also been seen that the efficacy of chitosan is affected by the concentration used, and even that depends on the species of microalgae for which it is used, although a taxonomic relationship has not been established (Heasman et al. 2000). There are other factors that determine this behavior, e.g., the culture medium used. Although chitosan is considered nontoxic, Heasman et al. (2000) described that the survival of oyster larvae fed microalgae flocculated with chitosan decreased drastically. This polysaccharide can also be used to generate a matrix to immobilize microalgal biomass in tertiary wastewater treatment (Kaya and Picard 1996).

Electroflocculation also modifies the surface charge, but does it by generating metal ions through a metal electrode rather than through the salts mentioned above (Dixon and Wilken 2018). In bioflocculation, bioflocculants are derived either from the microalgae themselves (autoflocculation) or from bacteria or fungi that can be added to the culture (Vandamme 2013; Dixon and Wilken 2018). For example, *Paenibacillus* sp. AM49 produces a bioflocculant that has been shown to be efficient for the collection of *C. vulgaris* (Hee-Mock et al. 2001). Autoflocculation is a process by which microalgae can generate spontaneous aggregation which leads to sedimentation. This can occur naturally or be induced by carbon limitation or other abiotic factors (Chen et al. 2011; Rawat et al. 2011). For example, autoflocculation caused by modification of the culture medium has been investigated in freshwater and saltwater systems (Cannell 1990; Lee et al. 1998).

Although flocculation can be efficient for the recovery of biomass, the residues that are generated make its use unfeasible when the objective is to obtain products of optimal purity for their application in food, health, and cosmetics (Rawat et al. 2011; Hernández-Pérez and Labbé 2014).

13.3.2.3 Cell Disruption

As mentioned in previous paragraphs, microalgae include prokaryotic cyanobacteria and eukaryotic photosynthetic organisms that use solar or light energy, nutrients, and CO_2 for their metabolism (Brasil et al. 2017; Salinas-Salazar et al. 2019). As a result, they biosynthesize various compounds, which are important for different industries. To obtain them, the biomass collected must be adapted to optimize the extraction, fractionation, and purification of the compounds.

The degree of residual humidity in the biomass can alter the chemical composition, by either reactions mediated by enzymes or different degradations that will also be favored by contamination. Therefore, intermediate steps are used for secondary dehydration prior to drying (Dixon and Wilken 2018). The concentrated biomass can be dried using air currents, solar energy, sprinkling, rotation, or vacuum, among others (Dixon and Wilken 2018). The drying or dehydration processes that are normally used are sun-drying, spray-drying, and freeze-drying or lyophilization (Show et al. 2013). Sun-drying, which is based on solar energy (De Farias Neves et al. 2019), is an uncontrollable process, which can lead to several problems, such as overheating, or change of texture, color, and taste of the microalgae (Show et al. 2013). In addition, it is limited by the weather conditions, the drying period, and the drying area. Spray-drying is a technique that produces dry powder from a fine spray of suspension droplets that is in continuous contact with hot air in a large vessel. The powder produced is very fine and the fast drying can maintain a good quality product (Mujumdar 2000; Nappa et al. 2016). This drying technique is usually chosen for high-value operations due to its efficiency, but it is high-cost and some algal components such as pigments can be significantly deteriorated (Show et al. 2013). Finally, freeze-drying or lyophilization is a process by which the moisture content of the microalgae is decreased slowly at low temperatures, which allows maintaining the solid structure and the quality of the product (Oyinloye and Yoon 2020). This technique is widely used at the laboratory scale only to dry microalgae because at large-scale production it can be very expensive (Molina Grima et al. 2003).

The dry material obtained by any of the methods indicated above must then be subjected to various disruption methods to break the cell walls and favor the entry of solvents and, therefore, the extractive process. This step is essential since it can condition the extraction yield, the metabolites obtained, and, therefore, the efficiency of the process, thereby increasing the production cost.

Thus, the first step to recovering biomolecules from microalgae is the disruption of the extracellular matrix (ECM). The ECMs of cells, which are involved in growth and development, serve to protect and defend the microalgal cell, and promote adhesion and interaction with other cells and substrates (Domozych et al. 2012). Since ECMs are highly diverse and varied depending on the species, when designing the production of a given species for metabolite production, its composition must be known to optimize this often costly step. In green microalgae, cell wall polymers found in ECMs include cellulose, xyloglucans, mannans, glucuronans, (1 → 3)-β-glucans, and ulvans (Popper et al. 2011). In some cases, the fibrillar wall is absent (e.g., *Dunaliella salina*, Polle et al. 2017) or very simple; in others, the fibrillar component prevails (consisting mainly of cellulose and mannans) and can form two or three layers (e.g., some species of *Chlorella*); and, in others, the constitution is extremely complex, forming several superimposed, stratified layers, made up of carbohydrates, proteins, glycoproteins, lipids, peptidoglycans, and phospholipids, among others. For example, species of *Chlamydomonas*, *Spirulina*, and *Chlorella* have variable and complex cell walls that provide properties such as rigidity, flexibility, and digestibility (Imam and Snell 1988; Ali and Saleh 2012; Gerken et al. 2013; Kim et al. 2016). Then, depending on the cellular characteristics of the microalgae, the disruption method selected should be the most appropriate and, as far as possible, the most economically profitable. The disruption methods can be mechanical or nonmechanical. Mechanical methods, which include bead beating, high pressure or high shear homogenization, ultrasonication, and pulsed electric field, are recommended for large-scale cell disruption because they allow complete disintegration with higher yield (Moraes et al. 2010; Salinas-Salazar et al. 2019). In general, mechanical methods are less specific to the type of microalgae but are more energy-intensive than nonmechanical methods (Dixon and Wilken 2018). Nonmechanical disruption methods, on the other hand, can be thermal, chemical, or biological. Thermal methods include freeze-thaw cycles and microwave; chemical methods include acid/alkali treatment and osmotic shock; and biological methods include the use of microorganisms and enzymes. Chemical methods do not consume a lot of energy, but imply the use of substances that can be toxic and/or degrade the metabolites (Show et al. 2015). Biological methods can be uneconomical and their use depends on the metabolites of interest since many can be degraded or modified during treatment (Günerken et al. 2015).

In this chapter only the main disruption methods have been listed; for a complete description, the reader may want to refer to Dixon and Wilken (2018).

13.3.3 Metabolite Production

The next step corresponds to the selection of the methodologies for the extraction, fractionation, and purification of the metabolites of interest. Once again, it is important to keep in mind the purpose of the application of the products obtained to consider all possible variables that may affect their quality and performance.

This chapter fundamentally considers the aspects related to the application of the products in health.

13.3.3.1 Extraction

Extraction is defined as the diffusion or transfer of matter from one phase to another. It consists of a set of operations aimed to obtain the substances contained at an initial phase, i.e., in the biomass. Figure 13.3 presents a summary of the steps of the extraction process considering the phases involved, the sample, and the resulting products.

The general classification of the process depends on:

- The state of the phases
- The total, partial, or no renewal of one or both phases; and
- The temperature to which the sample is exposed

The process involves considering the following critical factors:

- Objectives pursued
- Raw material, sample (= biomass)
- Solvent (unitary, mix)
- Procedure or methodology
- Cell diffusion control
- Interferences.

The choice of the extraction procedure depends on the nature of the sample and the compounds to be isolated.

Prior to choosing a method, it is necessary to establish the objectives pursued, since you may wish to extract:

- An unknown bioactive compound
- A known compound present in an organism
- A group of compounds within an organism that are structurally related
- All primary or secondary metabolites produced by one natural source under specified conditions; i.e., in the case the same species grown under different conditions can biosynthesize different metabolites, or

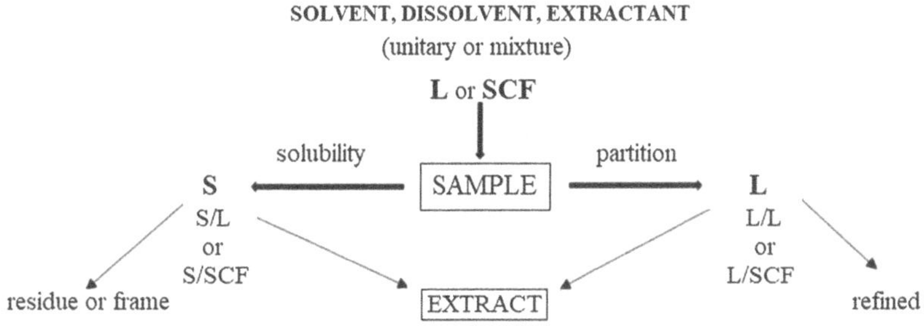

Fig. 13.3 Extraction scheme

- You may wish to identify all primary and/or secondary metabolites present in an organism for chemical fingerprinting or metabolomics study

Bioactive compounds from natural resources such as microalgal biomass can be extracted by classical extraction techniques based on different polar and unipolar unitary solvents (such as ethanol, methanol, water, dichloromethane, and hexane) or a mix of solvents, associated or not with heat and other techniques to enhance the extraction yield. Compounds can also be extracted by means of sequential extraction by nonpolar and polar solvents to generate a series of samples (Patras et al. 2018; Stirk et al. 2020).

Selecting the proper extraction solvent needs to consider the polarity of the solvents as well as the polarity of the compounds to be extracted (Table 13.1), their molecular affinity, mass transfer, possible use of co-solvents, environmental safety, human toxicity, and financial feasibility.

Traditional extraction methods have been extensively used to obtain metabolites from natural resources. Many of them continue to be used, especially at a laboratory scale; however, at an industrial level, they are not always efficient and can consume an excess of time and energy, and generate large amounts of waste, which is not profitable.

Simple extraction procedures are infusion, decoction, maceration, digestion, hydrodistillation, and semi-continuous or continuous extraction (e.g., Soxhlet), whereas other methods are percolation, liquid-liquid extraction, and sonication.

Some of the most used procedures in natural products are briefly described below:

- *Infusion* consists of an extraction with freshly boiled water that is poured over the material to be extracted and left in contact for 20 mins (FA 7° ed. 2013).
- *Decoction* also uses water, but, in this case, it is boiled together with the material to be extracted for 20 mins. In both systems, the ratio is 5% P/V (FA 7° ed. 2013).
- *Maceration* is also an inexpensive method to obtain compounds from crushed or ground material with a suitable solvent, at room temperature. Extraction can be optimized by magnetic, mechanical, or orbital stirring.
- *Digestion* is a hot maceration, where the working temperature must be lower than the boiling temperature of the solvent used. Also, in this case, the extraction can be optimized by stirring.

In all these cases, once the extraction is complete, the extract will be recovered by centrifugation or filtration.

- *Hydrodistillation* uses either water or steam; water distillation consists of a direct distillation which involves the hydrodiffusion of polar metabolites, whereas steam distillation is used for compounds of a volatile nature.
- *Soxhlet* extraction is generally used to extract substances of low or medium polarity such as pigments, lipids, steroids, and terpenes. Semi-continuous extraction equipment (or continuous when the extraction is performed at an industrial scale) use volatile solvents such as *n*-hexane, chloroform, dichloromethane, acetone, and methanol.

Modern extraction methodologies for different natural products, including microalgae, have been described (Patras et al. 2018). The main ones are supercritical fluid extraction, pressurized-liquid extraction, microwave-assisted extraction, ultrasound-assisted extraction, solid phase extraction, and their combination. They are described as follows:

- *Supercritical fluid extraction* is ideal for the extraction of thermolabile compounds. The most used solvent is CO_2 because of its favorable critical temperature and pressure (T_c = 31.1 °C and P_c = 73.9 bar). CO_2 also has low viscosity, low surface tension, high diffusivity and good density, and is nontoxic, nonflammable, inexpensive, widely available, chemically inert under several conditions, and gaseous at normal pressure and temperature, which allows eliminating the step of solvent evaporation after extraction. This solvent provides a nonoxidizing atmosphere, thus preventing extracts from degradation. It is used, e.g., to extract carotenoids and chlorophylls from *Scenedesmus obliquus*. However, the use is somewhat limited due to the cost of the equipment.
- *Pressurized liquid extraction*, also known as *subcritical fluid extraction*, uses temperatures of 50–200 °C and pressures in the range of 35–200 bar, lower than the T_c

Table 13.1 Eluotropic series of solvents and main chemical groups to be extracted

Solvent	Main chemical groups
Petroleum ether, *n*-hexane, cyclohexane	Fixed oils, fats, waxes, lignans
Benzene, toluene, carbon tetrachloride, chloroform, dichloromethane	Pigments, neutral alkaloids, alkaloids as bases, volatile oils, monomeric and dimeric quinone genins, terpenes, steroids, low-hydroxylated methoxylated flavanoid genins
Ethyl ether, ethyl acetate, acetone, *n*-butanol	Steroid glycosides, flavonoid genins, coumarins, monoglycosylated quinones
Ethanol, methanol	Glycosides, tannins
Hydroalcoholic mixtures (aqueous ethanol at 80–50%)	Flavonoid glycosides, oligosaccharides, monosaccharides, peptides, amino acids
Water	Amides, alkaloid N-oxides, quaternary ammonium salts, xanthines, saponins, carbohydrates in general, proteins, highly hydroxylated flavonoid glycosides, sulfated flavonoids
Acidified water	Quaternary ammonium salts, alkaloids as salts, anthocyanins
Alkalinized water	Saponins, lactones

and P_c of the solvents, which are kept in the liquid state. This increases the solubility and mass transfer rate of the solvents and reduces their viscosity and surface tension. In this process, the most widely used solvent is water, followed by propane and dimethyl ether. This methodology has been applied, e.g., for the extraction of antiviral polysaccharides from *C. vulgaris*.

- *Microwave-assisted extraction* is an efficient method that uses microwave irradiation to accelerate the removal of several compounds from natural matrices, by concomitant heat and mass gradients. Microwaves are electromagnetic waves with frequencies between 300 MHz and 300 GHz, which are generated by electric and magnetic fields. They induce dipole rotation and ion migration in solvents and structural molecules, a fact that results in heating, cleavage of chemical linkages, increased pressure and rupture of the natural material from inside, and final release of the product to the surroundings (Miazek et al. 2017). This methodology has been used for the extraction of metabolites from *Isochrysis galbana*, *Chlorella* sp., *Chlorella sorokiniana*, *Nannochloropsis salina*, *Cylindrotheca closterium*, and *Phaeodactylum tricornutum*, with a great improvement in product extraction yield (Miazek et al. 2017; Patras et al. 2018).
- *Ultrasound-assisted extraction* is a process by which high-frequency sounds (>20 kHz) are transmitted to a liquid, generating regions of alternating pressure, which lead to the formation of gas bubbles, in a process called cavitation (Miazek et al. 2017). When these bubbles come into contact with the surface of the natural material and implode, the high pressure and temperature created locally destroy the structure of the material. This ability of high-frequency sounds to form bubbles depends on the characteristics of the ultrasound wave and the solvent as well as on the ambient conditions. This methodology significantly reduces the extraction time and increases the extraction yields and has been applied, e.g., to extract lipids from *Nannochloropsis oculata* (Miazek et al. 2017).
- *Solid phase extraction* is a technique that allows obtaining, concentrating, and separating analytes (e.g., metabolites) present in a complex matrix (e.g., biomass), by means of a stationary solid phase. As a result, interferences are separated from the analyte and the latter will be obtained by desorption and can be analyzed with greater sensitivity using the appropriate analytical technique, e.g., ICP-OES (Inductively Coupled Plasma–Optical Emission Spectrometry) or ICP-MS (Inductively Coupled Plasma–Mass Spectrometry. This methodology is mainly used in three modalities or devices: cartridges, discs or filters, and fibers; the latter is called *solid phase microextraction*. One advantage is that it groups the extraction and concentration steps in a single device, which significantly reduces the handling and work time. It basically consists in the extraction of a metabolite (analyte) of interest through solid–liquid systems, using stationary phases that can be nonpolar (C8, C18), polar (silica alone or with cyano, amino, or diol groups) and resins of ion exchange (such as Amberlite, PorapaqQ). The retention capacity depends on the type of packing material, length, type of analyte, and extraction conditions. It uses less solvent than traditional solid–liquid and liquid–liquid extractions. The stationary phases offer a high surface area, reversible adsorption, low leaching of particles or impurities that may interfere with the analysis, and efficient contact surface with the sample (Fritz and Macka 2000; Ulrich 2000).

Of these, the most used methodologies to obtain microalgal products are solvent extraction favored with stirring, digestion, Soxhlet, enzyme-assisted extraction, microwave-assisted extraction, ultrasound extraction, solid-phase extraction, and, more recently, extraction with supercritical fluids (Moraes et al. 2010). The main factors to be taken into account for the extraction design include the disruption method, the type of solvent, the proportion of solvent with respect to the biomass, and the extraction time. These factors must be investigated together, so that the extraction process can be optimized, minimizing costs and maximizing yield (Moraes et al. 2010; Salinas-Salazar et al. 2019).

Regarding the type of solvent, organic solvents are carbon-based solvents that include methanol, ethanol, chloroform, acetone, isopropanol, butanol, ethyl acetate, and hexane, each of which has different boiling point, density, polarity and miscibility in water. These solvents can be applied in the pharmaceutical, food, oil, and petrochemical industries. On the other hand, Ionic liquids are organic salts that remain in a liquid state below 100 °C, with properties such as low volatility, high solvency, and nonflammability. These liquids can potentially replace the traditional organic solvents used in the industry (Miazek et al. 2017).

Although the new extraction systems use alternative solvents such as ionic liquids and surfactants, they must be still thoroughly evaluated. Solvent formulations must be treated carefully to avoid the use of highly volatile or toxic organic substances. It is also increasingly important to try to optimize the use of aqueous solutions.

In addition, the correlation between the chemical structures of the compounds of interest and the solvents and technologies to be used in relation to each specific compound must be adequately investigated. For example, the use of techniques involving temperature should be avoided to extract thermolabile metabolites. In this sense, the correlation between the compounds to be extracted, their properties and main structures, the solvents and the extraction techniques must be taken into account to improve the proper development of the process, as well as to be able to implement more efficient, environmentally friendly and economically viable procedures. Obtaining new products and creating new markets must be the focus of not only researchers, who should promote the optimization of methodologies, but also industries, which should collaborate with them (Ventura et al. 2017).

The excessive use of hazardous organic solvents and the growing environmental concern have led to the exploration of new extraction routes. Among these, innovative green solvents, such as natural deep eutectic solvents (NaDES), have been studied. However, the difficulty of fractionating NaDES microalgal extracts in separate streams rich in lipids, proteins and carbohydrates still represents a major drawback for a possible industrial application. In this regard, Ventura et al. (2017) explored the feasibility of using aqueous two-phase systems for the fractionation of the main bioactive compounds (lipids, proteins, carbohydrates) contained in an NaDES extract from the green microalga *C. vulgaris*.

13.3.3.2 Fractionation

Depending on the type of extraction performed and whether it was general or directed to a specific group of substances, an extract may contain a significant mixture of metabolites belonging either to various chemical groups or to a particular chemical group. In both cases, the next step will be the fractionation of the extract for the subsequent isolation of the metabolites.

The components of an extract can be divided into groups of compounds that share similar physicochemical features. This process, called fractionation, can be achieved in different ways, each of which separates compounds based on one or more specific features, such as size, shape, solubility, and electric charge. The methodologies can be combined to optimize the separation of chemical groups and/or compounds; e.g., the first fractionation may be based on solubility differences, whereas the next may be based on molecular size.

It should also be noted that, in some situations, the extract will contain large amounts of unwanted compounds that must be removed, e.g., pigments present in extracts targeting other metabolites. The procedures applied to separate the unwanted compounds are called clean-up procedures and are similar to those used in fractionation.

Selecting the fractionation method that will be used in each particular situation depends on several factors, including:

- *The nature of the compounds present in the extract*. Probably one of the most important points is to consider the solubility, ionic charge, thermostability, and photostability of the components. A general principle that should be considered is that the fractionation should be performed under the mildest conditions possible, i.e., minimizing temperature, protecting against light and avoiding solvents, reagents, and other chemical substances that can cause degradation.
- *The immediate destination of the separate fractions*. If the fractions are to be used for bioassays, solvents that are toxic and/or incompatible with the assays should be avoided. If the fractions are to be used for the isolation of individual chemicals, solvents that can be easily separated from the fraction by evaporation should be used. In all cases, inert, volatile solvents with low toxicity and, as far as possible, recyclable solvents must be selected to generate the least possible amount of waste.
- *Availability and cost of the equipment and materials required.*
- *Security*. Considering that most chemical manipulations carry some degree of risk for the personnel due to toxicity or possible damage of the material due to factors such as fire or corrosion, it is important to choose techniques and materials that reduce these risks as much as possible. Therefore, nonflammable solvents should be preferred to flammable ones.

The most common methods used to achieve fractionation (or clean-up) of extracts include precipitation, solvent–

solvent extraction, distillation, dialysis, electrophoresis, and chromatography. They are described briefly in the following sections.

13.3.3.2.1 Precipitation

When the concentration of a substance in a solution exceeds its maximum solubility, it precipitates. Precipitation can be used to remove the chemical group or compound of interest as a precipitate or to eliminate unwanted material and keep the material of interest in a solution.

Precipitation may be simply achieved by lowering the temperature of the extract solution. In this case, less soluble components will precipitate and can then be separated by centrifugation or filtration. If a substance is present as the main component, it can precipitate as crystals. The technique is widely used as a final step in the isolation of individual compounds from aliquots derived from the last fractionation step. In the case of aqueous solutions, the utility of precipitation by cooling is somewhat limited; if the aqueous extract is obtained at room temperature, only a small drop in temperature is possible before the entire solution freezes.

Another way in which precipitation can be affected is to modify the polarity of the solvent by adding a miscible solvent of different polarity. If the original extract is obtained with water, aliquots of methanol, 96% ethanol, or acetone can be added so that, with each addition, the solvent becomes somewhat less polar and more polar solutes precipitate. Conversely, if a nonpolar solvent such as hexane is used for the initial extraction, the addition of a more polar miscible solvent, such as acetone, will allow precipitating the less polar components.

Another way to precipitate aqueous extracts is to add a strong solution of an electrolyte that is very soluble in water, such as ammonium sulfate. By doing so, the less soluble nonionic compounds present in the extract will be displaced and then precipitate; this technique is widely used in the separation of proteins and peptides from extracts.

Another precipitation technique used is the formation of complexes between a compound present in the extract and a compound present in the solution. Examples of this are picric acid for alkaloids, 10% lead subacetate solution for phenols, and 5% gelatin solution for tannins. This technique can be applied for the separation of constituents, but it should be considered that any complex formed must be easily reversible to release the compound of interest.

It should be noted that when precipitates form, very fine particles are often obtained, a fact that causes problems in filtration as the filter pores are easily clogged. Filtration can be improved by using a filter bed of diatomaceous earth washed with the solvent used to prepare the filtration grade extract, mixing more diatomaceous earth with the extract and precipitate, and then pouring over the filter bed. The precipitate can then be dissolved out of the filter bed with an appropriate solvent. Another alternative is to remove the precipitate by centrifugation.

13.3.3.2.2 Solvent–Solvent Extraction

Immiscible liquids. The addition of a solvent to an extract dissolved in another solvent that is immiscible with the first will lead to the formation of two layers or phases. Any component of the mixture will be soluble in each of the two phases, and, after some time, will reach equilibrium. The time it takes to reach equilibrium is usually shortened by stirring the two phases in a stoppered flask. In any case, care must be taken when stirring, since if the solvents are very volatile, they can cause the stopper to lose and liquids to be lost. On the other hand, some organic phases such as chloroform and dichloromethane readily form emulsions with aqueous solutions; for this reason, it is recommended that the agitation be gentle. If emulsions do occur, they often form a third opaque layer at the interface of the two transparent phases.

The solubility of an uncharged substance in either phase at a given temperature depends on whether it has a similar polarity to the liquid phase. Charged molecules have a high affinity for solvents that contain a large number of oppositely charged ions. The ratio of the concentrations of a substance in the two phases is called the partition coefficient (K). Since different substances will have different K, if a compound is fairly polar, its K relative to the more polar phase will be much larger than that of a nonpolar compound. Thus, when an extract is exposed to two immiscible solvents, the different solutes will be distributed differently. If the partition coefficient is greater than 100 for any substance, a large proportion of it will be present in only one of the phases and can be separated from others.

To break emulsions, the emulsion must be transferred to a clean separating flask to separate it from the other phases, and then be treated as follows:

1. Gently rotate the separating flask in the vertical plane and gently tap on the sides with your fingers or a blunt, non-brittle rod.
2. Add a small amount of methanol or ethanol followed by the treatment indicated in the first step. The alcohol reduces the interfacial tension between the two phases and thus allows them to separate more easily.
3. Filter under reduced pressure through silicone-treated filter paper. This allows the passage of the nonpolar phase.
4. If there are small particles, filter under reduced pressure. Since the particles are often very small, they can easily block the pores in ordinary filter paper; thus, it is recommended to filter them by using a bed of diatomaceous earth.

Multiple extraction. Successive treatment of the original extract or phase with relatively small amounts of the second phase rather than by using the entire volume of the latter at one time will lead to greater fractionation. Multiple sequential extractions can also be performed. For example, an aqueous extract can be successively extracted with small volumes of each of the following solvents: hexane, dichloromethane, ethyl acetate (or dichloromethane-ethanol, 3:1), and butanol. A petroleum ether or dichloromethane extract can be extracted with a series of solvents of decreasing polarity, such as water, water-ethanol 3:1, water-ethanol 2:1, and water-ethanol 1:1. It is usual to extract each phase several times with the other and to combine the similar steps.

Salting out. The method known as salting out, i.e., adding a large amount of soluble inorganic salt to the solution, is used to improve the separation of moderately polar compounds from aqueous extracts by somewhat immiscible solvents such as butanol or methyl ethyl ketone. This greatly reduces the solubility of many nonionic compounds in water. If no other solvent is present, the nonionic compound may precipitate, but will generally dissolve in the presence of another moderately polar solvent. The salt often used for this purpose is sodium chloride because it is inexpensive and highly available; other salts such as ammonium sulfate are also widely used.

Partitions by changes in pH. Acidic and basic metabolites can be obtained from extracts by changes in pH. The ionic forms of certain compounds are soluble in aqueous media while the unionized forms are soluble in less polar solvents. For example, natural nitrogenous bases (fundamentally alkaloids) can be obtained from a nonpolar extract (e.g., dichloromethane) by means of an aqueous solution of an acid because the nitrogen of the bases is protonated, generating an electrical charge and giving it solubility in water. However, since the acidic aqueous fraction will also contain other water-soluble compounds, an additional step consisting of alkalinizing it to the acid fraction and partitioning it with a new addition of dichloromethane must be carried out. This step releases the protonated bases which have an affinity for dichloromethane and leaves the other polar components in the alkaline aqueous phase.

Elimination of traces of water from organic solvents. It should be noted that the organic solvent used for partitioning may also contain small amounts of the other solvent. When the latter is water and the other is a less polar solvent, the resulting organic phase often contains small amounts of water after separation from the aqueous phase. The organic layer is usually cloudy in appearance and can be dried by the addition of anhydrous sodium sulfate. Each molecule of anhydrous sodium sulfate occupies ten molecules of water for crystallization and thus this process absorbs water and leaves the organic phase dry. Salt should be added in successive moderate amounts until no more crystalline material is visible. The solid residue can then be filtered by using silicone-treated filter paper (which does not allow water to pass through) or by filtration through absorbent cotton where the cellulose in the cotton fibers absorbs large amounts of water.

13.3.3.2.3 Distillation

A mixture of volatile compounds can be separated by fractional distillation. This process is based on the temperature gradient established along a cooling column placed over a boiling liquid. The more volatile compounds condense far from the boiling chamber, whereas the less volatile ones condense much closer.

13.3.3.2.4 Dialysis

Another method to separate the components of a mixture is dialysis. This process, which occurs naturally across cell membranes, allows the separation of the components according to their molecular size and is very important in many physiological processes. It consists of the passage of molecules through a thin and semi-permeable membrane, made up of a polymeric material with regular pores. Under pressure or an osmotic gradient, this membrane allows the passage of molecules smaller than the pore, while retaining the larger ones inside the membrane. Membranes can have different pore sizes, which makes it a system of great application in the fractionation and purification of metabolites.

Dialysis is widely used in the purification of polysaccharides and proteins. The aqueous extract containing the large molecules is placed in a membrane-reinforced bag and immersed in water. Compounds of a low molecular size such as salts, small carbohydrates, amino acids, phenols, and others that are present in the extract pass through the pores of the membrane, while water molecules pass backward due to the osmotic gradient. Since large molecules are retained in the membrane, this process allows purifying them from small components. If small molecules are the focus of interest, the same process can also be used to remove substances of large molecular mass. An important issue to take into account during the dialysis procedure is that water should be renewed to prevent the process from stopping (for having reached an equilibrium of concentrations); therefore, it is recommended to renew the water every 8–12 h.

13.3.3.2.5 Electrophoresis

Electrophoresis is a method used primarily to separate substances that have an electrical charge. Under the influence of an electric field, the constituent molecular species will move at different speeds according to their size, shape, and total electric charge. It is primarily used as an analysis method for small sample mixtures of molecules, particularly proteins, peptides, and amino acids, rather than as a fractionation procedure.

More recently, there has been a considerable revival of interest in analytical electrophoresis due to the introduction of the technique known as capillary zone electrophoresis.

This technique allows analyzing charged and uncharged molecules.

13.3.3.2.6 Chromatography

According to the present definition, chromatography is the group or set of separative processes based on the migration of zones (Giddings). Chromatographic systems are the most diverse and widely used methods in the fractionation of extracts and in the purification and isolation of compounds. Chromatography is used mainly as an analytical method for the determination of one or more compounds present in a sample. However, it is also used for preparative purposes to isolate substances for later structural characterization.

Chromatography is based on the differential distribution of compounds between two phases, one of which moves relative to the other. These phases are called the mobile phase and the stationary phase, respectively. In this way, the components of a mixture (extract, fraction) may have a different tendency to remain in either of the phases involved. The more times the compounds interact between one phase and the other, the better the separation. The mobile phase is a fluid that can be liquid, gas, or supercritical fluid, being liquid the most used, whereas the stationary phase is generally a solid made up of very fine particles. In some cases, the solid material is in fact the phase involved in the chromatographic process; in other cases, the stationary phase is a liquid monomolecular film attached to the solid support material or a thicker film of liquid coated on it.

Chromatography is based on the fact that a dynamic equilibrium is established between the concentrations of a solute in two phases, where the equilibrium consists in the movement of solute molecules passing continuously between the two phases. At any time, the concentration ratio, i.e., the ratio of the number of molecules in each phase, is constant. However, each molecule spends time in both phases, and the proportion of time it spends in one phase or the other depends on the relative attraction of the substance to the two phases. The concentration ratio at equilibrium is called the distribution coefficient and the name for the attraction for any phase is called affinity. A slower-moving compound is said to have higher affinity for the stationary phase than a faster-moving compound.

Chromatography is thus a succession of equilibria established in a series of very small units of both phases along the path traveled by the mobile phase. These small units are called "theoretical plates," a term that has arisen from the concept of fractional distillation. The size of the theoretical plate is defined by the height equivalent theoretical plate. The smaller the size, the higher the number of theoretical plates produced in a column of a given length, a fact that results in a better separation.

The affinity of a compound for one of the phases depends on the attractive forces between the molecules of the compound and those of the phase. These forces can be Van der Waals interactions, dipolar attractions, hydrogen bonds, and ionic attractions, and are influenced by the structure, shape, and size of the molecules and their functional groups. The type of force that plays the most important role in the interaction varies according to the chromatography system used.

Components of the Chromatography System

The chromatography system includes the following:

- *Stationary phase*: it can be solid, liquid, polar, nonpolar, or ionic.
- *Mobile phase*: it can be liquid, gas, or supercritical fluid.
- *Support*: it can be active: paper; inactive: glass plate, plastic, or aluminum (for thin layer chromatography, TLC); glass tube; instrumental chromatography: stainless steel tube.
- *Sample*: it contains the solutes or substances to be separated; e.g., extract or fraction.
- *Process*: it depends on the temperature, humidity, pressure, centrifugal force, type of development, and equipment.
- *Developer or detector* (as appropriate): it can be *Universal*, for TLC iodine; for high performance liquid chromatography (HPLC): refractive index detector, diode array. *Specific*, for TLC UV light at 254 and 365 nm, visible light; reagents such as Dragendorff for alkaloids, Kedde for cardiotonics, ninhydrin for amino acids, ferric chloride for phenols; for HPLC: UV-Vis detector, fluorometer, mass spectrometer, electrochemical; for gas chromatography (GC): flame ionization detectors (FID), mass spectrometer, among others.

Main Chromatography Systems and Their General Characteristics:

The general characteristics are as follows:

- *Conventional laboratory column chromatography*: This type of chromatography uses glass columns (inactive support) filled with polar substances (stationary phase) such as alumina (Al_2O_3), silica, magnesium oxide, and cellulose. The polarity can also be changed using modified phases (in this case they will be called reversed phase), stationary phases made up of gels for permeation chromatography or gel filtration (according to molecular weight and shape) can be used; cationic or anionic resins also can be used as stationary phases.
- *TLC*: This type of chromatography uses a glass or plastic plate (inactive supports called chromatoplates) or a piece of aluminum (chromatofoils), coated with a stationary phase (generally of the type described in *Conventional Laboratory Column Chromatography*, with some

variants, especially the smaller particle size for this case), maintaining a small constant thickness throughout the plate. This is placed in a chromatographic tank, which must be previously saturated with the eluent (liquid mobile phase). The eluent will pass through the plate and drag the components along it, producing spots or bands or zones corresponding to the pure components (or mixed if the system does not achieve their correct separation). These areas will be observed once the process is finished, through the development that may be carried out using one of the developers indicated in previous paragraphs.

- *Paper chromatography*: This system is similar to the previous one, except that strips of chromatographic quality paper are used where the samples are seeded and developed in the chromatographic tank. The support is active and the main phenomenon is that of partition. The polarity of the paper can be decreased, e.g., by embedding it in paraffin or silicone (reverse phase paper).
- *HPLC*: This system is similar to *conventional laboratory column chromatography*, except that the flow is applied under pressure (between 1500 and 2200 psi), the particle size is generally less than 3 microns, the length of the column is between 5 and 25 cm, and the equipment can be expensive.

Detectors
Detectors are devices used to monitor the composition of the liquid that is coming out of the column (or eluted), that is, to record how the composition of the eluate varies as a function of time. As in any chromatographic system, it is necessary to see how the components of a sample have been separated, and, in an instrument, this could not be done directly. For this reason, it is necessary to have a device that gives a signal as the separation progresses, for which detectors are used. The detector is a device that continuously measures some physicochemical properties of the compounds and generates a signal proportional to the concentration of the compounds as they exit the column.

There are two groups or types of detectors:

- *Universal or general*: These detectors measure a general property that indicates that some substance is coming out, i.e., they distinguish the mobile phase alone from that which carries a substance. The most common are the refractive index (which is not very stable, does not allow the use of a solvent gradient, and has lower sensitivity), and the conductivity index.
- *Selective*: These detectors measure a characteristic property of a given substance. The most common are UV-Vis detectors (which determine the absorbance at a wavelength and can have a fixed or variable length), fluorescence detectors (which are very selective and stable), electrochemical detectors [which measure an acid-base or redox reaction; an example of this type of detector is the amperometric detector used in the high-performance anion-exchange chromatography/pulsed amperometric detection (HPAEC-PAD)], radiometric detectors (which measure gamma radiation), and mass detectors.

Types of HPLC Depending on the Stationary Phase
Stationary phases can be either common or modified.

Common stationary phases can be:

- *Partition phases*, such as liquid stationary phases, which are generally used for substances with molecular weight < 1500.
- *Adsorption phases* consist of a porous material up to 10 microns, generally less than 3 microns, or film when they are covering the internal walls of the columns; in general, silica gel is used.

Modified stationary phases can be:

- *Chemically bonded phases*: These phases can in turn be either *normal phases* or *reversed phases*. In the former, the silica gel increases the polarity of the phase by adding more polar groups such as diols, cyano groups, or amino groups; thus, since a more polar phase is achieved, the mobile phase will be used from the least polar to the most polar, and the less polar substances will have lower retention time. In the *reversed phases*, the silica gel decreases its polarity by the addition of groups or carbon chains of different lengths, such as C18 or RP18, C8, C4, or C2; thus, the mobile phase will be used from the most polar to the least polar, and less polar substances will have higher retention time.
- *Ion-exchange*: In this type of chromatography, the stationary phase is made up of silica or polystyrene resins linked or cross-linked with divinylbenzene (the latter are the most widely used). They can be cationic (in which case they expose a negative charge, and are thus used to separate cations), or anionic (in which case they expose a positive charge, and are thus used to separate anions; in the event that the detector is of the amperometric type, the chromatography is called HPAEC-PAD). In general, the mobile phase has water and another solvent, and a perfectly controlled pH.
- *Exclusion by gels*: This type of chromatography is used to separate substances with a molecular weight greater than 2000, and the separation is carried out by molecular size and shape. The smallest molecules penetrate the pores of the gel while the largest pass externally, and therefore the largest will come out first and the smallest last. They are

divided into two types: *gel filtration* (for biopolymers such as proteins and carbohydrates, where the stationary phase is hydrophilic, made up of dextran or agarose derivatives, and is used as mobile phase in general aqueous solutions), or *gel permeation* (for synthetic polymers, where the stationary phase is made up of neutral polystyrene resins cross-linked with divinylbenzene and the mobile phase is generally not very polar, e.g., toluene).

- *Affinity*: In this type of chromatography, the stationary phase is made up of materials that have the characteristic of temporarily and reversibly binding certain substances. For example, lectins bind certain carbohydrates, whereas antibodies bind certain antigens.
- *Chiral*: This type of chromatography is used to separate enantiomers (D and L). It allows separating the components of a racemic mixture, i.e., the chiral isomers. It is important to separate the active components from the nonactive ones, or to separate the more toxic ones. This can be done by using special stationary phases, such as a chiral reagent attached to the common stationary phase. When doing this, the isomers then associate with the attached reagent to give dissociable diastereomers that will have different retention time. The most common phases used for chiral chromatography are amino acid derivatives, e.g., dinitrobenzoylphenylglycine, or the phenylethylamine derivatives of valine. Another way is by adding additives to the mobile phase, in which case, the optical isomers react with a chiral reagent to form the diastereomers, and then an additional chiral center is achieved and therefore the geometric structure will be different, allowing separation. In other words, in this case, the chiral or selectant reagent is added to the mobile phase at low concentrations so that they form reversible complexes that will adsorb to the stationary phase with different affinities, allowing separation. Further isolation of the pure isomers requires the removal of the chiral reagent. Another more recent way is to use cyclodextrins made up of 6, 7, or 8 glucose units linked in 1,4 forming a cyclic structure or ring of different diameters, and according to the geometry of the chiral isomers only one can enter the cavity of the ring and the other will be excluded, achieving a separation by differential retention.
- *Gas Chromatography (GC)*: In this type of chromatography, the mobile phase is a gas called carrier gas, the most used of which is He, N_2, or Ar. The compounds that can be separated must be either naturally volatile, as volatile oils, some other terpenes or alkaloids, or made volatile by derivatization, e.g., carbohydrates, fatty acids, and steroids, and must be thermally stable.

In GC, columns are usually made of stainless steel, but can also be of Cu, Al, Ni, and glass. Columns can be *normal* or *packed*, *capillaries*, or *intermediate*. *Normal* or *packed* columns are 2–4 mm in internal diameter and up to 30 m long (for analytical chromatography), or 1–2 cm × 60 m (for preparative chromatography). In general, they are spiral or U-shaped, and have a stationary phase throughout the light of the column. *Capillaries* can have coated walls or coated supports. Their internal diameter is less than 1 mm (0.3–0.1 mm) and their length ranges from 60 to 300 m. They are made of stainless steel, borosilicate glass, or fused silica. They have a high efficiency and less bleeding than normal or packed columns. *Intermediate* columns are usually between 2 and 5 m long and 0.25–0.5 mm in internal diameter. They have an intermediate efficiency and can be used as the normal or packed ones, working with higher mobile phase flows (15–20 ml/min) or as the capillary ones (1–5 ml/min), although they never reach the efficiency of the latter.

In GC, the most common detectors are thermal conductivity detectors (TCD) and FID. Others are electron capture (ECD), nitrogen-phosphorus (NPD), atomic emission (AED), chemiluminescence (CLD), flame photometry (FPD), and photoionization (PID). In addition, the chromatograph can be coupled to a mass spectrometer (EMS).

The GC can be of two types:

- *Adsorption gas chromatography*: In this type of chromatography, the stationary phase is a finely powdered adsorbent solid (such as silica gel, alumina, coated glass beads, or activated carbon). Substances are separated by electrostatic forces on the surface of the solid.
- *Partition gas chromatography*: In this type of chromatography, the stationary phase is a liquid embedded in an inert porous (pore < 1 μ) or nonporous (glass microbeads) support that goes inside the column. This material can be diatomaceous earth (Chromosorb), siliceous earth, or porous polymers. The liquid stationary phase must be nonvolatile at operating temperature and can be very polar (propylene glycols and their esters), moderately polar to separate alcohols and esters (ethylene glycol polymers), or slightly polar (silicone or paraffin polymers, gums). So, if the stationary phase is a highly or moderately polar liquid, the elution occurs in order of increasing polarity, and if the stationary phase is a slightly polar liquid, the elution occurs in order of increasing molecular weight. In partition gas chromatography, solute molecules are retained in the liquid stationary phase according to their partition coefficient between it and the gaseous mobile phase.

Finally, other chromatography types include high-speed countercurrent chromatography (HCCC), centrifugal countercurrent chromatography, and supercritical fluids chromatography (general references: Houghton and Raman 1998;

Dixon and Wilken 2018; Patras et al. 2018; Corrêa et al. 2021).

13.3.3.3 Structural Characterization of the Metabolites

As described in the previous sections, several extraction and purification methods are used to recover the targeted bioactive compounds from microalgal biomass. Once the main metabolites have been purified, different analytical methods are used to characterize the chemical structures. They basically include determinations of physical constants, chromatographic properties, and spectral properties.

13.3.3.3.1 Determination of Physical Constants

The physical constants usually determined include the melting point, the rotational power, and the refractive index.

13.3.3.3.2 Determination of Chromatographic Properties

The chromatographic properties usually determined are the Rf values [by means of TLC or high-performance TLC (HPTLC)] and retention times (by HPLC or GC) (Patras et al. 2018).

The different types of chromatography techniques described above can be used to characterize different types of metabolites, as described below:

- *TLC and HPTLC*: These are the oldest and most widespread analytical chromatographic technique for the screening and characterization of microalgae and natural products in general. These methods are usually used for the analysis of the pigments and phenols from microalgae.
- *HPLC*: This technique can be used coupled with UV or photodiode array (PDA) detection. The best choice for the analytical separation and characterization of metabolites from their retention times is to use HPLC with reversed-phase columns. This technique allows efficiently identifying carbohydrates, phenols, mycosporine-like amino acids, pigments, and other active compounds from microalgae. For example, HPLC-ESI-MS has been used to study the carotenoid profiles of *Chlorella sorokiniana* and other microalgae, either in aqueous suspensions or powders obtained by freeze-drying. Triglycerides, phospholipids, and tocopherols can be analyzed by using liquid chromatography integrated with evaporative light-scattering detector; terpenes can be characterized by using HPLC-UV-MS; and carotenoids can be analyzed using HPLC-UV/Vis or HPLC-DAD (High Performance Liquid Chromatography-Diode Array Detector) (Hassan et al. 2022). Supercritical fluid extraction extracts make use of capillary electrophoresis coupled with PDA detection for less application time, high efficiency, and selectivity when compared to HPLC. The Chiral HPLC method coupled with 2D NMR spectral analysis can be used for structural elucidation of lipopeptides.
- *GC*: GC can be used for the chemical characterization of lipids (after hydrolysis and derivatization) and volatile compounds. GC-MS is the analytical method most commonly used to determine the fatty acid profile and can also be used to characterize terpenes (Hassan et al. 2022). GC-MS or GC-FID is usually used to analyze carbohydrates after hydrolysis and derivatization to the corresponding peracetylated nitriles or alditols (Flores et al. 2000).

13.3.3.3.3 Determination of Spectral Properties

The spectral properties of the metabolites obtained are usually determined by UV-Visible (UV/Vis) spectrophotometry, FTIR spectroscopy, Fluorescent spectroscopic analysis, mass spectroscopy (MS), and nuclear magnetic resonance (NMR) spectroscopy (Hassan et al. 2022).

- UV/Vis spectrophotometry is generally used to quantitatively determine the chlorophylls and carotenoids in whole pigment extracts from green microalgae and different natural products. This depends on the sample characteristics, the solvent system, and the spectrophotometric parameters. The absorption spectrum of an extract or fraction generally contains mixtures of chlorophyll *a* and *b* and total carotenoids. Chlorophyll *a* absorbs mainly at A428 (blue) and A661 (red), whereas chlorophyll *b* and carotenoids absorb mainly in the blue region (400 to 500 nm). However, by increasing the polarity of the solvent, the red absorption maximum of chlorophyll *a* shifts from 660 to 665 nm and the blue absorption maximum shifts from 428 to 432 nm (Patras et al. 2018).
- UV/Vis spectrophotometry is also used for the quantification of carbohydrates, using phenol–sulfuric acid, and measuring spectrophotometrically at 490 nm for hexoses and at 480 nm for pentoses and uronic acids (Flores et al. 2000). For mycosporine-like amino acids, the quantification needs absorption values at 330 and 310 nm. For phenols and derivatives, characteristic UV/Vis spectra are determined; e.g., typical flavonoids present two characteristic bands, I from 300 to 380 nm and II from 220 to 290 nm (Patras et al. 2018).
- Liquid chromatography coupled with MS (LC-MS) combines the high separation power of HPLC with the structural information provided by mass spectrometry. Both the sensitivity and the specificity of LC-MS are

much higher than those of the traditional UV/PDA detection and allow very fast chromatographic separations with high peak purity values. This technique can also use electron spray ionization (ESI) or atmospheric pressure chemical ionization (APCI), both of which transfer the molecules from the solution to the gas phase, and make them suitable for fragmentation in the mass spectrometer. LC-ESI/APCI-MS has been used to identify carotenoid esters present in extracts of pigmented freshwater microalgae and to identify and quantify glycolipids from microalgae (Hassan et al. 2022).

- Another high-performance technique used for protein fingerprinting and characterization of the phenotype (genus, species) is matrix-assisted laser desorption ionization–time of flight–mass spectrometry (MALDI-TOF-MS). In some cases, this technique can also be used to characterize microalgal strains, comparing the spectra of unknown microorganisms to reference spectra found in international databases. This provides reproducible and unique spectra covering a wide m/z range. The reproducibility of the spectra should then be confirmed by using composite correlation index calculation and main spectra library dendrogram creation, available with the MALDI Biotyper software (Hassan et al. 2022).
- NMR spectroscopy comprises 1D and 2D techniques such as ^{1}H, ^{13}C NMR, HMBC, HSQC, COSY, and NOESY. The most important 2D-NMR techniques used to identify natural products are DQF-COSY, TOCSY, NOESY, HSQC or HMQC, and HMBC, combined with HSQC and TOCSY, and INADEQUATE (Bross-Walch et al. 2005; Hassan et al. 2022).

13.3.3.4 Molecular Docking

Molecular docking is a technique that allows depicting the optimized interaction between the ligand (bioactive compound) and its target. This technique determines different binding modes of the ligand in the groove of the target molecule, thereby determining the potential drug candidate (Hassan et al. 2022).

For leading compounds, the physicochemical properties, fundamentally lipophilicity, acid-base character (pKa), and solubility, are determined, and the interpretation of these thermodynamic constants, at the molecular level, are useful tools to establish relationships between the chemical structure and the biological activity of drugs. The partition coefficient of the isolated substances can be determined using the ampoule method and/or HPLC; the pKa by potentiometry or by spectrophotometry; and the solubility by the method of saturation and spectrophotometric quantification or by HPLC.

13.4 Study of Two Promising Species

Microalgae are microscopic photosynthetic organisms found in both marine and freshwater environments. They have important applications, e.g., in food, aquaculture, and the production of alternative energies; some species also have applications in cosmetics. However, although they are a source of bioactive compounds with diverse biotechnological properties, few microalgal species have so far been explored and exploited industrially, especially in relation to the pharmaceutical industry, e.g., in therapeutic formulations (Priyadarshani and Rath 2012).

Among the most required groups of metabolites for the industry, pigments (astaxanthin, β-carotene), essential fatty acids, sterols, phenols, phycobiliproteins, and carbohydrates (mannitol, glucose, fructose, oligosaccharides, polysaccharides) stand out. Supercritical fluid chromatography has been used to recover astaxanthin, PUFAs, and other compounds. Hexane, ethanol, chloroform, and diethyl ether can extract fatty acids such as EPA, DHA, and arachidonic acid (AA) from various microalgae. Extraction with aqueous buffers has been used to obtain phycobiliproteins and lutein, ethanol for phenols, and aqueous extractions for carbohydrates (Molina Grima et al. 2003).

In commercial production, the main species of microalgae used belong to the genera *Isochrysis*, *Chaetoceros*, *Chlorella*, *Arthrospira*, and *Dunaliella*. However, other species are under investigation, and the results are showing significant industrial and commercial potential to meet the growing demands for this type of natural resource.

Among them, in this section, the authors highlight some of the main advances of two promising species of *Euglena* and *Scenedesmus*, some of whose studies have been carried out at the laboratories.

13.4.1 *Euglena gracilis*

The genus *Euglena* comprises unicellular freshwater flagellates with photoautotrophic, heterotrophic, or photoheterotrophic modes of nutrition. A total of 337 *Euglena* species names, 164 of which have been taxonomically accepted, can be found in the database AlgaeBase (http://www.algaebase.org). *Euglena* not only is used in bioremediation but also is promising as a cultivable microorganism for large-scale production of various important metabolites with potential application, including carbohydrates, proteins, lipids, carotenoids, phycobiliproteins, sterols, hormones, phenols, and mycosporin-like amino acids with important pharmaceutical value, mainly because of their neuroprotective, antioxidant, antimicrobial, anti-inflammatory, anticancer, and angiogenic properties. Unlike bacteria and protozoa, *Euglena* poses no pathological risk to humans and is safe to cultivate in large amounts. Besides, it can be grown heterotrophically on various carbon sources and over a wide range of pH values, and in wastewater containing chemical and organic pollutants (Kottuparambil et al. 2019). However, *Euglena* has been limitedly exploited for industrial production.

Euglena gracilis can grow autotrophically, heterotrophically, or myxotrophically and has been extensively studied, mainly for the production of primary metabolites. This species has a wide range of nutritional requirements, suggesting the existence of diverse physiological patterns, which can generate different metabolites and/or variations in the proportion in which they are biosynthesized. Among the Euglenophyceae, *E. gracilis* has long been used in basic research as a model organism in physiology and biochemistry (Kottuparambil et al. 2019).

In 2013, the authors published the results of a preliminary study of *E. gracilis* cultured in vitro, in its photosynthetic and bleached forms, on its exponential and stationary growth phase (Tolivia et al. 2013). Two strains were used in the study: the *E. gracilis* (Klebs) commercial strain (UTEX-753) and a wild-type strain (MAT) isolated from Río Matanza (Province of Buenos Aires, Argentina). The research was performed on the photosynthetic strains and their bleached counterparts, obtained by treatment with streptomycin. The cultures were grown in a growth chamber at 24 ± 1 °C, with 12:12 cool-white fluorescent light (150 mE m^{-2} s^{-1} irradiance) in EGM medium (*Euglena gracilis* medium). Cells were quantified with Neubauer's chambers and biomass was obtained by centrifugation at 4 °C after 72 h (exponential phase) and 144 h of growth (stationary phase). Biomass was washed four times with distilled water at 4 °C, and then dried by lyophilization. Extracts of ethanol 96° from both strains were screened for preliminary identification of chemical constituents. Pigments, lipids, and flavonoids were chromatographically analyzed. Besides antioxidant activity (DPPH method), growth inhibition of *Triticum sativum* seeds, and cytotoxicity by the brine shrimp, lethality test (*Artemia salina* test) were tested in vitro. The qualitative chemical analysis showed the presence of carbohydrates of low molecular weight, lipids, and steroids in all samples. Cardenolids were only present in the exponential phase samples and triterpenes only in the exponential phase samples of the bleached strains. With the exception of the photosynthetic MAT strain, tannins were present in the exponential phase of all the other samples. In contrast, flavonoids were only detected in the stationary phase samples of photosynthetic strains.

The presence in all photosynthetic samples of chlorophylls *a* and *b*, β-carotenes, diadinoxanthin, and neoxanthin was verified by TLC. Pigment analysis performed by HPLC in reverse phase with diode array detector (RP-HPLC-DAD) allowed yields between 33% (UTEX-b-ST, commercial strain, bleanched stationary phase) and 68.8% (MAT-ph-ST, strain isolated from Matanza river, photosyntetic exponential phase).

The chromatographic profiles showed that chlorophyll content decreased while carotenoids content increased in the stationary phase of both photosynthetic strains, and revealed the presence of flavonols derived from quercetin. Several authors have described pigment variation in *Euglena*. The relationships found in the results suggest that carotenoids may be involved in the formation of chlorophylls. Studies indicate that the same porphyrin-like molecule may influence the synthesis of both pigments.

Regarding flavonoids, there are some reports about this metabolite in algae, but the authors' work was the first to report this chemical group in Euglenoids. When nutrients become scarce, *E. gracilis* cells enter a non-growth phase known as stationary phase and develop a multiple-stress resistance response. The presence of flavonoids in the stationary phase may be associated with that response.

In concordance with the presence of polyphenols, the fractions with the highest polarity showed antioxidant activity against DPPH and growth inhibition of the seed activity in vitro even in the absence of paramylon. Paramylon has been previously reported to have anti-tumoral properties. Since paramylon precipitates in ethanol, the authors' work possibly demonstrated that the extracts may have antioxidant and possible anti-tumoral in vitro effect.

Before this study, antioxidant activity in *Euglena* had been previously reported in different strains, especially in relation to the presence of vitamins E and C and β-carotene; nevertheless, this activity had not been related to the polyphenols. This study showed that the fractions of higher polarity have the highest radical scavenging activity with $SC_{50} = 147.7$ μg/ml and 157.2 μg/ml (MAT-ph-ST and UTEX-ph-EX, respectively). This work was the first report about polyphenol (flavonoids and tannins) production in Euglenoids, which allows a first assessment of the potential of *E. gracilis* as a source of these bioactive products.

Later works, some of which cited the authors' work from 2013, analyzed the potential of the species to obtain products of industrial value, in particular related to bioactive metabolites. Kottuparambil et al. (2019) highlighted that species of the genus *Euglena* are particularly attractive due to their easiness to be grown in cultures and the ability to produce numerous bioactive compounds through extremely complex metabolic pathways. The plastids in *Euglena* cells have been recognized as a site for the production of proteins, fatty acids, and many other value-added metabolites.

The discovery of the high protein content and amino acid profiles of *E. gracilis* led the genus to be recognized as having commercial potential, mainly for its biotechnological applications of the specific type of β-1-3-glucan called paramylon. *Euglena* has been reported to be rich in carbohydrate-active enzymes and, despite being a single-celled organism, its high capacity to synthesize complex carbohydrates is exceptional. *Euglena* cells contain nutritionally important fatty acids, DHA, EPA, and vitamins, which corroborates the industrial potential of the genus as a valuable resource (Kottuparambil et al. 2019).

In the past years, works on the secondary bioactive metabolites of *Euglena*, aimed at the identification of new compounds, their potential properties, and promising applications have increased. In particular, *E. gracilis* has emerged as an interesting candidate for application-driven research and commercialization (Gissibl et al. 2019). Among these metabolites, paramylon is already marketed as an immunostimulatory agent in nutraceuticals. Future challenges include achieving the economy of large-scale cultivation. Recent insights into the complex metabolism of *E. gracilis* have highlighted unique metabolic pathways, which could provide new leads for product enhancement by genetic modification of the organism. Also, the development of molecular tools for strain improvement is emerging rapidly, making *E. gracilis* a noteworthy challenger for microalgae such as *Chlorella* sp. and their products on the market.

In a study, by using an untargeted metabolomic approach, Bernard and Guéguen (2022) showed how carbon sources can influence the phenolic compound production by *E. gracilis*. This species can tolerate high concentrations of metals via the excretion of organic metabolites, including phenols. This is important because industrial development and urbanization have led to the presence of diverse metals in wastewater, which are often improperly treated. Therefore, it is important to evaluate how carbon sources stimulate the production of phenolic compounds by this species. Bernard and Gueguen (2022) showed that each of the four carbon-amended cultures studied (i.e., glutamic acid, malic acid, glucose, and reduced glutathione) led to significant differences in the number, relative intensity, and molecular composition of the phenolic compounds during the logarithmic phase. Phenolic compounds were mainly produced during the minimum growth rate, probably as a response to stressful conditions. Further insights into the influence of the growth conditions on the production of phenolic compounds by *E. gracilis* will help identify conditions that favor the production of certain phenolic compounds for dietary, metal chelation, and pharmaceutical applications.

The presence of metals in aquatic ecosystems is a growing global problem because they can cause adverse and toxic effects on living organisms. Since estimates suggest that 40% of lakes and rivers are already contaminated, it is important to find ways to properly treat industrial and municipal effluents. In this context, several physical, chemical, and biological methods have been developed. One of them is phycoremediation using algal biomass that produces chelating ligands able to recover significant amounts of metals. The metabolic flexibility of *E. gracilis* suggests the existence of several physiological pathways affecting the biosynthesis and proportion of different metabolites. *E. gracilis* also shows high adaptation capacities for a broad range of environmental conditions through the plasticity of its atypical and unique metabolism, including its ability to tolerate high metal concentrations via the production of chelating ligands. The production of secondary metabolites is less known. These metabolites, which are produced in response to environmental changes, play a key role in the defense mechanism against abiotic and biotic stress. Among secondary metabolites, phenolic compounds act as reducing agents and hydrogen donors able to minimize oxidative stress. In addition, these compounds have shown excellent metal chelating potential due to the high number of oxygen functional groups in their structure.

Despite the industrial potential of *E. gracilis*, few reports have evaluated the production and bioactive functions of phenolic compounds and other secondary metabolites in this species. The progress in biotechnological techniques has encouraged a new trend of research on the potential of algal bioproducts. In this context, *Euglena* possesses an extremely complex metabolic capacity comparable to that of multicellular organisms, offering wide opportunities for metabolic engineering to produce value-added biomolecules.

13.4.2 *Scenedesmus obliquus*

In recent years, more and more species of microalgae are being investigated to identify, isolate, and industrially exploit the bioactive metabolites that they produce. Among the most studied are those of the genus *Scenedesmus* (Chlorophyta), although further chemical research remains to be done on most of its species, especially in relation to promising biological activities for health.

The genus is cosmopolitan and is commonly found in all types of freshwater. It plays a fundamental role in the food chain and contributes to the purification of eutrophic waters, characterized by its high growth rate and resistance to high concentrations of pollutants. Potentially, the species of this genus could be used for biotechnological applications in the remediation of wastewater, aquaculture, agriculture, and the production of biofuels (Badwy et al. 2008), and, more recently, in cosmetics and therapeutics. Regarding the latter, Guedes et al. (2011) reported that *Scenedesmus* is one of the few green algal genera with activity against pathogenic bacteria such as *Salmonella* sp., *Pseudomonas aeruginosa*, *Escherichia coli,* and *Staphylococcus aureus*. Other biological activities that position the genus as a promising source of pharmaceutical application have also been reported. *Scenedesmus* is characterized by its abundance of lutein, astaxanthin, and β-carotene (Ishaq et al. 2016), and by the production of essential fatty acids, phenols, carbohydrates, and terpenes.

As described in the previous sections, environmental factors, such as temperature, light, pH, CO_2 supply, the physiology and growth phase of the microalgae, as well as the composition of the wastewater, can largely modify the biochemical composition of the biomass. Likewise, seasonal variations due to ecological and geographical factors can lead to metabolic variations and, consequently, to variations in the metabolites produced and their biological properties among species of the same genus (Stengel et al. 2011). For this reason, the environmental conditions of growth as well as all the steps involved in the investigation of the species must be optimized to maximize the production of specific chemical compounds.

According to several reports, the cost of developing microalgae-based products remains relatively high. One alternative to reduce production costs is to use the nutrients present in wastewater and/or effluents. To study this issue, Fuentes-Grünewald et al. (2021) produced *C. vulgaris* and *S. obliquus* using anaerobically digested animal manure as a nutrient source and found that the protein content of the harvested biomass exceeded 45%, turning out to be an excellent source for animal feed. Ferreira et al. (2020) determined the biochemical composition, on a dry basis, of *S. obliquus* cultured in poultry effluent and found the following values: proteins (19.1%), lipids (19.8%), carbohydrates (36.2%), and ashes (4.2%).

As part of the joint investigations carried out by the authors of this chapter, in particular by Jiménez-Veuthey for her Ph.D. thesis, a strain of *S. obliquus* (IOAC081F) was isolated from the waters of Embalse Salto Grande, in the province of Entre Ríos, Argentina, and studied. This water reservoir has been contaminated by agro-industrial effluents from Argentina and Uruguay for several decades (Jiménez-Veuthey et al. 2019). Monitoring studies of its waters have recorded the presence of phytoplankton organisms belonging to the group Bacillariophyta, Chlorophyta, Cyanophyta, Euglenophyta, Chrysophyta, and Dinophyta (Bordet et al. 2011).

The isolation of IOAC081F is described in another chapter of this book ("Phycoremediation of Industrial Wastewater"; Jiménez-Veuthey et al.). According to the specific goals, the strain was isolated from the Embalse Salto Grande, generating and optimizing the appropriate operational conditions for its growth at laboratory scale using poultry effluent as the only culture medium, to maximize biomass production (Figs. 13.4 and 13.5). The ability of this strain for poultry effluent phycoremediation was evaluated using a pilot-scale raceway (Fig. 13.6). On the other hand, the chemical composition and the antimicrobial, antioxidant, and cytotoxic activity of the biomass were evaluated.

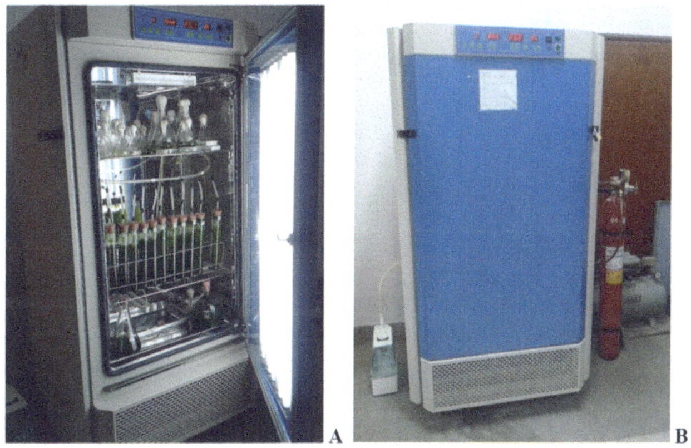

Fig. 13.4 (**a**) Phytotron (culture chamber) with photoperiod, temperature, and humidity automatic control; (**b**) Phytotron with air and CO_2 injection system on demand (Jiménez-Veuthey, LAMAS, Universidad Nacional de Entre Ríos, Entre Ríos, Argentina)

At microscopic observation, *S. obliquus* cells show an intense green color (Fig. 13.7). They are nonmotile and ovoid, ellipsoid, fusiform or spherical in shape with a well-defined apical crest, a narrowing central section in small individuals and very small or no apical crest in larger cells. Some cells are found as single cells (2–7 μm wide and 8-5 μm long) whereas others are found as four- to eight-celled coenobia (16–21 μm wide and 17–27 μm long, and 24–34 μm wide and 36–52 μm long, respectively). The identity of the isolated strain showed 99% similarity with *S. obliquus* through taxonomic keys and molecular techniques. The suitable growth conditions were determined

Fig. 13.5 *Scenedesmus obliquus* production in bubble column bioreactors inside the phytotron. pH control with CO_2 on demand (Photograph: Mariana Jiménez-Veuthey, LAMAS, Universidad Nacional de Entre Ríos, Entre Ríos, Argentina)

Fig. 13.6 Sampling of *S. obliquus* culture, during the phycoremediation process, using an outdoor raceway bioreactor (Photograph: Mariana Jiménez-Veuthey)

by simulating, at a laboratory scale, the environmental conditions from which the microalgae was isolated. In this sense, the Allen and Arnon culture medium enriched with $NaNO_3$ was formulated and the culture chamber conditions were 25 ± 1 °C, RH 65%, lighting with six fluorescent lamps of 36 W each and a 12 h light:12 h darkness cycle. The poultry effluent used as a culture medium for microalgal cultivation came from the poultry industry Las Camelias S. A., located in San José, Entre Ríos, Argentina. The effluent was characterized according to physicochemical and microbiological tests. After that, the phycoremediation ability of *S. obliquus* was evaluated at a laboratory scale. Subsequently, the process was scaled to a smaller raceway bioreactor (6 m^3) and the 60 m^3-capacity raceway bioreactor located in the Las Camelias S.A. Effluent Treatment Plant. The poultry effluent samples that fed the bioreactor were collected every 3 months, for a year, to simulate the four annual seasons. The other above-mentioned chapter of this book describes the experimental conditions and the results obtained in that part of the study. The results of the semi-industrial scale phycoremediation process showed an important influence of the seasons of the year on the tertiary treatment and biomass productivity.

The biomasses obtained from each season were lyophilized, subjected to disruption in a mortar, and extracted with 96° ethanol for 24 h, at room temperature and protected from light. The crude extracts obtained were fractionated in solvents of different polarities, obtaining a fraction with water and another with chloroform. The pellet remaining from the alcoholic extraction was extracted with water at room temperature. The fractions and extracts were analyzed by phycochemical screening, quantification of the main metabolites, and planar and instrumental chromatography (RP-HPLC-DAD and GC-MS). In addition, biological activities were determined (antibacterial by the agar inhibition method, antioxidant by bioautography, and cytotoxicity using the *Artemia salina* test).

The extraction yields were very important, standing out the spring and autumn ethanolic ones (43.5 and 39.5%, respectively) and the autumn and winter aqueous ones (40.8 and 43.6%, respectively).

The results of the study showed a great variability of metabolites, with carbohydrates, flavonoids, tannins, quinones, triterpenes, and cardenolides being the main chemical groups. The presence and abundance of such groups showed dependence on the culture medium used and the season of the year. Thus, of the four annual seasons, summer was the one that showed the greatest abundance and chemical variability, whereas autumn showed the opposite. Carbohydrates and tannins showed little seasonal variation; flavonoids stood out in summer, quinones, and triterpenes in spring; and cardenolides were present in all seasons, standing out in summer.

High levels of proteins, lipids, and carbohydrates, as well as high concentrations of minerals (mainly iron and calcium), phenolic compounds, and carotenoids, were quantified. Carbohydrates showed seasonal variations, from 9% in autumn to 25% in spring. Other authors such as Pereira et al. (2017) carried out studies of *S. obliquus* BR003 cultivated in oil wastewater in Brazil, to determine the productivity of biomass, lipids, and carbohydrates, and found

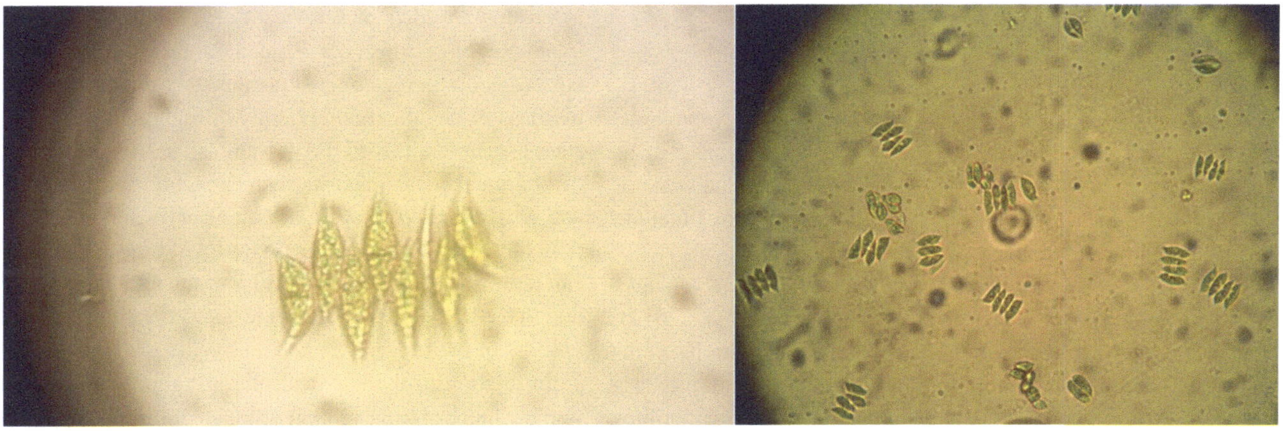

Fig. 13.7 Morphology of *S. obliquus* microalgae observed using an inverted light microscope (40x, Leica, DMIL; Photograph: Mariana Jiménez Veuthey)

between 3 and 18% of total carbohydrates. On the other hand, Singh and Gu (2010) reported that the content of these metabolites for *Scenedesmus* sp. was between 10 and 17%. The authors were also able to verify the influence of the season of the year on the profile of monosaccharides by means of GC-MS, after hydrolysis and derivatization of the hydrolysates. The main monosaccharides identified in all extracts were rhamnose and fucose, followed by mannose and galactose. However, in spring, rhamnose represented 66% of the total carbohydrates expressed in molar %; in summer, fucose prevailed (36%); in autumn, fucose and rhamnose (42 and 25%), and in winter rhamnose stood out with 36%. In addition, mannose, galactose, arabinose, glucose, and xylose were detected. The species most commonly used in the production of polysaccharides are *Tetraselmis* sp., *Isochrysis* sp., *Porphyridium cruentum*, *Porphyridium purpureum*, *Chlorella* sp., and *Rhodella reticulata* (Levasseur et al. 2020). The results position *S. obliquus* as a potential source of carbohydrates.

The study also showed that spring was the season with highest abundance of phenols (3.70%), followed by summer, winter, and autumn (3.10, 2.50, and 0.90%, respectively). El-Tablawy et al. (2020) showed similar results, highlighting that *S. obliquus* produced 2.9% of total phenols using the same methodology as the authors' work. They also analyzed the presence of flavonoids and pigments in various microalgae in relation to their antioxidant capacity. Other studies carried out with the *S. obliquus* 8610 strain isolated in Bulgaria showed similar amounts of total phenols (around 2%), and differences when the microalgae were grown under different light conditions (Zaharieva et al. 2022). The phenolic profile was analyzed by RP-HPLC-DAD. The results showed that gallic, chlorogenic, and coumaric acids, and flavonoids (flavonols, flavones, flavanones, and their glycosides) stood out. Jerez-Martel et al. (2017) also identified flavonoid compounds in *Ankistrodesmus* sp., *Spirogyra* sp., *Euglena cantabrica*, *Caespitella pascheri*, and in cyanobacteria, with catechins, chlorogenic acid, and gallic acid being the most abundant.

The amount of protein found was 20–33%, standing out in spring. Total lipids fluctuated between 10 and 16.80%; α-linolenic fatty acid (ALA) was the main omega-3 fatty acid identified. Lutein was the main carotenoid identified, followed by β-carotene and violaxanthin.

Cytotoxicity on *Artemia salina*, a monitor bioassay for the search of antitumor substances, was important in all seasons, with an LD_{50} between 19.3 and 63.6 μg/ml. In addition, in autumn, intercalation in the DNA of a dose-dependent type was highlighted. Studies on monosaccharides indicated that L-fucose and L-rhamnose are able to inhibit the activity of lymphokines in vitro, and suppress the manifestations of cellular immunity in vivo. In addition, they showed the ability to block the soluble suppressor factor of lymphocytes (Sánchez et al. 2002). In this context, the cytotoxicity results from the authors' work are related to the concentrations of rhamnose and fucose determined in different seasons.

Finally, studies on the antimicrobial activity against *Staphylococcus aureus*, *Escherichia coli*, *Pseudomonas aeruginosa*, and *Enterococcus faecalis* were negative in all seasons. The antioxidant capacity was also demonstrated by bioautography, coinciding with the presence of phenolic acids and flavonoids.

Therefore, the results of studies on *S. obliquus* show a solid basis for the production of metabolites by this species using a semi-industrial scale raceway. In addition, the nutrient recovery from poultry effluents allows obtaining a high-value biomass as well as treating industrial effluents, under the concept of circular economy.

13.5 Conclusions, Perspectives, and Projections

Although most pharmacognostic studies focus on plants, some are considering other types of organisms regarded as pharmacognostically interesting with possible added value of their products for industrial application. These include microalgae, whose potential is considerably high.

Microalgae have characteristics that constitute advantages to undertake large-scale production (Ventura et al. 2017). They require relatively simple cultivation conditions (temperature, nutrients, light), and can achieve high biomass yields and accumulation of bioactive metabolites (Hernández-Pérez and Labbé 2014). Gómez Luna (2007) summarized the benefits of microalgal cultivation in three points: (i) microalgae are efficient biological systems that use solar energy to produce organic matter, (ii) they grow faster than land plants, and (iii) they allow obtaining higher annual biomass yields than land plants. On the other hand, the biochemical composition of the metabolites produced can be modified by varying the environmental conditions and/or the composition of the culture medium. In addition, under certain conditions, many species of microalgae can accumulate high concentrations of compounds of commercial interest, such as proteins, lipids, carbohydrates, glycerol, natural pigments, and phenols. These molecules are thus attractive candidates for the pharmaceutical, cosmetic, and food industries, and are important for the production of biodiesel and biogas. In addition to this, these microorganisms are used in phycoremediation, which is a feasible solution to mitigate the environmental impact of industries with high waste production. An example of this is the biomass production of *Spirulina platensis*, *S. maxima*, and *Chlorella* sp. from wastewater from meat industries.

However, for microalgae to extend beyond traditional markets, it is necessary to plan production strategies to diversify products. This can be done by using a biorefinery model whereby biomass is converted into products of interest, maximizing the use of biomass and minimizing residual waste. To increase the concentration of certain compounds, the candidate strains can also be genetically manipulated, which implies that producers must also incorporate methodologies to evaluate the effects of manipulation.

In the quest to optimize microalgal production, to make it attractive to related industries and economically profitable, cultivation and harvesting operations receive a lot of attention (Dixon and Wilken 2018). Culture in photobioreactors is a more expensive capital investment than outdoor cultivation, but when processing costs are considered, the higher the cell density achievable in photobioreactors, the more economically advantageous they are. At a commercial level, mass microalgal cultures in open ponds are viable to obtain low-value products or services such as biofuels, wastewater treatment, biofertilizers, and biostimulants. Instead, the use of closed photobioreactors is justified for the production of biomass or chemical compounds of high commercial value, such as carotenoids, PUFAs, phenols, and other compounds of pharmaceutical interest, cosmetics, and food for humans and animals, among others. As for harvesting and dehydration, these are steps that consume a lot of time and energy and can represent between 10 and 50% of the total operating costs due to the high dilution of the algal suspensions and the large volume of the process (Dixon and Wilken 2018).

In relation to the harvest, microfiltration, and centrifugation are more expensive than sedimentation by gravity. However, the operation most appropriate for the biomass obtained should be considered, for which research carried out prior to the design of industrial development is crucial.

There is not much information regarding the economic study of the steps involved in obtaining quality products from microalgae (cell disruption, extraction, fractionation, purification, and characterization of the metabolites of interest). However, each step of the process can be decisive for the general sustainability of the production related to energy costs and environmental impacts.

Regarding extractive processes, as described in the corresponding section of this chapter, different alternatives, such as enzymatic hydrolysis, use of green extractions, and ways to balance costs and minimize environmental impact, are being evaluated.

In summary, microalgae can biosynthesize, metabolize, accumulate, and secrete a wide variety of primary and secondary metabolites with high value due to their great potential for industrial application: proteins (40–70%), carbohydrates (12–30%), lipids (4–20%), carotenoids (8–14%), and substantial amounts of vitamins, phenols, terpenes, steroids, and hormones. However, since the chemical composition varies with the species, geography, environmental factors, and seasonality, it is important to deepen the comprehensive studies of the species of interest to select the most suitable sources for each group of metabolites, seeking to achieve the highest yields.

In this sense, the study of the metabolism of native and genetically modified strains and the variations that occur under certain conditions can be decisive. Thus, new developments are being made in the use of microalgal biomass. Special emphasis is being made on the production of bioplastics from polyhydroxyalkanes, recombinant proteins for pharmacology, and increased productivity of lipids and carbohydrates by modifying metabolic pathways or structurally modifying cells (by reducing antenna pigments to reduce photoinhibition phenomena or increasing the efficiency of the Rubisco enzyme) (Borowitzka 2013; Kamalanathan and Quigg 2019). The global market of microalgal products has been predicted to reach 1143 million dollars by 2024. However, significant challenges remain in quantifying these benefits, understanding the possible adverse effects related to toxic bioproducts and ensuring the sustainable availability of algal biomass for large-scale industrial production. In addition, several microalgal species have not yet been thoroughly investigated, which constitutes an important challenge for the discovery of new sources of metabolites of pharmaceutical, cosmetic, nutraceutical, and related importance (Kamalanathan and Quigg 2019).

This chapter has described the main steps that make up the research strategies aimed at the integral study of microalgae and has highlighted two promising species whose research is advancing in different parts of the world. The results obtained will determine the potentiality of the resource to face an industrial development.

Acknowledgments The authors thank the financial support for their research from ANPCYT – FONCYT (PICTO GSJ, PICT 2019 03865) Argentina; SPU (PME 2015) Argentina; CONICET and Universidad Nacional de la Patagonia San Juan Bosco, Argentina. They also acknowledge the Universidad Nacional de Entre Ríos, Argentina, and Universidad de Almería, Spain, for the MJV mobility scholarship and collaboration in research, and the Asociación Universitaria Iberoamericana de Postgrado (AUIP).

References

Abd El-Baky HH, El Baz FK, El-Baroty GS (2009) Production of phenolic compounds from *Spirulina maxima* microalgae and its protective effects. Afr J Biotechnol 8(24):7059–7067. https://doi.org/10.5897/AJB2009.000-9548

Abdel-Raouf N, Al-Homaidan AA, Ibraheem IBM (2012) Microalgae and wastewater treatment. Saudi J Biol Sci 19:257–275

Ali SK, Saleh AM (2012) Spirulina-an overview. Int J Pharm Pharm Sci 4(3):9–15

Badwy TM, Ibrahim EM, Zeinhon MM (2008) Partial replacement of fish meal with dried microalgae (*chlorella sp.* and *Scenedesmus sp.*) In Nile tilapia (*Oreochromis niloticus*) diets. 8th international symposium on tilapia in aquaculture. Cairo, Egypt

Balasubramaniam V, Gunasegavan RD, Mustar S, Lee JC, Mohd Noh MF (2021) Isolation of industrial important bioactive compounds from microalgae. Molecules 26:943. https://doi.org/10.3390/molecules26040943

Barsanti L, Coltelli P, Evangelista V, Frassanito AM, Passarelli V, Vesentini N, Gualtieri P (2008) Oddities and curiosities in the algal world. In: Evangelista V, Barsanti L, Frassanito AM, Passarelli V, Gualtieri P (eds) Algal toxins: nature, occurrence, effect and detection. Springer, Dordrecht, pp 353–391

Bechet Q, Laviale M, Arsapin N, Bonnefond H, Bernard O (2017) Modeling the impact of high temperatures on microalgal viability and photosynthetic activity. Biotechnol Biofuels 10:136

Becker EW (1994) Microalgae: biotechnology and microbiology, vol 10. Cambridge University Press

Becker W (2004) Microalgae in human and animal nutrition. In: Richmond A (ed) Handbook of microalgal culture. Blackwell, Oxford, pp 312–351

Bernard E, Guéguen C (2022) Influence of carbon sources on the phenolic compound production by *Euglena gracilis* using an untargeted Metabolomic approach. Biomol Ther 12:795. https://doi.org/10.3390/biom12060795

Bordet F, López C, Victorino P (2011) Monitoreo emergencia de floración de cianobacterias: Áreas recreativas-Embalse de Salto Grande. p 15–34. En: Área de Gestión Ambiental Salto Grande (eds) Jornada de difusión. Programas Ambientales de Salto Grande. Salto Grande, Argentina-Uruguay

Borowitzka MA (1999) Commercial production of microalgae: ponds, tanks, tubes and fermenters. J Biotechnol 70:313–321

Borowitzka MA (2013) High-value products from microalgae—their development and commercialisation. J Appl Phycol 25:743–756. https://doi.org/10.1007/s10811-013-9983-9

Bross-Walch N, Kühn T, Moskau D, Zerbe O (2005) Strategies and tools for structure determination of natural products using modern methods of NMR spectroscopy. Chem Biodivers 2:147–177. https://doi.org/10.1002/cbdv.200590000

Burlew JS (ed) (1953) Algal culture: from lab to pilot plant, 1st edn. Carnegie Inst of Washington, Washington, D.C

Cannell RJP (1990) Algal biotechnology. Appl Biochem Biotechnol 26: 85–105

Chen C-Y, Yeh K-L, Aisyah R, Lee D-J, Chang J-S (2011) Cultivation, photobioreactor design and harvesting of microalgae for biodiesel production: a critical review. Bioresour Technol 102:71–81

Chen C-L, Chang J-S, Lee D-J (2015) Dewatering and drying methods for microalgae. Dry Technol 33:443–454. https://doi.org/10.1080/07373937.2014.997881

Chisti Y (2007) Biodiesel from microalgae. Biotechnol Adv 25(3): 294–306

Chisti Y, Moo-Young M (1991) Fermentation technology, bioprocessing, scale-up and manufacture. In: Moses M, Cape RE (eds) Biotechnology: the science and the business. Harwood Academic Publishers, New York, pp 167–209

Chojnacka K, Marquez-Rocha F-J (2004) Kinetic and stoichiometric relationships of the energy and carbon metabolism in the culture of microalgae. Biotechnology 3:21–34

Clomburg JM, Qian S, Tan Z, Cheong S, Gonzalez R (2019) The isoprenoid alcohol pathway, a synthetic route for isoprenoid biosynthesis. Proc Natl Acad Sci U S A 116(26):12810–12815. https://doi.org/10.1073/pnas.1821004116

Contreras-Flores C, Peña-Castro J, Flores-Cotera L, Cañizares-Villanueva R (2003) Avances en el diseño conceptual de fotobiorreactores para el cultivo de microalgas. Interciencia 28(8): 450–456

Converti A, Casazza AA, Ortiz EY, Perego P, Del Borghi M (2009) Effect of temperature and nitrogen concentration on the growth and lipid content of *Nannochloropsis oculata* and *Chlorella vulgaris* for biodiesel production. Chem Eng Process 48:1146–1151

Corrêa PS, Morais Júnior WG, Martins AA, Caetano NS, Mata TM (2021) Microalgae biomolecules: extraction, separation and purification methods. Processes 9:10. https://doi.org/10.3390/pr9010010

Dai J, Mumper RJ (2010) Plant phenolics: extraction, analysis and their antioxidant and anticancer properties. Molecules 15(10):7313–7352. https://doi.org/10.3390/molecules15107313

Das P, Aziz SS, Obbard JP (2011) Two phase microalgae growth in the open system for enhanced lipid productivity. Renew Energy 36(9): 2524–2528

Davies J (2007) Small molecules: the lexicon of biodiversity. J Biotechnol 129:3–5

Davies FK, Jinkerson RE, Posewitz MC (2015) Toward a photosynthetic microbial platform for terpenoid engineering. Photosynth Res 123(3):265–284. https://doi.org/10.1007/s11120-014-9979-6

De Farias Neves F, Demarco M, Tribuzi G (2019) Drying and quality of microalgal powders for human alimentation. In: Vítová M (ed) Microalgae-from physiology to application. IntechOpen, London, UK

Dillard CJ, German JB (2000) Phytochemicals: nutraceuticals and human health. J Sci Food Agric 80(12):1744–1756. https://doi.org/10.1002/1097-0010(20000915)

Dixon C, Wilken LR (2018) Green microalgae biomolecule separations and recovery. Bioresour Bioprocess 5:14. https://doi.org/10.1186/s40643-018-0199-3

Domozych DS, Ciancia M, Fangel JU, Mikkelsen M, Ulvskov P, Willats W (2012) The cell walls of green algae: a journey through evolution and diversity. Front Plant Sci 3:82. https://doi.org/10.3389/fpls.2012.00082

Brasil B d SAF, de Siqueira FG, TFC S et al (2017) Microalgae and cyanobacteria as enzyme biofactories. Algal Res 25:76–89. https://doi.org/10.1016/j.algal.2017.04.035

Drews P, Colares RG, Machado P, de Faria M, Detoni A, Tavano V (2013) Microalgae classification using semi-supervised and active learning based on Gaussian mixture models. J Braz Comput Soc 19(4):411

El-Tablawy NH, Mansour HA, Shaaban AEM (2020) Antioxidant activities of some edaphic algae in Egypt. Beni-Suef Uni J Basic Appl Sci 9(1). https://doi.org/10.1186/s43088-020-00060-0

Elmaleh S, Coma J, Grasmick A, Bourgade L (1991) Magnesium induced algal flocculation in a fluidized bed. Water Sci Technol 23:1695–1702

Farmacopea Argentina (2013) 7th ed. ANMAT-FA, Argentina

Ferreira AF, Ferreira A, Dias APS, Gouveia L (2020) Pyrolysis of *Scenedesmus obliquus* biomass following the treatment of different wastewaters. Bioenergy Res 13:896–906

Flores ML, Stortz CA, Cerezo AS (2000) Studies on the skeletal cell wall of the cystocarpic stage of the red seaweed *Iridaea undulosa* B.-part II - fractionation of the cell wall and methylation analysis of the inner core-fibrillar polysaccharides. Int J Biol Macromol 27:21–27

Fritz J, Macka M (2000) Review. Solid-phase trapping of solutes for further chromatographic or electrophoretic analysis. J Chromatogr A 902:137–166

Fuentes-Grünewald C, Gayo-Pelaez J, Ndovela V, Wood E, Vijay Kapoore R, Anne LC (2021) Towards a circular economy: a novel microalgal two-step growth approach to treat excess nutrients from digestate and to produce biomass for animal feed. Bioresour Technol 320:124349

Gerken HG, Donohoe B, Knoshaug EP (2013) Enzymatic cell wall degradation of *Chlorella vulgaris* and other microalgae for biofuels production. Planta 237:239–253. https://doi.org/10.1007/s00425-012-1765-0

Gissibl A, Sun A, Care A, Nevalainen H, Sunna A (2019) Bioproducts from *Euglena gracilis*: synthesis and applications. Front Bioeng Biotechnol 7:108. https://doi.org/10.3389/fbioe.2019.00108

Golueke CG, Oswald WJ (1965) Harvesting and processing sewage grown planktonic algae. J Water Pollut Control Fed 37:471–498

Gómez Luna LM (2007) Microalgas: aspectos ecológicos y biotecnológicos. Revista Cubana de Química XIX(2):3–20

Grima EM, Belarbi E-H, Fernández FA, Medina AR, Chisti Y (2003) Recovery of microalgal biomass and metabolites: process options and economics. Biotechnol Adv 20:491–515

Grobbelaar JU (2004) Algal nutrition. In: Richmond A (ed) Handbook of microalgal culture: biotechnology and applied phycology. Blackwell, Oxford, pp 97–115

Guedes AC, Catarina RB, Helena MA, Claudia IP, Francisco XM (2011) Microalgal and cyanobacterial cell extracts for use as natural antibacterial additives against food pathogens. Int J Food Sci Technol 46(4):862–870

Günerken E, Dhondt E, Eppink M, Garcia-Gonzalez L, Elst K, Wijffels RH (2015) Cell disruption for microalgae biorefineries. Biotechnol Adv 33:243–260. https://doi.org/10.1016/j.biotechadv.2015.01.008

Harun R, Singh M, Forde GM, Danquah MK (2010) Bioprocess engineering of microalgae to produce a variety of consumer products. Renew Sust Energ Rev 14:1037–1047

Hassan S, Meenatchi R, Pachillu K, Bansal S, Brindangnanam P, Arockiaraj J, Seghal Kiran G, Selvin J (2022) Identification and characterization of the novel bioactive compounds from microalgae and cyanobacteria for pharmaceutical and nutraceutical applications. J Basic Microbiol 62(9):1–31. https://doi.org/10.1002/jobm.202100477

Heasman M, Diemar J, O'Connor W, Sushames T, Foulkes L, Nell JA (2000) Development of extended shelf-life microalgae concentrate diets harvested by centrifugation for bivalve molluscs-a summary. Aquac Res 31(8–9):637–659

Hee-Mock O, Lee SJ, Park MH, Kim HS, Kim HC, Yoon JH et al (2001) Harvesting of *Chlorella vulgaris* using a bioflocculant from *Paenibacillus* sp. AM49. Biotechnol Lett 23:1229–1234

Hernández-Pérez A, Labbé JI (2014) Microalgae, culture and benefits. Rev Biol Mar Oceanogr 49(2):157–173. https://doi.org/10.4067/S0718-19572014000200001

Ho S-H, Chen C-Y, Lee D-J, Chang J-S (2011) Perspectives on microalgal CO_2-emission mitigation systems – a review. Biotechnol Adv 29:189–198

Ho SH, Huang SW, Chen CY, Hasunuma T, Kondo A, Chang JS (2013) Bioethanol production using carbohydrate-rich microalgae biomass as feedstock. Bioresour Technol 135:191–198

Houghton P, Raman A (1998) Laboratory handbook for the fractionation of natural extracts. Springer, New York, NY. https://doi.org/10.1007/978-1-4615-5809-5

Huang PW, Wang LR, Geng SS, Ye C, Sun XM, Huang H (2021) Strategies for enhancing terpenoids accumulation in microalgae. Appl Microbiol Biotechnol 105:4919–4930. https://doi.org/10.1007/s00253-021-11368-x

Imam SY, Snell WJ (1988) The Chlamydomonas cell wall degrading enzyme, lysin, acts on two substrates within the framework of the wall. Appl Environ Microbiol 53:1701–1704

Ishaq AG, Matias-Peralta HM, Basri H (2016) Bioactive compounds from green microalga – scenedesmus and its potential applications: a brief review. Pertanika J Trop Agric Sci 39(1):1–16

Jaramillo-Madrid AC, Abbriano R, Ashworth J, Fabris M, Pernice M, Ralph PJ (2020) Overexpression of key sterol pathway enzymes in two model marine diatoms alters sterol profiles in *Phaeodactylum tricornutum*. Pharmaceuticals (Basel) 13(12):481. https://doi.org/10.3390/ph13120481

Jerez-Martel I, García-Poza S, Rodríguez-Martel G, Rico M, Afonso-Olivares C, Gómez-Pinchetti JL (2017) Phenolic profile and antioxidant activity of crude extracts from microalgae and cyanobacteria strains. J Food Qual 2017(4):1–8

Jiménez-Veuthey M, Zapata LM, Flores ML (2019) Phycoremediation and microalgae production using poultry effluent from anaerobic digestion as a culture medium. J Environ Biotechnol Res 8:52–64

Kajikawa M, Kinohira S, Ando A, Shimoyama M, Kato M, Fukuzawa H (2015) Accumulation of squalene in a microalga *Chlamydomonas reinhardtii* by genetic modification of squalene synthase and squalene epoxidase genes. PLoS One 10(3):e0120446. https://doi.org/10.1371/journal.pone.0120446

Kamalanathan M, Quigg A (2019) Microalgae-from physiology to application. Physiological limitations and solutions to various applications of microalgae. https://doi.org/10.5772/intechopen.90206

Kaya VM, Picard G (1996) Stability of chitosan gel as entrapment matrix of viable *Scenedesmus bicellularis* cells immobilized on screens for tertiary treatment of wastewater. Bioresour Technol 56:147–155

Khan MI, Shin JH, Kim JD (2018) The promising future of microalgae: current status, challenges, and optimization of a sustainable and renewable industry for biofuels, feed, and other products. Microb Cell Factories 17:36. https://doi.org/10.1186/s12934-018-0879-x

Kim D-Y, Vijayan D, Praveenkumar R, Han J-I, Lee K, Park J-Y, Chang W-S, Lee J-S, Oh Y-K (2016) Cell-wall disruption and lipid/

astaxanthin extraction from microalgae: *chlorella* and *Haematococcus*. Bioresour Technol 199:300–310. https://doi.org/10.1016/j.biortech.2015.08.107

Kitaya Y, Azuma H, Kiyota M (2005) Effects of temperature, CO_2/O_2 concentrations and light intensity on cellular multiplication of microalgae, *Euglena gracilis*. Adv Space Res 35(9):1584–1588

Kobayashi J, Kubota T (2007) Bioactive macrolides and polyketides from marine dinoflagellates of the genus *Amphidinium*. J Nat Prod 70:451–460

Kottuparambil S, Thankamony RL, Agusti S (2019) *Euglena* as a potential natural source of value-added metabolites. A review. Algal Res 37:154–159. https://doi.org/10.1016/j.algal.2018.11.024

Lam MK, Lee KT (2012) Potential of using organic fertilizer to cultivate *Chlorella vulgaris* for biodiesel production. Appl Energy 94:303–308

Lee SJ, Kim SB, Kim JE, Kwon GS, Yoon BD, Oh HM (1998) Effects of harvesting method and growth stage on the flocculation of the green alga Botryococcus braunii. Lett Appl Microbiol 27:14–18

Levasseur W, Perré P, Pozzobon V (2020) A review of high value-added molecules production by microalgae in light of the classification. Biotechnol Adv 41:107545. https://doi.org/10.1016/j.biotechadv.2020.107545

Li JW-H, Vederas JC (2009) Drug discovery and natural products: end of an era or an endless frontier? Science 325:161–165

López A, Rico M, Santana-Casiano JM, González AG, González-Dávila M (2015) Phenolic profile of *Dunaliella tertiolecta* growing under high levels of copper and iron. Environ Sci Pollut Res 22(19): 14820–14828. https://doi.org/10.1007/s11356-015-4717-y

Lubian LM (1989) Concentrating cultured marine microalgae with chitosan. Aquac Eng 8:257–265

MacKay D, Salusbury T (1988) Choosing between centrifugation and crossflow microfiltration. Chem Eng (Lond) 477:45–50

Márquez-Rocha FJ, Palma-Ramírez D, García-Alamilla P, López-Hernández JF, Santiago-Morales IS, Flores-Vela AI (2019) Microalgae - physiology to application. Microalgae Cultivation for Secondary Metabolite Production. IntechOpen. https://doi.org/10.5772/intechopen.88531

Marrez DA, Naguib MM, Sultan YY, Higazy AM (2019) Antimicrobial and anticancer activities of *Scenedesmus obliquus* metabolites. Heliyon 5:e01404. https://doi.org/10.1016/j.heliyon.2019.e01404

Maschek JA, Baker BJ (2007) The chemistry of algal secondary metabolism. In: Amsler CD (ed) Algal chemical ecology. Springer, Berlin, pp 1–24

Mata TM, Martins AA, Caetano NS (2010) Microalgae for biodiesel production and other applications: a review. Renew Sust Energ Rev 14:217–232

Miazek K, Kratky L, Sulc R, Jirout T, Aguedo M, Richel A, Goffin D (2017) Effect of organic solvents on microalgae growth, metabolism and industrial bioproduct extraction: a review. Int. J. Mol. Sci. 18: 1429. https://doi.org/10.3390/ijms18071429

Milledge JJ (2010) Commercial application of microalgae other than as biofuels: a brief review. Rev Environ Sci Biotechnol 10:31–41. https://doi.org/10.1007/s11157-010-9214-7

Mishraa N, Gupta E, Singh P, Prasad R (2021) Application of microalgae metabolites in food and pharmaceutical industry. In: Preparation of phytopharmaceuticals for the management of disorders https://doi.org/10.1016/B978-0-12-820284-5.00005-8

Mohn FH (1980) Experiences and strategies in the recovery of biomass from mass cultures of microalgae. In: Shelef G, Soeder CJ (eds) Algae biomass. Elsevier, Amsterdam, pp 547–571

Moraes CC, De Medeiros Burkert JF, Kalil SJ (2010) C-phycocyanin extraction process for large-scale use. J Food Biochem 34:133–148. https://doi.org/10.1111/j.1745-4514.2009.00317.x

Morales J, de la Noue J, Picard G (1985) Harvesting marine microalgae species by chitosan flocculation. Aquac Eng 4:257–270

Mujumdar AS (2000) Classification and selection of industrial dryers. In: Sakamon D (ed) Mujumdar's practical guide to industrial drying: principles, equipment and new developments. Exergex Corporation, Montreal, QC, Canada, pp 23–77

Nappa M, Teir S, Sorsamäki L, Karinen P (2016) Energy requirements of microalgae biomass production; technical report for CCSP deliverable D606 Espoo 2016. Research Output, Helsinki, Finland

Narala RR, Garg S, Sharma KK et al (2016) Comparison of microalgae cultivation in photobioreactor, open raceway pond, and a two-stage hybrid system. Front Energy Res 4:1–10. https://doi.org/10.3389/fenrg.2016.00029

Onofrejová L, Vasickova J, Klejdus B, Stratil P, Misurcova L, Kracmar S, Kopecky J, Vacek J (2010) Bioactive phenols in algae: the application of pressurizaed-liquid and solid-phase extraction techniques. J Pharm Biomed Anal 51(2):464–470. https://doi.org/10.1016/j.jpba.2009.03.027

Oyinloye TM, Yoon WB (2020) Effect of freeze-drying on quality and grinding process of food produce: a review. PRO 8:354

Park JH, Yoon JJ, Park HD, Kim YJ, Lim DJ, Kim SH (2011) Feasibility of biohydrogen production from *Gelidium amansii*. Int J Hydrog Energy 36(21):13997–14003

Patras D, Moraru DV, Socaciu C (2018) Screening of bioactive compounds synthesized by microalgae: a progress overview on extraction and chemical analysis. Studia UBB Chemia LXIII(1): 21–35. https://doi.org/10.24193/subbchem.2018.1.02

Pereira EG, Martins MA, Mendes MDSA, Mendes LBB, Nunes Nesi A (2017) Outdoor cultivation of *Scenedesmus obliquus* BR003 in stirred tanks by airlift. Eng Agríc Jaboticabal 37(5):1041–1055. https://doi.org/10.1590/1809-4430-Eng.Agric.v37n5p1041-1055/2017

Petrusevski B, Bolier G, Van Bremen AN, Alaerts GJ (1995) Tangential flow filtration: a method to concentrate freshwater algae. Water Res 29:1419–1424

Polle JEW, Barry K, Cushman J et al (2017) Draft nuclear genome sequence of the halophilic and beta-carotene-accumulating green alga *Dunaliella salina* strain CCAP19/18. Genome Announc 5: e01105–e01117. https://doi.org/10.1128/genomea.01105-17

Popper ZA, Michel G, Hervé C, Domozych DS, Willats WGT, Tuohy MG, Kloareg B, Stengel DB (2011) Evolution and diversity of plant cell walls: from algae to flowering plants. Annu Rev Plant Biol 62: 567–590. https://doi.org/10.1146/annurev-arplant-042110-103809

Priyadarshani I, Rath B (2012) Commercial and industrial applications of micro algae – a review. J Algal Biomass Utln 3(4):89–100

Pulz O, Scheibenbogen K, Gross W (2001) Biotechnology with cyanobacteria and microalgae. In: Rehm H-J, Reed G (eds) Biotechnology, vol 10. Wiley-VCH, Weinheim, pp 105–136

Rawat I, Ranjith-Kumar R, Mutanda T, Bux F (2011) Dual role of microalgae: Phycoremediation of domestic wastewater and biomass production for sustainable biofuels production. Appl Energy 88: 3411–3424

Richmond A (2004) Biological principles of mass cultivation. Handbook of Microalgal Culture: Biotechnology and Applied Phycology, pp 125–177

Rizwan M, Mujtaba G, Memon SA, Lee K, Rashid N (2018) Exploring the potential of microalgae for new biotechnology applications and beyond: a review. Renew Sust Energ Rev 92:394–404. https://doi.org/10.1016/j.rser.2018.04.034

Salinas-Salazar C, Garcia-Perez JS, Chandra R, Castillo-Zacarias C, Iqbal HMN, Parra-Saldívar R (2019) Methods for extraction of valuable products from microalgae biomass. In: Alam MA, Wang Z (eds) Microalgae biotechnology for development of biofuel and wastewater treatment. Springer Nature Singapore. https://doi.org/10.1007/978-981-13-2264-8_11

Sánchez C, Gupta M, Santana AI (2002) Actividad inmunomoduladora de las plantas (I). Revista de Fitoterapia 2(2):151–163

Sasso S, Pohnert G, Lohr M, Mittag M, Hertweck C (2012) Microalgae in the postgenomic era: a blooming reservoir for new natural products. FEMS Microbiol Rev 36:761–785. https://doi.org/10.1111/j.1574-6976.2011.00304.x

Schmitz AC, Hartline CJ, Zhang F (2017) Engineering microbial metabolite dynamics and heterogeneity. Biotechnol J 12(10):1700422. https://doi.org/10.1002/biot.201700422

Shelef G (1978) Photosynthetic biomass production from sewage. Arch Hydrobiol Beih 11:3–14

Show K-Y, Lee D-J, Chang J-S (2013) Algal biomass dehydration. Bioresour Technol 135:720–729

Show K-Y, Lee D-J, Tay J-H, Lee T-M, Chang J-S (2015) Microalgal drying and cell disruption-recent advances. Bioresour Technol 184: 258–266. https://doi.org/10.1016/j.biortech.2014.10.139

Singh J, Gu S (2010) Commercialization potential of microalgae for biofuels production. Renew Sust Energ Rev 14(9):2596–2610. https://doi.org/10.1016/j.rser.2010.06.014

Singh SP, Singh P (2015) Effect of temperature and light on the growth of algae species: a review. Renew Sust Energ Rev 50:431–444

Sosik HM, Olson RJ (2007) Automated taxonomic classification of phytoplankton sampled with imaging-in-flow cytometry. Limnol Oceanogr Methods 5(6):204–216

Stengel DB, Connan S, Popper ZA (2011) Algal chemodiversity and bioactivity: sources of natural variability and implications for commercial application. Biotechnol Adv 29(5):483–501

Stirk WA, Balint P, Vambe M, Lovasz C, Molnar Z, van Staden J, Ördög V (2020) Effect of cell disruption methods on the extraction of bioactive metabolites from microalgal biomass. J Biotechnol 307(2020):35–43

Talapatra SK, Talapatra B (2015) Chemistry of plant natural products: stereochemistry, conformation, synthesis, biology, and medicine. Springer. https://doi.org/10.1007/978-3-642-45410-3-7

Tan XB, Lam MK, Uemura Y et al (2018) Cultivation of microalgae for biodiesel production: review on upstream and downstream processing. Chin J Chem Eng 26:17–30. https://doi.org/10.1016/j.cjche.2017.08.010

Tolivia A, Conforti V, Córdoba O, Flores L (2013) Chemical constituents and biological activity of *Euglena gracilis* extracts. J Pharm Res 7:209–214

Ulrich S (2000) Review. Solid-phase microextraction in biomedical analysis. J Chromatogr A 902:167–194

Vandamme D (2013) Flocculation based harvesting processes for microalgae biomass production. Dissertation, KU Leuven

Venkataraman LV (1978) New possibility for microalgae production and utilisation in India. Arch Hydrobiol Beih 11:199–210

Ventura SPM, Nobre BP, Ertekin F, Hayes M, Garcia-Vaquero M, Vieira F, Kocjj M, Gouveia L, Aires-Barros MR, AMF P (2017) Extraction of value-added compounds from microalgae. In: Microalgae-based biofuels and bioproducts. Elsevier. https://doi.org/10.1016/B978-0-08-101023-5.00019-4

Worms I, Simon DF, Hassler CS, Wilkinson KJ (2006) Bioavailability of trace metals to aquatic microorganisms: importance of chemical, biological and physical processes on biouptake. Biochimie 88(11): 1721–1731. https://doi.org/10.1016/j.biochi.2006.09.008

Xu W, Chai C, Shao L, Yao J, Wang Y (2016) Metabolic engineering of *Rhodopseudomonas palustris* for squalene production. J Ind Microbiol Biotechnol 43(5):719–725. https://doi.org/10.1007/s10295-016-1745-7

Zaharieva MM, Zheleva-Dimitrova D, Rusinova-Videva S, Ilieva Y, Brachkova A, Balabanova V, Gevrenova R, Chan Kim T, Kaleva M, Georgieva A, Mileva M, Yoncheva K, Benbassat N, Najdenski H, Kroumov AD (2022) Antimicrobial and antioxidant potential of *Scenedesmus obliquus* microalgae in the context of integral biorefinery concept. Molecules 27:519. https://doi.org/10.3390/molecules27020519

Zhu L (2015) Biorefinery as a promising approach to promote microalgae industry: an innovative framework. Renew Sust Energ Rev 41:1376–1384. https://doi.org/10.1016/j.rser.2014.09.040

Microalgae and Cyanobacteria Antioxidant Capacity under Stress Conditions

14

Mónica Bettina Félix-Castro, Fredy Hernández-Uribe, Karla María Cota-Quintero, Víctor Alonso Galván-Álvarez, Mario Rojas, Gracia A. Gómez-Anduro, and Bertha Olivia Arredondo-Vega

Abstract

Microalgae and cyanobacteria are photosynthetic organisms that produce a wide range of bioactive compounds that act as antioxidants such as carotenoids, chlorophylls, enzymes, fatty acids, and phycobiliproteins. This indicates that they have efficient protection mechanisms. Therefore, its use in human health and nutrition is of great interest. Different stress conditions modify cell metabolism either by increasing or modulating the activities of antioxidant enzymes of superoxide dismutase, catalase, peroxidase, as well as nonenzymatic compounds such as vitamin E and C, carotenoids, flavonoids, phycobiliproteins, among others. In this chapter, culture strategies are presented for cyanobacteria and microalgae for the evaluation of their antioxidant capacity. In *Arthrospira maxima* and *Phormidium persicinum*, the effect of nitrogen starvation and UV radiation was analyzed. In *Ettlia oleoabundans*, two independent experiments were carried out, nutrient limitation and the effect of marine culture media. In *Dunaliella salina*, the effect of total nitrogen deficiency, high salinity, light intensity, and temperature was evaluated. Nitrogen starvation or UV in *A. maxima* and *P. persicinum* showed an improvement in enzymatic antioxidant activity. The limitation of nutrients in *E. oleoabundans* increased the antioxidants and other bioactive compounds; in the marine culture media, an accumulation of up to 56% of lipid was obtained and gain of monounsaturated and polyunsaturated fatty acids was observed. Finally, *D. salina* increases the concentration of β-carotene, lycopene, and lutein. Accordingly, these results suggest that the culture stress conditions can be stimulated for the commercial production of antioxidants in the cyanobacteria and microalgae species.

Keywords

Antioxidant capacity · Stress conditions · Fatty acids · Pigments · Catalase · Superoxide dismutase

14.1 Introduction

Microalgae and cyanobacteria are microorganisms that produce oxygen by photosynthesis and contain molecules with antioxidant capacity. An antioxidant is defined as compounds that delay or prevent oxidation of lipids, nucleic acids, and proteins (Finkel y Holbrook 2000; Sies and Jones 2007). Under stress conditions, such as nutrient limitation (−N), high irradiance, temperature, salinity, metals, and ultraviolet radiation (UV), the concentration of reactive oxygen species (ROS) increase, which are the causative agents of oxidative damage (Halliwell 1993; Ballester 1996).). Resistance against ROS is based on the protective action of enzymatic and nonenzymatic compounds with intracellular or extracellular mode of action (Rice-Evans and Diplock 1993). The enzymatic antioxidant systems are essential for the maintenance of photosynthetic activity and other metabolic functions (Allen 1997; Latifi et al. 2009). The enzymatic compounds are given by enzymes superoxide dismutase (SOD), catalase (CAT), glutathione peroxidase (GPx), glutathione reductase (GR), ascorbate peroxidase (APX), as well as nonenzymatic antioxidants such as carotenoids, α-tocopherol (vitamin E), phenol oxidase, polyunsaturated fatty acids (PUFAs), and glutathione (GSH) (Polle 1996).

In photosynthetic organisms, ROS are produced by electron transport chain (Fig. 14.1). Light is essential for photosynthesis but at the same time, it can be a source of stress.

M. B. Félix-Castro · F. Hernández-Uribe · K. M. Cota-Quintero · M. Rojas · G. A. Gómez-Anduro · B. O. Arredondo-Vega (✉)
Centro de Investigaciones Biológicas del Noroeste (CIBNOR), Av. Instituto Politécnico Nacional, La Paz, Mexico
e-mail: kitty04@cibnor.mx

V. A. Galván-Álvarez
Centro Universitario de Ciencias Biológicas y Agropecuarias, Universidad de Guadalajara, Zapopan, Mexico

Centro de Investigaciones Biológicas del Noroeste (CIBNOR), Av. Instituto Politécnico Nacional, La Paz, Mexico

A. de Jesús Martínez-Roldán (ed.), *Biotechnological Processes for Green Energy, and High Value Bioproducts by Microalgae, and Cyanobacteria Cultures*, Developments in Applied Phycology 13, https://doi.org/10.1007/978-3-031-43969-8_14

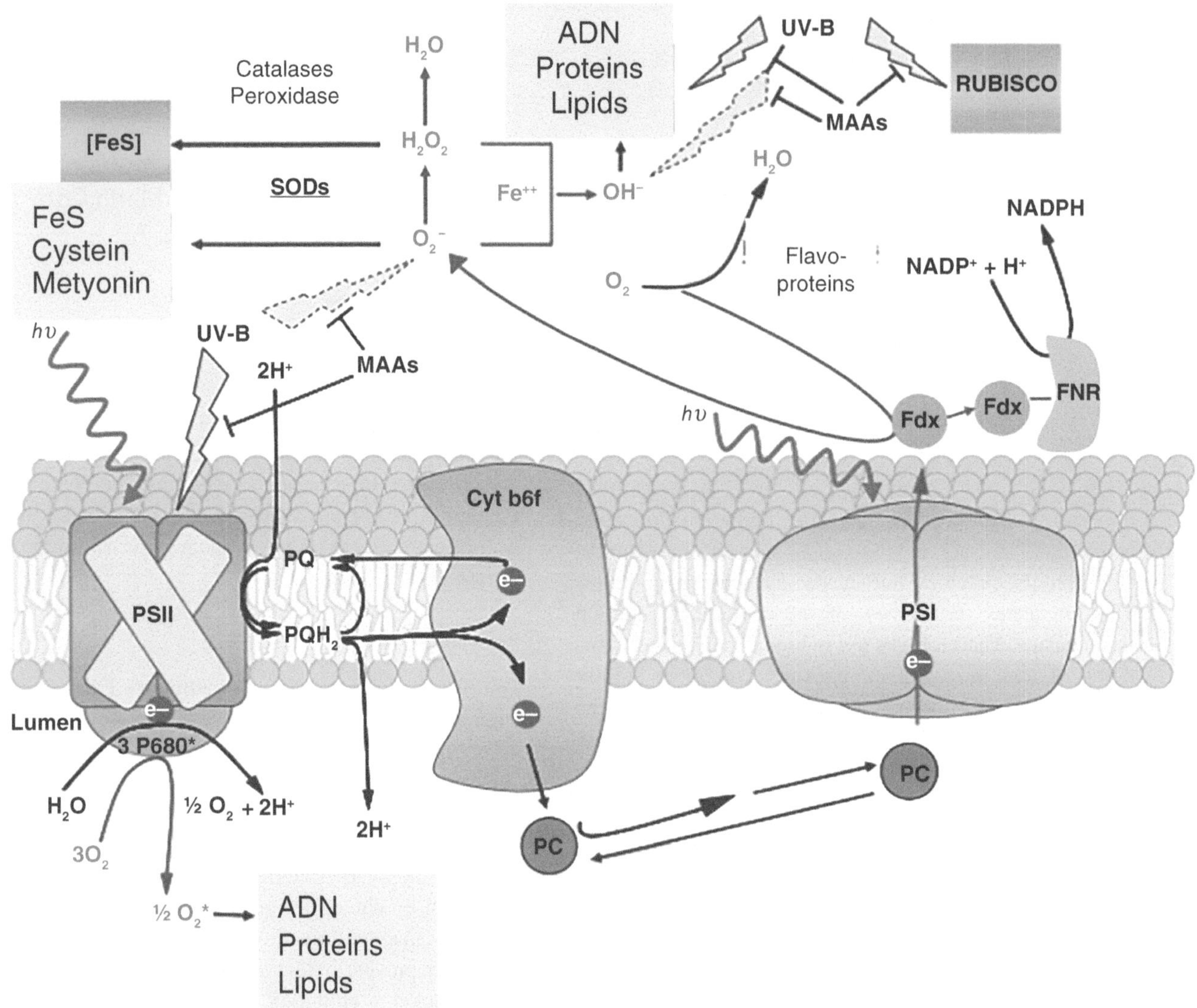

Fig. 14.1 Production of reactive oxygen species (ROS) and targets. Oxygen (O_2) leads to the formation of ROS. In photosystem II (PSII), singlet oxygen ($^1O_2$) is produced by energy input to oxygen by photosensitized chlorophyll. In photosystem I (PSI), the univalent reduction of O_2 using electrons from PSII generates superoxide radical ($O_2^{\bullet -}$). $O_2^{\bullet -}$ is dissociated to hydrogen peroxide (H_2O_2) and O_2 by superoxide dismutase (SOD) and H_2O_2 is reduced to water by catalases and peroxidases (modified from Latifi et al. 2009)

Singlet oxygen ($^1O_2$) is produced by an energy input to oxygen by photosensitized chlorophyll. If the light intensity is much higher than that is normally used by the capacity of the photosynthetic electron flow, it not only increases the production of $^1O_2$ but also the formation of other ROS, causing inactivation of the photosynthetic apparatus (Latifi et al. 2009).

Oxidation is a natural cell process that generates ROS and free radicals. ROS are reactive chemical entities that serve as molecular signals when they are at low concentrations (Finkel 2011); however, when ROS are highly produced, they can induce dangerous effects. ROS differ from each other in the source from which they come, the effects they have, and their chemical reactivity.

ROS are classified in free radicals and non-radical species (Beckman et al. 1993). Free radicals include the following ions: superoxide ($O_2^{\bullet -}$), hydroxyl ($HO^{\bullet}$), peroxyl ($RO_2^{\bullet}$), alkoxy ($RO^{\bullet}$), hydroperoxyl ($HO_2^{\bullet}$), and transition metals. The non-radical species are hydrogen peroxide (H_2O_2), hypochlorous acid (HOCl), ozone (O_3), singlet oxygen ($^1O_2$), and peroxynitrile ($ONOO^-$) (Halliwell 1993; Gutteridge et al. 1982).

Although the biological half-life of a free radical is microseconds, it has the ability to react with everything around it and provoking damage to molecules and cell membranes (Sieck 2004).

Free radicals are the result of the physiological processes such as respiration, photosynthesis, or environmental factors such solar and UV radiations, pollution, and other factors

(Halliwell 1993; Ballester 1996; Abdollahi et al. 2004). The increase in free radicals causes oxidative stress. Oxidative stress is the imbalance between oxidants and antioxidants compounds in favor of the oxidants that leads to an interruption of the redox signaling and control and/or damage to lipids, DNA, and proteins (Finkel y Holbrook 2000; Sies and Jones 2007). When the production of free radicals increases, antioxidant mechanisms are activated to control and stabilize the intra- or extracellular redox environment (Mátes 2000).

Antioxidants are classified as primary antioxidants: they prevent the formation of new free radicals, convert them into less dangerous molecules before they can react, or prevent the formation of free radicals from other molecules. Examples are SOD, CAT, and GPx (Velázquez-Paniagua et al. 2004). Secondary antioxidants capture free radicals, so preventing the chain reaction. They can be membrane associated (vitamin E, carotenoids, phycobiliproteins), dispersed in the cytoplasm (mycosporin-like amino acids), or bound to water-soluble reducing compounds such as GSH and vitamin C (Miller et al. 2002; Carletti et al. 2003). Other classes of antioxidants are simple phenolic compounds such as caffeic acid, chlorogenic acid, complex flavonoids, and anthocyanins (Aqil et al. 2006; Turyanitza et al. 2006; Moon and Shibamoto 2009; Ahmad et al. 2019).

Some secondary antioxidants are pigments (chlorophyll, carotenes, phycobiliproteins). These pigments have potential as natural colorants in the cosmetic, pharmaceutical, and food industries and have been extracted from species such as *Arthrospira maxima*, *Spirulina platensis*, *Porphyridium cruentum*, and *Rhodosorus marinus* (Básaca-Loya et al. 2009; Serval et al. 1994).

14.2 *Spirulina (Arthrospira) maxima* and *Phormidium persicinum*

Cyanobacteria and microalgae are particularly attractive as food source for animals and human consumption since they grow fast, can use nonarable land, brackish, marine, or wastewater, and the quality of protein is equal or better than standard vegetal or meat proteins (Mata et al. 2010). The proteins that have the highest economic potential are the phycobiliproteins, which are of three types: phycocyanin, allophycocyanin and phycoerythrin. The protein fraction may contain up to 20% phycocyanin, a blue water-soluble pigment (Ciferri 1983). Phycocyanin can be used as a natural pigment for food, medicine research, and in the cosmetic industry to replace synthetic pigments (Vonshak 1997).

Spirulina (Arthrospira) maxima is a planktonic cyanobacteria that forms massive populations in tropical and subtropical water bodies, characterized by high levels of carbonate or bicarbonate and high pH (Iltis 1980). This strain is the main component of phytoplankton of Texcoco Lake in Mexico (Durand-Chastel 1980). The protein content in *Spirulina* ranges between 50 and 70% of its dry weight (Becker and Venkataraman 1982) and contain all the essential amino acids, which represent 47% of their total weight (Tiboni and Ciferri 1985).

The genus *Phormidium* includes a large number of species from marine and fresh water environments, with different cell sizes and degrees of constriction in the transverse walls (Kutzing 1984).

This cyanobacterium has been little studied. Pintner and Provasoli (1958) cultivated under artificial conditions in order to get information about the range of temperature, preferential nitrogen source, and concentrations of vitamins. Palinska et al. (2011) used different approaches such as analysis of their morphology, pigments profile, and molecular markers and managed to differentiate other species of the same genera.

In the natural environment, some stress conditions can occur such as an increased dose of photosynthetically active radiation (PAR), UV radiation, or nutrient deficiency such as inorganic nitrogen; therefore, the antioxidant activity of the phytoplankton is modified as a response to ROS protection. For the scaling up and biorefinery of microalgae and cyanobacteria, it is important to evaluate those environmental conditions such as nitrogen deficiency and UV radiation in *Spirulina (Arthrospira) maxima* and *Phormidium persicinum*.

The strains of *S. maxima* (LB 2342) and *P. persicinum* (LB 2425) were obtained from the UTEX Culture Collection of Algae. The culture media selected for cell growth was as follows: at pH 10.5 for *S. maxima*, f/2 (Jourdan 2006); at pH 7.8 for *P. persicinum* (Guillard and Ryther 1962). Laboratory conditions were: temperature 24 1 ± 1 °C, irradiance 130 μmol photon m^{-2} s^{-1} for *S. maxima* and 30 μmol photon m^{-2} s^{-1} for *P. persicinum*, photoperiod 12:12 h light:dark, and continuous aeration. The treatments tested were: control: (+N + PAR), +N + PAR + UVR and total nitrogen deficiency: −N + PAR, −N + PAR + UVR.

Table 14.1 shows the results obtained for both cyanobacteria under the treatments tested (Hernández-Uribe 2016).

In *S. maxima* the most relevant dates were as follows: In +N + PAR, higher values of total antioxidant capacity and superoxide radical were reached, as well as antioxidant compounds such as carotenoids, phycobiliproteins, proteins, total phenolic compounds, and saturated fatty acids. In +N + PAR + UV treatments, the highest values of phycobiliproteins, proteins, and lipids were reached, as well as TBARS, CAT, and Chl *a*.

On the other hand, in the treatment −N + PAR + UV, there were an increase of total carbohydrates and the

Table 14.1 Parameters of *S. maxima* and *P. persicinum* under control conditions and treatments under nitrogen deficiency and UV radiation (Hernández-Uribe 2016)

	Spirulina (Arthrospira) maxima				*Phormidium persicinum*				
Analyses	+N + PAR	−N + PAR	+N + PAR + UVB	−N + PAR + UVB	+N + PAR	−N + PAR	+N + PAR + UVB	−N + PAR + UVB	Method
Protein (% dry weight)	53.35 ± 5.81^{b}	27.84 ± 3.72^{c}	72.8 ± 2.32^{a}	23.46 ± 1.09^{c}	25.75 ± 0.15^{c}	20.83 ± 3.32^{d}	47.81 ± 1.85^{a}	38.25 ± 4.49^{b}	Lowry et al. (1951)
Carbohydrates (% dry weight)	9.49 ± 0.79^{c}	65.1 ± 1.23^{a}	10.41 ± 0.43^{c}	60.34 ± 0.76^{b}	16.16 ± 1.46^{a}	4.85 ± 0.51^{d}	6.73 ± 0.35^{c}	11.68 ± 0.28^{b}	Dubois et al. (1956)
Lipids (% dry weight)	15.06 ± 0.75^{a}	6.88 ± 1.04^{c}	15.55 ± 1.74^{a}	8.97 ± 0.43^{b}	7.02 ± 0.56^{c}	6.6 ± 1.14^{c}	12.04 ± 0.28^{a}	10.18 ± 0.5^{b}	Marsh and Weinstein (1966)
Chl *a* (μg mg^{-1} dry weight)	12.17 ± 0.64^{b}	5.25 ± 0.56^{d}	16.6 ± 0.75^{a}	6.89 ± 0.23^{c}	5.57 ± 0.57^{c}	5.05 ± 0.38^{c}	10.44 ± 0.72^{a}	7.54 ± 0.74^{b}	Siegelman and Kycia (1978)
Carotenoids (μg mg^{-1} dry weight)	3.83 ± 0.94^{a}	1.29 ± 0.31^{b}	1.18 ± 0.41^{b}	0.49 ± 0.11^{b}	0.47 ± 0.11^{b}	1.25 ± 0.12^{a}	1.63 ± 0.34^{a}	0.28 ± 0.03^{b}	Vidussi et al. (1996)
Phycocyanin (μg mg^{-1} dry weight)	91.72 ± 8.78^{a}	30.72 ± 4.04^{b}	91.58 ± 13.11^{a}	15.75 ± 6.18^{c}	12.85 ± 1.8^{a}	9.8 ± 1.94^{b}	14.52 ± 1.9^{a}	12.46 ± 3.08^{ab}	Siegelman and Kycia (1978)
Allophycocyanin (μg mg^{-1} dry weight)	30.62 ± 2.27^{b}	15.30 ± 1.90^{c}	41.36 ± 2.77^{a}	32.22 ± 7.64^{ab}	15.65 ± 4.1^{a}	13.44 ± 2.3^{a}	17.67 ± 2.30^{a}	12.26 ± 0.34^{a}	
Phycoerythrin (μg mg^{-1} dry weight)	6.30 ± 0.53^{b}	2.06 ± 0.31^{c}	11.85 ± 6.49^{a}	7.15 ± 3.86^{ab}	48.52 ± 1^{b}	10.86 ± 3^{c}	81.69 ± 2.61^{a}	68.26 ± 12.33^{a}	Lawrenz et al. (2011)
Total phenolic compounds (μg mg^{-1} dry weight)	4.31 ± 0.46^{a}	1.55 ± 0.24^{b}	0.76 ± 0.33^{c}	0.55 ± 0.12^{c}	5.44 ± 0.65^{a}	2.8 ± 0.06^{b}	1.17 ± 0.04^{c}	1.41 ± 0.15^{bc}	Singleton and Rossi (1965)
Saturated FA (μg mg^{-1} dry weight)	18.33 ± 1.76^{a}	8.44 ± 0.74^{c}	20.09 ± 1.57^{a}	11.28 ± 0.27^{b}	7.7 ± 0.78^{b}	5.61 ± 0.71^{c}	10.11 ± 0.53^{a}	8.5 ± 0.21^{b}	Sato and Murata (1988)
Monounsaturated FA (μg mg^{-1} dry weight)	2.78 ± 0.22^{b}	1.47 ± 0.2^{d}	4.17 ± 0.6^{a}	1.92 ± 0.1^{c}	8.5 ± 1.09^{a}	7.25 ± 1.02^{b}	8.85 ± 0.66^{a}	9.23 ± 0.98^{a}	
Biunsaturated FA (μg mg^{-1} dry weight)	7.22 ± 0.81^{b}	1.99 ± 0.43^{d}	8.45 ± 1.13^{a}	3.53 ± 0.13^{c}	1.99 ± 0.25^{a}	1.29 ± 0.60^{b}	2.22 ± 0.22^{a}	1.85 ± 0.53^{ab}	
Polyunsaturated FA (μg mg^{-1} dry weight)	6.87 ± 0.41^{b}	4.60 ± 0.51^{c}	9.14 ± 0.91^{a}	4.94 ± 0.26^{c}	1.61 ± 0.26^{b}	0.9 ± 0.19^{c}	6.43 ± 0.32^{a}	1.79 ± 0.62^{b}	
Total antioxidant capacity (μM gallic acid)	370.45 ± 46^{a}	315.22 ± 7.3^{b}	56.34 ± 3.75^{c}	52.76 ± 14.63^{c}	467.4 ± 11^{a}	417.6 ± 15^{b}	351.14 ± 9.35^{c}	341.75 ± 19.81^{c}	Brand-Williams et al. (1995)
SOD (U mg^{-1})	20.2 ± 6.8^{c}	38 ± 16.8^{bc}	53.9 ± 5.9^{b}	158.5 ± 48.5^{a}	51.6 ± 10.6^{b}	530 ± 100^{a}	66.1 ± 11.1^{b}	102.5 ± 9.9^{b}	Suzuki (2000)
CAT (U mg^{-1})	7.1 ± 4.3^{c}	12.4 ± 8.3^{bc}	35.6 ± 20.3^{ab}	33.7 ± 17.9^{a}	10.8 ± 7^{c}	12.1 ± 4.8^{c}	28.8 ± 6.4^{b}	55 ± 5.3^{a}	Aebi (1984)
POX (U mg^{-1})	0.011 ± 0.003^{b}	0.045 ± 0.012^{a}	0.0006 ± 0.0002^{b}	0.014 ± 0.002^{b}	0.13 ± 0.02^{b}	0.25 ± 0.05^{a}	$0.008 \pm 1E\text{-}3^{c}$	$0.007 \pm 4E\text{-}4^{c}$	Kar and Mishra (1976)
Superoxide radical (nmol mg^{-1})	0.05 ± 0.018^{a}	0.014 ± 0.05^{b}	0.016 ± 0.05^{b}	0.014 ± 0.008^{b}	0.038 ± 0.011^{c}	0.107 ± 0.3^{b}	0.175 ± 0.011^{b}	0.157 ± 0.005^{a}	Drossos et al. (1995)
TBARS (nmol mg^{-1})	0.44 ± 0.12^{c}	1.03 ± 0.3^{b}	1.80 ± 0.11^{a}	0.75 ± 0.06^{b}	2.54 ± 0.15^{a}	0.98 ± 0.27^{b}	1.57 ± 0.27^{b}	2.43 ± 0.18^{a}	Persky et al. (2000)

Different letters indicate significant differences at $\alpha < 0.05$ (Fisher's *F* test) activities of the enzymes SOD and CAT and a decrease of proteins and lipids.

In the treatment +N + PAR, in *P. persicinum*, there was obtained the higher total antioxidant capacity and TBARs, as well an accumulation of phycocyanin and allophycocyanin, total phenolic compounds, carbohydrates, and mono- and biunsaturated fatty acids.

In +N + PAR + UV treatments, there were an increase in the levels of all photosynthetic pigments (Chl *a*, carotenoids, phycoerythrin, phycocyanin, allophycocyanin) as well as proteins, lipids, and fatty acids and a decrease in the levels of total carbohydrates.

In the treatment −N + PAR + UV, the level of superoxide radical was high, as well as CAT activity, TBAR, phycobiliproteins, and mono- and biunsaturated fatty acids.

14.3 *Ettlia oleoabundans*, Could Be a Source to Produce High-Value Antioxidants Compounds?

Ettlia oleoabundans (taxonomic synonym of *Neochloris oleoabundans*) is a unicellular member of the Neochloridaceae family (Chlorophyta; Deason et al. 1991). *Neochloris* was assigned by Starr (1955) in the *Chlorococcaceae*. Deason et al. (1991) suggested reclassification as *E. oleoabundans*. Finally, Garibay-Hernandez et al. (2017) verified the identification of the *N. oleoabundans* strain (UTEX 1185) using the molecular marker (18S ribosomal DNA), reclassifying it as *E. oleoabundans*. These microalgae are spherical, uninucleate with a diameter of 3–6 μm, containing a cup-shaped chloroplast with one or more pyrenoids (Deason et al. 1991). It is an edaphic organism, isolated from the dunes of Saudi Arabia (Chantanachat and Bold 1962). It was described as a freshwater organism and is commonly cultured on Bristol GR media, or similar mineral media (Bold 1949; Lopes da Silva et al. 2009). It is an oleaginous microalgae with 25–30% of total lipids by dry weight, and under stress conditions such as nitrogen deficiency, it can accumulate up to 56% dry weight (Garibay-Hernandez et al. 2013). Moreover, they are capable to accumulate more than 80% of their dry weight in triacylglycerols (Tornabene et al. 1983). *E. oleoabundans* is capable of growing in concentrations of 35 g L^{-1} of NaCl, modifying its biochemical composition in comparison to growing in freshwater culture (Arredondo-Vega et al. 1995). The specific physiological and metabolic process of *E. oleoabundans* grown under stress conditions promotes lipids production and compartmentalization (Matich et al. 2018), increases concentration of carbohydrates (Garibay-Hernández et al. 2013), and activates an efficient antioxidant machinery (de Jaeger et al. 2018), allowing it to establish itself under unfavorable environment conditions such as nutrients deficiency, high salinity, high temperature, irradiance among others.

Due to their metabolic versatility and lipid-rich biomass, these microalgae are classified within the oleaginous category and are considered promising candidates for the production of biofuels. However, there are still some technical barriers to overcome so that microalgae can be considered an economically viable source for biodiesel production (Radakovits et al. 2010), one of them being the cost of biomass production and processing. The cost of medium is a significant component of the total cost of microalgal production. Consequently, facing the economic and technological challenges represented by the exploitation of microalgae, biotechnological alternatives are proposed to develop a biorefinery approach, which allows an integral use of various metabolites and bioactive compounds of high value, allowing to counteract the high costs associated with maintenance and production of microalgae cultures. In addition, research focused on the exploration of raw material alternatives (nutrients from culture medium) that reduce production costs and promote the production of high-value compounds are being developed (Farooq et al. 2015; Olguín et al. 2015). However, supplying media with fluctuating nutrients affects the homogeneity in the cultures, making it necessary to establish culture systems that provide a medium that allows homogeneity in terms of the quantity and quality of high-value metabolites.

Different experiments on the effect of nutrient limitation in *E. oleoabundans*, specifically nitrogen deficiency, suggest that molecular regulation involved in stress response (Sturme et al. 2018) includes synthesis of molecules such as carbohydrates and lipids (Sun et al. 2014; Baldisserotto et al. 2014; Matich et al. 2018:). The production of ROS in microalgae occurs continuously since they are organisms that have an energy metabolism based on oxidation–reduction reactions (Mano 2002). If the optimal culture conditions are modified, the physiology of the organism can be altered, promoting ROS production. A change in balance between oxidants and antioxidants could cause stress and then oxidative damage (Sies 1986). Moreover, it has been documented that some species of microalgae during nitrogen deficiency stress promote the activation of nonenzymatic and enzymatic antioxidant machinery as the first barrier of defense against ROS, including superoxide dismutase (SOD), catalases (CAT), and peroxidases (Chokshi et al. 2017; Al-Rashed et al. 2016). On the other hand, the amount and source of nitrogen in the microalgae culture have also been shown to influence the composition and accumulation of lipids (Rodolfi et al. 2017; Wan Afifudeen et al. 2021).

For all of the aforementioned reasons, in this chapter, a cultivation platform for *E. oleoabundans* that enables production costs reduction through the use of an alternative culture medium is proposed, providing the capacity to produce high value-added compounds of interest by taking advantage of the production of biomass with a biorefinery approach (Cota-Quintero 2019). *E. oleoabundans* was cultivated in two stages. In the first stage, *E. oleoabundans* was cultivated for 5 days in f/2 medium (Guillard y Ryther. 1962) at 25 °C, light intensity of 62 μmol photons m^{-2} s^{-1}, and a 12:12 h light:dark photoperiod. In the second stage, the effect of nutrient limitation was evaluated using seawater (35 PSU) and an agricultural organic fertilizer (Fertiplus®) at two concentrations as an alternative to f/2 medium. In control treatment, the nitrogen concentration of Fertiplus® was equal to that of f/2 medium (882 μM, Guillard and Ryther 1962). Growth capacities, as well as the production of compounds of commercial interest such as lipids, carbohydrates, and proteins, were evaluated, highlighting some with antioxidant capacities (Table 14.2).

Under the above-described conditions, we evaluated parameters with the aim to evidence the potential of *E. oleoabundans* to produce high-value compounds though the antioxidant capacity of this specie. Total protein analysis showed a decrease in the proportion of the cell under stress conditions. Therefore, the results suggest that carbon from photosynthesis is used for the synthesis and storage of compounds including lipids and carbohydrates, identifying a greater accumulation for these two compounds under stress conditions. Similar results have been reported in stress culture conditions, including increased synthesis of lipids, carbohydrates, and compounds from secondary metabolism (Popovich et al. 2012; Rodolfi et al. 2017), whereas protein composition was decreased.

Under nitrogen limitation, *E. oleoabundans* shows greater antioxidant activity. SOD and CAT increases their activity under stress treatment compared to control conditions. This result agrees with the higher production rate of $O_2^{\bullet-}$ caused by nutritional stress. When there is a greater production of ROS (such as $O_2^{\bullet-}$ and H_2O_2), the activity of enzymes responsible for balancing ROS increases. The same effect has been reported in studies with other microalgae such as *Chlorella sorokiniana* (Zhang et al. 2013) and *Acutodesmus dimorphus* (Chokshi et al. 2017) cultivated under nitrogen deficiency. In the latter, it was reported that SOD activity increased by more than 300% in the nitrogen deficiency treatment, values similar to those found by Cota-Quintero. On the contrary, the activity of the GPx enzyme responsible for catalyzing the reduction of H_2O_2 and lipoperoxides using reduced glutathione (GSH) as a reducing agent, acting preferentially at low concentrations of H_2O_2 (Gill and Tuteja 2010), showed a decrease in antioxidant activity under control culture conditions. The rate of $O_2^{\bullet-}$ production was higher, being related to a higher concentration of TBARS in the nutrient-limited treatment, compared to the control treatment, similar results have been documented in *Dunaliella salina* (Yilancioglu et al. 2014), *Chorella sorokiniana* (Zhang et al. 2013), and *Chlorella pyrenoidosa*, (Fan et al. 2014). Although the limitation of nutrients increased the activity of antioxidant enzymes CAT and SOD, total antioxidant capacity does not show differences with respect to the control. This could suggest that these two enzymes provide the greatest contribution to the total antioxidant capacity under nutrient limitation. Probably, some compounds not quantified in this work, such as vitamin C and E or total phenolic compounds, could have decreased as reported in *Phaeodactylum tricornutum*, *Tetraselmis suecica*, and *Chlorella vulgaris* (Goiris et al. 2015). Regarding the concentration of lipids, in the present study, the limitation of nutrients caused an increase in the production of fatty acids 18:1ω9 and 18:2ω6 with respect to the control. Although the concentration of fatty acid 18:3ω3 decreased, this was one of the most abundant. In addition to enzymatic antioxidants, unsaturated fatty acids 18:1ω9 (oleic acid), 18:2ω6 (linoleic acid), 18:3ω3 (α-linolenic acid), among others, have documented antioxidant capacity (Henry et al. 2002), so the components of unsaturated fatty acids in microalgae could constitute an antioxidant contribution through their consumption in the diet because of protection against lipid peroxidation due to its low polarity and easy dissolution (Huang and Wang 2004). Finally, with the culture medium (seawater 35 UPS + Fertiplus), it was possible to obtain growth rates similar to those of conventional culture mediums used in other works. The organic fertilizer turns out to be cheaper and the use of seawater instead of freshwater (a scarce resource in many regions) makes this culture medium an option for scaling up microalgal biomass production. The use of this culture medium and the limitation of nutrients in a two-phase culture are proposed; in the first phase, nutrients are supplied without limitation to obtain maximum biomass production, and in a second phase, nutrients are limited to increase production of lipids, carbohydrates, and fatty acids. There are two objectives: (1) decrease production costs and (2) maximize the obtaining of compounds of biotechnological interest. Taking as background that incorporation of some of these compounds (polyunsaturated fatty acids) for use in human diet prevents the development of vascular diseases and inflammatory processes (Ahmad et al. 2019). The use of enzymatic antioxidants such as SOD, CAT, and GPx are used in pharmaceutical and cosmetology industries for protection against skin damage, prevention of lipid peroxidation derived from stress, as anti-inflammatories, to increase cellular oxygenation, reduce the risk of diabetes mellitus, as immune system stimulants, as well as in the food and textile industries (Roy et al. 2021).

Table 14.2 Parameters of *E. oleoabundans* under control and limitation nutrients culture conditions

Ettlia oleoabundans			
Analysis	Control (882 μM N)	NL (20 μM N)	Method
Macromolecules (%)			
Carbohydrates	15.0 ± 1.55 [b]	26.0 ± 1.48 [a]	Dubois et al. (1956)
Lipids	15.0 ± 3.00 [b]	28.4 ± 0.17 [a]	Marsh and Weinstein (1966)
Protein	27.0 ± 1.28 [a]	17.0 ± 0.50 [b]	Lowry et al. (1951)
Antioxidant activity enzymes (U mg^{-1} dry weight)			
SOD	527.54 ± 66.63 [b]	2096.26 ± 249.14 [a]	Suzuki (2000)
CAT	189.81 ± 66.63 [b]	797.07 ± 249.14 [a]	Aebi (1984)
GPx	92.67 ± 13.95 [a]	21.51 ± 3.76 [b]	Folhé and Gunzler (1984)
Antioxidant activity			
TBARS (nmol mg^{-1} protein)	0.139 ± 0.03 [b]	3.81 ± 1.20 [a]	Persky et al. (2000)
$O_2^{\bullet -}$ (nmol mg^{-1} protein min^{-1})	0.71 ± 0.02 [a]	0.139 ± 0.03 [b]	Drossos et al. (1995)
Total antioxidant activity (μM gallic acid)	68.68 ± 11.84 [a]	54.34 ± 8.49 [a]	Brand-Williams et al. (1995)
Fatty acid (%)			
18:1ω9 (oleic acid)	2.29 ± 0.16 [b]	16.03 ± 3.04 [a]	Parrish et al. (2009)
18:2ω6 (linoleic acid)	16.68 ± 1.05 [b]	20.14 ± 1.71 [a]	
18:3ω3 (α-linolenic acid)	30.58 ± 2.22 [a]	23.38 ± 3.43 [b]	

NL: nutrient limitation
Different letters indicate significant differences at $\alpha < 0.05$ (Fisher's *F* test)

14.4 *Dunaliella salina* β-Carotene Accumulation under Stress Conditions

Dunaliella salina is a type of unicellular and halophilic green biflagellate microalga without a rigid cell wall structure. It is well known for its ability to accumulate β-carotene, which can account for up to 14.0% of the total dry weight and has important antioxidant effects. Other carotenoids synthesized by *D. salina*, also considered high-efficient antioxidants, are chlorophylls, zeaxantin, lutein, α-carotene. Other compounds present in the organic extract of *D.* salina such as flavonoids, tannins, alkaloids, saponins, steroids, linoleic acid (ω-6), linolenic acid (ω-6), and other monounsaturated and polyunsaturated fatty acids also have antioxidant activity (Rammuni et al. 2019; Morais et al. 2020; da Silva et al. 2021). In *D. salina*, β-carotene composition is found mainly as two stereoisomers, all-*trans* β-carotene and 9-*cis* β-carotene, in an almost 50–50 ratio. 9-*cis* β-carotene is considered to have greater antioxidant capacity than synthetic β-carotene, whose production is limited to all-*trans* isomer (Rammuni et al. 2019). Both the amount of accumulated β-carotene and the 9-*cis* to all-*trans* ratio depend on light intensity and on the algal division time, which is determined by growth conditions. Any growth stress, which will slow down the rate of cell division under light, will in turn increase β-carotene production in *Dunaliella*. These microalgae have developed unique cellular adaptations that allow them to thrive in harsh environmental conditions.

Certain strains, isolated from highly saline environments (e.g., salt ponds) and low nutrient and high solar radiation habitats (environments with extreme concentrations of salt, exposed to high solar radiation in desert lands), are able to synthesize massive amounts of β-carotene, converting from green motile cells to red cells in the process and predominating other organisms (Ben-Amotz et al. 2009). Native strain *D. salina* CIBA29 isolated from salt pods in Guerrero Negro, B.C.S. Mexico, has great biotechnological potential due to its high accumulation of β-carotene under stress conditions. Galván-Álvarez (2018) evaluated β-carotene accumulation in the native strain *D. salina* CIBA29 and found that high salinity, in combination with nitrogen deficiency and high irradiance, increases pigment concentration, particularly β-carotene, and, to a lesser extent, lycopene and lutein (Table 14.3). *D. salina* CIBA29 was collected from crystallization glass number 33 in ESSA, Guerrero Negro, Mexico (N27°53.831′ W114°01.937′) and deposited in Centro de Investigaciones Biológicas del Noroeste S. C. (CIBNOR) Microalgae Collection. *D. salina* CIBA29 was cultivated in 2 L Erlenmeyer flasks with f/2 marine media (Guillard and Ryther 1962), at 25 °C, under cool-light fluorescent lighting (150 μmol photons m^{-2} s^{-1}), a 12:12 h light:dark photoperiod, and a salinity of 0.6 M NaCl. Filter-sterile air was pumped continuously to supply proper aeration. These conditions served as control. *D.* salina was submitted to nitrogen deficiency (0 mM $NaNO_3$) and an increase of irradiance (300 μmol photons m^{-2} s^{-1} + 0.6 M), salinity (150 μmol photons m^{-2} s^{-1} + 4 M NaCl), and an increase in both parameters (300 μmol photons m^{-2} s^{-1} + 4 M NaCl) to evaluate β-carotene accumulation and antioxidant capacity. Highest levels of β-carotene were obtained under nitrogen deficiency (0 mM $NaNO_3$) in

Table 14.3 Parameters of *D. salina* CIBA29 under control conditions and treatments

D. salina **CIBA29**					
Analysis	Control	ND + LS + HI	ND + HS + LI	ND + HS + HI	Method
Macromolecules (pg cell^{-1})					
Carbohydrates	47.29^{c}	41.40^{c}	753.01^{a}	235.26^{b}	Dubois et al. (1956)
Lipids	32.86^{c}	85.76^{a}	427.86^{b}	337.88^{a}	Marsh and Weinstein (1966)
Proteins	24.60^{c}	60.31^{b}	212.76^{a}	243.59^{a}	Lowry et al. (1951)
Carotenoids (pg cell^{-1})					
β-Carotene	0.67^{a}	1.17^{a}	12.46^{b}	16.34^{b}	Lamers et al. (2010)
Lycopene	0.51^{a}	2.04^{a}	8.80^{c}	8.39^{b}	
Lutein	0.14^{a}	0.21^{a}	2.67^{c}	2.48^{b}	
DPPH inhibition (%)					
Antioxidant capacity	78.16^{a}	77.94^{a}	70.97^{ab}	65.34^{b}	Brand-Williams et al. (1995))

ND: nitrogen deficiency (0 mM $NaNO_3$); LS: low salinity (0.6 M NaCl); HS: high salinity (4 M NaCl). LI: low irradiance (150 μmol photons m^{-2} s^{-1}); HI: high irradiance (300 μmol photons m^{-2} s^{-1})

Different letters indicate significant differences at $\alpha < 0.05$ (Fisher's *F* test)

combination with increased salinity (4 M NaCl) and irradiance (300 μmol photons m^{-2} s^{-1}), followed by nitrogen deficiency in combination with increased salinity and low irradiance; low salinity cultures (0.6 M NaCl) presented the lowest β-carotene levels with no significant difference regardless of irradiance (Table 14.3). Lycopene and lutein levels were highest under nitrogen deficiency (0 mM $NaNO_3$) in combination with increased salinity (4 M NaCl) and low irradiance (150 μmol photons m^{-2} s^{-1}) (Table 14.3). Coesel et al. (2008) evaluated β-carotene accumulation in *D. salina* (Dunal) Teodoresco (1905) strain CCAP 19/30 cultivated in round 5-L flasks with modified Walne medium at 23 °C under continuous illumination at low (65 μmol photons m^{-2} s^{-1}) or medium-low light intensity (120 μmol photons m^{-2} s^{-1}). Salt shits from 9% to 18% (w/v) were carried out. An additional light intensity upshift of 500 μmol photons m^{-2} s^{-1} was also evaluated. Total carotenoid content of cells was quantified photometrically. Under medium-low light conditions, carotenoid accumulation was significantly enhanced in cells that are nutrient limited. Largest carotenoid accumulation occurred in cells under constant nutrient limitation at 9% NaCl, as well as in cells under both nutrient and salt stress. Cells suffering nutrient stress and exposed to high light accumulated β-carotene faster than cells under nutrient stress alone, suggesting that high light promotes cell growth when nutrients are available and promotes carotenogenesis when nutrients become limiting. Lamers et al. (2010) evaluated carotenoid metabolisms of *D. salina* strain CCAP 19/18 cultivated in an aseptic panel photobioreactor with a culture volume of 2.275 L. Incident light intensity was set at 200 μmol photons m^{-2} s^{-1} for 15 days and supplied with nitrogen-containing medium and then it was replaced with nitrogen-free medium to study the initial response to nitrogen starvation. For both high light (500 μmol photons m^{-2} s^{-1}) and nitrogen stress, β-carotene accumulation was negatively correlated with the degree of unsaturation of the total fatty acid pool and positively correlated with oleic acid biosynthesis, indicating that oleic acid may be a key component of the lipid-globule-localized triacylglycerols and β-carotene accumulation. Wu et al. (2016) evaluated the effects of light, temperature, and nutrition on pigment accumulation of three *D. salina* strains isolated in Thailand, cultivated in a column photobioreactor with a volume of 250 mL in Modified Johnsons Medium, salinity of 2 M NaCl, a 12:12 h light: dark photoperiod, and a temperature of 22 °C. Three irradiance levels (68.5, 135.3, and 245.6 μmol photons m^{-2} s^{-1}) in combination with temperature (15, 22, and 30 °C) were evaluated. Irradiance of 245 μmol photons m^{-2} s^{-1} and a temperature of 22 °C induced the highest levels of β-carotene. The highest β-carotene rate per cell was obtained in cultures under high light intensity and low temperature, and the lowest total β-carotene content appeared at the lowest temperature. Gómez and González (2005) compared the physiological response of seven *D. salina* strains (three from Chile and one each from Australia, Mexico, China, and Israel) grown in Provasoli-enriched seawater medium (PES) supplemented with 12.5% (w/v) NaCl and under a combination of temperature and irradiance: 26 °C and 40 μmol photon m^{-2} s^{-1}; 26 °C and 110 μmol photon m^{-2} s^{-1}; 15 °C and 40 μmol photon m^{-2} s^{-1}; and 15 °C and 110 μmol photon m^{-2} s^{-1}. The Australian strain exhibited the maximum carotenoid accumulation per culture volume unit at 15 °C and 110 μmol photon m^{-2} s^{-1}, while the Chilean strain CONC-007 showed the maximum amount of carotenoids per cell at 26 °C and 40 μmol photon m^{-2} s^{-1}. The antioxidant capacity was also evaluated in native strain *D. salina* CIBA29 (Galván-Álvarez 2018) in the previously mentioned culture conditions. Although higher β-carotene was synthetized under nitrogen deficiency (0 mM $NaNO_3$) in combination with high salinity (4 M NaCl) and irradiance (300 μmol photons m^{-2} s^{-1}), lower percentage of radical 2,2-Diphenyl-1-picrylhydrazyl (DPPH) inhibition was observed in this culture conditions (Table 14.3). Hu et al. (2008) analyzed the antioxidant capacity of *D. salina* spray-dried powder cultivated in Taiwan. The antioxidant activities of the algal carotenoid extract were determined using DPPH radical scavenging assay, showing that the algal carotenoid extract had significantly higher antioxidant activity in comparison to all-*trans* forms of α-carotene, β-carotene, lutein, and zeaxanthin tested individually.

14.5 Conclusions

Environmental stress conditions can be exploited for the production of high-value compounds. Cyanobacteria and microalgae have grand potential for this purpose, especially when approached as biorefineries. *A. maxima* and *P. persicinum* are sources to get antioxidants such as phycocyanin and β-carotene. The quality of ammino acids that conform the proteins of *A. maxima* constitutes an added value product. Two-stage cultivation system could be applied in oleaginous microalgae such as *E. oleoabundans* for the exploitation of antioxidant compounds, using organic fertilizers as source of nutrients to lower overall costs of production, making it suitable for the recovery of high-value compounds and biofuels. Native strains such as *D. salina* present high potential as stress conditions under which they thrive can be easily replicated within the local area in which they were isolated. Identifying strains that naturally produce more of certain high-value compounds, as well as studying the environmental factors that trigger their accumulation, is important not only from the biotechnological point of view and the possibility of new compounds discovery but also for a better understanding of the stress adaptation mechanisms involved, and the possibility of improving high-value compound production and recovery using a biorefinery approach.

References

Abdollahi M, Ranjar A, Shadnia S, Nikfar S, Rezaie A (2004) Pesticides and oxidative stress: a review. Med Sci Monit 10(6):141–147

Aebi H (1984) Catalase in vitro. In: Packer L (ed) Methods in enzymology, oxygen radicals in biological systems, vol 105. Academics Press Inc, EUA, pp 121–126

Ahmad TB, Rudd D, Kotiw M, Liu L, Benkendorff K (2019) Correlation between fatty acid profile and anti-inflammatory activity in common Australian seafood by-products. Mar Drugs 17:155

Allen JF (1997) Superoxide and photosynthetic reduction of oxygen. In: Michelson AM et al (eds) Superoxide and superoxide dismutases. Academic Press, New York, pp 417–436

Al-Rashed SA, Ibrahim MM, El-Gaaly GA, Al-Shehri S, Mostafa A (2016) Evaluation of radical scavenging system in two microalgae in response to interactive stresses of UV-B radiation and nitrogen starvation. Saudi J Biol Sci 23:706–712

Aqil F, Ahmad I, Mehmood Z (2006) Antioxidant and free radical scavenging properties of twelve traditionally used Indian medical plants. Turk J Biol 30:177–183

Arredondo-Vega BO, Band-Schmidt CJ, Vazquez-Duhalt R (1995) Biochemical composition of *Neochloris oleoabundans* adapted to marine medium. Cytobios 83:201–206

Baldisserotto C, Giovanardi M, Ferroni L, Pancaldi S (2014) Growth, morphology and photosynthetic responses of *Neochloris oleoabundans* during cultivation in a mixotrophic brackish medium and subsequent starvation. Acta Physiol Plant 36:461–472

Ballester M (1996) Antioxidants, free radicals, and health. A chemical, organic, and physical approach. Med Clinic 107:509–515

Básaca-Loya GA, Valdez MA, Enríquez-Guevara EA, Gutiérrez-Millán LE, Burboa MG (2009) Extraction and purification of B-phycoerythrin from the red microalga *Rhodosorus marinus*. Cienc Mar 35(4):359–368

Becker FW, Venkataraman LY (1982) Biotechnology and exploitation of alga. The Indian Approach. Central Food Technological Research Institute, Mysore, 216

Beckman JS, Carson M, Smith CD, Koppenol WH (1993) ALS, SOD, and peroxynitrite. Nature 364(6438):584

Ben-Amotz A, Polle JEW, Subba Rao DV (eds) (2009) The Alga *Dunaliella*. Biodiversity, Physiology, Genomics and Biotechnology. Science Publishers, p 555

Bold HC (1949) The morphology of *Chlamydomonas chlamydogama* sp. nov. Bull Torrey Bot Club 76:101–108

Brand-Williams W, Cuvelier ME, Berst C (1995) Use of a free radical method to evaluate antioxidant activity. LWT-Food Sci Technol 28: 25–30

Carletti P, Masi A, Wonisch A, Grill D, Tausz M, Ferretti M (2003) Changes in antioxidants and pigment pool dimensions in UV-B irradiated maize seedlings. Environ Exp Bot 50:149–157

Chantanachat S, Bold HC (1962) Phycological studies III. Some algae from arid soils. The University of Texas Publication. Publication No. 6218, p 54

Chokshi K, Pancha I, Ghosh A, Mishra S (2017) Nitrogen starvation-induced cellular crosstalk of ROS-scavenging antioxidants and phytohormone enhanced the biofuel potential of green microalga *Acutodesmus dimorphus*. Biotechnol Biofuels 10:60

Ciferri O (1983) *Spirulina*, the edible microorganism. Microbiol Rev 47: 551–578

Coesel SN, Baumgartner AC, Teles LM, Ramos AA, Henriques NM, Cancela L, Varela JC (2008) Nutrient limitation is the main regulatory factor for carotenoid accumulation and for Psy and Pds steady state transcript levels in *Dunaliella salina* (Chlorophyta) exposed to high light and salt stress. J Mar Biotechnol 10:602–611

Cota-Quintero KM (2019) Efecto de la limitación de nutrientes en la composición bioquímica de *Ettlia oleoabundans* y su potencial en la producción de compuestos de alto valor agregado. MSc Thesis. Centro de Investigaciones Biológicas del Noroeste S.C. (CIBNOR), La Paz Baja California Sur, México

da Silva MRO, Moura B, Converti YAS, Porto A, Marques ALF, Daniela V, Bezerra RP (2021) Assessment of the potential of *Dunaliella* microalgae for different biotechnological applications: a systematic review. Algal Res 58:102396

de Jaeger L, Carreres BM, Springer J, Schaap PJ, Eggink G, Martins Dos Santos VAP, Wijffels RH, Martens DE (2018) *Neochloris oleoabundans* is worth its salt: transcriptomic analysis under salt and nitrogen stress. PLoS One 13(4):e0194834. https://doi.org/10.1371/journal.pone.0194834

Deason T, Silva PC, Watanabe S, Floyd GL (1991) Taxonomic status of the species of the green algal genus *Neochloris*. Plant Syst Evol 177: 213–219

Drossos G, Lazou A, Panagopoulos P, Westaby S (1995) Deforexamine cardioplegia reduces superoxide radical production in human myocardium. Ann Thorac Surg 59:169–172

Dubois M, Gilles K, Hamilton J, Rebers P, Smith F (1956) Colorimetric method for determination of sugar and related substances. Anal Chem 28:350–356

Durand-Chastel H (1980) Production and use of *Spirulina* in Mexico. In: Shelef G, Soeder CJ (eds) Algae Biomass. Elsevier, North Holland Biomedical Press, Amsterdam, 39

Fan J, Cui Y, Wan M, Wang W, Li Y (2014) Lipid accumulation and biosynthesis genes response of the oleaginous *Chlorella pyrenoidosa* under three nutrition stressors. Biotechnol Biofuels 7: 17

Farooq W, Suh WI, Park MS, Yang JW (2015) Water use and its recycling in microalgae cultivation for biofuel application. Bioresour Technol 184:73–81

Finkel T (2011) Signal transduction by reactive oxygen species. JCB 194(1):7–15

Finkel T, Holbrook NJ (2000) Oxidants, oxidative stress and the biology of ageing. Nature 408(6809):239–247

Folhé L, Günzler WA (1984) Assays for glutathione peroxidase. In: Packer L (ed) Methods in enzymology. Oxygen radicals in biological systems. Academic Press, New York, NY, pp 105, 114–105, 120

Galván-Álvarez VA (2018) Efecto de la deficiencia de nitrógeno, elevada salinidad e intensidad luminosa como inductores de la carotenogénesis en la microalga *Dunaliella salina*: rendimiento fotosintético, composición bioquímica y capacidad antioxidante. Bachelor Thesis. Universidad de Guadalajara, Jalisco, Mexico

Garibay-Hernández A, Vazquez-Duhalt R, Serrano-Carreón L, Martinez A (2013) Nitrogen limitation in *Neochloris oleoabundans*: a reassessment of its effect on cell growth and biochemical composition. Appl Biochem Biotechnol 171:1775–1791

Garibay-Hernández A, Barkla BJ, Vera-Estrella R, Martinez A, Pantoja O (2017) Membrane proteomic insights into the physiology and taxonomy of an oleaginous green microalga 1. Plant Physiol 173: 390–416

Gill SS, Tuteja N (2010) Reactive oxygen species and antioxidant machinery in abiotic stress tolerance in crop plants. Plant Physiol Biochem 48:909–930

Goiris K, Van Colen W, Wilches I, León-Tamariz F, De Cooman L, Muylaert K (2015) Impact of nutrient stress on antioxidant production in three species of microalgae. Algal Res 7:51–57

Gómez P, González M (2005) The effect of temperature and irradiance on the growth and carotenogenic capacity of seven strains of *Dunaliella salina* (Chlorophyta) cultivated under laboratory conditions. Bio Res 38:151–162

Guillard RRL, Ryther JH (1962) Studies of marine planktonic diatoms. I. *Cyclotella nana* Hustedt and *Detonula confervacea* Cleve. Can J Microbiol 8:229–239

Gutteridge JM, Rowley DA, Halliwell B (1982) Superoxide-dependent formation of hydroxyl radicals and lipid peroxidation in the presence of iron salts. Detection of catalytic iron and antioxidant activity in extracellular fluids. Biochemist 206(3):605–609

Halliwell B (1993) The chemistry of free radicals. Toxicol Ind Health 9(1–2):1–21

Henry GE, Momin RA, Nair MG, Dewitt DL (2002) Antioxidant and cyclooxygenase activities of fatty acids found in food. J Agric Food Chem 50:2231–2234

Hernández-Uribe F (2016) Efecto de la deficiencia de nitrógeno y la radiación UV en la actividad fotosintética y en los compuestos antioxidantes de las cianobacterias *Spirulina (Arthrospira) maxima* y *Phormidium persicinum*. MSc Thesis. Centro de Investigaciones Biológicas del Noroeste S.C. (CIBNOR), La Paz Baja California Sur, México

Hu CC, Lin JT, Lu FJ, Chou FP, Yang DJ (2008) Determination of carotenoids in *Dunaliella salina* cultivated in Taiwan and antioxidant capacity of the algal carotenoid extract. Food Chem 109:439–446

Huang HL, Wang BG (2004) Antioxidant capacity and lipophilic content of seaweeds collected from the Qingdao coastline. J Agric Food Chem 52:4993–4997

Iltis A (1980) Ecologie de *Spiruline platensis* dans les milieux natronés d'Afrique sahélienne. In: Materassi R (ed) Prospecttive della Coltura di *Spirulina* in Italia, Roma, p 41

Jourdan JP (2006) Grow your *Spirulina*. Manual artisan culture for the production of *Spirulina*., p 146

Kar M, Mishra D (1976) Catalase, peroxidase and polyphenoloxidase activities during rice leaf senescence. Plant Physiol 57:315–319

Kutzing TF (1984) Species algarum. F.A. Brockhaus, Leipzing, p 922

Lamers P, Van de Laak C, Kaasenbrood P, Lorier J, Janssen M, De Vos R, Bino R, Wijffels R (2010) Carotenoid and fatty acid metabolism in light- stressed *Dunaliella salina*. Biotechnol Bioeng 106: 638–648

Latifi A, Ruiz M, Zhang C (2009) Oxidative stress in cyanobacteria FEMS. *Microbiol Rev* 33:258–278

Lawrenz E, Fedewa EJ, Richardson TL (2011) Extraction protocols for the quantification of phycobilins in aqueous phytoplankton extracts. J Appl Phycol 23:865–871

Lopes da Silva TA, Reis R, Medeiros AC, Oliveira GL (2009) Oil production towards biofuel from autotrophic microalgae semicontinuous cultivations monitorized by flow cytometry. Appl Biochem Biotechnol 159:568–578. https://doi.org/10.1007/s12010-008-8443-5

Lowry OH, Rosenbrough NJ, Farr AL, Randall RJ (1951) Protein measurement with the Folin phenol reagent. J Biol Chem 193:265–272

Mano J (2002) Early events in environmental stresses in plants-induction mechanisms of oxidative stress. In: Inzé D, Van Montagu M (eds) Oxidative stress in plants, induction mechanisms of oxidative stress. Taylor and Francis Group, London, UK, pp 217–246

Marsh JB, Weinstein DB (1966) Simple charring method for determination of lipids. J Lipid Res 7:574–576

Mata TM, Martins AA, Caetano NS (2010) Microalgae for biodiesel production and other applications: a review. Renew Sust Energ Rev 14(1):217–232

Matés JM (2000) Effects of antioxidants enzymes in the molecular control reactive oxygen species toxicology. Toxicology 153(1–3): 83–104

Matich EK, Ghafari M, Camgoz E, Caliskan E, Pfeifer BA, Haznedaroglu BZ, Atilla-Gokcumen GE (2018) Time-series lipidomic analysis of the oleaginous green microalga species *Ettlia oleoabundans* under nutrient stress. Biotechnol Biofuels 11:29

Miller SR, Martin M, Touchton J, Castenholz RW (2002) Effects of nitrogen availability on pigmentation and carbon assimilation in the cyanobacterium *Synechococcus* sp. strain. Arch Microbiol 177(5): 392–400

Moon JK, Shibamoto T (2009) Antioxidant assays for plant and food components. J Agric Food Chem 57:1655–1666

Morais WGJ, Gorgich M, Corrêa PS, Martins AA, Mata TM, Caetano NS (2020) Microalgae for biotechnological applications: cultivation, harvesting and biomass processing. Aquaculture 528:735562

Olguín EJ, Dorantes E, Castillo OS, Hernandez-Landa VJ (2015) Anaerobic digestates from vinasse promote growth and lipid enrichment in *Neochloris oleoabundans* cultures. J Appl Phycol 27:1813–1822

Palinska KA, Deventer B, Hariri K, Lotocka M (2011) A taxonomic study on *Phormidium*–group (cyanobacteria) based on morphology, pigments, RAPD molecular markers and RFLP analysis of the 16S rRNA gene fragment. Fottea 11(1):41–55

Parrish CC, Deibel D, Thompson RJ (2009) Effect of sinking spring phytoplankton blooms on lipid content and composition in suprabenthic and benthic invertebrates in a cold ocean coastal environment. Mar Ecol Prog Ser 391:33–51

Persky AM, Green PS, Stubley L, Howell CO, Zaulyanov L, Brazeau GA, Simpkins JW (2000) Protective effect of estrogens against oxidative damage to heart and skeletal muscle in vivo and in vitro. Proc Soc Exp Biol Med 223:59–66

Pintner IJ, Provasoli L (1958) Artificial cultivation of a red-pigmented marine blue-Green alga, *Phormidium persicinum*. J Gen Microbiol 18:190–197

Polle A (1996) Protection of oxidative stress in trees as affected by elevated CO_2 and environmental stress. In: Koch G, Mooney H (eds) Terrestrial ecosystem response to elevated CO_2. Academic Press, New York, pp 299–315

Popovich CA, Damiani C, Constenla D, Martínez AM, Freije H, Giovanardi M, Leonardi PI (2012) *Neochloris oleoabundans* grown in enriched natural seawater for biodiesel feedstock: evaluation of its growth and biochemical composition. Bioresour Technol 114:287–293

Radakovits R, Jinkerson RE, Darzins A, Posewitz MC (2010) Genetic engineering of algae for enhanced biofuel production. Eukaryot Cell 9:486–501

Rammuni MN, Ariyadasa TU, Nimarshana P, Attalage RA (2019) Comparative assessment on the extraction of carotenoids from microalgal sources: Astaxanthin from *H. pluvialis* and β-carotene from *D. Salina*. Food Chem 277:128–134

Rice-Evans CA, Diplock AT (1993) Current status of antioxidant therapy. Free Rad Biol Med 15:77–96

Rodolfi L, Biondi N, Guccione A, Bassi N, D'Ottavio M, Arganaraz G, Tredici MR (2017) Oil and eicosapentaenoic acid production by the diatom Phaeodactylum tricornutum cultivated outdoors in Green Wall Panel (GWP®) reactors. Biotechnol Bioeng 114:2204–2210

Roy UK, Nielsen BV, Milledge JJ (2021) Antioxidant production in *Dunaliella*. *Appl Sci* 11:3959

Sato N, Murata N (1988) Membrane Lipids. Methods Enzymol 167: 251–259

Serval MO, Claire C, Darrien A, Coiffard L, Roeck-Holtzhauer Y (1994) Fatty acid composition of some marine microalgae. Phytochemistry 36:691–697

Sieck LW (2004) Reactions of $(CO_2)_2$ + $(CO_2)_2$ + association ions. Int J Chem Kin 10:335–366

Siegelman H, Kycia J (1978) Algal bili-proteins: handbook of phycological method. Cambridge University Press, Cambridge, pp 71–79

Sies H (1986) Biochemistry of oxidative stress. *Angewandte Chem* 25: 1058–1071

Sies H, Jones DP (2007) Encyclopedia of stress, vol 3. Elsevier, pp 45–48

Singleton VL, Rossi JA (1965) Colorimetry of total phenolic with phosphomolybdic phosphotungstic acid agent. Am J Enol Vitic 16: 144–158

Starr RC (1955) A comparative study of *Chlorococcum Meneghini* and other spherical, zoospore producing genera of the *Chlorococcales*. Science series. No. 20. Indiana University Press, Bloomington, pp 87–90

Sturme MH, Gong Y, Heinrich JM, Klok AJ, Eggink G, Wang D, Wijffels RH (2018) Transcriptome analysis reveals the genetic foundation for the dynamics of starch and lipid production in *Ettlia oleoabundans*. Algal Res 33:142–155

Sun X, Cao Y, Xu H, Liu Y, Sun J, Qiao D, Cao Y (2014) Effect of nitrogen-starvation, light intensity and iron on triacylglyceride/carbohydrate production and fatty acid profile of *Neochloris oleoabundans* HK-129 by a two-stage process. Bioresour Technol 155:204–212

Suzuki K (2000) Measurements of Mn-SOD and Cu, Zn-SOD. In: Taniguchi N, Gutteridge J (eds) Experimental protocols for reactive oxygen and nitrogen species. Oxford University Press, pp 91–95

Tiboni O, Ciferri O (1985) The biochemistry and industrial potential of *Spirulina* sp. Annual Rev Microbiol 39:503–526

Tornabene TG, Holzer G, Lien S, Burris N (1983) Lipid composition of the nitrogen starved green alga *Neochloris oleoabundans*. Enzym Microb Technol 5:435–440

Turyanitza I, Baloghova M, Paulovicsova B, Burdova M, Hudec J, Balint L (2006) Antioxidant and antiradical activity of plant resources as indicators of food value. Ann Univ Maria E. Curie-Sklodowska. Lublin-Polonia 19(1):91–94

Velázquez-Paniagua M, Prieto-Gómez B, Contreras-Pérez R (2004) El envejecimiento y los radicales libres. Cienc Mar 75:36–43

Vidussi F, Claustre H, Bustillos-Guzman J, Cailliau C, Marty JC (1996) Determination of chlorophylls and carotenoids of marine phytoplankton: separation of chlorophyll *a* from divinyl-chlorophyll a and zeaxanthin from lutein. J Plankton Res 18:2377–2382

Vonshak A (1997) Spirulina platensis (Arthrospira): physiology, cell-biology and biotechnology, 1st edn. Taylor y Francis, London, p 233

Wan Afifudeen CL, Loh SH, Aziz A, Takahashi K, Mohd Effendy AW, Cha TS (2021) Double-high in palmitic and oleic acids accumulation in a non-model green microalga, *Messastrum gracile* SE-MC4 under nitrate-repletion and -starvation cultivations. Sci Rep 11:38

Wu Z, Duangmanee P, Zhao P, Juntawong N, Ma C (2016) The effects of light, temperature, and nutrition on growth and pigment accumulation of three *Dunaliella salina* strains isolated from saline soil. Jundishapur J Microbiol 9:e26732

Yilancioglu K, Cokol M, Pastirmaci I, Erman B, Cetiner S (2014) Oxidative stress is a mediator for increased lipid accumulation in a newly isolated *Dunaliella salina* strain. PLoS One 9(3):e91957. https://doi.org/10.1371/journal.pone.0091957

Zhang YM, Chen H, He CL, Wang Q (2013) Nitrogen starvation induced oxidative stress in an oil-producing green alga *Chlorella sorokiniana* C3. PLoS One 8(7):e69225. https://doi.org/10.1371/journal.pone.0069225

15 Biogas Obtained from the Anaerobic Digestion of Microalgal Biomass

Alfredo de Jesús Martínez-Roldán, María Dolores Josefina Rodríguez Rosales, and Sergio Valle Cervantes

Abstract

Microalgae field is a very efficient biomass generator because it has a simple cellular structure and a very fast growth rate. In addition, it is important to highlight the fact that the biomass composition of the microalgae is very similar to the vegetal biomass. Carbon is the major component and can reach values close to 50% of the dry biomass, and the proximal composition can vary from one group to another; however, many microalgae have higher amounts of total carbohydrates (cellulose, hemicellulose, starch, glucose, etc.) and there can be harness to develop biotechnological processes to produce biofuels as the biogas. The biogas can be used as fuel if the amount of methane is higher than 60% and produced by the anaerobic digestion of organic matter. In this chapter, we analyze the potential of the microalgal biomass to be used in biogas production by anaerobic digestion.

Keywords

Biogas · Anaerobic digestion · Microalgae · Methane

15.1 Introduction

Currently, the world has focused on developing clean energies since the processes used in energy production using fossil fuels that, in addition to their depletion, are highly polluting, contributing to the deterioration of air quality and therefore to climate change. Using natural resources such as wind, water, solar energy, and biomass has been the focus of research studies aimed at producing clean energy. Europe has been a leader in using its waste or biomasses in biofuel production, as in Germany, Sweden, Holland, and Italy, among others. Biomasses as an energy source in the future are a potential that exists in all countries around the world and must be used for renewable energy production.

In recent years, studies on the production of biofuels from available and abundant biomasses have increased, showing that some substrates have the potential for conversion, and others need the help of other substrates to improve their performance in bioenergy production.

One of the most studied processes for clean energy generation is anaerobic digestion, a technology for handling organic materials that play an essential role in the present and future of renewable energy production (Appels et al. 2011; Surendra et al. 2014). For example, the stabilization of various organic compounds by anaerobic digestion has been used on a large scale for methane production (Carrere et al. 2016; Cavinato et al. 2013).

This anaerobic digestion process transforms biodegradable organic matter into biogas with the help of microorganisms. This biogas is composed mainly of methane and carbon dioxide. Several process variables are involved in the process that influences the high or low yield of the biogas. These factors are temperature, retention time, and organic load. In addition, the geometry of the bioreactors and their agitation components are of great importance so that the organic content with the active biomass is mixed. Then, the reaction is carried out in the entire volume of the bioreactor. Anaerobic digestion is carried out in four stages: hydrolysis, acidogenesis, acetogenesis, and methanogenesis, the by-products in each stage are produced because of the growth of microorganisms, mainly bacteria and archaea (Fig. 15.1).

In the hydrolysis stage, organic matter is composed of carbohydrates (cellulose, hemicellulose xylanase, and

A. de Jesús Martínez-Roldán
Investigador Nacional Nivel I, Catedrático CONAHCYT - TecNM/ITD, Departamento de Ingenierías Química y Bioquímica, División de Estudios de Posgrado, Maestría en Sistemas Ambientales, Durango, Mexico
e-mail: adjmartinezro@conacyt.mx

M. D. J. Rodríguez Rosales · S. Valle Cervantes (✉)
Tecnológico Nacional de México/IT de Durango. Maestría en Sistemas Ambientales, Durango, Mexico
e-mail: svalle@itdurango.edu.mx

A. de Jesús Martínez-Roldán (ed.), *Biotechnological Processes for Green Energy, and High Value Bioproducts by Microalgae, and Cyanobacteria Cultures*, Developments in Applied Phycology 13, https://doi.org/10.1007/978-3-031-43969-8_15

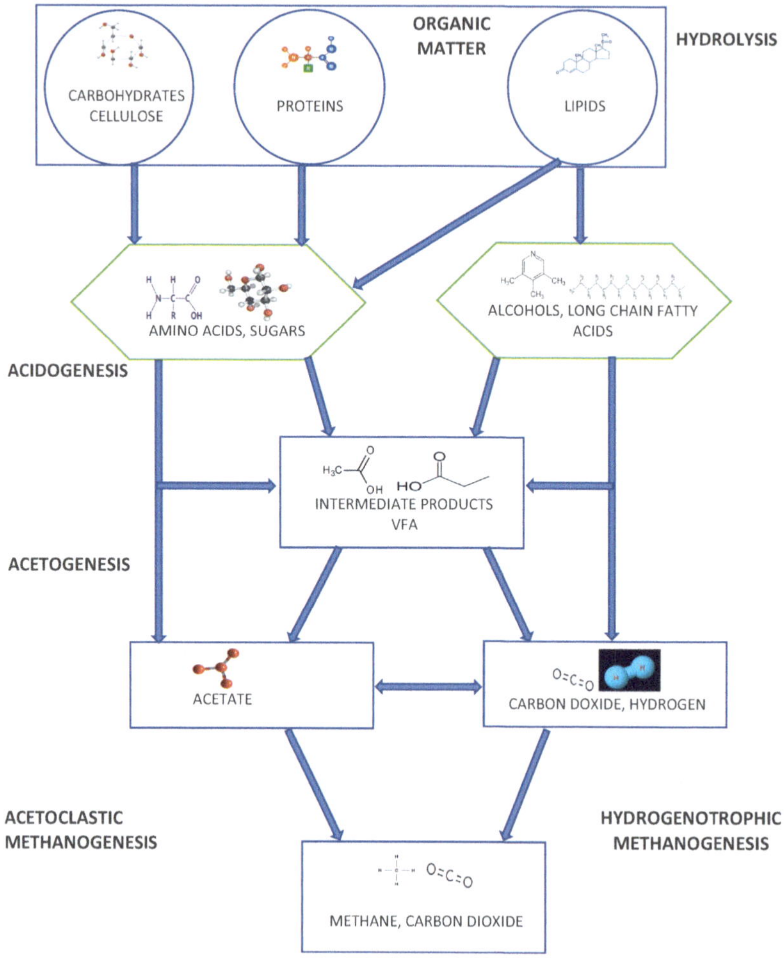

Fig. 15.1 Diagram of pathways in an anaerobic digestion process to obtain biogas (adapted by Gonzalez-Fernandez et al. 2015)

amylase), proteins, and lipids, which are broken down by hydrolytic bacteria into amino acids, sugars, and long-chain volatile fatty acids (LCFA) and alcohols. In the acidogenesis stage, these products are converted to intermediate products such as volatile fatty acids (VFA) by acidogenic bacteria. Acetate, hydrogen, and carbon dioxide are produced in the acetogenesis stage. Finally, in the methanogenesis stage, acetate, hydrogen, and carbon dioxide are converted to methane and carbon dioxide by archaea microorganisms.

Hydrolysis as a limiting step is considered when the feedstock or substrates contain large organic particles such as cellulose and fats (Parkin and Owen 1986). Hydrolysis of complex organic matter such as carbohydrates, proteins, and lipids is carried out by extracellular enzymes excreted by different bacteria. Hydrolyzed molecules include amino acids, sugars, long-chain volatile fatty acids (LCFA), and alcohol. The molecules are converted into volatile fatty acids (VFA), hydrogen, and carbon dioxide in the acidogenic and acetogenic stages. Degradation of LCFA results in acetate and hydrogen by hydrogen-producing acetogenic bacteria. The substrates for methanogenic microorganisms are acetate, hydrogen, and carbon dioxide. As the growth rate of archaea is slow, there is the problem of VFA accumulation and hence inhibition of the activity of these microorganisms. Acetoclastic methanogenesis is generally the preferred catabolic route, accounting for 70% of the methane produced. Methanogenesis is the limiting phase in the anaerobic digestion (AD) of easily degradable substrates.

The anaerobic digestion process efficiency depends on the type and composition of the substrates or biomasses to be digested. These substrates are composed of organic and inorganic materials degraded by the microorganisms to simpler compounds through the stages of the anaerobic digestion process (Deublein and Steinhauser 2008). Hence, it is important to know the properties of biomasses to determine the feasibility of using these substrates in the production of biofuels through an anaerobic digestion process. The characterization of the substrates, biodegradability, bioaccessibility, and bioavailability must be known to optimize an anaerobic digestion process. In addition to the operational control variables such as temperature, organic load, pH, and retention time (Amon et al. 2007).

Since the 2010s, there has been increasing research in using cosubstrates such as fats and oils, algae, and sludge found in wastewater treatment plants. Combining diverse substrates results in the balance between micro- and macronutrients and the dilution of toxic or inhibitory compounds (Mata-Alvarez et al. 2011). Some of the characteristics of each of the substrates studied are as follows: sewage sludge characterized by low organic loads, animal manure by low organic loads and high nitrogen concentrations, municipal organic solid waste that has high concentrations of heavy metals, agro-industrial waste that may lack nitrogen, and slaughterhouse waste (SHW) that may have high concentrations of nitrogen and long-chain fatty acids.

15.2 Microbiology in the Anaerobic Digestion Process

In anaerobic digestion, the dominant microbial consortium is bacteria and archaea. The bacteria degrade organic matter to VFA, carbon dioxide, and hydrogen, and the archaea are responsible for converting the bacterial products to methane and carbon dioxide. The bacteria act in the stages of hydrolysis, acidogenesis, and acetogenesis. The archaeal consortium (methanogens) is responsible for methane production during the methanogenic stage. The archaea are divided into acetoclastic methanogens and hydrogenotrophic methanogens.

The two main groups of bacteria in AD are acidogenic bacteria and methanogens. The former breaks down or decompose smaller organic molecules to produce hydrogen and acetic acid, and the methanogens use hydrogen, carbon dioxide, and acetic acid and convert them to methane. During acetogenesis, different organic acids, such as propionic and butyric, and some alcohols and ketones can be formed (Cooney et al. 2007).

The composition of the raw material or substrates is a factor that affects the diversity of bacteria in the process (Wang et al. 2010), finding a different bacterial consortium (Ziganshin et al. 2013). Factors such as temperature, organic load, and retention time affect the AD process. Temperature affects bacteria and the genus archaea; these microorganisms are susceptible to changes in temperature (Goberna et al. 2010; Sundberg et al. 2013). Methanogens are the most sensitive microorganisms to changes in environmental and operational conditions in the AD process.

Methanogens or methane archaea are methane-forming microorganisms and utilize organics such as acetate, carbon dioxide/hydrogen, methanol, and formate (De Bok et al. 2004; Gujer and Zehnder 1983; Thauer 1998). Short-chain volatile fatty acids and alcohols are produced from the biodegradation of complex organic compounds (De Bok et al. 2004; Shin et al. 2010). These reduced organic compounds are usually degraded by syntrophic microorganisms to acetate, hydrogen, and carbon dioxide and consumed by methanogens. Acetoclastic methanogens utilize acetate; hydrogenotrophic methanogens consume hydrogen for methane production via carbon dioxide reduction. In addition, these microorganisms maintain low hydrogen partial pressures, which provide conditions for fermentative bacteria to continue the fermentation of reduced organic compounds (De Bok et al. 2004; Gujer and Zehnder 1983; Shin et al. 2010).

Several authors found that the most common strains in anaerobic digestion are *Methanospirillum hungatei*, *Methanosarcina barkeri*, *Methanobacterium thermoautotrophicum*, and *Methanobacterium formicicum*. They found that the inhibitory effects are more significant in acidogenic bacteria than in methanogenic bacteria (Chen et al. 2008; Angelidaki and Ahring 1993; Bhattacharya and Parkin 1989; Borja et al. 1996; Koster and Lettinga 1984; Robbins et al. 1989; Kayhanian 1994). In addition, they found that methanogens that consume acetate have more excellent resistance to high concentrations of ammonia nitrogen (Chen et al. 2008; Kayhanian 1999; Wiegant and Zeeman 1986; Zeeman et al. 1985). It is important to note that anaerobic digestion will be successful depending on the bacterial consortium present. The bacteria involved in anaerobic digestion need access to the microalgae cells for the digestible components to convert to methane biogas (Ward et al. 2014). Some strains found in the digestion process are more sensitive to ammonia nitrogen concentrations, such as *M. hungatei*, which was inhibited at a concentration of 4200 mg L^{-1} (Jarrell et al. 1987).

The importance of developing research on the microorganisms developing within the digestion process should be a priority to determine the microbial indicators responsible for achieving anaerobic digestion to pass the hydrolysis barrier due to the accumulation of toxins during the process. When working with microalgae, the hydrolysis phase is the one that determines the success of methane production (González-Fernández et al. 2012b).

Bacteria involved in anaerobic digestion need access to the microalgae cells that digestible components can be converted to methane biogas (Ward et al. 2014). During acetogenesis, different organic acids such as propionic, butyric, and some alcohols and ketones can be formed (Cooney et al. 2007).

Sulfur is toxic to methanogenic bacteria at concentrations greater than 200 mg L^{-1} (Parkin and Owen 1986). Sulfate-reducing bacteria compete with methanogenic bacteria for acetate. The former has a higher affinity for acetate and competes with each other even at low acetate concentrations (Bitton 2005). At COD/SO_4^{2-} ratios of 1.7 to 2.7, both are competitive, favoring an increase in this ratio to methanogens and a decrease in sulfate reducers (Metcalf and Eddy 2006; Bitton 2005). Aspe et al. (1997) found that at a ratio lower than 0.5, methane fermentation was inhibited.

15.3 Microalgae Properties

Each microalgae species has different properties, so it is advisable to treat them with cosubstrates that increase their effectiveness in anaerobic digestion. In case these biomasses are present in a biorefinery process, the production of biofuels with the digestion process would be viable by treating their residues after having extracted the bioproducts and thus achieving that this microalgal biomass has a future in producing clean energy. Microalgae generally contain proteins, lipids, carbohydrates, nucleic acids, pigments, and vitamins. The cell wall of microalgae is composed of biopolymeric compounds, hence they have a low biodegradability (Passos et al. 2014).

Microalgal biomass has low C/N ratios in the range of 5–9, resulting from their high protein concentrations (Oh-Hama and Miyachi 1988; Yen and Brune 2007; Lardon et al. 2009; Zhong et al. 2012). For anaerobic digestion processes, it is detrimental to have low carbon/nitrogen ratios during fermentation as ammonium will be produced, which can result in high pH levels and form free ammonia that is highly toxic to methanogenic microorganisms (Chen et al. 2008; Kayhanian 1994).

Microalgae such as *Chlorella* and *Scenedesmus* possess robust and rigid cell walls, and in *Chlorella*, microalgae contain cellulose and cellulose intermediates and mannan cellulose with mannan copolymers, in addition to pectin, chitosan, and chitin-like glycans (Chitin) (Gerken et al. 2013; Kapaun and Reisser 1995; Mihara 1961). Several authors have observed the correlation between cell wall degradation with the quantity and quality of biogas in an anaerobic digestion process. Therefore, it is essential to submit microalgae to pretreatments to break the cell wall and increase bacterial hydrolysis before submitting them to AD (Mussgnug et al. 2010; Golueke and Oswald 1959; Chen and Oswald 1998; Zamalloa et al. 2011; Samson and Leduy 1983; Sialve et al. 2009, Gonzalez-Fernandez et al. 2012).

Microalgal biomass is a protein-rich substrate, and the by-products in anaerobic digestion are propionate and butyrate, which are degraded to amino acids and converted to ammonium (nitrogen mineralization). High nitrogen mineralization can mediate the inhibition of anaerobes by high ammonium/ammonia (ammonium/ammonia) concentration (Mahdy et al. 2015). These compounds affect archaea the most (Yenigün and Demirel 2013). Protein-rich substrates will be formed during DA propionate and butyrate (Nagase and Matsuo 1982).

Lipids and proteins are essential in anaerobic digestion as they are broken down during hydrolysis (Chen et al. 2008). High protein concentrations result in the formation of ammonia compounds (Chen et al. 2008; McCarty 1964). Ammonia results from the breakdown of the biological decomposition of nitrogenous matter (Buswell and Boruff 1932).

Lipids in microalgae are an essential property in anaerobic digestion since they have a high theoretical methane potential compared to proteins and carbohydrates (Zamalloa et al. 2011). However, they can inhibit anaerobic digestion due to their low buffering capacity and intermediate products such as long-chain fatty acids (LCFAs) and volatile fatty acids (VFAs) (Park and Li 2012). When the total lipid content is less than 40%, the suggestion is to submit them to AD and then extract the lipids (Sialve et al. 2009). Ward et al. (2014) observed that at concentrations of 31, 40, and 47%, there was inhibition due to high lipid fraction. Cirne et al. (2007) reported that there was inhibition at 5, 10, and 18%. In case of removal when it comes to liquid biofuel production, this residual microalgal biomass can be subjected to anaerobic digestion (Sialve et al. 2009; Park and Li 2012; Cirne et al. 2007).

C/N ratios below 20 result in an imbalance in anaerobic digestion due to a high concentration of total ammonia-nitrogen (TAN) and an accumulation of volatile fatty acids (VFAs) in the digester. These two intermediate products of anaerobic digestion inhibit the process (Parkin and Owen 1986).

This inhibition is due to the production of decomposition intermediates such as long-chain fatty acids and volatile fatty acids. Microalgae have low C/N ratios between 4 and 8. Therefore, it is necessary to use cosubstrates that have a high carbon content (animal wastes, fatty oils rich in lipids and fats, and cellulose) to counteract the accumulation of N-NH_3 (Ehimen et al. 2009; Gonzalez-Fernandez et al. 2011; Saxena et al. 1984; Wang et al. 2009; Yen and Brune 2007).

Therefore, to have high productivity in the anaerobic digestion system, it is necessary that the C/N ratio is maintained above 20, and knowing that this ratio depends on the species of microalgae, it will be necessary to look for a suitable secondary cosubstrate to balance this ratio (Ward et al. 2014). One of the meaningful properties of microalgae is the lipid content, making them significant in the production of biogas, mainly methane, as opposed to proteins and carbohydrates. However, a high concentration of these can also inhibit the digestion process, due to the chemical association of lipids, with low buffering capacity and alkalinity (Park and Li 2012; Ward et al. 2014).

There is still much controversy in the literature about the optimum C/N ratio in anaerobic digestion; an acceptable range is 20/1–30/1 (Parkin and Owen 1986). When the C/N ratio is less than 20, it is said that there is no balance required by the microbial consortium and results in the release of nitrogen in the form of ammonia, which is an inhibitor for methanogenic bacteria and results in the accumulation of volatile fatty acids during anaerobic digestion (Sialve et al. 2009). Both ammonia nitrogen and volatile fatty acids are intermediate products formed during anaerobic digestion and are potential inhibitors that accumulate during AD (Parkin and Owen 1986).

Parkin and Owen (1986) and Yen and Brune (2007) indicate that the optimum C/N ratios should be 20:1–30:1 and 20:1–25:1, respectively. Increasing this ratio reduces the effect of inhibition by ammonia nitrogen. A concentration of 50–200 mg L^{-1} ammonia nitrogen should be kept at a minimum for the requirements of the microbial community in anaerobic digestion. Ammonia nitrogen is toxic at 1500–3000 mg L^{-1}; above 3000 mg L^{-1} is a potent inhibitory effect associated with nitrogen-ammonium (Parkin and Owen 1986), which causes biogas production to decline. According to several investigations, it has been shown that maintaining NH_3-N concentrations between 50 and 200 mg L^{-1} is beneficial for the bacterial population because ammonia nitrogen is an essential nutrient required by the microbial community (Parkin and Owen 1986). The release of ammonia nitrogen (ammonia-nitrogen) during anaerobic digestion is due to high nitrogen and protein concentrations in microalgae (Sialve et al. 2009).

There is a need to study the cell wall composition of microalgae to increase the rate of anaerobic hydrolysis. Low C/N ratios result in high ammonia nitrogen levels that inhibit biomethane production, which has been tested at very high temperatures (Wang et al. 2013; Olsson et al. 2014; Caporgno et al. 2015; Passos et al. 2013; Schwede et al. 2013). The micronutrients in microalgae are key factors that can prevent digestion failure when (Organic Loading Rate) OLRs are increased above 2 gVS L^{-1} day^{-1} (Schwede et al. 2013). An alternative in microalgae is to recover high-value compounds such as lipids, proteins, antioxidants, and pigments and subsequently treat the microalgae waste through anaerobic digestion (Ramos-Suárez and Carreras 2014; Mahdy et al. 2015; Prajapati et al. 2014; Keymer et al. 2013; Astals et al. 2011; Barontini et al. 2016) reported that the methane yield of microalgae residues is between 20 and 100% higher than the methane yield of natural microalgae.

15.4 Microalgae and Cosubstrates in Anaerobic Digestion

Microalgae are found in all bodies of water where there are nutrients where they can grow and develop. When microalgae are derived from a production process, they offer alternatives for later use in the production of biofuels. Microalgae can also be used in wastewater treatment to remove nutrients. Algae are significant microorganisms in the wastewater treatment ecosystem as they are essential in supplying oxygen to the bacteria for degrading biodegradable organic materials. Bacteria also supply carbon dioxide to algae for their growth and development.

Anaerobic digestion has been studied in terms of the use of different biomasses of high production, such as animal waste, agricultural waste, municipal solid waste, and sludge from wastewater treatment. Some of these wastes need to be combined with other wastes to adjust or raise the C/N ratio, thus avoiding the production of toxic substances such as ammonia nitrogen and volatile fatty acids. Waste subjected to anaerobic digestion individually does not generate high methane yields, hence the need to combine one or more substrates to increase methane yields in anaerobic digestion. The search for substrates with high energy potential should be a primary objective in all countries around the world to take advantage of them and generate wealth with bioenergy production. Anaerobic digestion for microalgal biomass and its residues plays a principal role in clean energy and sustainability.

The study of microalgae in the production of biofuels has been thought of from a clean energy point of view for a sustainable future. In a biorefinery process, the microalgae and all their remnants resulting from the process are used or exploited to have zero discharges and take advantage of everything produced. One of the most promising processes in bioenergy is anaerobic digestion or fermentation where the residual microalgal biomass converts to methane-rich biogas (Chisti 2007; De Schamphelaire and Verstraete 2009).

Due to the photosynthesis reaction, the solar energy stored in algal biomass can release methane through anaerobic digestion (Oswald and Golueke 1960). Anaerobic digestion is a process that transforms organic matter into renewable energy, and this technology has capitalized for the recovery of microalgae biomasses. The advantages of using microalgae are the mobilization of nitrogen and phosphorus nutrients and the availability of CO_2. In addition, the benefit of codigestion is the dilution of potentially toxic ammonia, which allows a higher loading rate and better biogas yield (Sosnowski et al. 2008). Golueke et al. (1957) were the first authors to report the anaerobic digestion of microalgal biomass and did so with *Chlorella vulgaris* and *Scenedesmus*. Codigestion helps cosubstrates increase the C/N ratio, balance micro- and macronutrients, rapidly degrade carbohydrates, and slowly degrade proteins and fats. The most studied substrates in anaerobic digestion are animal wastes such as pig manure and cow manure, besides sewage and activated sludge.

The authors studied the use of cosubstrates to increase the C/N ratio. They tested different microalgae species with pig manure, cow manure, corn stalks, sewage sludge, municipal wastewater solids, lipid-rich fats, oil and grease, domestic, municipal wastewater, and soybean oil. The result obtained was an increase in methane production in the anaerobic digestions (Gonzalez-Fernandez et al. 2011; Wang et al. 2009; Saxena et al. 1984; Ramamoorthy and Sulochana 1989; Peng et al. 2012; Samson and Leduy 1983; Yuan et al. 2012; Park and Li 2012; Salerno et al. 2009).

Microalgae are substantial biomasses in biogas production because of their utilization after extracting all the valuable bioproducts from them or their raw harvest. One of the essential sources of microalgae growth is the use of treated effluents from municipal wastewater treatment plants. Due to the biological treatment applied, the resulting effluents contain nutrients such as nitrogen and phosphorus, among others, and these nutrients are a source of microalgae growth. The study of anaerobic digestion with microalgae was accomplished by Golueke et al. (1957), whom they worked with *Chlorella vulgaris* and *Scenedesmus*.

There are several studies on anaerobic digestion for different species of microalgae. According to different microalgae species, the experiment's speed or rate of organic loading results in different methane yields, even for the same species. Different loading rates or velocities have been worked on, resulting in different yields for the same initial process conditions (Moheimani et al. 2015). There is a need to standardize the terminology for reporting methane yield as the volume of methane produced as a function of volatile solids loading or total solids added or removed to the process because some studies report it as the volume of methane as a function of COD removed. Even though microalgae have a low methane potential, they cosubstrate in anaerobic digestion. Therefore, microalgae biomasses can have a high biogas potential compared to corn silage commonly used for methane production generating 345 ± 7 mL g^{-1} VS (Bauer et al. 2010). Residual microalgae biomass can be converted into biogas by dark fermentation, and this is of great importance in bioenergy production.

Golueke et al. (1957) found that feeding an anaerobic digester with a concentrated algal biomass mixture resulted in a lower yield than that with raw sewage sludge, resulting in 0.5 m^3 of biogas per kg of dry algal organic matter (VS) from algae with 62.5% methane. Yen and Brune (2007) and Zhong et al. (2012) determined that to achieve an optimal C/N ratio, codigestion with carbon-rich substrates should be carried out, resulting in biogas productivity and decreased ammonia nitrogen during fermentation. Biogas production using microalgae has several barriers such as dense and thick cell walls for hydrolysis, its low C/N ratio, and high protein content (Gonzalez-Fernandez et al. 2015; Show et al. 2015). In their studies, Thorin et al. (2018) show the characteristics of some substrates such as proteins, carbohydrates, total solids lipids, volatile solids, and C/N ratio. This information is of interest for the search for cosubstrates that can help to increase methane yield during anaerobic digestion.

15.5 Microalgae Pretreatments and Cosubstrates

Microalgae and cosubstrates should be pretreated to make the substrate more biodegradable during digestion to increase methane production. Before undergoing the anaerobic digestion process, microalgae should be pretreated to break the cell wall, make their intracellular content more available, and improve biodegradability in optimal anaerobic digestion and thus increase bacterial hydrolysis (Chen and Oswald 1998; Golueke and Oswald 1959; Mussgnug et al. 2010; Sialve et al. 2009; Zamalloa et al. 2011).

Currently, the most used pretreatments are thermal, mechanical, chemical, and biological. These have been used to solubilize biomass and improve methane yields (Passos et al. 2014). Emphasizing that if the same treatment is applied to different substrates, the substrates respond differently (González-Fernández et al. 2012a, b, c; Mahdy et al. 2015). In addition to the energy costs that some of them imply in their application, such as the thermal method, it is necessary to submit highly concentrated biomasses to the anaerobic digestion process so that it is economically feasible to produce biogas (Mendez et al. 2014a, b; Passos et al. 2013). In addition, the results obtained are different depending on the mode in which the experiments are performed, either in batch mode or semicontinuous, resulting in different yields (Gonzalez-Fernandez et al. 2015).

To be able to treat microalgae through anaerobic digestions, it is necessary to submit to chemical, enzymatic, physical, and chemical pretreatment methods (Alzate et al. 2012; Keymer et al. 2013; Mahdy et al. 2014a, 2014b; Mendez et al. 2014a, b; Schwede et al. 2013). Figure 15.2 shows the different pretreatments for microalgae and cosubstrates.

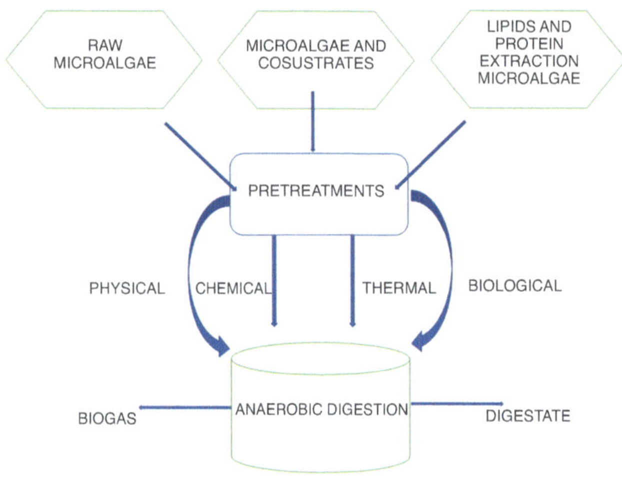

Fig. 15.2 Pretreatments in microalgae and cosubstrates in an anaerobic digestion

Mendez et al. (2014a, b) concluded that different methane yields will be obtained in the anaerobic digestion process depending on the microalgae species and the pretreatment applied. Jankowska et al. (2017) and Passos et al. (2014) extensively studied pretreatments.

In microalgae, *C. vulgaris* and *Chlorella* sp. uronic acids and a variety of sugars, such as glucosamine, were found once their cell walls were broken (Gerken et al. 2013; Huss et al. 1999). They indicated the presence of cellulose, hemicellulose, chitin, chitosan, and pectin in the cell wall (Gerken et al. 2013; Popper et al. 2011). Some microalgae contain cellulose and hemicellulose (Bisalputra and Weier 1963; Takeda 1996). Chen and Oswald (1998) found that productivity increased by 33% when microalgae were exposed to a temperature of 100 °C for 8 h. In this study, 60% of the untreated biomass remained undigested because the cell wall remained intact during the digestion. In addition, Gonzalez-Fernandez et al. (2012) found that methane production was increased by 9% with pretreatment at 70 °C. Alzate et al. (2012) showed that a thermal treatment increased methane productivity from 46 to 62%.

The thermochemical, mechanical, and ultrasonic pretreatments are also treatments that Samson and Leduy (1983) applied, giving similar results with the ultrasonic and thermal methods, with the difference being the time of each pretreatment, the ultrasonic being shorter, 10 min compared with 1 h at 150 °C. The chemical pretreatment of sodium hydroxide addition was not efficient for anaerobic digestion. Gonzalez-Fernandez et al. (2012) concluded that the ultrasonic method applied to microalgae resulted in a higher methane production than those without pretreatment, obtaining 44% of biodegradability in sonicated biomasses compared to 23% of biomasses in non-sonicated samples.

Lu et al. (2013) found that adding *Clostridium thermocellum* bacteria to *C. vulgaris* biomass increased biogas production by 17–24%. Mendez et al. (2013) found that applying a chemical pretreatment to the microalgal biomass resulted in higher solubilization of organic matter than the thermal method; however, the difference in biogas production was minimum. Thermal pretreatment improved methane yield by 62% for *C. vulgaris* (Gutierrez et al. 2015). Codigestion of microalgal biomass and other substrates improves methane productivity by providing easy hydrolysis and hence methanogenesis. Biological pretreatment using biocatalysts is an alternative to avoid the high energy consumption that can be applied in the pretreatment of substrates. Furthermore, intrinsic cell enzymes have been introduced to induce the autolysis of microalgae (Demuez et al. 2015).

Enzymatic hydrolysis can be implemented by

1. a dual-stage anaerobic digestion system (a hydrolytic–acidogenic stage and a methanogenic stage),
2. bioaugmentation sludge addition of bacterial and fungal cultures to accelerate the degradation rate, and
3. genetic modification of anaerobic microorganisms.

Ometto et al. (2014) concluded that methane production increased using proteases in anaerobic digestion. Physical pretreatments were studied by Ruffino et al. (2015), Lindner et al. (2015), and Tsapekos et al. (2015) as mechanical and thermal, using high shear mixers (high shear mixer) and ball-milling grinding, sandpapers/mesh grating, where they found that the biogas improved after these pretreatments.

For substrates with lignin and cellulose, organosolv pretreatment with acetic acid, ethanol, or methane at a temperature of 190 °C and time of 1 h can be used. Enzymes and energy supply are other pretreatments needed to increase methane productivity during AD (Lin et al. 2015a, b). Few studies exist on the codigestion of microalgae with sludge. The information is scarce to perform data processing for understanding the process. More information is needed about other parameters such as alkalinity, ammonia nitrogen, and volatile fatty acids (Thorin et al. 2018). The lower VFA concentration in a codigestion indicates that adding microalgae with animal manure may result in a more stable process.

15.6 Methane Yield in Anaerobic Codigestion with Microalgae

Biochemical methane potential (BMP) is tested to estimate methane productivity for substrates of interest in anaerobic digestion. These tests are performed at the laboratory level to evaluate in a fast way the methane yields with different proportions of substrates, either pretreated or untreated, to mark the differences that may exist in different pretreatments in methane production. These BMP tests are influenced by the capacities of the inoculums and substrates (Koch et al. 2017; De Vrieze et al. 2015).

Sialve et al. (2009) found methane yields in microalgae in the range of 0.1–0.4 $LCH_4\ g^{-1}$ VS (0.07–0.27 $LCH_4\ g^{-1}$ COD) equivalent to conversion efficiencies of 17–76% at loading rates of 1.5–5 g COD L^{-1} day^{-1}. The HRT hydraulic retention time for anaerobic digestion is 10–30 days for conventional digestion.

It has been demonstrated that, at thermophilic conditions in the digestion process, microalgae have reached conversion efficiencies of 54%, reaching organic loading of up to 2.9 kg SVm^{-3} day^{-1}. For example, for *Spirulina maxima*, a yield of

0.3–0.37 m^3 biogas kg SV^{-1} with 70% methane and conversion efficiencies higher than 48% were obtained (Samson and Leduy 1982, 1983, 1986).

Yen and Brune (2007) increased the C/N ratio from 6.7 to 36.4 by adding wastepaper with a combination of *Scenedesmus* sp. and *Chlorella* sp. microalgae. Showing that combining 50% paper and 50% microalgae was the best codigestion, resulting in a C/N ratio of 18 with 1170 ± 75 mL day^{-1} biogas produced, compared to the digested microalgae alone, which resulted in 573 ± 28 mL day^{-1} of biogas.

In 1982, Samson and Leduy studied the cyanobacterium *Spirulina maxima* that lacks a rigid cell wall, and hydrolysis is not hindered in the production of biogas in anaerobic digestion. Reporting results at a mesophilic temperature of 0.26 L g $VSin^{-1}$ at an OLR of 0.97 g VS L^{-1} day^{-1} and an HRT of 33 days. These cyanobacteria have low C/N ratios so it is advisable to use carbon-rich cosubstrates (Mahdy et al. 2015; Park and Li 2012). Markou (2012) showed that feeding the bioreactors with carbohydrates increased methane yield, for example, feeding 60% carbohydrate biomass resulted in 0.20 ± 0.01 L CH_4 g (chemical oxygen demand) $CODin^{-1}$ compared to feeding 20%, which resulted in lower methane yield, 0.12 ± 0.01 L CH_4 g $CODin^{-1}$.

Authors such as Passos et al. (2013) reported for non-pretreated microalgal biomass of 0.17 L CH_4 g $VSin^{-1}$. Wieczorek et al. (2015) obtained higher methane yields of 0.24–0.30 L CH_4 g $VSin^{-1}$, emphasizing that it depends on the substrate/inoculum ratio. In addition to harvest, time is also attributable to producing lower or greater methane yields. Table 15.1 shows the results of methane production at mesophilic temperature for some microalgae species. It can be observed that organic loading, retention time, and type of reactor influence on methane yield.

Reactor configuration and bioprocess parameters affect methane production (Mendez et al. 2013; Passos and Ferrer 2015). Some of the pretreatments have been performed at batch scale to quickly compare substrates and operating conditions such as HRT, OLR, and temperature. The most commonly used reactors are CSTRs, where the mixture is suspended, and OLRs range from 1 to 6 g COD L^{-1} day^{-1} with retention times from 10 to 30 days. The HTR hydraulic retention time in a typical digester range from 10 to 30 days (Sialve et al. 2009). Ras et al. (2011) worked with *C. vulgaris*, and the methane yield was higher operating at a lower organic loading and more extended retention time (0.11 L CH_4 g $CODin^{-1}$ at HRT of 16 day and OLR of 1 g COD L^{-1} day^{-1}).

The results that many researchers have reported make the differences due to biomass composition, variability, and cell wall hardness of the microalgae strains. Microalgae work better at mesophilic than thermophilic temperatures because high temperatures result in high concentrations of ammonia nitrogen and VFA. For example, Kinnunen et al. (2014) studied *Nannochloropsis* sp. at the two temperatures and obtained 48% more at mesophilic conditions operating at OLR less than 2 kg VS m^{-3} day^{-1}.

It has been shown that under mesophilic conditions, anaerobically digesting microalgae has resulted in high methane yields compared to thermophilic temperatures.

Combining microalgae with septic sludge, biogas production increased by 400% compared to digesting microalgae alone. Lu and Zhang (2016) concluded that it favors this cosubstrate in increasing the C/N ratio, thereby reducing the possibility of ammonia nitrogen formation, improving the digestibility of microalgal biomass, and decreasing H_2 concentration, which helps to increase methane quality and quantity.

Some studies of codigestion with activated sludge as cosubstrates have increased biogas yields (Yen and Brune 2007). In addition, sewage sludge cosubstrate with microalgae has been studied to improve biogas production (Cecchi et al. 1996; Sosnowski et al. 2008; Wang et al. 2013). Finally, some authors have worked on various cosubstrates with microalgae to improve the C/N ratio.

Table 15.1 Methane production by microalgae using different types of microalgae

Microalgae	Reactor	OLR g COD L^{-1} day^{-1}	OLR g VS L^{-1} day^{-1}	HRT (days)	Methane L CH_4 g COD in^{-1}	Methane L CH_4 g $VSin^{-1}$	Author
C. vulgaris	CSTR	1		16	0.11		Ras et al. (2011)
C. vulgaris	CSTR	1.5		15	0.05		Mahdy et al. (2015)
Scenedesmus sp.	CSTR	1		15	0.03		González-Fernández et al. (2013)
Oocystis sp.	CSTR	2.3		20	0.08		Passos and Ferrer (2015)
Scenedesmus sp.	UASB		2.25	3.8		0.22	Tartakovsky et al. (2015)
Scenedesmus sp.	CSTR		0.65	16		0.22	Tartakovsky et al. (2013)
Scenedesmus sp.	UASB		2.7	2.2		0.11	Zamalloa et al. (2012)

Table 15.2 Production of methane in anaerobic digestion with microalgae and cosubstrates

Substrates	Mode	Ratio mass	CN	*T* (°C)	OLR (kg m^{-3} day^{-1})	Biogas mL g^{-1} VS	Methane yield CH_4 m^{-3} kg^{-1}	CH_4 mL day^{-1}	References
Microalgae + pig manure	Batch	50:50 (COD)		35			0.22		Gonzalez-Fernandez et al. (2011)
Scenedesmus spp. + *Chlorella* spp.	SC		6.7	35	4			573 ± 28	Yen and Brune (2007)
Scenedesmus spp. + *Chlorella* spp. + waste paper	SC	50:50 (VS)	18	35	4			1170 ± 75	Yen and Brune (2007)
Spirulina maxima + sewage sludge	C	50:50 (VS)		35	3.9		0.36		Samson and Leduy (1983)
Scenedesmus sp. + sewage sludge	SC	50:50 (VS)		35	0.5		0.39		Peng and Colosi et al. (2016)
Microalgal biomass + sewage sludge	SC	25:75 (VS)		35	0.5		0.51		Peng and Colosi et al. (2016)
Microalgal biomass + primary sludge	SC	25:75 (VS)		35	1.2		0.46		Solé-Bundó et al. (2017)
Scenedesmus sp. + *Chlorella* sp. + sewage sludge	SC	40:60 (VS)		37	2.4		0.20		Olsson et al. (2018)
Scenedesmus sp. + paper sludge	C	74:26 (VS)		37			0.173		Ramos-Suárez and Carreras (2014)
Scenedesmus sp. + pig manure	C	15:85 (VS)		37			0.319		Astals et al. (2015)

C, continuous; SC semicontinuous. Data from Solé-Bundó et al. (2019)

The agro-industrial waste is an example in which the codigestion of microalgae with this type of waste improves biogas quality. Agro-industrial wastes such as papaya, corn, and wheat straw, corn cob and silage, cheese whey, coffee husks, and cellulose recognized for having high C/N values >45 and codigestion with microalgae overcome the limitation of low C/N ratio (Mata-Alvarez et al. 2014; Fernández-Rodríguez et al. 2014). Table 15.2 summarizes the results of several studies of anaerobic digestion with microalgae and cosubstrates at mesophilic temperature. The methane yield depends on the microalgae type, the mixture proportion, organic loading, pretreatment, and mode operation.

Microalgae and microalgae wastes are characterized by low methane yields (150–300 mL CH_4 gVS^{-1}) and degradation rates (0.05–0.25 day^{-1}) compared to other substrates such as sewage sludge (200–350 mL CH_4 gVS^{-1}, 0.20–0.40 day^{-1}), animal manure (200–400 mL CH_4 gVS^{-1}, 0.10–0.30 day^{-1}), and food waste (400–550 mL CH_4 gVS^{-1}, 0.30–0.70 day^{-1}) (Ramos-Suárez and Carreras 2014; Gunaseelan 1997; Nasir et al. 2012; Astals et al. 2013; Braguglia et al. 2018; Raposo et al. 2012; Li et al. 2018; Dębowski et al. 2017; Komilis et al. 2017; Olsson et al. 2018; Carminati et al. 2018). Both microalgae and some animal wastes have low C/N ratios; some examples are pig, cattle, and poultry manure, which have been studied in BMP tests with different ratios in v/v, SV, and COD. For example, pretreated *Chlorella vulgaris* was digested with cattle manure and given 431 mL CH_4 gSV^{-1} (Mahdy et al. 2017).

15.7 Methane and Codigestate as End Products of Anaerobic Digestion

The end products of anaerobic digestion are biogas and digestate, which are valuable by-products of the process because, depending on the success of the process, the quality of the subproducts will be high. Therefore, in all treatment processes, the products derived from these processes should be applicable and economically affordable, the biogas should be reused as a source of clean energy, and the digestate should be reused in the agricultural sector to improve the nutrients in the soil. The importance of determining the quality of the digestate as the concentration of nutrients, heavy metals, pathogens, phytotoxicity, and stability of the organic matter is relevant for the subsequent use of the product. The concentration of heavy metals increases when digested with microalgae and can impact the reuse of the digestate on agricultural land (Ramos-Suárez and Carreras 2014; Olsson et al. 2018; Solé-Bundó et al. 2017). It is relevant to mention that compounds such as heavy metals and trace organic compounds can be found in anaerobic digestions where this depends on the origin of the organic substrates (Markou et al. 2013).

A digestate can have concentrations of 2940 mg L^{-1} ammonia-nitrogen, 390 mg L^{-1} total phosphorus, and 320 mg L^{-1} potassium (Collet et al. 2011). Zamalloa et al. (2012) found a clear liquid digestate with high concentrations of nutrients derived from anaerobic digestion of microalgal biomass of *P. tricornutum* with total ammonia-nitrogen (TAN) and phosphate concentrations of 546 ± 48 mg L^{-1} and 141 ± 41 mg L^{-1}, respectively.

Although phosphorus has been investigated little compared to other nutrients (Vaccari 2009), it is necessary because it is a nutrient in most substrates or organic biomasses. The advantages of a digestion process are stable biosolid production and a liquid with properties that can be reused, such as a biofertilizer and a nutrient-rich liquid with nitrogen and phosphorus. The nutritional contents of both ammonia-nitrogen and phosphorus in the anaerobic digestion process of microalgae are 546–2940 mg L^{-1} and 141–390 mg L^{-1}, respectively (Collet et al. 2011; Erkelens et al. 2014; Ward et al. 2014; Zamalloa et al. 2012).

Solé-Bundó et al. (2017) concluded that codigestion between primary sludge and microalgae was better for reuse in agriculture than when digested only with microalgae.

15.8 Conclusions

Microalgae benefit the environment because they grow in aquatic environments and are CO_2 capturers, besides being used for cleaning treated wastewater in terms of nitrogen and phosphorus. The properties of microalgae must be utilized to anaerobically digest them and obtain biogas rich in methane so that it can be reused as a source of energy. There is a global need to generate clean energy based on exploring biomasses that are generated as waste and have the potential to be used through anaerobic digestion. Microalgae are a potential source to be used as a substrate or cosubstrate in anaerobic digestion. Increasing the methane yield through anaerobic digestions searches substrates that counteract the low C/N ratio and make the hydrolysis phase not the obstacle to achieving methanogenesis. Different authors have studied different pretreatments that have helped to increase methane yields in the AD process. However, they emphasize that the pretreatment of microalgae generates an extra economic expense that can increase the cost of the process. Some recommend that lipids and proteins be recovered and their residues treated in AD; however, these treatments must be safe because any remaining chemical residues can affect the AD process.

There is a need to standardize methane yield results, basing reports on the volume of methane produced as a function of volatile solids loading, total solids, or COD. A necessity in showing results is what percentage of SV represents COD. More research is necessary on reactor configurations tested with microalgae and pretreated cosubstrates to see if methane yields are affected.

References

Alzate ME, Munoz R, Rogalla F, Fernandez-Polanco F, Perez-Elvira SI (2012) Biochemical methane potential of microalgae: influence of substrate to inoculum ratio, biomass concentration and pretreatment. Bioresour Technol 123:488–494

Amon T, Amon B, Kryvoruchko V, Zollitsch W, Mayer K, Gruber L (2007) Biogas production from maize and dairy cattle manure - influence of biomass composition on the methane yield. Agric Ecosyst Environ 118:173–182

Angelidaki I, Ahring BK (1993) Thermophilic digestion of livestock waste, the effect of ammonia. Appl Microbiol Biotechnol 38:560–564

Appels L, Lauwers J, Degreve J, Helsen L, Lievens B, Willems K et al (2011) Anaerobic digestion in global bio-energy production: potential and research challenges. Renew Sust Energ Rev 15:4295–4301

Aspe E, Marti MC, Roeckel M (1997) Anaerobic treatment of fishery wastewater using a marine sediment inoculum. Water Res 31(9): 2147–2160

Astals S, Ariso M, Galí A, Mata-Alvarez J (2011) Co-digestion of pig manure and glycerine: experimental and modelling study. J Environ Manag 92:1091–1096

Astals S, Esteban-Gutiérrez M, Fernández-Arévalo T, Aymerich E, García-Heras JL, Mata-Alvarez J (2013) Anaerobic digestion of seven different sewage sludges: a biodegradability and modelling study. Water Res 47:6033–6043

Astals S, Musenze RS, Bai X, Tannock S, Tait S, Pratt S, Jensen PD (2015) Anaerobic co-digestion of pig manure and algae: impact of intracellular algal products recovery on co-digestion performance. Bioresour Technol 181:97–104. https://doi.org/10.1016/j.biortech.2015.01.039

Barontini F, Biagini E, Dragoni F, Corneli E, Ragaglini G, Bonari E, Tognotti L, Nicolella C (2016) Anaerobic digestion and co-digestion of oleaginous microalgae residues for biogas production. Chem Eng Trans:91–96

Bauer A, Leonhartsberger C, Bösch P, Amon B, Friedl A, Amon T (2010) Analysis of methane yields from energy crops and agricultural by-products and estimation of energy potential from sustainable crop rotation systems in EU-27. Clean Techn Environ Policy 12 (2):153–161. https://doi.org/10.1007/s10098-009-0236-1

Bhattacharya SK, Parkin GF (1989) The effect of ammonia on methane fermentation process. J Water Pollut Control 61(1):55–59

Bisalputra T, Weier T (1963) The cell wall of *Scenedesmus quadricauda*. Am J Bot 50:1011–1019

Bitton G (2005) Wastewater microbiology. Wiley

Borja R, Sanchez E, Duran MM (1996) Effect of the clay mineral zeolite on ammonia inhibition of anaerobic thermophilic reactors treating cattle manure. J Environ Sci Health A31(2):479–500

Braguglia CM, Gallipoli A, Gianico A, Pagliaccia P (2018) Anaerobic bioconversion of food waste into energy: a critical review. Bioresour Technol 248:37–56

Buswell AM, Boruff CS (1932) The relationship between chemical composition of organic matter and the quality and quantity of gas produced during sludge digestion. Sewage Works J 4(3):454–460

Caporgno MP, Trobajo R, Caiola N, Ibanez C, Fabregat A, Bengoa C (2015) Biogas production from sewage sludge and microalgae co-digestion under mesophilic and thermophilic conditions. Renew Energy 75:374–380

Carminati P, Gusmini D, Pizzera A, Catenacci A, Parati K, Ficara E (2018) Biogas from mono- and co-digestion of microalgal biomass grown on piggery wastewater. Water Sci Technol 78:103–113

Carrere H, Antonopoulou G, Affes R, Passos F, Battimelli A, Lyberatos G, Ferrer I (2016) Review of feedstock pretreatment strategies for improved anaerobic digestion: from lab-scale research to full-scale application. Bioresour Technol 199:386–397

Cavinato C, Bolzonella D, Pavan P, Fatone F, Cecchi F (2013) Mesophilic and thermophilic anaerobic co-digestion of waste activated sludge and source sorted biowaste in pilot-and full-scale reactors. Renew Energy 55:260–265

Cecchi F, Pavan P, Mata-Alvarez J (1996) Anaerobic co-digestion of sewage sludge: application to the macroalgae from the Venice lagoon. Resour Conserv Recy 17(1):57–66

Chen PH, Oswald WJ (1998) Thermochemical treatment for algal fermentation. Environ Int 24(8):889–897

Chen Y, Cheng JJ, Creamer KS (2008) Inhibition of anaerobic digestion process: a review. Bioresour Technol 99:4044–4064

Chisti Y (2007) Biodiesel from microalgae. Biotechnol Adv 25:294–306

Cirne DG, Paloumet X, Björnsson L, Alves MM, Mattiasson B (2007) Anaerobic digestion of lipid-rich waste—effects of lipid concentration. Renew Energy 32:965–975

Collet P, Hélias A, Lardon L, Ras M, Goy RA, Steyer JP (2011) Life-cycle assessment of microalgae culture coupled to biogas production. Bioresour Technol 102:207–214

Cooney M, Maynard N, Cannizzaro C, Benemann J (2007) Two-phase anaerobic digestion for production of hydrogen-methane mixtures. Bioresour Technol 98:2641–2651

De Bok FAM, Plugge CM, Stams AJM (2004) Interspecies electron transfer in methanogenic propionate degrading consortia. Water Res 38:1368–1375

De Schamphelaire L, Verstraete W (2009) Revival of the biological sunlight to biogas energy conversion system. Biotechnol Bioeng 103:296–304

De Vrieze J, Raport L, Willems B, Verbrugge S, Volcke E, Meers E, Angenent L, Boon N (2015) Inoculum selection influences the biochemical methane potential of agro-industrial substrates. Microb Biotechnol 8:776–786

Dębowski M, Zieliński M, Kisielewska M, Krzemieniewski M (2017) Anaerobic co-digestion of the energy crop *Sida hermaphrodita* and microalgae biomass for enhanced biogas production. Int J Environ Res 11:243–250

Demuez M, Mahdy A, Tomas-Pejo E, Gonzalez-Fernandez C, Ballesteros M (2015) Enzymatic cell disruption of microalgae biomass in biorefinery processes. Biotechnol Bioeng 112:1955–1966

Deublein D, Steinhauser A (eds) (2008) Biogas from waste and renewable resources-an introduction. Wiley, Weinheim, p 443

Ehimen EA, Connaughton S, Sun Z, Carington GC (2009) Energy recovery from lipid extracted, transesterified and glycerol codigested microbial biomass. GCB Bioenergy 1:371–381

Erkelens M, Ward AJ, Ball AS, Lewis DM (2014) Microalgae digestate effluent as a growth medium for Tetraselmis sp. in the production of biofuels. Bioresour Technol 167:81–86. https://doi.org/10.1016/j.biortech.2014.05.126

Fernández-Rodríguez MJ, Rincón B, Fermoso FG, Jiménez AM, Borja R (2014) Assessment of two-phase olive mill solid waste and microalgae co-digestion to improve methane production and process kinetics. Bioresour Technol 157:263–269

Gerken HG, Donohoe B, Knoshaug EP (2013) Enzymatic cell wall degradation of *Chlorella vulgaris* and other microalgae for biofuels production. Planta 237:239–253

Goberna M, Gadermaier M, García C, Wett B, Insam H (2010) Adaptation of methanogenic communities to the cofermentation of cattle excreta and olive mill wastes at 37 °C and 55 °C. Appl Environ Microbiol 76:6564–6571

Golueke CG, Oswald WJ (1959) Biological conversion of light energy to the chemical energy of methane. Appl Microbiol 7:219–227

Golueke CG, Oswald WJ, Gotaas HB (1957) Anaerobic digestion of algae. Appl Microbiol 5:47–55

Gonzalez-Fernandez C, Molinuevo-Salces B, Garcia-Gonzalez MC (2011) Evaluation of anaerobic codigestion of microbial biomass and swine manure via response surface methodology. Appl Energy 88:3448–3453

González-Fernández C, Sialve B, Bernet N, Steyer JP (2012a) Impact of microalgae characteristics on their conversion to biofuel. Part II: focus on biomethane production. Biofuels Bioprod Biorefin 6:205–218

González-Fernández C, Sialve B, Bernet N, Steyer JP (2012b) Thermal pretreatment to improve methane production of *Scenedesmus* biomass. Biomass Bioenergy 40:105–111

González-Fernández C, Sialve B, Bernet N, Steyer JP (2012c) Comparison of ultrasound and thermal pretreatment of *Scenedesmus* biomass on methane production. Bioresour Technol 110:610–616

González-Fernández C, Sialve B, Bernet N, Steyer (2013) Effect of organic loading rate on anaerobic digestion of thermally pretreated Scenedesmus sp. biomass. Bioresour Technol 129:219–223. https://doi.org/10.1016/j.biortech.2012.10.123

Gonzalez-Fernandez C, Sialve B, Molinuevo-Salces B (2015) Anaerobic digestion of microalgal biomass: challenges, opportunities and research needs. Bioresour Technol 198:896–906

Gujer W, Zehnder AJ (1983) Conversion processes in anaerobic digestion. Water Sci Technol 15:127–167

Gunaseelan VN (1997) Anaerobic digestion of biomass for methane production: a review. Biomass Bioenergy 13:83–114

Gutierrez R, Ferrer I, Garcia J, Uggetti E (2015) Influence of starch on microalgal biomass recovery, settleability and biogas production. Bioresour Technol 185:341–345

Huss VAR, Frank C, Hartmann EC, Hirmer M, Kloboucek A, Seidel BM, Wenzeler P, Kessler E (1999) Biochemical taxonomy and molecular phylogeny of the genus *Chlorella sensu lato* (Chlorophyta). J Phycol 35:587–598

Jankowska E, Sahu AK, Oleskowicz-Popiel P (2017) Biogas from microalgae: review on microalgae's cultivation, harvesting and pretreatment for anaerobic digestion. Renew Sust Energ Rev. https://doi.org/10.1016/j.rser.2016.11.045

Jarrell KF, Saulnier M, Ley A (1987) Inhibition of methanogenesis in pure cultures by ammonia, fatty acids, and heavy metals, and protection against heavy metal toxicity by sewage sludge. Can J Microbiol 33:551–555

Kapaun E, Reisser W (1995) A chitin-like glycan in the cell-wall of a *Chlorella* sp. (Chlorococcales, Chlorophyceae). Planta 197:577–582

Kayhanian M (1994) Performance of a high solids anaerobic digestion process under various ammonia concentrations. J Tech Biotechnol 59:349–352

Kayhanian M (1999) Ammonia inhibition in high-solids biogasification, an overview and practical solutions. Environ Technol 20:355–365

Keymer P, Ruffell I, Pratt S, Lant P (2013) High pressure thermal hydrolysis as pre-treatment to increase the methane yield during anaerobic digestion of microalgae. Bioresour Technol 131:128–133

Kinnunen HV, Koskinen PEP, Rintala J (2014) Mesophilic and thermophilic anaerobic laboratory-scale digestion of *Nannochloropsis* microalga residues. Bioresour Technol 155:314–332

Koch K, Lippert T, Drewes JE (2017) The role of inoculum's origin on the methane yield of different substrates in biochemical methane potential (BMP) tests. Bioresour Technol 243:457–463

Komilis D, Barrena R, Grando RL, Vogiatzi V, Sánchez A, Font X (2017) A state of the art literature review on anaerobic digestion of food waste: influential operating parameters on methane yield. Rev Environ Sci Biotechnol 16:347–360

Koster IW, Lettinga G (1984) The influence of ammonia nitrogen on the specific activity of pelletized methanogenic sludge. Agric Waste 9: 215–216

Lardon L, Helias A, Sialve B, Steyer JP, Bernard O (2009) Life-cycle assessment of biodiesel production from microalgae. Environ Sci Technol 43:6475–6481

Li L, He Q, Zhao X, Wu D, Wang X, Peng X (2018) Anaerobic digestion of food waste: correlation of kinetic parameters with operational conditions and process performance. Biochem Eng J 130:1–9

Lin JH, Lee DJ, Chang JS (2015a) Lutein production from biomass: marigold flowers versus microalgae. Bioresour Technol 184:421–428

Lin JH, Lee DJ, Chang JS (2015b) Lutein in specific marigold flowers and microalgae. J Taiwan Inst Chem Eng 49:90e94

Lindner J, Zielonka S, Oechsner H, Lemmer A (2015) Effects of mechanical treatment of digestate after anaerobic digestion on the degree of degradation. Bioresour Technol 178:194–200

Lu D, Zhang XJ (2016) Biogas production from anaerobic codigestion of microalgae and septic sludge. J Environ Eng 142(10). https://doi.org/10.1061/(ASCE)EE.1943-7870.0001126

Lu F, Shao L, He P (2013) Bacterial bioaugmentation for improving methane and hydrogen production from microalgae. Biotechnol Biofuels 6:92

Mahdy A, Mendez L, Ballesteros M, Gonzalez-Fernandez C (2014a) Autohydrolysis and alkaline pretreatment effect on *Chlorella vulgaris* and *Scenedesmus* sp. methane production. Energy 78:48–52

Mahdy A, Mendez L, Ballesteros M, Gonzalez-Fernandez C (2014b) Enhanced methane production of *Chlorella vulgaris* and *Chlamydomonas reinhardtii* by hydrolytic enzymes addition. Energy Convers Manag 85:551–557

Mahdy A, Mendez L, Ballesteros M, González-Fernández C (2015) Algaculture integration in conventional wastewater treatment plants: anaerobic digestion comparison of primary and secondary sludge with microalgae biomass. Bioresour Technol 12:236–244

Mahdy A, Fotidis IA, Mancini E, Ballesteros M, González-Fernández C, Angelidaki I (2017) Ammonia tolerant inocula provide a good base for anaerobic digestion of microalgae in third generation biogas process. Bioresour Technol 225:272–278

Markou G (2012) Alteration of the biomass composition of *Arthrospira* (*Spirulina*) platensis under various amounts of limited phosphorus. Bioresour Technol 116:533–535

Markou G, Angelidaki I, Georgakakis D (2013) Carbohydrate-enriched cyanobacterial biomass as feedstock for bio-methane production through anaerobic digestion. Fuel 111:872–879. https://doi.org/10.1016/j.fuel.2013.04.013

Mata-Alvarez J, Dosta J, Macé S, Astals S (2011) Codigestion of solid wastes: a review of its uses and perspectives including modeling. Crit Rev Biotechnol 31:99–111

Mata-Alvarez J, Dosta J, Romero-Güiza MS, Fonoll X, Peces M, Astals S (2014) A critical review on anaerobic co-digestion achievements between 2010 and 2013. Renew Sust Energ Rev 36:412–427

McCarty PL (1964) Anaerobic waste treatment fundamentals: part 3. Public Works 95:91–94

Mendez L, Mahdy A, Timmers RA, Ballesteros M, González-Fernández C (2013) Enhancing methane production of *Chlorella vulgaris* via thermochemical pretreatments. Bioresour Technol 149:136–141

Mendez L, Mahdy A, Ballesteros M, González-Fernández C (2014a) Methane production of thermally pretreated *Chlorella vulgaris* and *Scenedesmus* sp. biomass at increasing biomass loads. Appl Energy 129:238–242

Mendez L, Mahdy A, Demuez M, Ballesteros M, Gonzalez-Fernandez C (2014b) Effect of high pressure thermal pretreatment on *Chlorella vulgaris* biomass: organic matter solubilisation and biochemical methane potential. Fuel 117:674–679

Metcalf L, Eddy HP (2006) Wastewater engineering treatment and reuse, 4th edn. McGraw-Hill Publishing Company Limited, New York

Mihara S (1961) Change in glucosamine content of *Chlorella* cells during the course of their life cycle. Plant Cell Physiol 2:25–29

Moheimani NR, McHenry MP, de Boer K, Bahri PA (2015) Biomass and biofuels from microalgae advances in engineering and biology. Springer International Publishing, Cham

Mussgnug JH, Klassen V, Schlüter A, Kruse O (2010) Microalgae as substrates for fermentative biogas production in a combined biorefinery concept. J Biotechnol 150:51–56

Nagase M, Matsuo T (1982) Interactions between amino-acid-degrading bacteria and methanogenic bacteria in anaerobic digestion. Biotechnol Bioeng 24:2227–2239

Nasir IM, Mohd Ghazi TI, Omar R (2012) Anaerobic digestion technology in livestock manure treatment for biogas production: a review. Eng Life Sci 12:258–269

Oh-Hama T, Miyachi S (1988) Chlorella. Cambridge University Press, Cambridge

Olsson J, Feng XM, Ascue J, Gentili FG, Shabiimam MA, Nehrenheim E, Thorin E (2014) Codigestion of cultivated microalgae and sewage sludge from municipal waste water treatment. Bioresour Technol 171:203–210

Olsson J, Forkman T, Gentili FG, Zambrano J, Schwede S, Thorin E, Nehrenheim E (2018) Anaerobic co-digestion of sludge and microalgae grown in municipal wastewater - a feasibility study. Water Sci Technol 77:682–694

Ometto F, Quiroga G, Pšenicka P, Whitton R, Jefferson B, Villa R (2014) Impacts of microalgae pre-treatments for improved anaerobic digestion: thermal pretreatment, thermal hydrolysis, ultrasound and enzymatic hydrolysis. Water Res 65:350–361

Oswald JW, Golueke CG (1960) Biological transformation of solar energy. Adv Appl Microbiol 2:223–261

Park S, Li Y (2012) Evaluation of methane production and macronutrient degradation in the anaerobic co-digestion of algae residue and lipid waste. Bioresour Technol 111:42–48

Parkin GF, Owen WF (1986) Fundamentals of anaerobic digestion of wastewater sludges. J Environ Eng 112:867–920

Passos F, Ferrer I (2015) Influence of hydrothermal pretreatment on microalgal biomass anaerobic digestion and bioenergy production. Water Res 68:364–373. https://doi.org/10.1016/j.watres.2014.10.015

Passos F, Solé M, García J, Ferrer I (2013) Biogas production from microalgae grown in wastewater: effect of microwave pretreatment. Appl Energy 108:168–175

Passos F, Uggetti E, Carrere H, Ferrer I (2014) Pretreatment of microalgae to improve biogas production: a review. Bioresour Technol 172:403–412

Peng S, Colosi LM (2016) Anaerobic digestion of algae biomass to produce energy during wastewater treatment ABSTRACT. Water Environ Res 88(1):29–39. https://doi.org/10.2175/106143014X14062131179195

Peng S, Hou C, Wang J, Chen T, Liu X, Yue Z (2012) Performance of anaerobic co-digestion of corn straw and algae biomass from lake Chaohu. Trans Chin Soc Agric Eng. https://www.ingentaconnect.com/content/tcsae/tcsae/2012/00000028/00000015/art00028

Popper ZA, Michel G, Herve C, Domozych DS, Willats WGT, Tuohy MG, Kloareg B, Stengel DB (2011) Evolution and diversity of plant cell walls: from algae to flowering plants. Annu Rev Plant Biol 62: 567–588

Prajapati SK, Choudhary P, Malik A, Vijay VK (2014) Algae mediated treatment and bioenergy generation process for handling liquid and solid waste from dairy cattle farm. Bioresour Technol 167:260–268

Ramamoorthy S, Sulochana N (1989) Enhancement of biogas production using algae. Current science (Bangalore). https://www.cabdirect.org/cabdirect/abstract/19901357375

Ramos-Suárez JL, Carreras N (2014) Use of microalgae residues for biogas production. Chem Eng J 242:86–95

Raposo F, De La Rubia MA, Fernández-Cegrí V, Borja R (2012) Anaerobic digestion of solid organic substrates in batch mode: an overview relating to methane yields and experimental procedures. Renew Sust Energ Rev 16:861–877

Ras M, Lardon L, Sialve B, Bernet N, Steyer JP (2011) Experimental study on a coupled process of production and anaerobic digestion of *Chlorella vulgaris*. Bioresour Technol 102:200–206

Robbins JE, Gerhard SA, Kappel TJ (1989) Effects of ammonia in anaerobic digestion and an example of digester performance from cattle manure protein mixtures. Biol Wastes 27:1–14

Ruffino B, Campo G, Genon G, Lorenzi E, Novarino D, Scibilia G, Zanetti M (2015) Improvement of anaerobic digestion of sewage sludge in a wastewater treatment plant by means of mechanical and thermal pre-treatments: performance, energy and economical assessment. Bioresour Technol 175:298e308

Salerno M, Nurdogan Y, Lundquist TJ (2009) Biogas production from algae biomass harvested at wastewater treatment ponds. Proceedings of the 2009 Bioenergy Engineering Conference. https://digitalcommons.calpoly.edu/cenv_fac/180

Samson R, Leduy A (1982) Biogas production from anaerobic digestion of *Spirulina maxima* algal biomass. Biotechnol Bioeng 24:1919–1924

Samson R, Leduy A (1983) Influence of mechanical and thermochemical pretreatments on anaerobic-digestion of *Spirulina-maxima* algal biomass. Biotechnol Lett 5:671–676

Samson R, Leduy A (1986) Detailed study of anaerobic digestion of Spirulina maxima algal biomass. Biotechnol Bioeng 28(7):1014–1023. 10.1002/(ISSN)1097-0290. 10.1002/bit.v28:7. https://doi.org/10.1002/bit.260280712

Saxena VK, Tandon SM, Singh KK (1984) Anaerobic digestion of green filamentous algae and water hyacinth for methane production. Nat Acad Sci Lett 7:283–284

Schwede S, Rehman Z-U, Gerber M, Theiss C, Span R (2013) Effects of thermal pretreatment on anaerobic digestion of *Nannochloropsis salina* biomass. Bioresour Technol 143:505–511

Shin SG, Han G, Lim J, Lee C, Hwang S (2010) A comprehensive microbial insight into two-stage anaerobic digestion of food waste-recycling wastewater. Water Res 44:4838–4849

Show KY, Lee DJ, Tay JH, Lee TM, Chang JS (2015) Microalgal drying and cell disruption e recent advances. Bioresour Technol 184:258–266

Sialve B, Bernet N, Bernard O (2009) Anaerobic digestion of microalgae as a necessary step to make microalgae biodiesel sustainable. Biotechnol Adv 27:409–416

Solé-Bundó M, Cucina M, Folch M, Tàpias J, Gigliotti G, Garfí M, Ferrer I (2017) Assessing the agricultural reuse of the digestate from microalgae anaerobic digestion and codigestion with sewage sludge. Sci Total Environ 586:1–9

Solé-Bundó M, Garfí M, Matamoros V, Ferrer I (2019) Co-digestion of microalgae and primary sludge: effect on biogas production and microcontaminants removal. Sci Total Environ 660:974–981. https://doi.org/10.1016/j.scitotenv.2019.01.011

Sosnowski P, Klepacz-Smolka A, Kaczorek K, Ledakowicz S (2008) Kinetic investigations of methane co-fermentation of sewage sludge and organic fraction of municipal solid wastes. Bioresour Technol 99:5731–5737

Sundberg C, Al-Soud WA, Larsson M, Alm E, Yekta SS, Svensson BH, Sørensen SJ, Karlsson A (2013) 454 pyrosequencing analyses of bacterial and archaeal richness in 21 full-scale biogas digesters. FEMS Microbiol Ecol 85:612–626

Surendra KC, Takara D, Hashimoto AG, Khanal SK (2014) Biogas as a sustainable energy source for developing countries: opportunities and challenges. Renew Sust Energ Rev 31:846–859

Takeda H (1996) Cell wall sugars of some *Scenedesmus* species. Phytochemistry 42:673–675

Tartakovsky B, Matteau-Lebrun F, McGinn PJ, O'Leary SJB, Guiot SR (2013) Methane production from the microalga Scenedesmus sp. AMDD in a continuous anaerobic reactor. Algal Res 2(4):394–400. https://doi.org/10.1016/j.algal.2013.07.002

Tartakovsky B, Matteau F, Guiot SR (2015) High-rate biomethane production from microalgal biomass in a UASB reactor. Algal Res 7:86–91. https://doi.org/10.1016/j.algal.2014.12.004

Thauer RK (1998) Biochemistry of methanogenesis: a tribute to Marjory Stephenson: 1998 Marjory Stephenson prize lecture. Microbiol 144: 2377–2406

Thorin E, Olsson J, Schwede S, Nehrenheim E (2018) Co-digestion of sewage sludge and microalgae – biogas production investigations. Appl Energ 227:64–72

Tsapekos P, Kougias PG, Angelidaki I (2015) Biogas production from ensiled meadow grass; effect of mechanical pretreatments and rapid determination of substrate biodegradability via physicochemical methods. Bioresour Technol 182:329e335

Vaccari DA (2009) Phosphorus: a looming crisis. Sci Am 300:54–59

Wang S, Yan Q, Miao H, Ruan W (2009) Effect of inoculum to substrate ratios on methane production in mixed anaerobic digestion of pig manure and blue-green algae. Trans Chin Soc Agric Eng 25:172–176

Wang H, Tolvanen K, Jaakko P, Tolvanen K, Lehtomäki A, Puhakka J, Rintala J (2010) Microbial community structure in anaerobic co-digestion of grass silage and cow manure in a laboratory continuously stirred tank reactor. Biodegradation 21:135–146

Wang M, Sahu KA, Björn R, Park C (2013) Anaerobic co-digestion of microalgae *Chlorella* sp. and waste activated sludge. Bioresour Technol 152:585–590

Ward AJ, Lewis D, Green FB (2014) Anaerobic digestion of algae biomass: a review. Algal Res 5:204–214

Wieczorek N, Kucuker MA, Kuchta K (2015) Microalgae-bacteria flocs (MaBFlocs) as a substrate for fermentative biogas production. Bioresour Technol 194:130–136

Wiegant WM, Zeeman G (1986) The mechanism of ammonia inhibition in the thermophilic digestion of livestock wastes. Agric Wastes 16: 243–253

Yen HW, Brune DE (2007) Anaerobic co-digestion of algal sludge and waste paper to produce methane. Bioresour Technol 98:130–134

Yenigün O, Demirel B (2013) Ammonia inhibition in anaerobic digestion: a review. Process Biochem 48:901–911

Yuan X, Wang M, Park C, Sahu AK, Ergas SJ (2012) Microalgae growth using high-strength wastewater followed by anaerobic co-digestion water environment. Research 84(5):396–404. https://doi.org/10.2175/106143011X13233670703242

Zamalloa C, Vulsteke E, Albrecht J, Verstraete W (2011) The techno-economic potential of renewable energy through the anaerobic digestion of microalgae. Bioresour Technol 102:1149–1158

Zamalloa C, Vrieze JD, Boon N, Verstraete W (2012) Anaerobic digestibility of marine microalgae *Phaeodactylum tricornutum* in a lab-scale anaerobic membrane bioreactor. Bioenerg Biofuels 93: 859–869

Zeeman G, Wiegant WM, Koster-Treffers ME, Lettinga G (1985) The influence of the total-ammonia concentration on the thermophilic digestion of cow manure. Agric Wastes 14:19–35

Zhong W, Zhang Z, Luo Y, Qiao W, Xiao M, Zhang M (2012) Biogas productivity by co-digesting Taihu blue algae with corn straw as an external carbon source. Bioresour Technol 114:281–286

Ziganshin AM, Liebetrau J, Proter J, Kleinsteuber S (2013) Microbial community structure and dynamics during anaerobic digestion of various agricultural waste materials. Appl Microbiol Biotechnol 97: 5161–5174

Microalgal Biomass-Derived Biochars 16

Eduardo Fuentes-Quezada, Alfredo de Jesús Martínez-Roldán, and Diana Cristina Martínez-Casillas

Abstract

Microalgal biomass is a promising alternative that can be transformed into other value-added products such as biochar. In the last decade, the preparation of biochar from microalgal biomass has been on the increase because of their cost, availability, high production rate, and eco-friendly character viewed as a circular economy if the microalgal biomass came from a wastewater treatment process. Additionally, microalgal biomass has a high content of nitrogen, phosphorus, and oxygen, which remain in the biochars as functional groups, improving its properties. This chapter summarizes the recent research progress on the synthesis of biochar derived from different microalgal biomasses and its applications in environmental field as carbon sequestrant, pollutants adsorbent, and soil additive, as well as in energy storage as electrode material in electrochemical devices.

Keyword

Microalgal biomass · Carbonization · Biochar applications · Pollutants removal · Solid fuel

E. Fuentes-Quezada
Tecnológico Nacional de México/IT de Durango. Maestría en Sistemas Ambientales, Durango, Mexico

A. de Jesús Martínez-Roldán
Investigador Nacional Nivel I, Catedrático CONAHCYT - TecNM/ITD, Departamento de Ingenierías Química y Bioquímica, División de Estudios de Posgrado, Maestría en Sistemas Ambientales, Durango, Mexico
e-mail: adjmartinezro@conacyt.mx

D. C. Martínez-Casillas (✉)
Tecnológico Nacional de México/IT de Durango. Maestría en Sistemas Ambientales, Durango, Mexico

Escuela Nacional de Estudios Superiores Unidad Juriquilla, Universidad Nacional Autónoma de México, Queretaro, Santiago de Queretaro, Mexico
e-mail: d.martinez@unam.mx

16.1 Introduction

Continuous world population growth leads to high energy consumption, and environmental issues arise from human activities. Today, the majority of energy generation systems are based on the combustion of fossil fuels, resulting in an increase in the levels of greenhouse gases in the atmosphere and depletion of fossil fuels. Moreover, a combination of various energy storage devices is required for an efficient exploitation of renewable energy sources. On the other hand, adsorption technique is one of the most used to remove water pollutants, and the main material used has been activated carbon. Therefore, materials to substitute fossil energy resources and/or to be used in energetic and environmental applications are needed.

Biochar is defined as a carbon-rich material generated after thermochemical conversion of biomass between 150 °C and 1000 °C, with or without oxygen (Lehmann and Joseph 2015; Ok et al. 2019; Thapar Kapoor et al. 2021). Usually, biochar contains 50–90% fixed carbon, 0–40% volatiles, 1–15% moisture, and 0.5–5% ash (Brewer and Brown 2012). The properties of biochar are mainly determined by the type of biomass used, the parameters of the carbonization process, and posttreatments. Therefore, the selection of biomass is one of the key steps in biochar production. Nonetheless, the precursor and carbonization process of biochar will have a variety of potential applications (Hyland and Sarmah 2014; Ok et al. 2019; Thapar Kapoor et al. 2021). For example, biochar can be used as an alternative solid fuel trough combustion; also due to its adsorption capacity, biochar can be considered in the removal of organic and mineral pollutants; furthermore, biochar is a source of nutrients and can be added to soils with environmental and agronomic benefits.

Biomass is any renewable organic material derived from plants and animals, as well as wastes generated from their transformations. Biomass can be classified depending on its nature or composition in two groups: lignocellulosic such as

A. de Jesús Martínez-Roldán (ed.), *Biotechnological Processes for Green Energy, and High Value Bioproducts by Microalgae, and Cyanobacteria Cultures*, Developments in Applied Phycology 13, https://doi.org/10.1007/978-3-031-43969-8_16

wood, plants, agricultural crops residues, and forestry residues; and non-lignocellulosic such as algae, livestock wastes, and food supply chain residues (food and farm residues). Biomass is a biopolymer made up by proteins, carbohydrates, cellulose, hemicellulose, lignin, lipids, starches, extractives, water, ash, and other trace components (Vassilev et al. 2010). Among the variety of biomasses used as feedstock in thermochemical technologies for biochar production, microalgal biomass is an attractive and novel source, which has been gained attention in recent years. The great potential of microalgal biomass as feedstock is because of its high growth rates (plentiful mass generation in matter of days), which lead to abundant availability, low production cost, high photosynthetic efficiency compared to terrestrial biomass, efficient CO_2 fixation, high lipid, nitrogen, phosphorus, and oxygen content, and environmentally friendly nature (Kumar et al. 2017; Yu et al. 2017). Microalgal biomass mainly contains proteins, carbohydrates, lipids, water, ash, and little or no lignin. These microalgal characteristics make them suitable as biochar precursor. Moreover, the main advantages of microalgal biomass are simple nutrient requirement such as CO_2 and sunlight, non-potable water requirement, non-arable land cultivation, non-herbicidal/pesticidal cultivation, carbon-neutral feedstock usage, and high-quality production of carbonized materials.

In this chapter, recent developments and ongoing trends in research of microalgal biomass-derived biochars (MB) are reviewed. First, characteristics and advantages of microalgal biomass are summarized, then carbonization by thermochemical processes are presented, and finally, some applications of MB are discussed (Fig. 16.1).

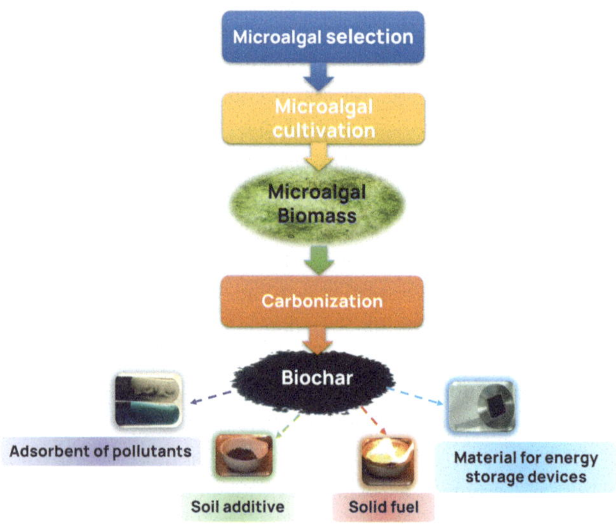

Fig. 16.1 Obtention process and applications of microalgal biomass-derived biochars

16.2 Microalgal Biomass Precursors

Microalgae are a group of highly diverse organisms, that have the ability to carry out photosynthesis, i.e., to growth they use the light as energy source and carbon dioxide as food source (Barsanti et al. 2008). Moreover, microalgae have the capacity to grow in a wide ranging of temperatures, and can be found in soil, and water bodies (Barsanti and Gualtieri 2006). In recent years, due to its large number of uses the development of microalgae's biotechnology has been motivated (Martínez-Roldán and Ibarra-Berumen 2019). Today a massive cultivation is possible using photobioreactors that are technological devices designed to meet the specific needs of the microorganism to be cultivated (Martínez-Roldán and Cañizares-Villanueva 2015). The possibility that microalgae can be cultivated in an easy way and that their nutritional requirements also are also simple, make them an excellent renewable resource for obtaining various high added value products such as pigments, fatty acids, proteins, and even compounds to be used as biofuels (Binda et al. 2020; Cherubini 2010).

Among the diversity of this microbial group with a highly specific metabolic characteristics, the most abundant and widely distributed show a lot of characteristics like the superior plants (Barsanti et al. 2008). Most eukaryotic microalgae have a cellular wall similar to that of plants, composed of polysaccharides like cellulose (Fu et al. 2010; Shokrkar et al. 2022). Regarding reserve substances, in normal conditions starch is the most common and sometime polysaccharides of similar structure are stored. These carbohydrates represent the majority fraction of the biomass when the culture is in its exponential phase of growth (Brányiková et al. 2011; Ran et al. 2019). Meanwhile, under nutrient limitations, some genus tend to accumulate large amounts of glycerol (Shetty et al. 2019), another kind tend to storage triacyl glycerol made up of long-chain (12–20 carbon atoms) fatty acids in a varied composition dependent on each microalgae specie (Arias-Peñaranda et al. 2013). Nonetheless, reserve substance accumulated the amount of carbon contained in the biomass remains high, usually around 40–50% wt. (Cordoba-Perez and de Lasa 2021; Raheem et al. 2015).

In this sense, microalgal biomass represents a raw material to produce carbonaceous materials through treatments similar to those applied to plant biomass, which have diverse applications (Nie et al. 2023; Wang et al. 2023). In addition, it is possible to carry out chemical modifications that change the structure and surface charge of the particulate material, increasing the potential of this type of biomass as raw material for the generation of biochars (Wang et al. 2023). Furthermore, the microalgal biomass used can be the residue from the production process of some high-value product such as pigments or fatty acids. Additionally, using residual microalgal biomass a value-added by-product that can contribute to the financial balance of process is generated (Kim et al. 2015; Sharma et al. 2022). Another advantage of produce biochar from microalgae, is that some biomass could no be used for any use, due to its growth conditions, such as biomass obtained using wastewater as a source of nutrients (Law et al. 2022; Yu et al. 2017).

Recently, biochar has been proposed as one of the high added value products that can be obtained from microalgae and cyanobacteria, derived from the fact that is necessary the generation of integral process without waste (Shokrkar et al. 2022; Yen et al. 2013). Moreover, in the last years, the number of publications about biochars derived from microalgal biomass has been increased. Nonetheless, most of the studies are related to the microalgal biomass conversion, and just few studies are devoted to the production and applicability of the produced biochars. Most of papers studied the effects of thermochemical process conditions in the yield and characteristic of biochar. Furthermore, several microalgae species have been used to produce biochar, such as *Arthrospira platensis* (Conti et al. 2016; Yao et al. 2016), *Chaetocerous muelleri* (Grierson et al. 2009), *Chlamydomonas sp.* (Chen et al. 2016), *Chlamydomonas reinhardtii* (Heilmann et al. 2010; Torri et al. 2011), *Chlorella like* (Grierson et al. 2009; Bolognesi et al. 2021), *Chlorella minutissima* (De Bhowmick et al. 2022), *Chlorella sp.* (Amin and Chetpattananondh 2019; Ashokkumar et al. 2019; Yang et al. 2021), *Chlorella vulgaris* (Grierson et al. 2009; Wang et al. 2013; Lee et al. 2017; Liberale et al. 2017; Yu et al. 2018; Binda et al. 2020; Viegas et al. 2021), *Chlorella sorokiniana* (Chen et al. 2014; Jaiswal et al. 2021), *Cladophora sp.* (Maddi et al. 2011), *Cyanidioschyzon merolae* (Dandamudi et al. 2021), *Desmodesmus communis* (Conti et al. 2016), *Dunaliella salina* (Heilmann et al. 2010), *Dunaliella tertiolecta* (Grierson et al. 2009), *Haematococcus lacustris* (Ashokkumar et al. 2021), *Isochrysis sp.* (Aysu et al. 2016), *Lyngbya sp.* (Maddi et al. 2011), *Nannochloropsis sp.* (Aysu and Sanna 2015; Lu et al. 2015; Binda et al. 2020), *Nannochloropsis oculata* (Levine et al. 2013), *Oscillatoria* (Kawale and Kishore 2019), *Pavlova sp.* (Aysu et al. 2017), *Scenedemus dimorphus* (Bordoloi et al. 2017), *Spirogyra sp.* (Chaiwong et al. 2012), *Spirulina platensis* (Nautiyal et al. 2016), *Scenedemus obliquus* (Viegas et al. 2021) *Spirulina sp.* (Heilmann et al. 2010; Leng et al. 2015; Binda et al. 2020; Choi et al. 2020; Yang et al. 2021), *Synechococcus* (Grierson et al. 2009), *Synechocystis spp.* (Heilmann et al. 2010), *Tetraselmis chui* (Grierson et al. 2009), and *Tetraselmis sp.* (Aysu et al. 2016).

16.3 Microalgal Biomass-Derived Biochars Production

Biochar is the solid product obtained from the thermochemical conversion of biomass; it can be produced from lignocellulosic and non-lignocellulosic biomasses. The structural components of biomass go through depolymerization, crosslinking, and fragmentation during carbonization, and the biomass gets transformed into biochar along with liquid and gas products. In most of thermochemical processes, temperature plays an important role over products' yields. Some of these processes are combustion, gasification, hydrothermal carbonization, torrefaction, and pyrolysis. During the carbonization, the moisture and the volatile matter of biomass are removed, while its structural components, i.e., proteins, cellulose, hemicellulose, lignin, carbohydrates, and lipids, are decomposed. Thus, the remaining biochar often displays properties different from the pristine biomass materials. This final biochar has demonstrated potential uses in various fields, including air pollution control, wastewater treatment, catalyst support, hydrogen storage, and energy conversion and storage. The characteristics, properties, and consequently the potential uses of biochar mainly depend on the original biomass and thermochemical conversion process. Thus, it is necessary to find the most suitable biomass processing strategy, which could include post-modification of biochar to increase its specific surface area (SSA) and pore fraction, or to form functional groups in it. Some post-modification processes are physical/chemical activation and surface functionalization. In the case of biochar derived from microalgal biomass, often only activation strategies are necessary, because the biochars produced have oxygen and nitrogen functional groups on their surface.

16.3.1 Thermochemical Processes for Microalgal Biomass Carbonization

Before processing, biomass is especially important to fully understand the carbonization method and mechanism. At present, microalgal biomass-derived biochars (MB) have been produced by hydrothermal carbonization, pyrolysis, and torrefaction (Heilmann et al. 2010; Chen et al. 2015; Yu et al. 2017; Gan et al. 2018; Yang et al. 2019). These techniques are presented in Fig. 16.2 with their main characteristics. The major difference between them is the operation temperature range at which each method proceeds.

Table 16.1 summarizes thermochemical processes used to produce MB from a broad range of microalgal biomasses, as well as the biochar's yield and operation conditions (temperature and residence time) are also included. Studies about each thermochemical conversion process will be discussed in

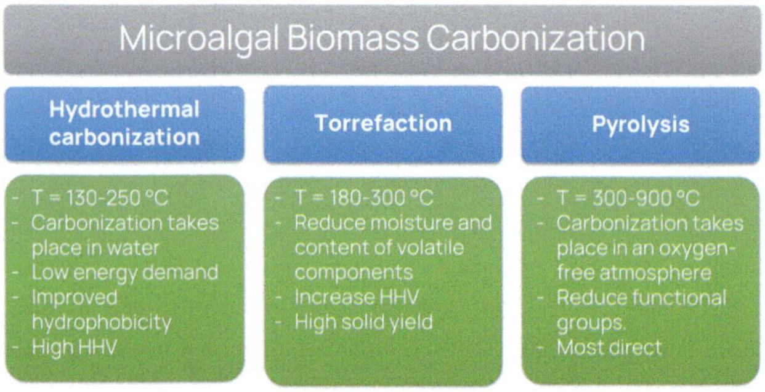

Fig. 16.2 Thermochemical processes for microalgal carbonization

the following sections in order to get a better understanding about factors such as microalgae specie, operation temperature, residence time, and heating rate, affecting microalgal biomass-derived biochar production and properties.

16.3.2 Torrefaction

Torrefaction is an emerging thermochemical conversion process in which moderate temperature conditions between 200 °C and 300 °C in an inert atmosphere are applied. It is an effective process to remove moisture and reduce the content of the volatile components from biomass, increasing its higher heat value (HHV) and carbon content and promoting the biochar production (Chen et al. 2015). Thermochemical conversion of microalgal biomass using torrefaction has been reported in the last decade, including biochar's applicability as solid fuel, biofertilizer, and adsorbent (Chen et al. 2014, 2016; Bach et al. 2017; Gan et al. 2018; Viegas et al. 2021).

Chen et al. (2014) investigated the torrefaction of *Scenedemus obliquus* CNW-N at five different temperatures and a residence time of 60 min. The MB's yields are 86.37%, 82.74%,76.45%, 67.19%, and 63.23% at 200, 225, 250, 275, and 300 °C, respectively. As can be seen, a higher torrefaction temperature leads to a lower solid yield, evidencing that increasing torrefaction temperature affects the conversion of microalgal biomass. This is due to higher degradation of carbohydrate and protein occurs at high torrefaction temperature (>250 °C) (Gan et al. 2018). From elemental analysis of the MB obtained at 200 °C, the weight percentages of C, H, N, and O are 39.26, 5.47, 7.38, and 23.19%, respectively, whereas for the MB obtained at 300 °C, the contents of C, H, N, and O are 36.93, 3.63, 6.53, and 28.21%, respectively. Thus, increasing torrefaction temperature decreases the relative contents of C, H, and N, whereas the relative O content increases.

Torrefaction of *Chlamydomonas* sp. was performed at five different temperatures (200 °C, 225 °C, 250 °C, 275 °C, and 300 °C) and three residence times (15, 30, and 60 min) to study microalgal biomass degradation under such conditions (Chen et al. 2016). Briefly, this study revealed that increasing torrefaction severity, i.e., increasing operation factors such as temperature or residence time, leads to lower solid yield and produces decarbonization, dehydrogenation, and deoxygenation. Furthermore, they demonstrate the impact of torrefaction on the decomposition of microalgal biomass and surface properties of MB produced by FTIR analyses.

Viegas et al. (2021) reported that torrefaction of *Chlorella vulgaris* and *Scenedesmus obliquus* at 250 °C for 60 min with a heating rate of 14 °C min^{-1} resulted in biochar yields of 73.7% and 70%, respectively. These yields are in good agreement with those reported in the previous studies described above under the same thermochemical conditions, indicating that the selection of microalgal biomass slightly affects the solid yield (>60%) in torrefaction (see Table 16.1). Figure 16.3 shows the MB's yields for torrefaction of different microalgal biomasses under different conditions. As can be seen, regardless of the microalgal specie biomass used in torrefaction process, performing this process at low temperature, short residence time, and low heating rate can result in a higher yield of the MB produced. Furthermore, as torrefaction severity increases, the C, H, and O content of MB decreases.

16.3.3 Hydrothermal Carbonization

Hydrothermal carbonization (HTC), also known as wet torrefaction, is an alternative and promising thermochemical process to produce biochar from microalgal biomass because of its low cost, mild synthesis conditions, and environmental friendliness. This process occurs at a low temperature of 130–250 °C in water under self-generated pressure (Titirici et al. 2012). HTC produces a solid product, usually named hydrochar, that has high calorific value, improved hydrophobicity due to oxygenated groups on its surface, and lower ash content. Likewise, torrefaction temperature and residence time are the main factors that affect the solid yield (see Table 16.1).

Heilmann et al. (2010) investigated the HTC of *Chlamydomonas reinhardtii*, *Dunaliella salina*, *Spirulina* sp., and *Synechocystis* spp. under a variety of experimental conditions such as temperature, residence time, and percentage of solid levels. Nonetheless, just one MB obtained from each specie at specific conditions was reported. Solid yields were 39%, 36%, 27%, and 18% for *Chlamydomonas reinhardtii*, *Dunaliella salina*, *Spirulina* sp., and *Synechocystis* spp., respectively. Higher solid yields (>30%) were obtained at 200 °C, whereas lower solid yields were obtained at 213 °C; evidencing that increasing HTC temperature decreases the solid yield. Moreover, the MB obtained from each specie at specific conditions is compared with the microalgal biomass used as precursor. For all the

Table 16.1 Microalgal biomass-derived biochar yields from different thermochemical process of several microalgae species at different conditions

Type of process	Temperature °C	Residence time minutes	Microalgal biomass	Biochar yield %	Reference
Hydrothermal carbonization (HTC)	190	120	Arthrospira platensis	37	Yao et al. (2016)
		180		34	
		240		32	
	200	120		31	
		180		29	
		240		27	
	210	120		27	
		180		24	
		240		21	
	200	120	Chlamydomonas reinhardtii	39	Heilmann et al. (2010)
	330	30	*Cyanidioschyzon merolae*	32	Dandamudi et al. (2021)
	160	10	*Chorella vulgaris ESP-31*	62	Bach et al. (2017)
	170	5		63	
		10		60	
		30		52	
	180	10		54	
	200	120	*Chorella vulgaris*	50	Chu et al. (2021)
	260			45	
	200	180	*Dunaliella salina*	36	Heilmann et al. (2010)
	200	120	*Microcystis sp.*	59	Chu et al. (2021)
	260			55	
	180	15	*Nannochloropsis sp.*	47	Lu et al. (2015)
		30		42	
	200	15		40	
		30		39	
	220	15		36	
		30		30	
	180	15	*Nannochloropsis oculate*	51	Levine et al. (2013)
		30		49	
	190	15		51	
		30		47	
	200	15		47	
		30		44	
	210	15		45	
		30		41	
	213	180	*Spirulina sp.*	27	Heilmann et al. (2010)
	213	120	*Synechocystis spp.*	18	Heilmann et al. (2010)
Pyrolysis	500	20	*Arthrospira platensis*	25	Conti et al. (2016)
	500	n.a.	*Chaetocerous muelleri*	53	Grierson et al. (2009)
	350	20	*Chlamydomonas reinhardtii*	44	Torri et al. (2011)
	500	n.a.	*Chlorella like*	34	Grierson et al. (2009)
	400	n.a.	*Chlorella sp.*	41	Ashokkumar et al. (2019)
	200	120		57	Yang et al. (2021)
	400			43	
	600			30	
	500	30	*Chlorella vulgaris*	27	Yu et al. (2018)
	350	60		35	Binda et al. (2020)
	500	n.a.		37	Grierson et al. (2009)
	600	20	*Cladophora sp.*	28	Maddi et al. (2011)
	500	20	*Desmodesmus communis*	34	Conti et al. (2016)
	500	n.a.	*Dunaliella tertiolecta*	63	Grierson et al. (2009)

(continued)

Table 16.1 (continued)

Type of process	Temperature °C	Residence time minutes	Microalgal biomass	Biochar yield %	Reference
	500	60	*Isochrysis sp.*	42	Ashokkumar et al. (2021)
	530	20	*Lyngbya sp.*	22	Maddi et al. (2011)
	600			17	
	650			13	
	400	60	*Nannochloropsis sp.*	34	Aysu and Sanna (2015)
	500			30	
	600			25	
	350	60		43	Binda et al. (2020)
	450	60	*Pavlova sp.*	49	Aysu et al. (2017)
	500			41	
	650			39	
	350	60	*Spirulina sp.*	36	Binda et al. (2020)
	200	120		53	Yang et al. (2021)
	500			38	
	600			28	
	710	n.a.	*Synechococcus*	44	Grierson et al. (2009)
	710	n.a.	*Tetraselmis chui*	37	Grierson et al. (2009)
	500	60	*Tetraselmis sp.*	20	Aysu et al. (2016)
Torrefaction	200	30	*Chlamydomonas sp.*	94	Chen et al. (2016)
	250	15		93	
		30		82	
		60		64	
	300	30		51	
	250	60	*Chlorella vulgaris*	74	Viegas et al. (2021)
			Scenedemus obliquus	70	
	200	60	*Scenedemus obliquus* CNW-N	87	Chen et al. (2014)
	225			83	
	250			77	
	275			67	
	300			63	

n.a. not available

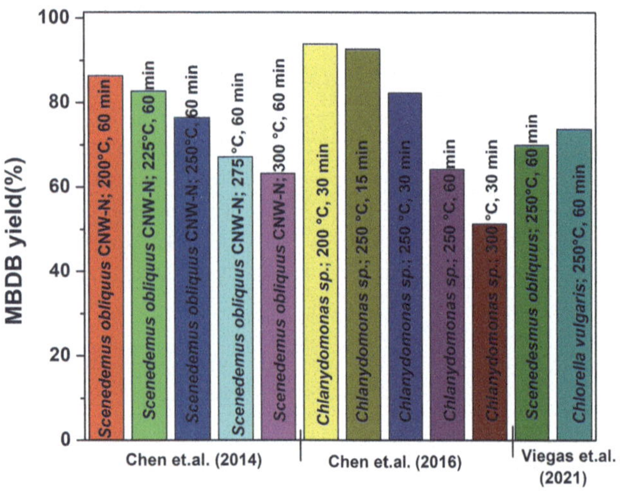

Fig. 16.3 MB yields obtained under different conditions by torrefaction of different microalgal species reported in the literature

microalgae species, hydrothermal carbonization increases the relative contents of C and H, whereas the relative N content decreases.

Levine et al. (2013) reported the production of MB from *Nannochloropsis oculata* by HTC at 180 °C, 190 °C, 200 °C, and 210 °C for 15, 30, and 45 min. The biochar yields obtained are in the range of 41–51% (see Table 16.1). The solid yield tends to decrease with increasing temperature and residence times, and this is due to an increased hydrolysis of biomass constituents (mainly carbohydrates, proteins, and lipids). On the other hand, the content of C in the MB increases as temperature and residence time increase.

Lu et al. (2015) examined the hydrothermal carbonization of *Nannochloropsis* sp. at temperatures ranging from 180 °C to 220 °C and residence times of 15 and 30 min. The biochar yields obtained were between 30% and 47%, and these results are similar to those previously observed for other microalgae species (see Table 16.1) in the studies mentioned above. Another interesting study was about the HTC of *Arthrospira platensis* under three temperatures (190 °C, 200 °C, and 210 °C) and residence times (2, 3, and 4 h), with MB's yields between 22% and 37% were reported by Yao et al. (2016). They found that increasing temperature or residence time leads to lower solid yield. Bach et al. (2017) used *Chlorella vulgaris ESP-31* for thermochemical conversion at three different temperatures (160 °C, 170 °C, and 180 °C) and residence times (5, 10, and 30 min). The results show that the solid yield and carbohydrate content decrease with increasing temperature and/or residence time where a wide range of biochar yields are reported.

Chu et al. (2021) investigated the HTC of *Chlorella vulgaris* and *Microcystis* sp. at 200 °C and 260 °C using two different solvents (deionized water and 1%wt. citric acid solution). They found that lower MB yields were obtained with increasing HTC temperature regardless of the solvent used. For example, using the citric acid solution, with the increase of temperature to 60 °C during HTC, MB yields decrease 4.3% and 3.6% for *Chlorella vulgaris* and *Microcystis* sp., respectively. Additionally, using deionized water as solvent, the same effect of reducing MB's yield with increasing temperature was observed for both microalgae species, as reported by previous HTC studies.

In general, the microalgal biomass-derived biochar yield tends to decrease as HTC's temperature and residence times increase (see Fig. 16.4a). This is in good agreement with the previous torrefaction studies mentioned above. On the other hand, as can be observed in Fig. 16.4b, the higher heat value (HHV) increases with increasing temperature. However, as can be seen in Table 16.1 and Fig. 16.4, considerable variation across microalgal species is observed in relation to MB's production by hydrothermal carbonization. Therefore, the inherent composition of biomass (proteins, carbohydrates, lipids, extractives, and minerals) also influences the biochar yields obtained by hydrothermal carbonization.

16.3.4 Pyrolysis

Pyrolysis is the oldest and most used thermochemical process for biochar production. This method proceeds in a wide range of temperatures varying from 300 °C to 900 °C in an oxygen-free atmosphere. It is derived from the original method of burning charcoal and is considered to be the most direct method for producing biochars. There are several types of pyrolysis based on their operating conditions such as conventional slow and fast pyrolysis; advanced pyrolysis methods have been also developed such as microwave-assisted pyrolysis and catalytic pyrolysis (Yang et al. 2019). Nonetheless, biochar production is significantly favored at moderate temperatures, slower heating rates, and longer residence time, which is representative of slow pyrolysis also known as conventional pyrolysis. As this process is the most widely used thermochemical conversion process, several pyrolysis studies have been performed on microalgal biomass (see Table 16.1).

The pyrolysis of *Lyngbya* sp. was investigated by Maddi et al. (2011) at three different temperatures. The MB's yields were 22%, 17%, and 13% at 530 °C, 600 °C, and 650 °C, respectively. As can be seen, pyrolysis at lower temperatures

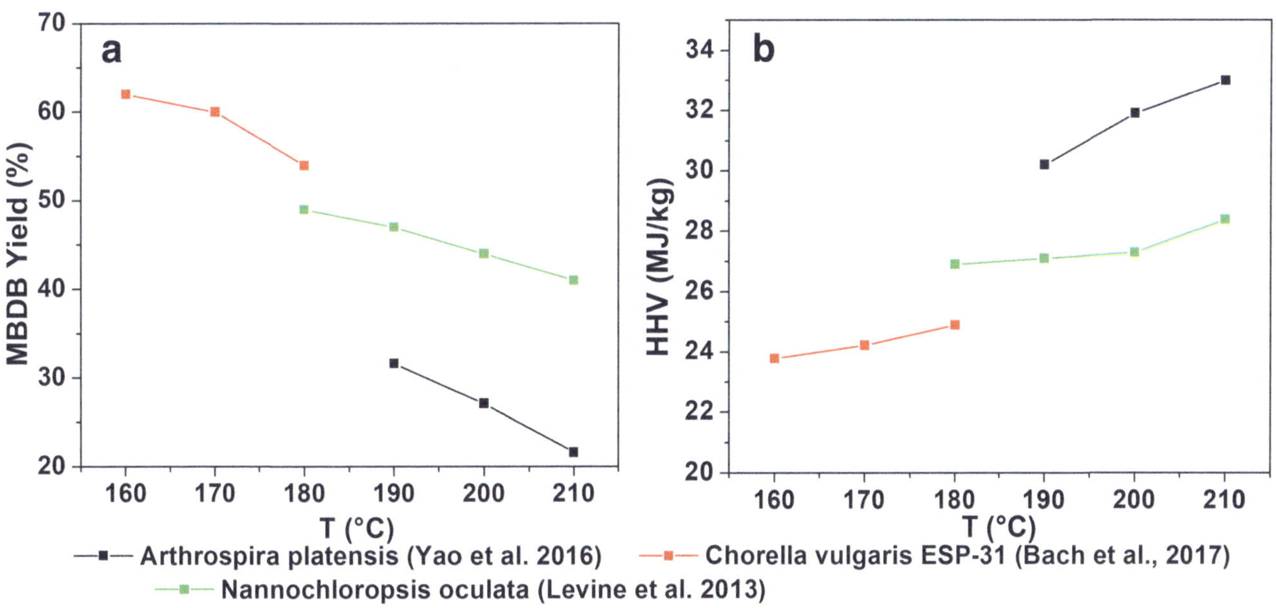

Fig. 16.4 (**a**) Yields and (**b**) HHV of microalgal biomass-derived biochars obtained at different temperatures

resulted in higher biochar's yield. Aysu and Sanna (2015) studied the effect of pyrolysis temperature on conversion of *Nannochloropsis* sp. into MB. Biochar' yields were 34%, 30%, and 25% at 400 °C, 500 °C, and 600 °C, respectively, indicating that biochar's yield tends to decrease as pyrolysis temperature increases. The pyrolysis of Pavlova at three different temperatures was reported by Aysu et al. (2017). Their MB's yields were 48.96%, 40.90%, and 38.97% at 450 °C, 500 °C, and 650 °C, confirming a higher biochar production at lower temperature. Yang et al. (2021) studied the effect of pyrolysis temperature on conversion of *Chlorella* sp. and *Spirulina* sp. into MB. For both microalgal species, the same trend in biochar yield with the pyrolysis temperature was observed, i.e., a decrease with increasing temperature (see Table 16.1).

Biochar yields obtained from these studies confirmed that temperature is the key parameter on the biochar production by pyrolysis. Additionally, similar to hydrothermal carbonization, significant variation across microalgal species in relation to biochar production is observed (Table 16.1) and reported in previous studies (Grierson et al. 2009; Maddi et al. 2011; Conti et al. 2016; Yang et al. 2021).

16.3.5 Activation

Activation is the most common method to increase the specific surface area (SSA) and adjust the pore structure of carbonaceous materials by opening the original occlusion holes, expanding original pores, and/or forming new ones. Moreover, activation could also modify the surface functional groups on the biochar. The obtained material after this process is known as activated carbon. Depending on the activation mechanism, the activation methods can be divided into physical activation and chemical activation. In physical activation, the biochar is exposed to the flow of an oxidizing gas at above 700 °C, generally are used oxidizing agents such as CO_2, H_2O steam, or mixture. On the other hand, chemical activation involves the addition of activators such as NaOH, KOH, K_2CO_3, H_3PO_4, H_2O_2, and $ZnCl_2$ during carbonization (in situ) or added to produced biochar (Lee et al. 2017; Anto et al. 2021; De Bhowmick et al. 2022).

Studies on activation of biochar produced from microalgal biomass are very limited in the literature, and recently is being investigated the effect of activation on MB. In this sense, microalgal biomass-derived activated carbon (MAC) was produced by pyrolysis of *Chlorella vulgaris* at 500 °C for 2 h and post-chemical activation using KOH as activating agent at 800 °C in a N_2 atmosphere (Lee et al. 2017). They found that higher SSA and pore volume are obtained at higher KOH concentration. De Bhowmick et al. (2022) investigated the production of MAC from the pyrolysis of *Chlorella minutissima* at 500 °C for 60 min and post-chemical activation using hydrogen peroxide. Results indicated that activation method increases the specific surface area of biochar 3.7 times and, additionally, help to increase the oxygen and nitrogen containing surface functional groups of the biochar. Another interesting study on the activation and doping of microalgal biomass-derived biochar was reported by González-Hourcade et al. (2022). They performed the activation with phosphoric acid and doping with melamine of a microalgae polyculture of *Coelastrum* ssp., *Desmodesmus* ssp., *Scenedesmus* ssp., and *Chlorella* ssp. before pyrolysis at 600 °C for 1 h in a nitrogen atmosphere. Results indicate an increase of 33.6% in the specific surface area related to the doping process with melamine as nitrogen source. Thus, an activation treatment to MB could result in a material with improved properties and performance for any application.

In general, biochar yield depends on the rate limiting factors such as temperature and heating rate whatever thermochemical conversion method being used. Temperature plays a major role in determining biochar quantity and quality in all thermochemical process. A few properties of biochar that are widely influenced by carbonization temperature are product yield, elemental composition, i.e., carbon, hydrogen, nitrogen, sulfur, and oxygen (CHNSO) contents, and higher heating value (HHV). In this regard, Fig. 16.5a shows MB's yields vs atomic H/C for microalgal biomass-derived biochars obtained at temperatures ranging from 190 °C to 750 °C by hydrothermal carbonization, pyrolysis, and torrefaction. There is a strong correlation between solid product yield and atomic H/C ratio, indicating that both properties tend to decrease with increasing temperature regardless of thermochemical process used. Higher MB yields are obtained by torrefaction, while MB produced by HTC exhibits a wide MB yield from 18% to 63%.

In addition to the MB's elemental composition, Fig. 16.5b presents a van Krevelen diagram to compare atomic H/C and O/C ratios of microalgal biomass-derived biochars produced at temperatures ranging from 190 °C to 750 °C by the thermochemical processes presented above. Atomic H/C and O/C ratios in biochar tend to lower with increasing process temperature due to decarbonization, dehydrogenation, and deoxygenation of microalgal biomass (Chen et al. 2016; Bach et al. 2017; Choi et al. 2020; Viegas et al. 2021). Furthermore, Fig. 16.6 shows the higher heating value versus carbon content of microalgal biomass-derived biochars produced by HTC, pyrolysis, and torrefaction. As can be seen, there is a strong correlation between MB's carbon content and HHV, as the carbon content is directly proportional to HHV, whatever the thermochemical process used. The carbon content and HHV of MB are ranging from 11% to 73% and from 6 MJ kg^{-1} to 33 MJ kg^{-1}, respectively. However, biochars derived from microalgal biomass produced by hydrothermal carbonization exhibit the highest higher

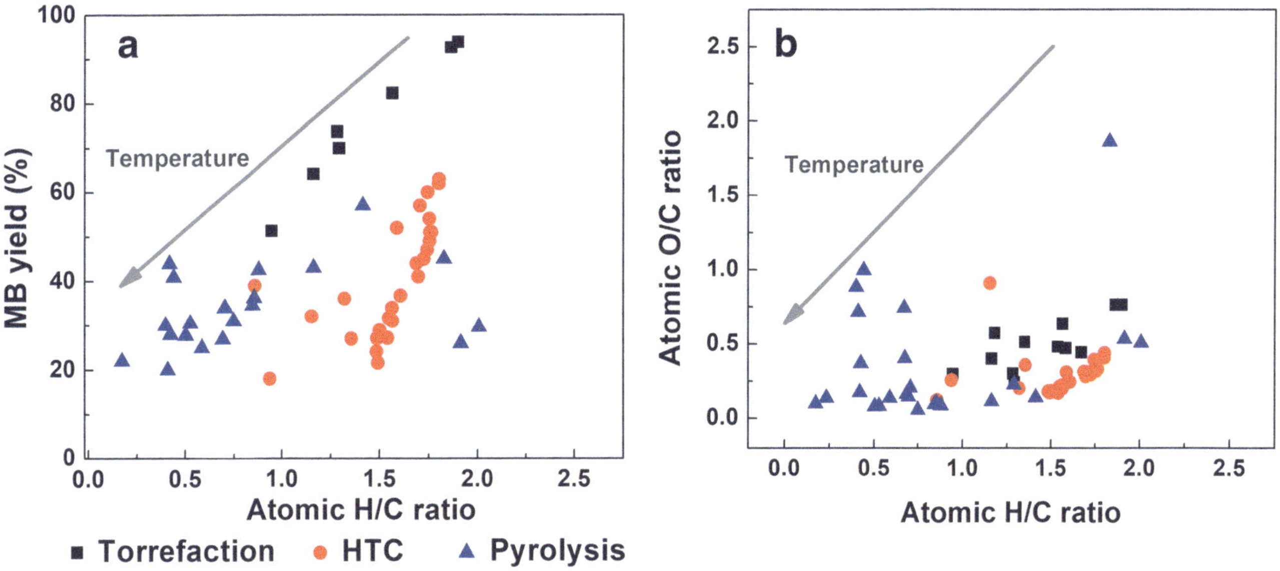

Fig. 16.5 MB yields vs atomic H/C ratio (**a**) and Van Krevelen plot (**b**) for microalgal biomass-derived biochars produced at different temperatures

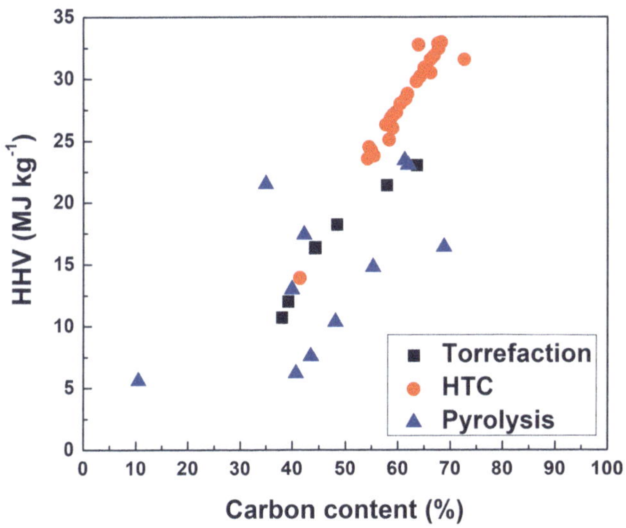

Fig. 16.6 HHV vs carbon content of microalgal biomass-derived biochars produced by different thermochemical processes

heating values and carbon content (red circles in Fig. 16.6). This effect on the HHV will be analyzed later in the application section since HHV is an important parameter to consider these MB as solid fuel.

16.4 Applications of Microalgal Biomass-Derived Biochars

As mentioned earlier, biochar has demonstrated potential in several fields such as soil remediation, wastewater treatment, catalyst, and energy conversion and storage. The efficiency of biochar in most of the applications in these fields significantly depends on biochar's properties such as carbon content, specific surface area, pore size distribution, elemental composition, alkalinity, and ion-exchange capacity. These properties are subject to variations depending on the biochar feedstock and production parameters such as temperature, heating rate, residence time, oxidation medium, and post-processing treatments.

Biochars derived from microalgal biomass are characterized by special pH and surface charge, porous structure, nutrient content, plentiful surface functional groups, and inorganic elements. These characteristics make MB suitable to be used in environmental applications such as carbon sequestration, biofertilizer, and adsorbent of emerging pollutants and dyes (Torri et al. 2011; Chaiwong et al. 2012; Leng et al. 2015; Nautiyal et al. 2016; Bordoloi et al. 2017; Zheng et al. 2017; Yu et al. 2018; Choi et al. 2020; Tan et al. 2020; Chu et al. 2021; Yang et al. 2021; Jaiswal et al. 2021; De Bhowmick et al. 2022). Nonetheless, MB had also been investigated in energy storage and conversion field (Lee et al. 2017; Liberale et al. 2017; Lee et al. 2017). Table 16.2 summarizes the studies about the usage microalgal biomass-derived biochars in several applications. This section would mainly focus on the application in energy, soil amendment, and water treatment of biochars derived from microalgal biomass.

Table. 16.2 Properties and application of biochars derived from microalgal biomass

Microalgal biomass	MB production	Elemental analysis wt%					HHV	Textural properties			Application	Reference
		C	H	N	S	O	MJ kg^{-1}	SSA $m^2\ g^{-1}$	V_P $cm^3\ g^{-1}$	$Size_P$ nm		
Chlamydomonas sp.	Pyr. T: 600 °C; Rt: 0.5 h	80	2	7		7		15.03	0.001	152.4	Adsorption of p-nitrofenols	Zheng et al. (2017)
Chlamydomonas reinhardtii	Pyr. T: 350 °C; HR: 20 min	40	1	5	<0.1	9	13				Biofertilizer	Torri et al. (2011)
Chlorella minutissima	Pyr. T: 500 °C; Rt: 1h; HR: 10 °C min^{-1}; Activated with H_2O_2	47	3	4	0.3	46		14.36	0.015	14.6	Adsorption of metformin	De Bhowmick et al. (2022)
Chlorella sp.	Pyr. T: 600 °C; Rt: 0.5 h	60	2	6		11		6.16	<0.001	151.38	Adsorption of p-nitrofenols	Zheng et al. (2017)
	Pyr. T: 450 °C; Rt: 1h	11	2	4	1	26	6	266	0.61		Adsorption of Cr (VI), Zn(II), and Ni (II)	Amin and Chetpattananondh (2019)
	Pyr. T: 400 °C; HR: 20 °C min^{-1}	35	9	3	2	38	22				Hematite iron ore reduction	Ashokkumar et al. (2019)
	Pyr. T: 600 °C; Rt: 2h	62	3	3		7		4.05	0.07	45.6	Adsorption of Pb (II)	Yang et al. (2021)
Chlorella vulgaris	Pyr. T: 500 °C; Rt: 0.5 h; HR: 10 °C min^{-1}	61	4	10		12	23				Solid fuel and carbon sequestration	Yu et al. (2018)
	Torr. T: 250 °C; Rt: 1h; HR: 14 °C min^{-1}	39	4	6	1	16	12				Adsorption of methylene blue	Viegas et al. (2021)
	HTC T: 260 °C; Rt: 2 h; HR: 3 °C min^{-1}; 1 wt% citric acid solution	59	4	6.2	0.8	30		6.4	0.04		Biofertilizer	Chu et al. (2021)
	Pyr. T: 500 °C; Rt: 2h; Activation with KOH							1785	1.084	2.4	Electrode in supercapacitor	Lee et al. (2017)
Chlorella sorokiniana	HTC; T: 400 °C; Rt: 6h; HR: 5 °C min^{-1}	10	0.4	1	2	87		14.76			Adsorption of heavy metals	Jaiswal et al. (2021)
Coelastrum sp.	Pyr. T: 600 °C; Rt: 0.5 h	64	2	5		9		2.12	0.006	15.3	Adsorption of p-nitrofenols	Zheng et al. (2017)
Haematococcus lacustris	Pyr. T: 450 °C										Hematite iron ore reduction	Ashokkumar et al. (2019)
Microcystis sp.	HTC T: 260 °C; Rt: 2 h; HR: 3 °C min^{-1}; 1 wt% citric acid solution	61	3	8	0.6	27		6.1	0.03		Biofertilizer	Chu et al. (2021)
Scenedemus dimorphus	Pyr. T: 500 °C; HR: 40 °C min^{-1}	50	8	36		36					Adsorption of Co (II)	Bordoloi et al. (2017)
Scenedemus obliquus	Torr. T: 250 °C; Rt: 1h; HR: 14 °C min^{-1}	38	4	6	1.2	12	10.7				Biofertilizer and adsorption of methylene blue	Viegas et al. (2021)

Spirogyra spp.	Pyr. T: 550 °C; Rt: 1h	62	0.4	2	0.5	4	22.96				Fuel, Carbon sequestration	Chaiwong et al. (2012)
Spirulina platensis	Pyr. T: 450 °C; Rt: 2h; HR: 20 °C min^{-1}	63	4	7	0.37	34					Adsorption of dye	Nautiyal et al. (2016)
Spirulina sp.	Pyr. T: 550 °C; Rt: 1h	45	1	3	0.07	0.28	15.78				Fuel, Carbon sequestration	Chaiwong et al. (2012)
	Torr.; T: 350°C; Rt: 20 min	70	4	9		16		1.56	0.014	32.4	Adsorption of methylene blue	Leng et al. (2015)
	Pyr. T: 750 °C; Rt: 2h	67	1	9		12		2.63			Adsorption of tetracycline	Choi et al. (2020)
	Pyr. T: 700 °C; Rt: 1.5h							133.19	0.20		Adsorption of dyes	Tan et al. (2020)
	Pyr. T: 600 °C; Rt: 2h	65	3	5		7		7.89	0.07	53.7	Adsorption of Pb (II)	Yang et al. (2021)
Polyculture of *Coelastrum ssp.*, *Desmodesmus ssp.*, *Scenedesmus ssp.*, *and Chlorella ssp.*	Pyr. T: 600 °C; Rt: 1h; HR: 10 °C min^{-1}; Activation with H_3PO_4.							324	0.25		Adsorption of acetaminophen	González-Hourcade et al. (2022)

Note: Pyr.-Pyrolysis; HTC-Hydrothermal carbonization; Torr.-Torrefaction; T-Temperature; Rt-Residence time; HR-Heating rate; HHV-Higher heating value; SSA-Specific surface area; V_P-Pore volume; $Size_P$-Pore size

16.4.1 Microalgal Biomass-Derived Biochars in Energy

Continuous growth of energy consumption, climate change, and depletion of fossil fuels leads society to move toward sustainable and renewable sources. This promotes ongoing research in the fields of fossil fuel substitutes and renewable energy sources. Regarding biochars, including those derived from microalgal biomass, they can be used in energy as a solid fuel (Chaiwong et al. 2012; Yu et al. 2018) to substitute fossil energy resources, particularly coal. In order to determine if an MB could be a potential fuel, it is necessary to investigate its fuel properties. There are some key indicators to analyze the fuel properties of MB such as higher heating value (HHV), energy yield, and H/C and O/C atomic ratios. As mentioned previously, these MB's properties depend on the carbonization method and parameters.

The HHV is the amount of heat released when one unit of fuel is burned. Higher heating values of microalgal biomass-derived biochars produced by the thermochemical process described in the previous section are also reported in Table 16.2. HHV of MB is different for every microalgal biomass used as biochar's precursor, and it is ranged between 6 MJ kg^{-1} and 23 MJ kg^{-1} (see Table 16.2). Furthermore, Fig. 16.7 presents the HHV vs H/C ratio of the MB produced from different microalgal biomass by hydrothermal carbonization, torrefaction, and pyrolysis. As can be seen in Fig. 16.7a, the MB produced by HTC can reach the range of coal fuels around 25–35 MJ kg^{-1} (United Nations and Statistical Division 2018), indicating that thermochemical

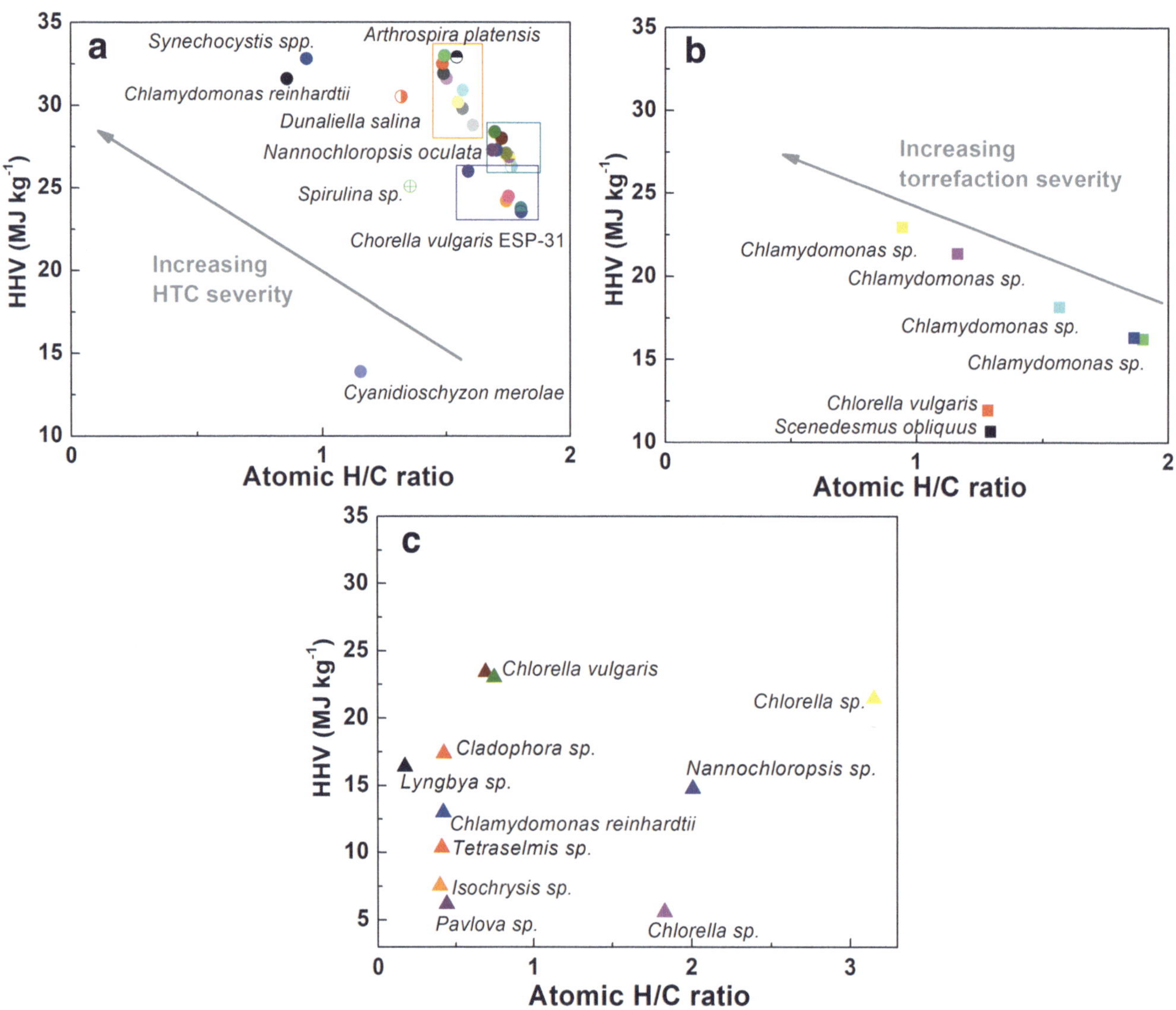

Fig. 16.7 HHV vs atomic H/C ratio of microalgal biomass-derived biochars produced by hydrothermal carbonization (**a**), torrefaction (**b**), and pyrolysis (**c**) processes

conversion of microalgal biomass by this process is a promising method for producing biochar with high energy density. The HHV of microalgal biomass-derived biochar increases with increasing process severity, while the H/C ratio of microalgal biomass-derived biochar decreases (see Fig. 16.7a and b). The H/C ratio decreases with increasing process temperature and duration time, which is because of increase in carbon elements. This indicates that better fuel properties of MB are achieved by carbonizing the microalgal biomass at higher temperatures and longer residence time.

On the other hand, the exploitation of renewable energies such as solar radiation, wind, and ocean wave is limited by their intermittent operation. Therefore, an efficient energy storage system is a major issue for harvesting the energy produced by these renewable sources. Biochar can also be used in energy technologies such as batteries and supercapacitors (SC), which are the most suitable electrochemical energy conversion devices (Shen et al. 2018; Escobar et al. 2021). SC are systems that store and deliver energy at very high power via electrosorption to enable high discharge rates, while batteries rely on more sluggish redox process such as intercalation or conversion reactions (Simon et al. 2014; Escobar et al. 2021). Most batteries have a higher specific energy and lower specific power compared to SC, and these characteristics make them suitable for mobile applications (Simon et al. 2014; Escobar et al. 2021). Whereas SC are utilized for portable and/or wearable electronics, miniaturized electronic devices, hybrid and/or electric vehicles, industrial power management, and renewable energy management (Béguin and Frąckowiak 2013; Simon et al. 2014). Batteries and SC have the same basic components: two electrodes, an electrolyte, and an ion permeable separator. The electrode material and electrolyte determine the energetic features of these electrochemical devices due to the energy storage mechanism takes place between the electrode and the electrolyte. Therefore, electrode material determines the intrinsic characteristics and performance of such energy storage devices. Biochars have emerged as very promising electrode materials because of their properties, which include good chemical stability and electrical conductivity, good specific surface areas and highly porous surface, availability, and low to moderate cost (Pandolfo and Hollenkamp 2006).

In this regard, the use of microalgal biomass-derived biochar as electrode material in energy conversion devices has been reported recently. MB derived from *Chlorella vulgaris* was tested as anode material in sodium-ion batteries (Liberale et al. 2017). Although its coulombic efficiency approached 100% for most cycles, its initial specific capacity of around 150 mAh g^{-1} was 60% lower than the performance of a synthetic hard carbon evaluated under the same conditions. Nonetheless, this study demonstrates that the use of low-cost MB as active materials for Na-ion battery electrodes is feasible, which could minimize the environmental impact and production costs of such devices. Another study using *Chlorella vulgaris* as precursor for electrode material was reported by Lee et al. (2017). They prepared activated carbon with a high specific surface area of 1785 m^2 g^{-1} by carbonizing and activating the microalgal biomass with KOH. Then, the obtained material was tested as electrode material in a supercapacitor, showing a high specific capacitance of 117 F g^{-1}, and this value is in the order of that reported for the activated carbons commonly used in SC. This study further reveals that the specific capacitance increases with the improvement of nitrogen level, indicating that nitrogen-rich microalgal biomass is a good material to use in the production of MB with high excellent performance as electrodes in supercapacitors.

16.4.2 Microalgal Biomass-Derived Biochars in Soil Amendment

The use of biochar onto soils is not a new concept; in fact, this was the first application of biochar. Several studies about the role of biochar in improving the physicochemical and biological properties of soil and increasing crop productivity have been carried out. Nevertheless, biochar composition and properties and, therefore, its performance as a soil amendment are highly dependent on the feedstock and thermochemical process conditions. Biochar has a porous structure, which gives it the ability to retain moisture and nutrients in soil. Additionally, biochars with high specific surface area can be used as a remediation tool in soil to adsorb organic and/or inorganic contaminants (Hyland and Sarmah 2014).

Since microalgal biomass-derived biochars are enriched in nitrogen, they could be used as a slow nitrogen-releasing source to soil. In this regard, MB produced from *Chlamydomonas reinhardtii* showed a nitrogen content of 5%, which is higher than those values reported for biochar derived from lignocellulosic biomass (Torri et al. 2011). Therefore, from an agronomical point of view, these materials have a great potential to be used as biofertilizer; moreover, some interesting results have been reported for MB in this application lately. Viegas et al. (2021) reported the fertilizing effect of an MB produced by torrefaction of *Scenedesmus obliquus* at 250 °C in germination tests of wheat and watercress seeds. MB had a positive effect on seed germination, it shows a significant increase in the germination index of 45% in watercress and 35% in wheat. They associated this good MB's performance as biofertilizer with its content of minerals and micronutrients such as calcium, magnesium, potassium, phosphorus, boron, copper, iron, manganese, sodium, and zinc. In another study on wheat fertilization with biochars derived from two different microalgae produced by hydrothermal carbonization at

260 °C (Chu et al. 2021), they reported gains in wheat yield grain of 14.5% and 21.6% for *Chlorella vulgaris* and *Microcystis* sp. derived biochars, respectively. These results could be related to the high N and P content of the obtained microalgal biomass-derived biochars.

16.4.3 Microalgal Biomass-Derived Biochars in Water Treatment

The major issue that the world faces today is the scarcity and contamination of water. The most frequent application of biochar is as a pollutant adsorbent in water treatment because of its special characteristics such as high stability, excellent surface properties, specific surface area, pore structure, and surface functional groups. Biochar has been used for wastewater treatment as low-cost and environmentally friendly adsorbent of dyes, heavy metals, and pharmaceuticals with high efficiency (Thapar Kapoor et al. 2021). As can be seen in Table 16.2 also biochars derived from microalgal biomass have been widely applied as adsorbent due to their unique properties. The adsorption capacity of biochar is closely related to physicochemical properties such as specific surface area, pore structure, and chemical composition. For example, high specific surface area and developed pore structure can provide more adsorption sites, whereas oxygen-containing groups and mineral components can improve the adsorption effect (Thapar Kapoor et al. 2021).

Dye industry has been rapidly expanded due to their wide consumption in several industries such as tanneries, food, cosmetic, textile, and medicine. Consequently, pollution of water by the discharging of dye-contaminated wastewater has escalated, causing destruction of the aquatic systems and hazards for human life. Therefore, eradication of dye molecules from wastewater is needed. Regarding microalgal biomass-derived biochars, some investigations have highlighted their use as competitive dye adsorbents removal from effluents (see Table 16.2). Leng et al. (2015) reported the adsorption of methylene blue onto MB produced from *Spirulina* sp. by liquefaction at 350 °C. They demonstrated that the maximum adsorption capacity of 128.6 mg g^{-1} found for the MB exceeded the values of other pyrolytic biochars and is comparable to the commercial activated carbon adsorbent reported in the literature, despite its lower specific surface area (1.56 m^2 g^{-1}). Another study on methylene blue remotion from aqueous solutions using MB produced by torrefaction at 250 °C of *Chlorella vulgaris* and *Scenedemus obliquus* found removals of 56.5% and 62.9% were reported, respectively (Viegas et al. 2021). Nautiyal et al. (2016) investigated the adsorption of Congo red dye onto MB from *Spirulina platensis*; they found a high removal of 82.6%, which is close to the 85.4% capacity adsorption of commercial activated carbon.

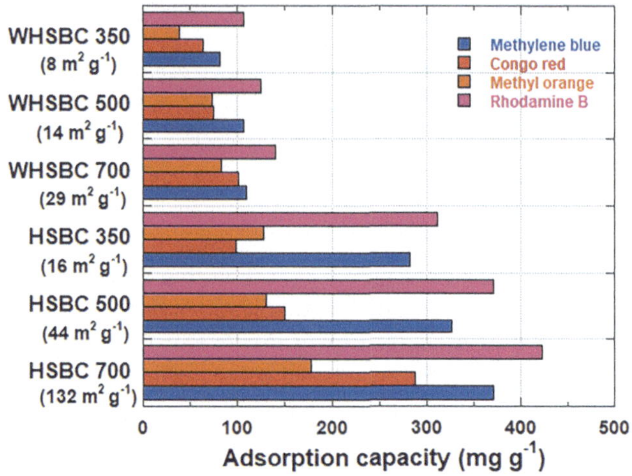

Fig. 16.8 Maximum adsorption capacities for different dyes onto MB

Recently, Tan et al. (2020) investigated the capacity adsorption of ionized dyes such as rhodamine B, methylene blue, Congo red, and methyl orange by six microalgal biomass-derived biochars produced by pyrolysis of *Spirulina* sp. under different conditions. Figure 16.8 shows the maximum adsorption capacity of each dye by sample, with specific surface areas are also indicated. The highest values were achieved by the MB with the highest SSA for the four dyes. As can be observed, here is a direct correlation between adsorption capacity and SSA, indicating that increasing specific surface area of MB increases its adsorption capacity regardless of the dye studied. Moreover, a difference in adsorption capacities between WHBCs and HSBCs is also seen, which is related to the content of inorganic minerals.

Another interesting example of using microalgal biomass-derived biochar for water treatment is like adsorbents of heavy metals such as Cd, Co, Cu, Cr, Ni, Pb, and Zn (see Table 16.2). These elements are of great concern due to their toxic and persistent nature in the environment. Usually, they are produced from various industrial operations such as mining, smelting, and electroplating. Bordoloi et al. (2017) investigated the copper (Co) removal efficiency of MB produced by pyrolysis at 500 °C of *Scenedemus dimorphus*. Despite the low adsorption capacity, their study shown that the obtained MB is an effective adsorbent for the removal of Co(II) from aqueous solutions. Amin and Chetpattananondh (2019) evaluated the adsorption capacity of heavy metals using an MB obtained from *Chlorella* sp. by pyrolysis at 450 °C. The removal was 45%, 95%, and 98% for Cr(IV), Zn(II), and Ni (II), respectively; these high removal efficiencies could be attributed to the high MB's specific surface area of 266 m^2 g^{-1}. Another study on Pb (II) removal by MB produced by pyrolysis of *Chlorella* sp. and *Spirulina* sp. at different temperatures was recently performed by Yang et al. (2021). The MB exhibited that adsorption capacities for Pb(II) are affected by the pyrolysis temperature. The adsorption capacity was as high as 131.41 mg g^{-1} for the MB derived from *Chlorella* sp. and 154.56 mg g^{-1} for the MB derived from *Spirulina* sp. pyrolyzed at 600 °C. In order to verify the feasibility of the practical application of microalgal biomass-derived biochars, the MB prepared from *Chlorella sorokiniana* was used to remove metal ions (Cu, Ni, Pb, and Zn) from a real municipal wastewater (Jaiswal et al. 2021). The result shows high removal efficiencies of 97–98% in 240 min for all the metal ions present.

Lately, a new class of compounds named "emerging pollutants" whose effects on the environment and human health are unknown have been found in wastewater and aquatic environments (Deblonde et al. 2011). These emerging pollutants include a wide variety of pharmaceutical and personal care products. In this regard, Zheng et al. (2017) investigated the adsorption of *p*-nitrophenols on MB derived from three different microalgal species by pyrolysis at 600 ° C. The adsorption capacities for the MB derived from *Chlamydomonas* sp., *Chlorella* sp., and *Coelastrum* sp. were 105.9 mg g^{-1}, 203.5 mg g^{-1}, and 109.9 mg g^{-1}, respectively. Choi et al. (2020) evaluated the adsorption of tetracycline using three biochars derived from *Spirulina* sp. biomass produced by pyrolysis at 350 °C, 550 °C, and 750 °C. Among the tetracycline adsorption capacity of MB ranged from 14.9 mg g^{-1} to 132.8 mg g^{-1}, the highest adsorption capacity was observed in the MB produced at the highest pyrolysis temperature. This result is similar to the one obtained in the study on the adsorption of heavy metals (Yang et al. 2021), suggesting that higher pyrolysis temperature leads to higher adsorption capacity regardless of the adsorbate and microalgae species used. Another interesting investigation using MB derived from *Chlorella minutissima* for metformin adsorption reported a removal of 62% (De Bhowmick et al. 2022). González-Hourcade et al. (2022) reported the adsorption of acetaminophen using a nitrogen doped MB produced from a microalgal polyculture, which include *Coelastrum* ssp., *Desmodesmus* ssp., *Scenedesmus* ssp., and *Chlorella* ssp. by pyrolysis at 600 ° C. The maximum adsorption capacity of acetaminophen was 120.7 mg g^{-1}, which was related to improved surface area and surface functionalities due to the nitrogen doping process.

Among all the studies described above, demonstrate the good performance and high potential of MB as adsorbents of dyes, heavy metals, and pharmaceutical pollutants in wastewater treatment. Adsorption evaluation protocols for each type of pollutant must be developed in order to be able to compare all the studies about adsorption on the microalgal biomass-derived biochars produced by different methods under a variety of operational conditions.

References

Amin M, Chetpattananondh P (2019) Biochar from extracted marine chlorella sp. residue for high efficiency adsorption with ultrasonication to remove Cr(VI), Zn(II) and Ni(II). Bioresour Technol 289:121578. https://doi.org/10.1016/j.biortech.2019.121578

Anto S, Sudhakar MP, Shan Ahamed T et al (2021) Activation strategies for biochar to use as an efficient catalyst in various applications. Fuel 285:119205. https://doi.org/10.1016/j.fuel.2020.119205

Arias-Peñaranda MT, Martínez-Roldán, AJ, Cañizares-Villanueva RO (2013) Biodiesel production from microalgae: cultivation parameters that affect lipid production. Acta Biologica Colombiana 18:43–68

Ashokkumar V, Chen W-H, Kamyab H et al (2019) Cultivation of microalgae chlorella sp. in municipal sewage for biofuel production and utilization of biochar derived from residue for the conversion of hematite iron ore (Fe2O3) to iron (Fe)–integrated algal biorefinery. Energy 189:116128. https://doi.org/10.1016/j.energy.2019.116128

Ashokkumar V, Chen W-H, Kumar G et al (2021) A biorefinery approach for high value-added bioproduct (astaxanthin) from alga Haematococcus sp. and residue pyrolysis for biochar synthesis and metallic iron production from hematite (Fe2O3). Fuel 304:121150. https://doi.org/10.1016/j.fuel.2021.121150

Aysu T, Abd Rahman NA, Sanna A (2016) Catalytic pyrolysis of Tetraselmis and Isochrysis microalgae by nickel ceria based catalysts for hydrocarbon production. Energy 103:205–214. https://doi.org/10.1016/j.energy.2016.02.055

Aysu T, Ola O, Maroto-Valer MM, Sanna A (2017) Effects of titania based catalysts on in-situ pyrolysis of pavlova microalgae. Fuel Process Technol 166:291–298. https://doi.org/10.1016/j.fuproc.2017.05.001

Aysu T, Sanna A (2015) Nannochloropsis algae pyrolysis with ceria-based catalysts for production of high-quality bio-oils. Bioresour Technol 194:108–116. https://doi.org/10.1016/j.biortech.2015.07.027

Bach Q-V, Chen W-H, Lin S-C et al (2017) Wet torrefaction of microalga Chlorella vulgaris ESP-31 with microwave-assisted heating. Energy Convers Manag 141:163–170. https://doi.org/10.1016/j.enconman.2016.07.035

Barsanti L, Gualtieri P (2006) General overview. In Barsanti L, Gualtieri P (eds) Algae. Anatomy, biochemistry, and niotechnology, 1st edn. Taylor & Francis

Barsanti L, Coltelli P, Evangelista V, Frassanito AM, Passarelli V, Vesentini N, Gualtieri P (2008) The world of algae. In Algal toxins: nature, occurrence, effect and detection. NATO science for peace and security series A: chemistry and biology. Springer Netherlands, Dordrecht, pp 1–15

Béguin F, Frąckowiak E (2013) Supercapacitors: materials, systems, and applications, 1st edn. Wiley

Binda G, Spanu D, Bettinetti R et al (2020) Comprehensive comparison of microalgae-derived biochar from different feedstocks: a prospective study for future environmental applications. Algal Res 52:102103. https://doi.org/10.1016/j.algal.2020.102103

Bolognesi S, Bernardi G, Callegari A et al (2021) Biochar production from sewage sludge and microalgae mixtures: properties, sustainability and possible role in circular economy. Biomass Conv Bioref 11:289–299. https://doi.org/10.1007/s13399-019-00572-5

Bordoloi N, Goswami R, Kumar M, Kataki R (2017) Biosorption of co (II) from aqueous solution using algal biochar: kinetics and isotherm studies. Bioresour Technol 244:1465–1469. https://doi.org/10.1016/j.biortech.2017.05.139

Brányiková I, Maršálková B, Doucha J, Brányik T, Bišová K, Zachleder V, Vítová M (2011) Microalgae-novel highly efficient starch producers. Biotechnol Bioeng. 108:766–776

Brewer CE, Brown RC (2012) Biochar. In: Comprehensive renewable energy. Elsevier, pp 357–384

Chaiwong K, Kiatsiriroat T, Vorayos N, Thararax C (2012) Biochar production from freshwater algae by slow pyrolysis. Maejo Int J Sci Technol 6:186–195

Chen W-H, Peng J, Bi XT (2015) A state-of-the-art review of biomass torrefaction, densification and applications. Renew Sust Energ Rev 44:847–866. https://doi.org/10.1016/j.rser.2014.12.039

Chen W-H, Wu Z-Y, Chang J-S (2014) Isothermal and non-isothermal torrefaction characteristics and kinetics of microalga Scenedesmus obliquus CNW-N. Bioresour Technol 155:245–251. https://doi.org/10.1016/j.biortech.2013.12.116

Chen Y-C, Chen W-H, Lin B-J et al (2016) Impact of torrefaction on the composition, structure and reactivity of a microalga residue. Appl Energy 181:110–119. https://doi.org/10.1016/j.apenergy.2016.07.130

Cherubini F (2010) The biorefinery concept: using biomass instead of oil for producing energy and chemicals. Energy Convers Manage 51:1412–1421

Choi Y-K, Choi T-R, Gurav R et al (2020) Adsorption behavior of tetracycline onto spirulina sp. (microalgae)-derived biochars produced at different temperatures. Sci Total Environ 710:136282. https://doi.org/10.1016/j.scitotenv.2019.136282

Chu Q, Lyu T, Xue L et al (2021) Hydrothermal carbonization of microalgae for phosphorus recycling from wastewater to crop-soil systems as slow-release fertilizers. J Clean Prod 283:124627. https://doi.org/10.1016/j.jclepro.2020.124627

Conti R, Fabbri D, Vassura I, Ferroni L (2016) Comparison of chemical and physical indices of thermal stability of biochars from different biomass by analytical pyrolysis and thermogravimetry. J Anal Appl Pyrolysis 122:160–168. https://doi.org/10.1016/j.jaap.2016.10.003

Cordoba-Perez M, de Lasa H (2021) CO2-derived carbon capture using microalgae and sodium bicarbonate in a photobioCREC unit: Kinetic modeling. Processes 9. https://doi.org/10.3390/pr9081296

Dandamudi KPR, Murdock T, Lammers PJ et al (2021) Production of functionalized carbon from synergistic hydrothermal liquefaction of microalgae and swine manure. Resour Conserv Recycl 170:105564. https://doi.org/10.1016/j.resconrec.2021.105564

De Bhowmick G, Briones RM, Thiele-Bruhn S et al (2022) Adsorptive removal of metformin on specially designed algae-lignocellulosic biochar mix and techno-economic feasibility assessment. Environ Pollut 292:118256. https://doi.org/10.1016/j.envpol.2021.118256

Deblonde T, Cossu-Leguille C, Hartemann P (2011) Emerging pollutants in wastewater: a review of the literature. Int J Hyg Environ Health 214:442–448. https://doi.org/10.1016/j.ijheh.2011.08.002

Escobar B, Martínez-Casillas DC, Pérez-Salcedo KY et al (2021) Research progress on biomass-derived carbon electrode materials for electrochemical energy storage and conversion technologies. Int J Hydrog Energy 46:26053–26073. https://doi.org/10.1016/j.ijhydene.2021.02.017

Fu CC, Hung TC, Chen JY, Su CH, Wu WT (2010) Hydrolysis of microalgae cell walls for production of reducing sugar and lipid extraction. Bioresour Technol 101:8750–8754

Gan YY, Ong HC, Show PL et al (2018) Torrefaction of microalgal biochar as potential coal fuel and application as bio-adsorbent. Energy Convers Manag 165:152–162. https://doi.org/10.1016/j.enconman.2018.03.046

González-Hourcade M, Simões dos Reis G, Grimm A et al (2022) Microalgae biomass as a sustainable precursor to produce nitrogen-doped biochar for efficient removal of emerging pollutants from aqueous media. J Clean Prod 348:131280. https://doi.org/10.1016/j.jclepro.2022.131280

Grierson S, Strezov V, Ellem G et al (2009) Thermal characterisation of microalgae under slow pyrolysis conditions. J Anal Appl Pyrolysis 1–2:118–123. https://doi.org/10.1016/j.jaap.2008.10.003

Heilmann SM, Davis HT, Jader LR et al (2010) Hydrothermal carbonization of microalgae. Biomass Bioenergy 34:875–882. https://doi.org/10.1016/j.biombioe.2010.01.032

Hyland C, Sarmah AK (2014) Advances and innovations in biochar production and utilization for improving environmental quality. In: Bioenergy research: advances and applications. Elsevier, pp 435–446

Kim SS, Ly HV, Kim J, Lee EY, Woo HC (2015) Pyrolysis of microalgae residual biomass derived from Dunaliella tertiolecta after lipid extraction and carbohydrate saccharification. Chem Eng J 263:194–199

Law XN, Cheah WY, Chew KW, Ibrahim MF, Park YK, Ho SH, Show PL (2022) Microalgal-based biochar in wastewater remediation: Its synthesis, characterization and applications. Environ Res 204. https://doi.org/10.1016/j.envres.2021.111966

Lehmann J, Joseph S (eds) (2015) Biochar for environmental management: science, technology and implementation, 1st edn. ROUTLEDGE, New York

Jaiswal KK, Kumar V, Verma R et al (2021) Graphitic bio-char and bio-oil synthesis via hydrothermal carbonization-co-liquefaction of microalgae biomass (oiled/de-oiled) and multiple heavy metals remediations. J Hazard Mater 409:124987. https://doi.org/10.1016/j.jhazmat.2020.124987

Kawale HD, Kishore N (2019) Production of hydrocarbons from a green algae (Oscillatoria) with exploration of its fuel characteristics over different reaction atmospheres. Energy 178:344–355. https://doi.org/10.1016/j.energy.2019.04.103

Kumar G, Shobana S, Chen W-H et al (2017) A review of thermochemical conversion of microalgal biomass for biofuels: chemistry and processes. Green Chem 19:44–67. https://doi.org/10.1039/C6GC01937D

Lee KS, Park M, Park CW, Kim J-D (2017) Sustainable fabrication of nitrogen activated carbon from chlorella vulgaris for energy storage devices. Colloids Surf A Physicochem Eng Asp 529:102–106. https://doi.org/10.1016/j.colsurfa.2017.05.051

Leng L, Yuan X, Huang H et al (2015) Characterization and application of bio-chars from liquefaction of microalgae, lignocellulosic biomass and sewage sludge. Fuel Process Technol 129:8–14. https://doi.org/10.1016/j.fuproc.2014.08.016

Levine RB, Sierra COS, Hockstad R et al (2013) The use of hydrothermal carbonization to recycle nutrients in algal biofuel production. Environ Prog Sustain Energy 32:962–975. https://doi.org/10.1002/ep.11812

Liberale F, Dossi C, Bettinetti R et al (2017) Algae derived electrodes for rechargeable Na-ion batteries: materials characterization and electrochemical performances. ECS Trans 80:349–355. https://doi.org/10.1149/08010.0349ecst

Lu Y, Levine RB, Savage PE (2015) Fatty acids for nutraceuticals and biofuels from hydrothermal carbonization of microalgae. Ind Eng Chem Res 54:4066–4071. https://doi.org/10.1021/ie503448u

Maddi B, Viamajala S, Varanasi S (2011) Comparative study of pyrolysis of algal biomass from natural lake blooms with lignocellulosic biomass. Bioresour Technol 102:11018–11026. https://doi.org/10.1016/j.biortech.2011.09.055

Martínez-Roldán AJ, Cañizares-Villanueva RO (2015) Photobioreactors: Improving the biomass productivity. In Torres-Bustillos L (ed) Microalgae and other phototrophic bacteria: culture, processing, recovery and new products, 1st edn. Nova, pp 145–170

Martínez-Roldán AJ, Ibarra-Berumen J (2019) Employment of wastewater to produce microalgal biomass as a biorefinery concept. In Microalgae biotechnology for development of biofuel and wastewater treatment. Springer Singapore, Singapore, pp 487–504

Nautiyal P, Subramanian KA, Dastidar MG (2016) Adsorptive removal of dye using biochar derived from residual algae after in-situ transesterification: alternate use of waste of biodiesel industry. J

Environ Manag 182:187–197. https://doi.org/10.1016/j.jenvman.2016.07.063

Nie Y, Zhao C, Zhou Z, Kong Y, Ma J (2023) Hydrochloric acid-modified fungi-microalgae biochar for adsorption of tetracycline hydrochloride: Performance and mechanism. Bioresour Technol 383. https://doi.org/10.1016/j.biortech.2023.129224

Ok Y, Tsang DCW, Bolan N, Novak JM (eds) (2019) Biochar from biomass and waste: fundamentals and applications. Elsevier, Amsterdam, the Netherlands

Pandolfo AG, Hollenkamp AF (2006) Carbon properties and their role in supercapacitors. J Power Sources 157:11–27. https://doi.org/10.1016/j.jpowsour.2006.02.065

Raheem A, Sivasangar S, Wan Azlina WAKG, Taufiq Yap YH, Danquah MK, Harun R (2015) Thermogravimetric study of Chlorella vulgaris for syngas production. Algal Res 12:52–59

Ran W, Wang H, Liu Y, Qi M, Xiang Q, Yao C, Lan X (2019) Storage of starch and lipids in microalgae: Biosynthesis and manipulation by nutrients. Bioresour Technol 291. https://doi.org/10.1016/j.biortech.2019.121894

Sharma AK, Ghodke P, Sharma PK, Manna S, Pugazhendhi A, Matsakas L, Patel A (2022) Holistic utilization of Chlorella pyrenoidosa microalgae for extraction of renewable fuels and value-added biochar through in situ transesteri-fication and pyrolysis reaction process. Biomass Convers Biorefinery. https://doi.org/10.1007/s13399-022-02713-9

Shen PK, Wang C-Y, Jiang SP et al (eds) (2018) Electrochemical energy: advanced materials and technologies, 0th edn. CRC Press

Shetty P, Gitau MM, Maróti G (2019) Salinity stress responses and adaptation mechanisms in eukaryotic green microalgae. Cells 8. https://doi.org/10.3390/cells8121657

Shokrkar H, Zamani M, Ebrahimi S (2022) Exploring strategies for the use of mixed microalgae in cellulase production and its application for bioethanol production. Biofuels, Bio Bioprod Biorefin 16:816–825

Simon P, Gogotsi Y, Dunn B (2014) Where do batteries end and supercapacitors begin? Science 343:1210–1211. https://doi.org/10.1126/science.1249625

Tan X, Zhu S, Show PL et al (2020) Sorption of ionized dyes on high-salinity microalgal residue derived biochar: electron acceptor-donor and metal-organic bridging mechanisms. J Hazard Mater 393: 122435. https://doi.org/10.1016/j.jhazmat.2020.122435

Thapar Kapoor R, Treichel H, Shah MP (eds) (2021) Biochar and its application in bioremediation. Springer Singapore, Singapore

Titirici M-M, RobinJ W, Falco C, Sevilla M (2012) Black perspectives for a green future: hydrothermal carbons for environment protection and energy storage. Energy Environ Sci 5:6796. https://doi.org/10.1039/c2ee21166a

Torri C, Samorì C, Adamiano A et al (2011) Preliminary investigation on the production of fuels and bio-char from Chlamydomonas reinhardtii biomass residue after bio-hydrogen production. Bioresour Technol 102:8707–8713. https://doi.org/10.1016/j.biortech.2011.01.064

United Nations, Statistical Division (2018) International recommendations for energy statistics (IRES)

Vassilev SV, Baxter D, Andersen LK, Vassileva CG (2010) An overview of the chemical composition of biomass. Fuel 89:913–933. https://doi.org/10.1016/j.fuel.2009.10.022

Viegas C, Gouveia L, Gonçalves M (2021) Aquaculture wastewater treatment through microalgal. Biomass potential applications on animal feed, agriculture, and energy. J Environ Manag 286: 112187. https://doi.org/10.1016/j.jenvman.2021.112187

Wang K, Brown RC, Homsy S et al (2013) Fast pyrolysis of microalgae remnants in a fluidized bed reactor for bio-oil and biochar production. Bioresour Technol 127:494–499. https://doi.org/10.1016/j.biortech.2012.08.016

Wang C, Bi L, Liu J, Huang B, Wang F, Zhang Y, Song M (2023) Microalgae-derived carbon quantum dots mediated formation of metal sulfide nano-adsorbents with exceptional cadmium removal performance. J Colloid Interface Sci 629:994–1002

Yang C, Li R, Zhang B et al (2019) Pyrolysis of microalgae: a critical review. Fuel Process Technol 186:53–72. https://doi.org/10.1016/j.fuproc.2018.12.012

Yang Z, Hou J, Wu J, Miao L (2021) The effect of carbonization temperature on the capacity and mechanisms of Pb(II) adsorption by microalgae residue-derived biochar. Ecotoxicol Environ Saf 225: 112750. https://doi.org/10.1016/j.ecoenv.2021.112750

Yao C, Wu P, Pan Y et al (2016) Evaluation of the integrated hydrothermal carbonization-algal cultivation process for enhanced nitrogen utilization in Arthrospira platensis production. Bioresour Technol 216:381–390. https://doi.org/10.1016/j.biortech.2016.05.110

Yen H-W, Hu I-C, Chen C-Y, Ho S-H, Lee D-J (2013) Microalgae-based biorefinery – From biofuels to natural products. Bioresour Technol 135:166–174

Yu KL, Show PL, Ong HC et al (2018) Biochar production from microalgae cultivation through pyrolysis as a sustainable carbon sequestration and biorefinery approach. Clean Techn Environ Policy 20:2047–2055. https://doi.org/10.1007/s10098-018-1521-7

Yu KL, Show PL, Ong HC et al (2017) Microalgae from wastewater treatment to biochar – feedstock preparation and conversion technologies. Energy Convers Manag 150:1–13. https://doi.org/10.1016/j.enconman.2017.07.060

Zheng H, Guo W, Li S et al (2017) Adsorption of p-nitrophenols (PNP) on microalgal biochar: analysis of high adsorption capacity and mechanism. Bioresour Technol 244:1456–1464. https://doi.org/10.1016/j.biortech.2017.05.025

Process Integration via a Sustainable Biorefinery Approach Using Agro-industrial Residues and Photosynthetic Consortia

17

Pablo A. López-Pérez, Dulce J. Hernández-Melchor, Lizeth Vanessa Hernández Quijano, Mónica Ivette Sánchez Contreras, and R. Icela Beltrán-Hernández

Abstract

Integration of processes is proposed in this chapter to describe the production of bioenergy and value-added co-products via a biorefinery approach. Bioprocessing of biofuels is based on recovery of waste (pretreatment), transformation thermochemistry–biochemistry, and finally the combined heat and power (CHP), also known as cogeneration (production of electricity or mechanical power via useful thermal energy). CHP can use a variety of fuels, both fossil and renewable-based. The proposed technology for coproducing bioenergy and bioproducts increases the benefit and sustainability relative to fossil-driven products. This will allow optimizing the energy consumption of the process in the future while maintaining production without affecting the environment and society. Likewise, the implementation and specialized software tools for the simulation and calculations of process and environmental intensification indicators are described. A novel diagram was proposed that describes process integration via a sustainable biorefinery approach using wastes and agro-industrial residues. The above reveals that the generated methodology can be considered for control, multi-scale-up, and optimization of bioprocesses purposes.

Keywords

Bioprocessing · Environmental intensification · Process integration · Bioenergy · Optimization

P. A. López-Pérez (✉) · L. V. Hernández Quijano
Universidad Autónoma del Estado de Hidalgo, Escuela Superior Apan, Apan, Hgo, México
e-mail: pablo_lopez@uaeh.edu.mx

D. J. Hernández-Melchor
Colegio de Postgraduados, Campus Montecillo. Departamento de Edafología e Hidrociencias, Montecillo, Estado de México, México

M. I. Sánchez Contreras · R. I. Beltrán-Hernández
Centro de Investigaciones Químicas, Universidad Autónoma del Estado de Hidalgo, Pachuca, México

17.1 Sustainability and Circular Economy on Bioprocesses

Currently, the circular economy seeks to generate economic prosperity, takes care of the environment, and prevents pollution so that, consequently, it generated a sustainable development. This is why the circular economy aims to produce resistant products based on recycled or reused raw materials, thus maintaining a social responsibility. In addition, it seeks to highlight the role that engineering has in the construction of a sustainable society based on a circular economy using an undeniable connection relationship (Prieto-Sandoval et al. 2017; Stefanakis and Nikolaou 2021). Sustainability is based on three pillars: social, economic, and environmental. It is common to visualize these pillars as circles related to sustainability as an improved product, as well as their production methods in the future, will be increased.

One way to adapt these products to a sustainable and promising future is through mathematical descriptions of cells and microbial consortia to better understand them, especially for in silico research and cell models. Biological engineering is capable of accelerating the growth and construction of these cells and microbial consortia to generate higher yields, also having zero emissions and good resources from bioprocess engineering, thus generating sustainable technological systems (Straathof et al. 2019).

There are great advances in the field of microbial biotechnology in terms of understanding the machinery of natural systems applicable to fields of science. Within these advances, an important aspect to take into account is sustainability in crop production improvement processes due to the depletion of natural resources, environmental degradation, the degradation of pollutants, nanomaterials, and their toxicity, and food safety problems. To deal with them, we try to apply the concept of sustainability in each stage of bioprocesses (Soumia et al. 2021).

Sustainability: Sustainability implies the invariance of the system, in relation to the disturbances that affect it, the

A. de Jesús Martínez-Roldán (ed.), *Biotechnological Processes for Green Energy, and High Value Bioproducts by Microalgae, and Cyanobacteria Cultures*, Developments in Applied Phycology 13, https://doi.org/10.1007/978-3-031-43969-8_17

reversibility of the deviations that emerge, in turn, the development in the classical sense a change characterized by the simultaneous presence of properties of irreversibility, direction, and regularity. Sustainability is manifested as immutability, variability, sustainability, and sustainable development.

Circular economy: The circular economy is an economic concept interrelated with sustainability. Its objective is that the value of products, materials, and resources is maintained in the economy for as long as feasible, and that the generation of waste can be reduced as much as possible (Melgarejo 2019).

17.1.1 Integration of Bioprocesses

Process integration allows tools such as simulation, software sensor, intensification, control, modeling, and optimization to be combined in order to systematically analyze processes, unit operations, and their interactions. The steps in the proposed methodology consider a metabolic pathway, process, environmental, and social integration. Various biotechnological solutions have been proposed for batch operations within a process synthesis. After this advance, the challenge of integrating unit operations that allows establishing a continuous manufacturing system takes place. To this end, all the unit operations must be linked and a continuous flow process connected, giving rise to nonstop, continuously operating bioprocessing (López et al. 2013; López-Pérez et al. 2015, 2016a, b, Hernández-Melchor et al. 2017). These methods are tools of biotechnology and process engineering. Because of the above, researchers employ the following three approaches to accomplish bioprocess integration:

1. *Modular approach*: It is about the independent optimization of each module or unit operation to be able to integrate it with another module without undergoing modifications with the help of facilitating technology.
2. *Adaptation approach*: Unit operations are adapted so that they can be integrated.
3. *Merge approach*: In this case, unit operations are merged into one general step.

The proposed sustainable, effective, and economic bioprocesses for microalgal and photosynthetic microbial consortium (MC) products require the integration of microalgal molecular biology, including strain selection, isolation of microbial consortium, process, and reactor design. Furthermore, the integration into mass and energy fluxes for integration features into several degrees of integration with respect to the classical upstream, bioreaction, and downstream steps (Rathore et al. 2018; Hernández Touset et al. 2021).

Bioprocesses: Bioprocesses are an important and basic part of the food, chemical, and pharmaceutical industries. Bioprocesses handle microbial, animal, and plant cells as well as certain cell components such as enzymes. In short, a bioprocess is a process capable of generating an environment for increased or improved use of cellular material (López et al. 2018, 2020).

17.1.2 Indicator: Intensification of the Process

To achieve an intensification of processes (IP), it is necessary to start from the intensification of unitary operations. For example, sample pretreatment is one of the most energetically demanding stages; this is why the parameters of the process must be adjusted and intensified (Castillo-Saldarriaga and Velásquez-Lozano 2018). Process intensification is the conceptual or existing design or redesign of a process to optimize it, making it more efficient and more effective by design. This is why those processes designed to be sustainable manage to be synthesized and intensified at the same time at unit operational, functional, and/or phenomena levels that can be obtained by integration of unit operations for a set of target operations. We understand intensification as any process modification (improvements) realizing higher efficiency, lower expenses, environmentally sociable operation, or any combination of the above. Process intensification is considered as an alternative for reducing cost since the kinds of improvements are oriented to the integration of operations, functions, and phenomena. Intensified alternatives also seem to have many other characteristics if the goal is to obtain a bioprocess (green chemistry) or biorefinery approach. If process intensification effectively implemented, it will lead to improvements in environmental suitability, energy efficiency, cost, and social influence (Jiménez-González et al. 2012; Bildea et al. 2015; Sánchez-Ramírez et al. 2017; Hernández-Melchor et al. 2020; Lopéz-Peréz et al. 2022; Segovia-Hernández et al. 2020).

17.1.3 Indicator: The LCA of the Process

Life cycle assessment (LCA) tools provide a comprehensive method to examine impact with a methodological protocol that includes environmental impact metrics and hotspots (Katakojwala and Mohan 2021). The objective of the LCA is to make decisions so that the development of biorefinery technologies from the initial stages contemplates the environmental impact (Koch et al. 2020). The LCA is responsible for evaluating the terms of equivalent carbon dioxide emissions, acidification, and ecotoxicity, in addition to the emergency such as human toxicity, carcinogens, and abiotic depletion

(Singh et al. 2018). Life cycle assessment can be carried out on bioenergy products from biomass of different denominations, for example, microalgae and lignocellulosic treated with thermochemical processes. These processes can be analyzed using a life cycle framework to identify possible challenges in bioenergy production (Ubando et al. 2019). Continuously, the life cycle analyses that are carried out are useful to evaluate the impacts that a product has. However, this type of analysis has limitations when it comes to accounting for problems related to services provided to ecosystems at the land and supply chain level. Life cycle assessments are standardized tools that are responsible for evaluating impacts generated in production processes (see Eq. 17.1). This characteristic allows it to be possible to evaluate the sustainability of biological products (D'amato et al. 2020). On the other hand, Eco-indicator 99 (see Eq. 17.2) is used to evaluate the environmental impact and sustainability of processes, providing a powerful and reliable tool to choose processes with low environmental impact considering design and synthesis stages.

Eco-indicator 95 is calculated as follow:

$$\mathcal{L}(\alpha)=\sum_{i=1}^{n}\frac{\delta i}{\mathrm{T}i}\cdot\frac{\mathcal{L}i(\alpha)}{\delta i-\mathcal{L}i^{TH}} \quad (17.1)$$

Here

$\mathcal{L}(\alpha) = \textit{Total environmental impact caused by the alternative } a.$

$\delta i = \textit{Normalization reference for impact category } i.$

$\mathrm{T}i = \textit{Benchmark for impact category } i.$

$\mathcal{L}i(\alpha) = \textit{Resulting impact indicator for the category } i$

$\textit{caused by alternative } \alpha.$

$\mathcal{L}i^{TH}$
$= \textit{Indicator of the reference threshold of the impact category} i.$

The Eco-indicator 99 is calculated by:

$$E_{99}=\sum_{i}\sum_{d}\sum_{e\in E}\Gamma_d\Theta_d\omega_i\Psi_{i,e} \quad (17.2)$$

Here

ω represents the damage category (the total amount of chemical i released per unit of reference flow due to direct emissions), while Γ and Θ are normalization and weighting factors for damage of category d, respectively. $\Psi_{i,\ e}$ is a parameter associated to the direct releases, raw materials production, and energy generation (Segovia-Hernández et al. 2020).

The use of LCA in the evaluation of the sustainability of biorefineries presents a series of difficulties, given its multi-product and multipurpose nature. LCA can be used in conjunction with optimization methods, for example, the tool the pareto curve considering greenhouse gas emissions as indicators (kg year^{-1}) related to net profit (USD year^{-1}) for a specific bioprocess (Andiappan et al. 2015).

LCA is a tool that allows:

- Identify opportunities to improve a production system.
- Characterize production systems and their production units over time.
- Relative comparisons between product systems based on indicators selected by category.
- Indicate environmental effects for decision-making.

During the life cycle analysis, the inventory data is grouped by categories of environmental impacts. Assessment methodologies allow these impacts to be derived into global categories: damage to human health, depletion of natural resources, and the ecosystem. A lower value of the indicators of these final categories indicates a greater sustainability of the process or product under analysis. Additionally, to evaluate the sustainability of a biorefinery, the following series of indicators are used to quantify the state of development of the technology:

- Economic indicators (production costs)
- Energy input and output or net energy ratio
- Fossil fuel replaced per hectare
- Carbon footprint: greenhouse gas emissions avoided (reduction of equivalent CO_2 emissions)
- Environmental impacts (indicators by impact categories)
- Carbon emissions due to changes in land use
- Renewability indicators (exergy)
- Social indicators

The main impact categories of LCA are as follows:

- WD: Water demand
- ADP: Abiotic depletion potential
- AP: Freshwater acidification potential
- POP: Photo-oxidant formation potential
- POCP: Photochemical ozone creation potential
- HTP: Human toxicity potential
- ODP: Ozone depletion potential
- AETP: Aquatic ecotoxicity potential
- TETP: Terrestrial ecotoxicity potential
- EP: Eutrophication potential
- GWP: Global warming potential
- CED: Cumulative energy demand

The LCA analyzes the effects on the environment by both the use of resources (inputs) and the emissions created by a given process (outputs).

- *Inputs*: raw materials, water, energy, and chemicals.
- *Outputs*: product, co-product, solid waste, air emissions, water emissions, and emissions to land (Goedkoop and Spriensma 2000; Turton 2003; Mariani and Vastola 2015; Muthu 2020).

17.1.4 Indicator: Economic Indicators

Some LCA indicators are Eco-indicator 99 and total annual cost (TAC), which evaluate the performance of certain processes that are intended to be sustainable and economical. Speaking specifically of the TAC (Eq. 17.3), it is about adding the costs of the necessary equipment, the operations, the resources, and the cost of the materials of each one of the stages of the process (Onishi et al. 2017). The process total cost comprises the contributions related in general to

1. Operational cost (OC):
   - electricity,
   - thermal services consumption,
   - cooling, and
   - heating fluids.
2. Capital cost (CC):
   - investment in equipment,
   - electric generators,
   - heaters and cosolers,
   - compressors heat exchangers, and
   - valves.

$$TAC_{min} = OC - CC \tag{17.3}$$

In addition, this methodology can be adjusted according to the properties of each bioprocess, for example, the cost of an industrial plant by means of equations using Eq. (17.4):

$$TAC = \frac{\sum_{i}^{n} CC_i}{PP_y} + \sum_{k}^{n} OC_k \tag{17.4}$$

where PP_y is the payback period of y years (Turton 2003).

17.2 Biorefinery and Co-products

17.2.1 The Concept of the Biorefinery

Conceptually, a biorefinery is an energy mechanism that has biomass from different origins as input and, on the other hand, output of optimal bioproducts for its comprehensive and sustainable use. In the same way, it is about alternatives to convert specialized raw material into sustainable bioproducts. Actually, they have as their objective seeking sustainability within the different stages of material conversion, generated biofuels, and value-added co-products. Before the fermentation stage, pretreatment and characterization techniques have been developed by physical, chemical, physicochemical, and biological methods in order to optimize yields according to the requirements of each bioprocessing (Kumar et al. 2020; Wang et al. 2019). The biorefinery concept is based on the production of marketable biological inputs analogous to those produced by fossil waste. These inputs are produced through different decarbonization pathways and replace certain products with biological ones with greater sustainability (Katakojwala and Mohan 2021).

A biorefinery that uses microalgae as a substrate seeks its conversion into biomass to obtain a wide variety of value-added products. These products can be biofuels, biochemicals, biomaterials, and bioactive substances. In this sense, biorefineries must consider several key aspects for their sustainability, for example, biomass can be used directly or indirectly through pretreatment, converting it into liquid or gaseous fuel (biodiesel, bioethanol, and biogas) and, in turn, into electrical energy (combustion, gasification, and pyrolysis) and co-products: biochar, biohydrogen, biofertilizers, glycerin, fodder, and biomethanol, among others (de Siqueira Castro et al. 2020). So far, there are no proposals for biorefineries that integrate the generation of energy based on these biofuels and allow regulating energy efficiency, environmental, social, and economic impacts, in order to give greater viability to these integrated processes known as biorefineries as shown in Fig. 17.1 (Ramos Miranda et al. 2018).

Figure 17.1 shows an integrated biorefinery scheme based on four general stages: stage (a) Pretreatment of the waste and agro-industrial residues as feedstocks, i.e., also known as valorization; stage (b) Process of thermochemical and biochemical transformation; stage (c) Integration of conversion operations of generation of energy and heat; and finally stage (d) Co-products that are generated along with the biofuels. In addition, this scheme makes it possible to relate the important

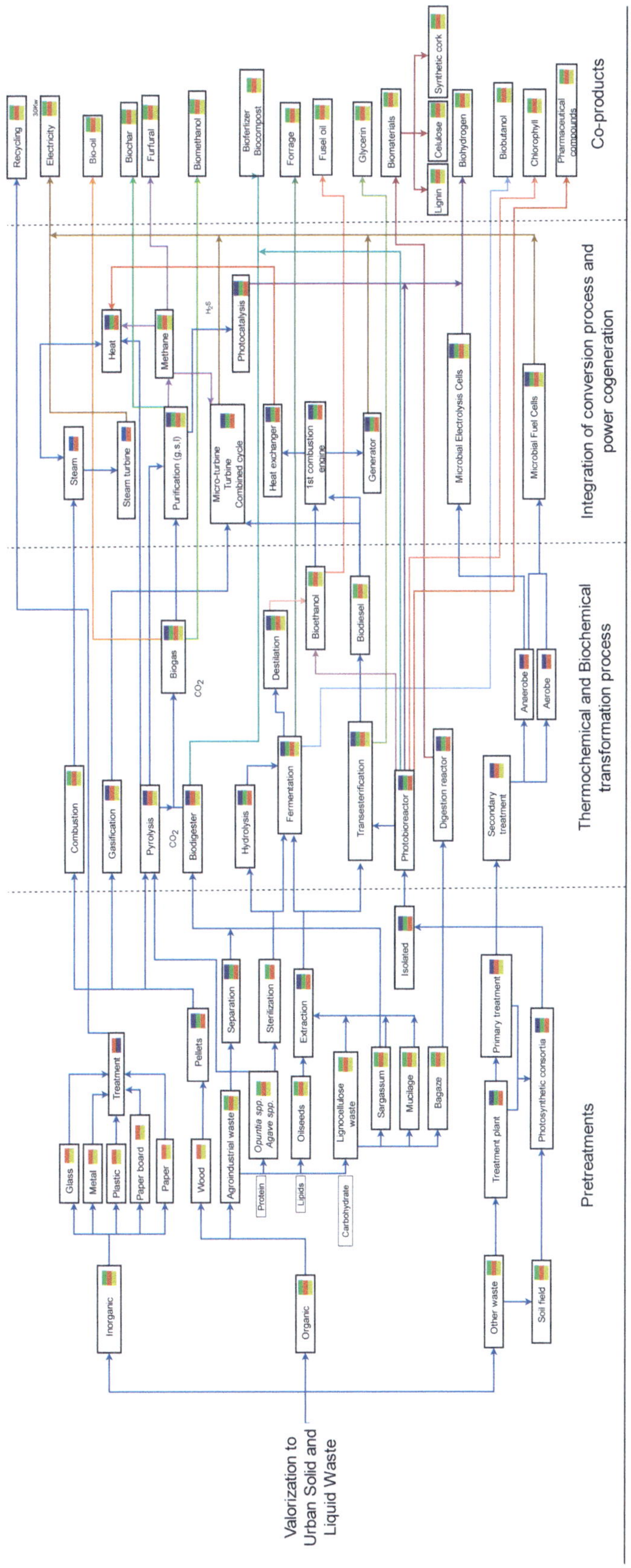

Fig. 17.1 Process integration via a sustainable biorefinery approach using wastes and agro-industrial residues

aspects of sustainability defined by the indicators at each stage and unit operation: intensification, LCA, social impact, as well as annual operating cost indicators. We will begin by describing the recovery of waste according to its solid or liquid properties as well as its origin, known as urban solid waste and wastewater treatment plants (Aguilar-López et al. 2010; Velázquez-Sánchez et al. 2017; Hernández-Melchor et al. 2018; Ramos Miranda et al. 2018; López Pérez et al. 2018, 2022).

At present, the production of biofuels through biomass has shown yields substrate-product (y_{sp}) in an interval of $y_{sp} = 40 - 70$ in the most optimal cases, even using genetically modified strains and intensification of industrial processes. In addition, we know that the efficiency (γ) of the machines and cycles of transformation of electrical energy and/or heat range of 20–60 in the most ideal cases. Consequently, under these operating parameters, a comprehensive vision is needed to give viability to the different production schemes for biofuels, energy, and co-products, in order to reduce economic, environmental, and social impacts (see Fig. 17.1). Therefore, its environmental, economic, and social assessment becomes important to assess its feasibility of installing biorefineries since they must be sustainable, competitive, and eco-efficient compared to energy processes based on fossil resources. In this sense, the analysis of sustainability is used as a decision-making tool for prospective technologies and the generation of new products based on waste recovery (Lopez et al. 2013).

- *Pretreatment of the waste*: The complete conversion of biomass implies that the pretreatment stage is aimed at increasing the accessibility of all carbohydrates within the crystalline structure of cellulose. Some are alkali or acid treatment, predigestion of substrate (sargassum, bagaze, mucilage, landfill leachate from food fermentation beans cocoa-coffee), enzymatic hydrolysis, enzymatic cellulose conversion, thermochemical pretreatment, hydrothermolysis, and ultrasonic pretreatment. Other reported pretreatment methods are steam explosion, ammonia and carbon dioxide pretreatment, use of ionic liquids, sulfite and inorganic salt pretreatment, and irradiation. In addition, one of the main sugars present on biomass is a C_5 sugar, such as xylan, which is harder to ferment than the C_6 sugar, such as glucose (see Fig. 17.1). Pretreatment comprises operations such as collection, segregation, adjustment, and decontamination that is applied to liquid and solid wastes (Bajpai 2012).
- *Process of thermochemical and biochemical transformation*: Fermentation and anaerobic digestion. *Thermochemical*: Applied to any biomass feedstock, multiple high-value products using Separations Unit Operations, independent of external conditions and reaction time <<<< (see Fig. 17.1). *Biochemical*: Involves the use of strain, CM, enzymes, and/or chemicals; yield is limited due to microbial conversion; mostly susceptible to external perturbations, production of secondary wastes, and reaction time >>>> (Singh et al. 2015).
- *Integration of conversion operations of generation of energy and heat*: Biomass gasification is the conversion of solid fuels (wood, wood waste, and agricultural waste, among others) into a mixture of fuel gases called syngas. A thermochemical process converts solid biomass into a fuel gas under high temperature conditions and in the presence of a gasifying agent (air, oxygen, water vapor, etc.). Depending on the type of gasifying agent, different gas qualities was obtained. For the generation of thermal or electrical energy, turbo-steam generators or co-generators, internal combustion engines or combined cycle engines are used. Downdraft gasifiers have the ability to convert biomass with high volatile content into a gas with low impurity content. In addition, the initial investment and the cost of producing electricity for this technology, on scales of up to 2 Megawatts (MW), make it competitive compared to other renewable energies (see Fig. 17.1).
- *Co-products that are generated along with the biofuels*: Including value-added co-products at the output of a biorefinery produces a greater economic improvement within all the unit operations involved in the process, mainly the third-generation biofuels (García Prieto 2018).

Co-products that have a high benefit can be replaced by alternatives of analogous products that come from technological processes derived from natural resource residues through the transformation of raw materials (Placeres Remiur et al. 2012). A co-product is defined as those intermediate and final products that are used as an alternative to chemical products with greater environmental impact; these co-products come from residues of agricultural, industrial, or urban process (López Ortega 2015).

17.2.2 Valorization of Organic By-products

Using by-products from residues and organic waste has many advantages. For example, they fulfill the functions of public services by taking advantage of alternative carbon sources, promoting the use of renewable energies. In this sense, biomass is a resource that can be exploited and that, on occasions, is not assigned a use. Despite this, the potential with which biomass can be used as a raw material or as a source of energy in broad terms has not been measured at a quantitative level. This is because there is no quality control that allows these terms to be established (Brosowski et al.

2016). The recovery of agro-industrial waste represents an alternative for the use of certain types of materials such as oils, animal fats, or vegetable residues, with high concentrations of sugar. Biomass offers a continuous and growing source for bioenergy production. Some of the advantages represented by the use of biomass for energy purposes are as follows:

- substrate for bioenergy production is already present;
- this carbon source can be found close to the place of transformation, so it does not need transport costs;
- a system is formed to reduce waste, eliminate pests, reduce the risk of forest fires and reduce bad odors, in addition to benefits for the air, water, and soil (Suárez et al. 2020).

The concept of valuation includes the conversion of waste and other materials paying attention to environmental and sustainability indicators. To achieve the objective of recovering some waste, it makes use of fields such as engineering, health, and safety. The nopal is a very attractive plant as feed for livestock because of its efficiency in converting water into biomass, and in addition to its digestible energy content, it is rich in carbohydrates and calcium. However, due to its poor protein value, it is necessary to combine it with other foods to complement the daily diet of cattle. It is estimated that only 15 species are used as forage (Díaz-Plascencia et al. 2012). Various authors point out that dehydrated nopal provided in concentrations of 20%, 30%, and 40% of the diet of fattening sheep produces daily weight gains, similar to diets based on grains and meat by-products. There are viable options to increase the protein value of nopal from 60% to 40% and are mainly based on semisolid or liquid fermentations with yeasts such as *Saccharomyces cerevisiae*, *Trichoderma harzianum*, and *Kluyveromyces lactis* (Díaz-Plascencia et al. 2012; Herrera et al. 2017). The proposed experimental methodology uses the nopal as substrate and prickly pear biomass to later convert it into an efficient and accessible livestock feed, allowing it to be used as the main protein source of forages (see Fig. 17.1).

17.2.3 Biofuels

Microalgae biomass is composed of carbohydrates, long-chain fatty acids, pigments, and proteins. In this sense, carbohydrates from microalgae are broken down into cellulose and starch without the presence of lignin, which is already ready for fermentation. The long-chain fatty acids produced by some algae can be used as supplements in the health area. Finally, the pigments and proteins produced by microalgae also have their application in the medical field (Yen et al. 2013). Microalgae provided a new perspective regarding their application toward ecological services and environmental and economic sustainability. Microalgae are capable of fixing CO_2 from the atmosphere; they can valorize desired products and generate bioproducts with them. As a result, they are a reliable source for the creation of multifaceted biorefineries (Javed et al. 2019). One of the most widely used sources of third-generation biofuels is microalgae (Pérez et al. 2019). These stand out for their high lipid content. Despite this, the use of microalgae for this purpose is still not considered viable in economic terms. Within the wide range of existing biofuels, biodiesel is the one with the shortest harvest time and has a continuous production period if it is in photosynthetic conditions (Ananthi et al. 2021a, b).

Macroalgae are an alternative for the production of biofuels classified as the third generation, replacing starch biomass from food crops and lignocellulosic biomass. The use of macroalgae is mainly due to its rapid growth rate, because it is not necessary for them to compete for agricultural land to fulfill a life cycle and because of its high carbohydrate content, in addition to simpler steps for its use compared to lignocellulosic biomass (Tan et al. 2020). The macroalgae used as an alternative, sustainable biomass resource is green, brown, and red marine algae. These are used to produce biofuels, chemicals, and food. The fact that its chemical composition is varied creates both new opportunities and challenges for bioenergy production. Numerous studies support the use of these carbon sources to generate bioethanol (Sudhakar et al. 2018).

Cyanobacteria are photosynthetic feeding microorganisms and can be found in various environments. This type of microorganisms serves as raw material for the production of biofuels classified as third and fourth generation. To generate a biofuel with species such as cyanobacteria, certain considerations must be taken into account, such as knowing the metabolic pathways by which they work, the structure of genetically engineered or modified cyanobacteria, and the scaling of processes (Farrokh et al. 2019).

17.2.3.1 Bioethanol

There are three generations of bioethanol production and they are divided depending on the raw material to be fermented. The first generation of bioethanol is produced through the fermentation of glucose from starch crops, cane sugar, and corn, among others (Cortes et al. 2018). Second-generation bioethanol is also known as one of those belonging to advanced biofuels, and it is obtained from lignocellulosic matter and forest and agro-industrial residues (Pérez et al. 2018). Third-generation bioethanol is the one that provides the greatest benefits compared to its predecessors (López-Pérez et al. 2016a, b). This generation obtains its fermentable carbon source from microalgae and cyanobacteria. These microorganisms are preferred for their high content of lipids and carbohydrates, in addition to culture techniques that do not require soil and their ability to absorb carbon dioxide

(Bastos 2018). The production of third-generation bioethanol is processed in three main ways: fermentation of previously treated microalgae biomass, dark fermentation of reserved carbohydrates, and direct photofermentation of CO_2 using light energy. Bioethanol represents a green resource to meet the energy demand for transportation. All methodologies are possible solutions, however, in a practical way; traditional fermentation is still the best accepted (Lakatos et al. 2019).

Second-generation bioethanol is produced through lignocellulosic biomass obtained from woody residues, pastures, or crops. Bioethanol production represents 3% of the total ethanol produced and reduces greenhouse gas (GHG) to a greater extent than first-generation bioethanol. Despite this, the environmental impact is still present and depends on how much availability there is the raw material and the technology used to transform it, since the latter must overcome challenges in technological and economic fields, including phases such as pretreatment, fermentation, hydrolysis, and separation of final products (Sharma et al. 2020). However, there are some drawbacks to this due to the food that must be converted and its occupation of the territories. Instead, the integration of the second-generation bioethanol process with the infrastructure of the first-generation process means that bioethanol production has great potential (Ayodele et al. 2020).

A reliable source of fermentable sugars that do not compete in the food field are those from agro-industrial lignocellulosic residues. These have a direct impact on waste disposal techniques by converting them into second-generation bioethanol. Despite the methodologies developed, the high economic costs and the great variation that exists in the composition of the substrate means that the procedure cannot be scaled for commercialization. Bioethanol is considered the most successful third-generation biofuel because of its performance and its microalgae-based production. The use of microalgae has advantages such as cultivation areas and low environmental impact in terms of CO_2 emissions. Microalgae and macroalgae have a starch composition similar to that found in certain higher plants, making them good candidates for biofuel production. One of the techniques for improving the yield of bioethanol production is to increase the concentration of carbohydrates in the biomass, which can be directed to biorefineries that generate high value-added products (da Maia et al. 2020).

Case Study: Ethanol Production Based on Rice Water Using Photosynthetic Consortium

Among the potential sources of biofuels, the most efficient producers of biomass are microalgae and cyanobacteria. These versatile organisms can be used for the production of bioethanol, biodiesel, biohydrogen, and biogas. In fact, most of the economic methods for the production of biofuels produced by algae combine an approach where multiple biofuels are produced from the same biomass source. The production of bioethanol from algae is limited by the content of carbohydrates contained in the microalgal biomass and the portion of carbohydrates that can be converted to fermentable sugars and therefore to ethanol; reporting an average amount of 0.13 L bioethanol kg^{-1} of algal biomass. Therefore, the objective of this case study was the production of ethanol from microbial consortia using residual substrates such as rice water as a carbon source.

A photosynthetic microbial consortium isolated from the "Sierra Hermosa" wastewater treatment plant in charge of the Decentralized Potable Water, Sewerage Agency Sanitation (ODAPAS) of the Municipality of Tecámac was used. The culture was propagated and maintained in the mineral medium BG11. For ethanol production, 250 mL photobioreactors were used in batch culture with 180 mL of BG11 medium, 10% rice water, and 10% inoculum (v/v). Operating conditions of the photobioreactors were the following: aeration of 0.5 vvm, an amount of light of 80 μmol photons m^{-2} s^{-1}, and a light:dark photoperiod of 12:12. The hydraulic residence time of the batch of culture was 10 days. Aliquots were taken every 24 h for determinations of biomass and ethanol. The ethanol concentration was determined by dry gas chromatography. The chromatographic conditions were determined by a previous program based on the volatility of alcoholic compounds, the maximum temperature of the detector was 250 °C, and the temperature of the injector was 220 °C. Each analytical determination was performed in triplicate. Gram stain and Schaeffer and Fulton stain for the identification of microorganisms present in the microalgae consortium: According to its morphology, microorganism could be *Nostoc comunne* (see Figs. 17.2 and 17.3) because it has a linear conformation, does not present an internal color, and is made up of circles of different sizes called heterocysts. In addition, within the staining, a large number of oval microorganisms apparently without external coloration were identified, with some green pigments and without flagella, which could be *Monodopsis subterranea*. The Shaeffer and Fulton technique shows that there are no photosynthetic sporulated microorganisms, only microalgae that are dyed green, those illuminated brown parts are *Botrycoccus braunii*, which is easy to identify due to its color and opacity that do not allow the passage of the entire light of the microscope. With the result of the staining, it was ruled out that the production of ethanol is not due to bacteria or sporulated microorganisms, but rather to a microalgal consortium, specifically *Monodopsis subterrane*, *Botryococcus braunii*, *Haematococcus pluvialis*, *Chlorella* spp., and *Crypthecodinium cohnii*. This case presents a batch fermentation for ethanol production, using starch-

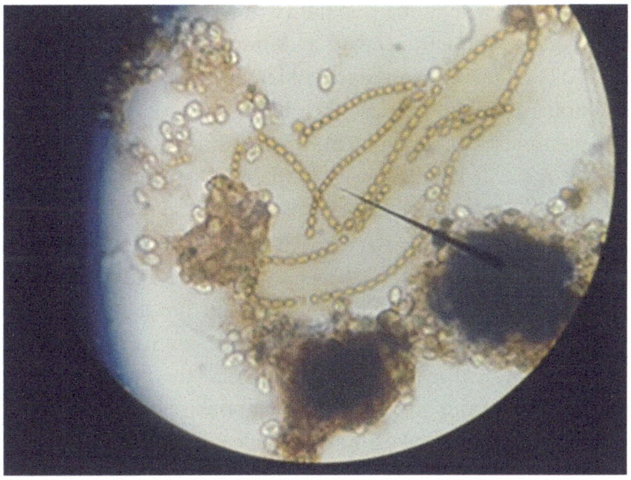

Fig. 17.2 Morphology and composition of the microalgae consortium in an optical microscope

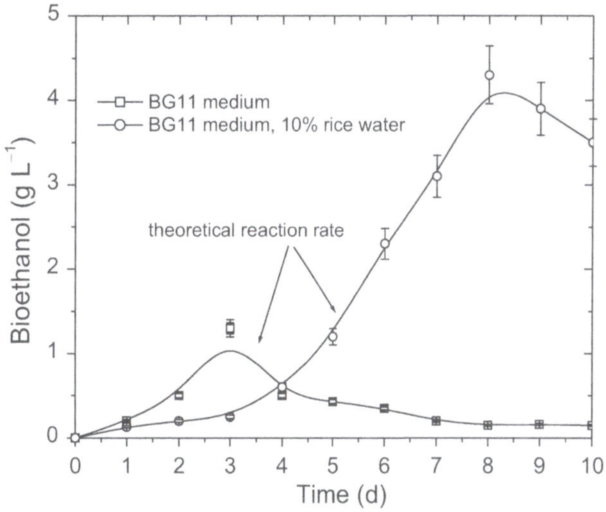

Fig. 17.4 Production of bioethanol considering the residue from the industrial processing of rice

containing residues from rice processing from the food industry. Furthermore, this experiment showed that the maximum concentration of ethanol was approximately 4.3 g L^{-1}, with a productivity of 0.53 g L^{-1} d^{-1} (Fig. 17.4).

The LCA of the production of bioethanol generated from microalgae was investigated to determine the CO_2 emissions of the process and the energy balance. It was shown that from 6 ha of land in the high seas, 220 tons of biomass were obtained for the conversion of bioethanol. The studies resulted in 218.86 tons of CO_2 per year. A favorable energy ratio was obtained, which represents a low water and land footprint. In summary, low concentrations of CO_2 emissions were obtained, little demand for land and water in the process and cultivation, and high energy generation, in addition to certain by-products, which represented a favorable impact on energy and environmental fields (Hossain et al. 2019).

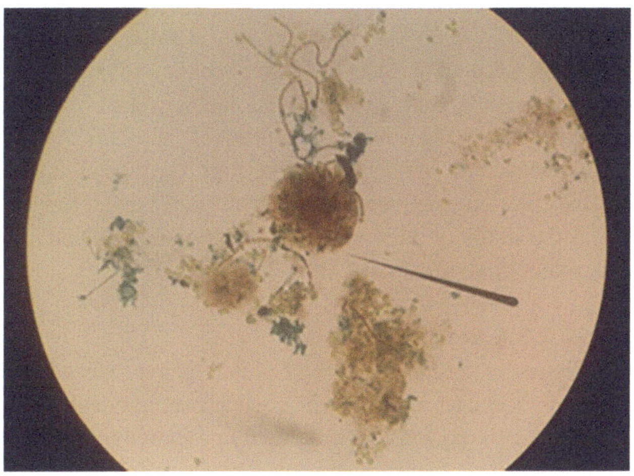

Fig. 17.3 Morphology and composition of the microalgae consortium with staining in an optical microscope

Currently, research on bioethanol as a bioenergy product focuses on the use of micro- and macroalgae instead of terrestrial substrates. Macroalgae, in particular, are an important source of carbohydrates. Bioethanol production is divided into three main steps: first mechanical processing, then biomass pretreatment, and finally microbial fermentation. A crucial step for bioconversion is a pretreatment for the saccharification process, which can be carried out in two steps or combined, varies the yield of reducing sugars for bioethanol production. Within the methods to carry out the pretreatment of the sample, there are physical and physicochemical, chemical, and biological methods (Dave et al. 2019).

17.2.3.2 Biohydrogen

A new form of energy is biohydrogen, which has certain advantages over other fossil fuels, such as recyclability, high conversion efficiency, nonpollution and high-energy efficiency (López et al. 2017). Currently, hydrogen releases energy and water as a by-product of combustion, which is why it is considered a source of environmental pollution. Instead, biohydrogen is considered as one of the cleanest biofuels and with a growing demand. Biohydrogen has a higher energy density, but obtaining it from fossil fuels has a negative effect on the environment. On the other hand, the production of biohydrogen by means of green algae and cyanobacteria generates a positive contribution as a source of renewable energy. Microalgae can generate other fuels such as bioethanol and biodiesel (Mona et al. 2020).

The production of biohydrogen has been carried out by photosynthesis and fermentation of algae, cyanobacteria, and anaerobic bacteria. The production of biohydrogen requires less energy, and it can be carried out at room temperature and pressure and has a long-term production. Biohydrogen has application for internal combustion engines or as a fuel cell electricity generation. One of the reasons for this is because the enzyme hydrogenase, which catalyzes the production of biohydrogen, is sensitive to molecular oxygen, hence the molecule inhibits the enzyme during photosynthesis (Romero 2020). Has been studied of modeling and multi-objective optimization for simulation of hydrogen production using a photosynthetic consortium. A Plackett–Burman design and a steepest ascent were used to screen the key factors to obtain the best hydrogen concentration. The optimal conditions, using the method of steepest ascent, were aeration (0.125 L min^{-1}), acetate (200 mg L^{-1}), and cysteine (498 mg L^{-1}). The combinations acetate + cysteine without aeration and cysteine without aeration increased hydrogen concentration. The simultaneous optimization of reaction time (RT) and cysteine improved hydrogen concentration to 28.19 mg L^{-1} (Rangel et al. 2016; Hernández-Melchor et al. 2020).

Biohydrogen has the highest specific energy content within biofuels, around 122 kJ g^{-1}. Biohydrogen applications range from transportation fuel to electricity generation. Biomass production and combustion for this biofuel is zero CO_2 emissions. Dark fermentation does not require light, requires anaerobic conditions, and uses organic waste as a substrate. In this technique, metabolic routes have been identified that discard acetic acid, butyric acid, butanol, ethanol, and propionic acid as a product. For the production of this type of biofuel, genera such as *Clostridium*, *Enterobacter*, and *Bacillus* are mainly studied. Biohydrogen generated by microalgae produces clean energy through carbon neutrality and generating bioenergy sustainability (Pérez et al. 2015). Dark fermentation of microalgae is, in general, better than other techniques such as photofermentation, biophotolysis, and microbial electrolysis due to its continuous generation of biohydrogen and flow of natural sunlight (Li et al. 2022). Biological fermentation is recommended for its biological and economic approaches analogous to thermochemical processes (Torres-Ortega et al. 2018). Despite this, the low yield of H_2 and high production cost are limiting in the scaling of processes for commercial purposes (Goswami et al. 2021).

17.2.3.3 Biodiesel

Biodiesel is one of the most widely used biofuels to partially replace fossil fuels. For this purpose, it is important that the production of raw material is vegetable oils. Bio-oils and biodiesel are alternatives to those petroleum products that have emerged in recent years. Currently, traditional fuels continue to be used; however, bioconversion fuels have been in increasing use for some time and their use is growing. The great current interest is the development of biodiesel from algae, which is why it is classified as a third-generation biofuel. Despite the popularity of beveling, there is a need to reduce production costs and improve productivity (Konur 2021). One of the main problems of the large-scale production of biodiesel is the high cost of production and manufacturing compared to the traditional production of fossil fuels. It is necessary that there is an improvement in the development of biodiesel production technologies with continuous improvement to make a competitive product (Tabatabaei et al. 2019). The latter enables us to quantify and monitor the limits of biofuels production when this information is used to implement a control loop. Biodiesel is ecological and nontoxic in nature, its production can be achieved from numerous raw materials, catalysts, and various technologies. For example, nanocatalyst technology is used with certain advantages such as its large surface area, the reuse it provides, and the high activity of the nanocatalyst used (Ambat et al. 2018).

Microalgae are one of the renewable raw materials with the greatest potential for the production of green energy. These microorganisms are used to generate qualitative biodiesel. After the recollection of the microalgae, a dehydration of the same is carried out, and these processes can be carried out by a combination of sequential operations (Intensification). There are certain drawbacks such as the high cost of production, which could be reduced as the process is automated (Ananthi et al. 2021a, b). Three estimators were applied to the Droop model to produce microalgae (lipids). Linear and nonlinear observers were assessed to estimate changes in process inputs and parameters. Biomass and nitrogen concentrations were estimated in spite of errors in the evaluation of the parameters using substrate concentration as measurable output (Aguilar-López et al. 2010). An important aspect for a cyanobacterial biorefinery is the amount of lipids in them. *Cyanobacteria* is preferred as an alternative to produce biodiesel because of their high lipid profiles. In addition to these quantitative results, it was found that strains of the *Cyanobacterium* genus are suitable for generating biodiesel due to their vast composition of saturated and monounsaturated fatty acids (Bolatkhan et al. 2020). The expectations in the production of biofuels based on cyanobacteria are based on the high content of lipids and the chemical composition of the fatty acids that it has, as well as the ability of microorganisms to grow in open and residual waters (Sadvakasova et al. 2021).

17.2.3.4 Biogas

Biogas is a colorless combustible gas produced by the biological decomposition of organic matter; occurring in the absence of oxygen. It is commonly generated from the anaerobic digestion of biodegradable materials such as biomass, manure, green waste, and agricultural residues such as cassava and sugar cane, among others. Biogas comprises a mixture of different gases, mainly methane (CH_4), carbon dioxide (CO_2), and about 1–5% of other gases, which makes it an attractive source of energy. Its use offers multiple benefits such as the improvement of agriculture in rural communities, the reduction of waste through the use of organic agricultural residues and urban solid waste for energy production, and the improvement of the quality of the environment through the reduction of CO_2 emissions (Sawyerr et al. 2018). Globally, biogas production has been boosted considerably in the last 15 years; however, there is significant variation between countries in terms of sector development and number of plants. While some countries, such as Germany and China, have shown tremendous growth over the last decade, the biogas industry in other countries is just emerging. As recorded, globally, biogas generation capacity reached 16.9 GW in 2017 compared to 6.7 GW in 2008; and in terms of installed electrical capacity, Europe leads the sector with approximately 17,662 biogas plants that contribute 9985 MWe (Megawatts electric) (Benato and Macor 2019), followed by the United States with more than 2200 digesters with an installed capacity of 977 MWe (American Biogas Council 2015). Importantly, the use of biogas varies around the world, ranging from the creation of millions of small-scale biogas plants providing cooking gas in China and India to electricity and improved biomethane as fuel for vehicles in Germany and Sweden. Generally, biomass of any type that contains carbohydrates (either starch, cellulose, or hemicellulose), proteins and fats as core components can be used for biogas production.

17.2.4 Co-products

Biofertilizer: Biofertilizers take part within the integrated management of nutrients, and these are known to increase the economic benefits of reducing expenses when acquiring a fertilizer of chemical classification; this results in comprehensive and sustainable agriculture (Atieno et al. 2020). The use of fertilizers causes the loss of nitrogen in crops, which generates diffuse pollution. Instead, a biofertilizer is considered as an alternative to synthetic nitrogen fertilizers (Azizoglu 2019). Biomass is largely made up of an important nutrient reservoir, which can be recovered through certain physicochemical and biochemical technologies that allow these nutrients to be converted into biofertilizers (de Siqueira Castro et al. 2020).

The yield of agricultural crops and their commercial value depends, to a large extent, on the level of fertilizers used throughout the process. There is a biological counterpart to fertilizers that claims to be of lower economic cost, more ecological, and broad-spectrum, which makes it an economically sustainable product. A clear example of this is cyanobacteria and algae. The former are used to fix atmospheric nitrogen and other trace elements in the soil. On the other hand, algae are responsible for producing secondary metabolites to inhibit certain pathogenic bacteria and fungi, which results in better growth of certain plant species (Dineshkumar et al. 2019).

Biobutanol: Biobutanol can be fermented from acetone–biobutanol–ethanol (ABE), and this is known as a liquid drop in biofuel with the ability to mix with gasoline. The production process of this biofuel has been innovated through the fermentation of ABE. To achieve good production, the selection of a raw material, such as lignocellulose biomass, is crucial. Because of its sustainable supply, it does not pose a threat to food. In addition, the source must go through a pretreatment filter for proper growth (Huzir et al. 2018). The raw material used as a substrate ranges from starch, lignocellulosic residues, or algae biomass, each with its advantages and disadvantages in the face of bioconversion. The production of this biofuel owes its success to the relationship between microbial behavior and the way in which bioconversion is carried out so that the process can be scaled (Ibrahim et al. 2018). On the other hand, microalgae are a good resource to deal with this problem because they do not need a large territorial extension and are capable of growing in both fresh and salt water. Microalgae use carbon from the atmosphere for their photosynthetic processes, thus reducing GHG emission levels. The use of microalgae is considered as a third-generation substrate for the production of biofuels due to its rapid growth rate and high carbohydrate content (Shanmugam et al. 2021).

Biochar: Biochar is known to be a carbon-dense material obtained from organic, agricultural, or urban biomass (Quintanar-Orozco et al. 2018; Cruz-Méndez et al. 2021). Due to its high carbon content and its cation exchange potential, its large specific surface area and its stable structure, biochar have gained more attention and tentative uses continue to be analyzed (Wang et al. 2019). Another use of biochar is wastewater treatment because of its adsorption capacity. Soil bioremediation and gas storage and separation are the uses of biochar because of its ability to remove toxic metals and organic contaminants found in wastewater (Xiang et al. 2020). The characteristics of biochar allow it to be used in the formation of nutrients based on carbon; these include anaerobic digestion, gasification, pyrolysis, and energy storage (Zhou et al. 2021). Biocarbon is characterized by having a vast number of surface functional groups, in addition to being porous, which allows it to easily store and convert energy. Biocarbon is a co-product of fermented biomass and has the option of becoming a sustainable carbon that replaces products based on organic carbon obtained from fossil waste (Liu et al. 2019).

Compost: The composition of the biocompost is based on carbon, potassium, and more micronutrients, as well as bacteria and fungi. Among its uses is its ability to maintain carbon within the soil, and it helps the fertilizer to be able to increase or decrease the mobility of pesticides, helping to manage soil contamination (Singh 2008). According to the European Commission 69, about 25% of the biowaste produced is an optimal source for the generation of biocompost. One of the co-products of a biorefinery is biocompost, and the product of a composting process made of minerals and other organic material has been difficult to degrade such as lignin and its derivatives. Among its uses are soil improvement, fungal growth substrate, or heat generation (van Groenestijn et al. 2020).

Furfural: Furfural has application as a fungicide and nematicide. It also functions as a raw material to produce furfuryl alcohol, furan heterocycles, methylfuran, acetylfuran, furfurylamine, and furoic acid (Zeitsch 2000). Furfural is an aldehyde derived from xylose, with its applications in the production of plastic, glass, and fungicides and insecticides.

Glycerin: One of the main by-products of biodiesel is glycerin. It is estimated that for every 10 kg of biodiesel generated, 1 kg of glycerin is recovered. In countries such as Colombia, for every 50 L of biodiesel generated, approximately 4 kg of crude glycerin is obtained. For glycerin to be considered a co-product of diesel, its degree of refining must be taken into account (Afanador et al. 2018).

Fusel oil: Fusel oil is a long-chain alcohol, i.e., a higher alcohol. This is obtained as a co-product of bioethanol production. It has applications as a fuel for compression- or spark-ignition engines if used as a blend with other fuels. To obtain fusel oil, a substrate fermentation must be carried out. The application that has had the most impact is its potential use as fuel for internal combustion engines. Its performance as a fuel has been found to be similar to that of ethanol (Ardebili et al. 2019).

17.2.5 Electric Energy Generation Based on Biofuels

Biomass gasification is the conversion of solid fuels (wood, wood waste, and agricultural waste, among others) into a mixture of fuel gasses called syngas. A thermochemical process converts solid biomass into a fuel gas under high temperature conditions and in the presence of a gasifying agent (air, oxygen, water vapor, etc.). Depending on the type of gasifying agent, different gas qualities were obtained. For the generation of thermal or electrical energy, turbo-steam generators or co-generators, internal combustion engines or combined cycle engines are used. Downdraft gasifiers have the ability to convert biomass with high volatile content into a gas with low impurity content. In addition, the initial investment and the cost of producing electricity for this technology, on scales of up to 2 Megawatts (MW), make it competitive compared to other renewable energies. Likewise, there are studies on technologies that use biofuels as an input. Among them is the solid oxide fuel cell (SOFC), which is based on the fact that the fuel cell is powered by bioethanol and allows electricity to be generated. The system has a catalytic reformer that releases the hydrogen contained in the biofuel; later, it is converted into electricity in an electrochemical device, i.e., the solid oxide fuel cell (Porras 2021). The biomass that are recovered can be used to extract the oil that can later be converted into biodiesel if the processes are separated. A fraction of the biomass has the capacity to be used as animal feed and as high value-added products. If an anaerobic digestion process is carried out, an important part of the biomass can produce biogas, which allows the generation of electrical energy and functions as a support for the maintenance of microalgae (Fig. 17.5).

If the bioproduct (biofuels generated) meets the quality specifications that exist at the international level, it can:

- form part of the current global transport infrastructure;
- use directly as a solid fuel to be able to produce energy in the form of electricity, heat, and steam;
- bioconvert into gaseous biofuels, e.g., biogas and biohydrogen;
- excrete starch with the capacity to be fermented to produce liquid biofuels such as bioethanol and biobutanol.

Another gaseous biofuel produced by gasification of solid biomass is the gasification gas obtained when the biomass is placed in the presence of very high temperatures (approximately 800–1000 °C) and low oxygen levels. Two different products are obtained depending on the purity of the gas used. If air is used, gasogen or producer gas is produced, i.e., the mixture of carbon monoxide and nitrogen, which can be used to obtain both electricity and steam. If pure oxygen is used, synthesis gas such as carbon monoxide and hydrogen can be obtained, which can be used as a direct fuel, as another source of hydrogen. Liquid biofuels are fuels obtained from biomass in a liquid state in the presence of normal conditions of pressure and temperature. Boilers are used to produce heat and electricity as well as in combustion engines. Among the most used are biodiesel, bioethanol, and pyrolysis oils. In addition, electricity can be generated through gas turbines or reciprocating engines that use fuels such as synthesis gas that comes from the gasification of biomass or biogas produced by anaerobic digestion (Perdices 2018).

- *Micro turbine*: Small number of moving parts, low emissions and mechanical efficiency, compact size and light weight, power stations: 30–250 kW.
- *Internal combustion engine*: Operate on low pressure gas (biogas), relatively low investment cost, high maintenance costs, power stations: <5 MW in distributed generation, high speed (1200 rpm): ≤ 4 MW.

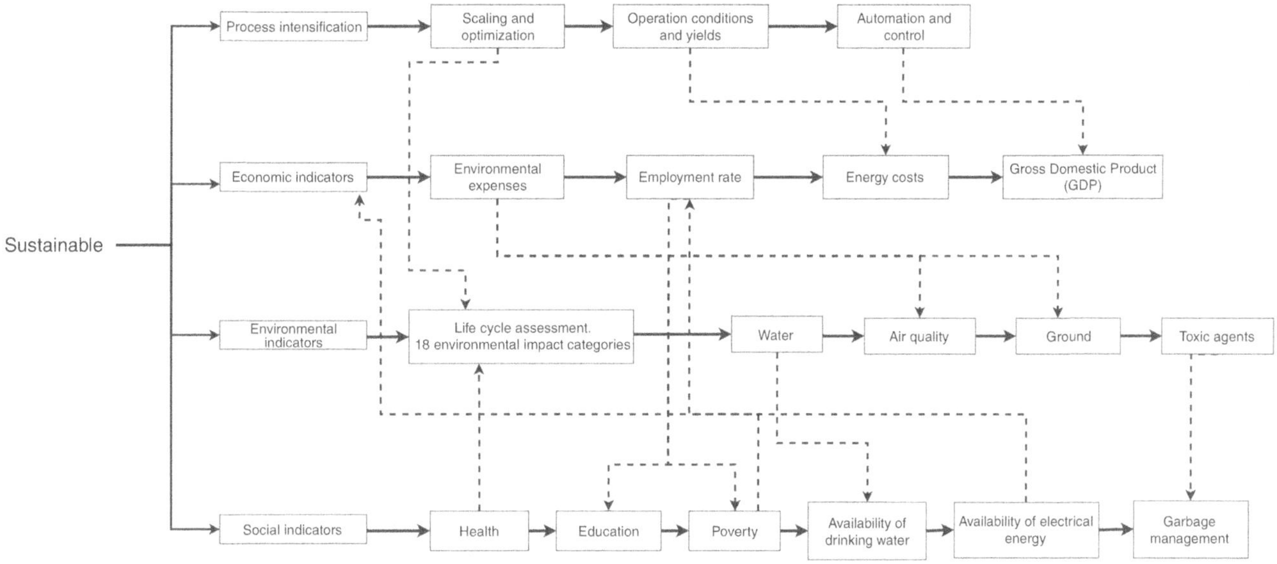

Fig. 17.5 Scheme of implementation in process of the general sustainability criteria

- *Steam turbine*: Any type of fuel can be used (biofuels), power stations: 50 kW–250 MW.
- *Fuel cells*: Low emission and low noise, fuels requiring process power stations: 5 kW–2 MW.

Combined heat and power (CHP) plants (co-generation plants), using biomass as fuel, can be an interesting alternative generated for this type power plant. Electrical efficiencies of biomass plants can reach 33–34%, and up to 40% if operated in electricity-only mode and recovery of heat from 75% to more than 90%, depending on technology plant (Abbas et al. 2020).

17.2.6 Free Plant Simulation Software

COCO-COFFEE: COCO simulation software is used to model processes. The COCO simulator (CAPE-OPEN to CAPEOPEN) is based on the CAPE-OPEN standards, which allows CAPE applications and components to operate on different applications within the CAPE-OPEN protocol seamlessly, such as COCO, DWSim, and Aspen HYSYS. The CAPE-OPEN protocol has been developed by COLaN in collaboration with Air Liquide, Chemical Company, and Shell Global Solutions. COCO is an open-source software that has been developed and maintained by amsterCHEM and it works on the same protocol. Initially, it was developed for use in chemical engineering to model steady state processes (van Baten and Szczepanski 2011; van Baten 2014; van Baten and Baur 2015; Rahman and Abir 2019).

Eco-indicator: BioGrace is a spreadsheet-style model used to calculate GHG emissions from various biofuels. A European cooperative has been formed with the effort in which the owners are from Germany, the Netherlands, Spain, and the United Kingdom, and these countries developed the calculator to implement EU-RED and FQD. Currently, the manager of the model is the Institute for Energy and Environmental Research of Germany (Pereira et al. 2019). BioGrace was developed for consultants and politicians to improve decision-making and implementation of national programs that seek to reduce GHG emissions. In addition, it is a reliable tool for other fields such as agriculture, to; for example, verify the carbon footprint of biofuel production (Głąb and Sowiński 2019; BIOGRACE 2020).

17.3 Conclusions

This chapter presents the development of a methodology to generate a sustainable biorefinery based on agro-industrial residues and photosynthetic consortia. The biorefinery approach reveals that processes can be integrated to generate biofuels, energy, and co-products considering sustainable indicators. Finally, this approach allows for the rapid determination whether a biorefinery is feasible for scaling up industrial from a laboratory study.

References

Abbas T, Issa M, Ilinca A, El-Ali A (2020) Biomass combined heat and power generation for Anticosti Island: A case study. Int J Power Energy Syst 8(03):64

Afanador G, Avellaneda Y, Mejía G, Mayorga O, García G, Pérez C, Ordóñez C, Rubiano A, Ramos Y, Ortiz R, Malagon K, Montañes D,

Loaiza A, Reina A, Téllez L, Rodríguez, Stella (2018) Glicerina y subproductos del biodiesel: alternativa energética para la alimentación de aves y cerdos
Aguilar-López R, López-Pérez PA, Neria-González MI, Domínguez-Bocanegra AR (2010) Observer based adaptive model for a class of aerobic batch bioreactor. Rev Mex Ing Quím 9(1):29–35
Ambat I, Srivastava V, Sillanpää M (2018) Recent advancement in biodiesel production methodologies using various feedstock: A review. Renew Sust Energ Rev 90:356–369
American Biogas Council (2015) Current and potential biogas production. https://www.americanbiogascouncil.org/pdf/biogas101.pdf
Ananthi V, Balaji P, Sindhu R, Kim SH, Pugazhendhi A, y Arun, A. (2021a) A critical review on differente harvesting techniques for algal based biodiesel production. Sci Total Environ 780:146467
Ananthi V, Raja R, Carvalho IS, Brindhadevi K, Pugazhendhi A, y Arun, A. (2021b) A realistic scenario on microalgae based biodiesel production: third generation ethanol. Fuel 284:118965
Andiappan V, Ko ASY, Lau VWS, Ng LY, Ng RTL, Chemmangattuvalappil NG, Ng DKS (2015) Synthesis of sustainable integrated biorefinery via reaction pathway synthesis: economic, incremental environmental burden and energy assessment with multiobjective optimization. AIChE J 61:132–146
Ardebili SMS, Solmaz H, Mostafaei M (2019) Optimization of fusel oil–gasoline blend ratio to enhance the performance and reduce emissions. Appl Therm Eng 148:1334–1345
Atieno M, Herrmann L, Nguyen HT, Phan HT, Nguyen NK, Srean P et al (2020) Assessment of biofertilizer use for sustainable agriculture in the great Mekong region. J Environ Manag 275:111300
Ayodele BV, Alsaffar MA, Mustapa SI (2020) An overview of integration opportunities for sustainable bioethanol production from first- and second-generation sugar-based feedstocks. J Clean Prod 245: 118857
Azizoglu U (2019) *Bacillus thuringiensis* as a biofertilizer and biostimulator: a mini-review of the little-known plant growth-promoting properties of Bt. Curr Microbiol 76(11):1379–1385
Bajpai P (2012) Biotechnology for pulp and paper processing. Springer, New York, pp 7–13
Bastos RG (2018) Biofuels from microalgae: bioethanol. In: Energy from microalgae. Springer, Cham, pp 229–246
Benato A, Macor A (2019) Italian biogas plants: trend, subsidies, cost, biogas composition and engine emissions. Energies 12(6):979
BIOGRACE (2020) Harmonized calculations of biofuel greenhouse gas emissions. Biograce.net. https://www.biograce.net/. Accesed 15 Apr 2022
Bildea CS, Győrgy R, Sánchez-Ramírez E, Quiroz-Ramírez JJ, Segovia-Hernandez JG, Kiss AA (2015) Optimal design and plantwide control of novel processes for di-n-pentyl ether production. J Chem Technol Biotechnol 90(6):992–1001
Bolatkhan K, Sadvakasova AK, Zayadan BK, Kakimova AB, Sarsekeyeva FK, Kossalbayev BD, Allakhverdiev SI et al (2020) Prospects for the creation of a waste-free technology for wastewater treatment and utilization of carbon dioxide based on cyanobacteria for biodiesel production. J Biotechnol 324:162–170
Brosowski A, Thrän D, Mantau U, Mahro B, Erdmann G, Adler P, Stinner W, Reinhold G, Hering T, Blanke C (2016) A review of biomass potential and current utilisation–status quo for 93 biogenic wastes and residues in Germany. Biomass Bioenergy 95:257–272
Castillo-Saldarriaga CR, Velásquez-Lozano ME (2018) Efecto del pretratamiento con ácido sulfúrico diluido sobre la hidrólisis enzimática del panicum maximum. Biotecnología en el Sector Agropecuario y Agroindustrial 16(1):68–79
Cortes TR, Cuervo-Parra JA, Robles-Olvera VJ, Cortes ER, Pérez PAL (2018) Experimental and kinetic production of ethanol using mucilage juice residues from cocoa processing. Int J Chem React Eng 16(11)
Cruz-Méndez AS, Ortega-Ramírez E, Lucho-Constantino CA, Arce-Cervantes O, Vázquez-Rodríguez GA, Coronel-Olivares C, Beltrán-Hernández RI (2021) Bamboo biochar and a nopal-based biofertilizer as improvers of alkaline soils with low buffer capacity. Appl Sci 11(14):6502
da Maia JL, Cardoso JS, da Silveira Mastrantonio DJ, Bierhals CK, Moreira JB, Costa JAV, de Morais MG (2020) Microalgae starch: A promising raw material for the bioethanol production. Int J Biol Macromol 165:2739–2749
D'amato D, Gaio M, Semenzin E (2020) A review of LCA assessments of forest-based bioeconomy products and processes under an ecosystem services perspective. Sci Total Environ 706:135859
Dave N, Selvaraj R, Varadavenkatesan T, Vinayagam R (2019) A critical review on production of bioethanol from macroalgal biomass. Algal Res 42:101606
de Siqueira Castro J, Calijuri ML, Ferreira J, Assemany PP, Ribeiro VJ (2020) Biofertilizante a base de microalgas: un enfoque de ciclo de vida. Ciencia del Medio Ambiente Total 724:138138
Díaz-Plascencia D, Rodríguez-Muela C, Mancillas-Flores P, Ruíz-Holguín N, Mena-Mungía S, Salvador-Torres F, Duran-Melendez L (2012) Fermentación *in vitro* de nopal forrajero con un inóculo de levadura Kluyveromyces lactis obtenida a partir de manzana de desecho. REDVET Rev Electron de Vet 13(1):1–11
Dineshkumar R, Subramanian J, Gopalsamy J, Jayasingam P, Arumugam A, Kannadasan S, Sampathkumar P (2019) The impact of using microalgae as biofertilizer in maize (*Zea mays L.*). Waste Biomass Valorization 10(5):1101–1110
Farrokh P, Sheikhpour M, Kasaeian A, Asadi H, Bavandi R (2019) Cyanobacteria as an eco-friendly resource for biofuel production: a critical review. Biotechnol Prog 35(5):e2835
García Prieto CV (2018) Diseño óptimo de biorrefinerías integradas basadas en microalgas. Tesis Doctoral. Universidad Nacional del Sur, Argentina
Głąb L, Sowiński J (2019) Sustainable production of sweet sorghum as a bioenergy crop using biosolids taking into account greenhouse gas emissions. Sustainability 11(11):3033
Goswami RK, Mehariya S, Obulisamy PK, Verma P (2021) Advanced microalgae-based renewable biohydrogen production systems: A review. Bioresour Technol 320:124301
Hernández Touset JP, de Armas Martínez AC, Espinosa Pedraja RO, Pérez Ones O, Guerra Rodríguez LE (2021) Procedimiento de análisis energético para la conversión de industrias de la caña de azúcar en biorrefinerías. Rev Univ Soc 13(5):277–288
Hernández-Melchor DJ, Camacho-Pérez B, Ríos-Leal E, Alarcón-Bonilla J, López-Pérez PA (2020) Modelling and multi-objective optimization for simulation of hydrogen production using a photosynthetic consortium. Int J Chem React Eng 18(7)
Hernández-Melchor DJ, Cañizares-Villanueva RO, Terán-Toledo JR, López-Pérez PA, Cristiani-Urbina E (2017) Hydrodynamic and mass transfer characterization of flat-panel airlift photobioreactors for the cultivation of a photosynthetic microbial consortium. Biochem Eng J 128:141–148
Hernández-Melchor DJ, López-Pérez PA, Carrillo-Vargas S, Alberto-Murrieta A, González-Gómez E, Camacho-Pérez B (2018) Experimental and kinetic study for lead removal via photosynthetic consortia using genetic algorithms to parameter estimation. Environ Sci Pollut Res 25(22):21286–21295
Herrera E, Murillo M, Berumen L, Soto-Cruz NO, Páez-Lerma JB (2017) Protein enrichment of Opuntia ficus-indica using *Kluyveromyces marxianus* in solid-state fermentation. Cienc Investig Agrar 44(2):113–120
Hossain N, Zaini J, Mahlia TMI (2019) Life cycle assessment, energy balance and sensitivity analysis of bioethanol production from microalgae in a tropical country. Renew Sust Energ Rev 115:109371
Huzir NM, Aziz MMA, Ismail SB, Abdullah B, Mahmood NAN, Umor NA, Muhammad SAFAS (2018) Agro-industrial waste to biobutanol

production: eco-friendly biofuels for next generation. Renew Sust Energ Rev 94:476–485

Ibrahim MF, Kim SW, Abd-Aziz S (2018) Advanced bioprocessing strategies for biobutanol production from biomass. Renew Sust Energ Rev 91:1192–1204

Javed F, Aslam M, Rashid N, Shamair Z, Khan AL, Yasin M, Tahir F, Ainy H, Fahad R, Muhammad S, Zakir K, Javed I, Bazmi A (2019) Microalgae-based biofuels, resource recovery and wastewater treatment: a pathway towards sustainable biorefinery. Fuel 255:115826

Jiménez-González C, Constable DJ, Ponder CS (2012) Evaluating the "greenness" of chemical processes and products in the pharmaceutical industry—a green metrics primer. Chem Soc Rev 41(4): 1485–1498

Katakojwala R, Mohan SV (2021) A critical view on the environmental sustainability of biorefinery systems. Curr Opin Green Sustain Chem 27:100392

Koch D, Paul M, Beisl S, Friedl A, Mihalyi B (2020) Life cycle assessment of a lignin nanoparticle biorefinery: decision support for its process development. J Clean Prod 245:118760

Konur O (2021) Producción de biomasa de algas para la producción de biodiesel: una revisión de la investigación. In: Combustibles biodiésel basados en materias primas comestibles y no comestibles, desechos y algas, pp 695–717

Kumar B, Bhardwaj N, Agrawal K, Chaturvedi V, Verma P (2020) Current perspective on pretreatment technologies using lignocellulosic biomass: an emerging biorefinery concept. Fuel Process Technol 199:106244

Lakatos GE, Ranglová K, Manoel JC, Grivalský T, Kopecký J, Masojídek J (2019) Bioethanol production from microalgae polysaccharides. Folia Microbiol (Praha) 64(5):627–644

Li S, Li F, Zhu X, Liao Q, Chang JS, Ho SH (2022) Biohydrogen production from microalgae for environmental sustainability. Chemosphere 291:132717

Liu WJ, Jiang H, Yu HQ (2019) Emerging applications of biochar-based materials for energy storage and conversion. Energy Environ Sci 12(6):1751–1779

López Ortega S (2015) Fermentación sólida de coproductos Y subproductos de la agroindustria de La caña de azúcar por *pleurotus ostreatus*. Tesis de maestría. Universidad Veracruzana, México

Lopez PP, Neria GM, Flores CL, Aguilar LR (2013) A mathematical model for cadmium removal using a sulfate reducing bacterium: Desulfovibrio alaskensis 6SR

López RA, Camacho BR, Neria-González MI, Rangel E, Santos O, Pérez PAL (2017) State estimation based on nonlinear observer for hydrogen production in a photocatalytic anaerobic bioreactor. Int J Chem React Eng 15(5)

López RA, Pérez PAL, Femat R (2020) Control in bioprocessing: modeling, estimation and the use of soft sensors. John Wiley & Sons

López-Pérez PA, Cuevas-Ortiz FA, Gómez-Acata RV, Aguilar-López R (2015) Improving bioethanol production via nonlinear controller with noisy measurements. Chem Eng Commun 202(11):1438–1445

López-Pérez PA, Neria-González MI, Aguilar-López R (2016a) Improvement of activated sludge process using a nonlinear PI controller design via adaptive gain. Int J Chem React Eng 14(1): 407–416

López-Pérez PA, Puebla H, Sánchez HV, Aguilar-López R (2016b) Comparison tools for parametric identification of kinetic model for ethanol production using evolutionary optimization approach Int. J Chem React Eng 14(6):1201–1209

López Pérez PA, Cortes TR, Santos O, Hernández Melchor D, Peña Caballero JV, Martínez Farías F, On Observability and Controllability for a Bioethanol Dynamical Model Obtained from Cocoa Industrial Waste (2018) IEEE 38th Central America and Panama Convention (CONCAPAN XXXVIII). San Salvador, El Salvador, pp 1–6. https://doi.org/10.1109/CONCAPAN.2018.8596444

Lopéz-Peréz PA, Rodriguez-Mata AE, Hernández-González O, Amabilis-Sosa LE, Baray-Arana R, Leon-Borges J (2022) Design of a Robust sliding mode controller for bioreactor cultures in overflow metabolism via an interdisciplinary approach. Open Chem 20(1):120–129

Goedkoop M, Spriensma R (2000) Eco-indicator 99 manual for designers. PRéConsult Amersfoort, Netherlands

Mariani A, Vastola A (2015) Sustainable winegrowing: current perspectives. Int J Wine Res 7(1):37–48

Melgarejo J (2019) Agua y economía circular. Congreso Nacional del Agua Orihuela

Mona S, Kumar SS, Kumar V, Parveen K, Saini N, Deepak B, Pugazhendhi A (2020) Green technology for the sustainable production of biohydrogen (waste to energy): a review. Total Environmental Science 728:138481

Muthu SS (2020) Assessing the environmental impact of textiles and the clothing supply chain. Woodhead Publishing

Onishi VC, Ravagnani MA, Jiménez L, Caballero JA (2017) Multi-objective synthesis of work and heat exchange networks: optimal balance between economic and environmental performance. Energy Convers Manag 140:192–202

Perdices MB (2018) El papel de la bioenergía en la bioeconomía. Mediterráneo económico 31:353–369

Pereira LG, Cavalett O, Bonomi A, Zhang Y, Warner E, Chum HL (2019) Comparison of biofuel life-cycle GHG emissions assessment tools: the case studies of ethanol produced from sugarcane, corn, and wheat. Renew Sust Energ Rev 110:1–12

Pérez PAL, Castillo-Baltazar OS, Aguilar-López R, Caballero VP, Castañeda EV (2019) Virtual sensors for biofuels production: a brief mathematical description for synthesis of algorithms. Comptes Rendus de l'Academie Bulgare des Sciences 72(10):1383–1392

Pérez PAL, Cortes TR, Santos O, Melchor DJH, Caballero VP, Farías FM (2018) On observability and controllability for a bioethanol dynamical model obtained from cocoa industrial waste. In: 2018 IEEE 38th Central America and Panama convention (CONCAPAN XXXVIII). IEEE, pp 1–6

Pérez PAL, Neria-González MI, López RA (2015) Increasing the bio-hydrogen production in a continuous bioreactor via nonlinear feedback controller. Int J Hydrog Energy 40(48):17224–17230

Pérez PAL, Yescas RM, Acata RVG, Caballero VP, López RA (2013) Software sensors design for the simultaneous saccharification and fermentation of starch to ethanol. Fuel 110:219–226

Placeres Remiur A, González Suárez E, Concepción Toledo DN, Mesa Garriga L, González Herrera I (2012) Gestión de ciencia y técnica para el desarrollo prospectivo de tecnologías obtención de etanol, coproductos y otros biocombustibles de la industria de la caña de azúcar. Centro azúcar 39(2):41–55

Porras CB (2021) El rol de los biocombustibles líquidos en la transición energética. C3-BIOECONOMY: Circular and Sustainable Bioeconomy 2:69–80

Prieto-Sandoval V, Jaca-García C, Ormazabal-Goenaga M (2017) Circular economy: relationship with the evolution of the concept of sustainability and strategies for its implementation

Quintanar-Orozco ET, Vázquez-Rodríguez GA, Beltrán-Hernández RI, Lucho-Constantino CA, Coronel-Olivares C, Montiel SG, Islas-Valdez S (2018) Enhancement of the biogas and biofertilizer production from Opuntia heliabravoana Scheinvar. Environ Sci Pollut Res 25(28):28403–28412

Rahman M, Abir MA (2019) A technical analysis regarding the potentiality of rice husk-based power plant in Bangladesh using Cape open to Cape open (COCO) simulator. In International Conference on Innovation in Engineering and Technology (ICIET) (Vol. 23, p. 24)

Ramos Miranda FE, Cuellar MA, Saez Jover D (2018) El desarrollo tecnológico y social de la industria azucarera. El caso de las biorefinerías. Rev Univ soc 10(5):295–300

Rangel E, Sansores E, Vallejo E, Hernández-Hernández A, López-Pérez PA (2016) Study of the interplay between N-graphene defects and small Pd clusters for enhanced hydrogen storage via a spillover mechanism. Phys Chem Chem Phys 18(48):33158–33170

Rathore AS, Kateja N, Kumar D (2018) Process integration and control in continuous bioprocessing. Curr Opin Chem Eng 22:18–25

Romero MG (2020) Biocombustibles y producción de biohidrógeno. MoleQla: Rev de Ciencias de la Universidad Pablo de Olavide 38(8)

Sadvakasova AK, Kossalbayev BD, Zayadan BK, Kirbayeva DK, Alwasel S, Allakhverdiev SI (2021) Potential of cyanobacteria in the conversion of wastewater to biofuels. World J Microbiol Biotechnol 37(8):1–22

Sánchez-Ramírez E, Quiroz-Ramírez JJ, Hernández S, Segovia-Hernández JG, Kiss AA (2017) Optimal hybrid separations for intensified downstream processing of biobutanol. Sep Purif Technol 185:149–159

Sawyerr N, Trois C, Workneh T, Okudoh V (2018) Comparison and modelling of biogas production from unpeeled and peeled cassava tubers at a mesophilic temperature. Int J Mech Eng Technol 9(12): 338–352

Segovia-Hernández JG, Sánchez-Ramírez E, Alcocer-García H, Quíroz-Ramírez JJ, Udugama IA, Mansouri SS (2020) Analysis of intensified sustainable schemes for biobutanol purification. Chem Eng Process 147:107737

Shanmugam S, Hari A, Kumar D, Rajendran K, Mathimani T, Atabani AE, Pugazhendhi A et al (2021) Recent developments and strategies in genome engineering and integrated fermentation approaches for biobutanol production from microalgae. Fuel 285:119052

Sharma B, Larroche C, Dussap CG (2020) Comprehensive assessment of 2G bioethanol production. Bioresour Technol 313:123630

Singh N (2008) Biocompost from sugar distillery effluent: effect on metribuzin degradation, sorption and mobility. Pest Manag Sci 64(10):1057–1062

Singh R, Prakash A, Balagurumurthy B, Bhaskar T (2015) Hydrothermal liquefaction of biomass. In: Recent advances in thermo-chemical conversion of biomass. Elsevier, pp 269–291

Singh V, Dincer I, Rosen MA (2018) Life cycle assessment of ammonia production methods. In: Exergetic, energetic and environmental dimensions. Academic Press, pp 935–959

Soumia PS, Krishna R, Jaiswal DK, Verma JP, Yadav J, Singh M (2021) Entomopathogenic microbes for sustainable crop protection: future perspectives. In: Yadav AN, Singh J, Singh C, Yadav N (eds) Current trends in microbial biotechnology for sustainable agriculture. Environmental and microbial biotechnology. Springer, Singapore, pp 469–497. https://doi.org/10.1007/978-981-15-6949-4_19

Stefanakis AI, Nikolaou I (2021) Circular economy and sustainability, vol 2. Elsevier Publishing, Amsterdam, The Netherlands

Straathof AJ, Wahl SA, Benjamin KR, Takors R, Wierckx N, Noorman HJ (2019) Grand research challenges for sustainable industrial biotechnology. Trends Biotechnol 37(10):1042–1050

Suárez L, Benavente-Ferraces I, Plaza C, de Pascual-T S, Suárez-Ruiz I, Centeno Teresa A (2020) Hydrothermal carbonization as a sustainable strategy for integral valorisation of apple waste. Bioresour Technol 309:123395. https://doi.org/10.1016/j.biortech.2020.123395

Sudhakar K, Mamat R, Samykano M, Azmi WH, Ishak WFW, Yusaf T (2018) An overview of marine macroalgae as bioresource. Renew Sust Energ Rev 91:165–179

Tabatabaei M, Aghbashlo M, Dehhaghi M, Panahi HKS, Mollahosseini A, Hosseini M, Soufiyan MM (2019) Reactor technology for biodiesel production and processing: a review. Prog Energy Combust Sci 74:239–303

Tan IS, Lam MK, Foo HCY, Lim S, Lee KT (2020) Advances of macroalgae biomass for the third generation of bioethanol production. Chin J Chem Eng 28(2):502–517

Torres-Ortega CE, Ramírez-Márquez C, Sánchez-Ramírez E, Quiroz-Ramírez JJ, Segovia-Hernandez JG, Rong BG (2018) Effects of intensification on process features and control properties of lignocellulosic bioethanol separation and dehydration systems. Chem Eng Process 128:188–198

Turton R (2003) Analysis, Synthesis, and Design of Chemical Processes, 7th edn. Prentice Hall, Upper Saddle River, N.J

Ubando AT, Rivera DRT, Chen WH, Culaba AB (2019) A comprehensive review of life cycle assessment (LCA) of microalgal and lignocellulosic bioenergy products from thermochemical processes. Bioresour Technol 291:121837

van Baten JM, Baur R (2015) COCO 3.0: some experiences with multithreading. In: Presented at the 2015 CAPE-OPEN Annual Conference, Amsterdam, the Netherlands. (slides and notes)

van Baten JM (2014) CAPE-OPEN COM Objects Without Fringes. Presented at the 2014 CAPE-OPEN Annual Conference, Mörfelden, Germany

van Baten JM, Szczepanski R (2011) A thermodynamic equilibrium reactor model as a CAPE-OPEN unit operation. Comput Chem Eng 35:1251–1256

van Groenestijn J, Harmsen P, Bos H (2020) Biomass for the circular economy: everything you wanted to know about biomass but were afraid to ask. Wageningen Food & Biobased Research

Velázquez-Sánchez HI, López-Pérez PA, Neria-González MI, Aguilar-López R (2017) Enhancement of bio-hydrogen production technologies by Sulphate-reducing bacteria. Int J Hydrog Energy:385–406

Wang P, Chen YM, Wang Y, Lee YY, Zong W, Taylor S, Wang Y et al (2019) Towards comprehensive lignocellulosic biomass utilization for bioenergy production: efficient biobutanol production from acetic acid pretreated switchgrass with *clostridium saccharoperbutylacetonicum* N1-4. Appl Energy 236:551–559

Xiang W, Zhang X, Chen J, Zou W, He F, Hu X, Gao B et al (2020) Biochar technology in wastewater treatment: a critical review. Chemosphere 252:126539

Yen HW, Hu IC, Chen CY, Ho SH, Lee DJ, Chang JS (2013) Microalgae-based biorefinery–from biofuels to natural products. Bioresour Technol 135:166–174

Zeitsch KJ (2000) The chemistry and technology of furfural and its many by products. Elsevier Book

Zhou Y, Qin S, Verma S, Sar T, Sarsaiya S, Ravindran B, Awasthi MK et al (2021) Production and beneficial impact of biochar for environmental application: A comprehensive review. Bioresour Technol 337:125451

Attached Biofilm Cultivation (ABC) of Mixotrophic Microalgae for the Sustainable Supply of Innovative New Bioproducts

18

Linda O'Higgins and Imen Hamed

Abstract

The industrial utility of microalgae is critically constrained by high-volume, monotrophic suspension approaches to biomass cultivation. Suspension cultivation suffers from high capital costs, labor-intensive culture maintenance, low biomass productivity, and prohibitively high energy demands for biomass mixing and harvesting. Attached Biofilm Cultivation (ABC) of aquatic mixotrophic microalgae is attracting increasing interest as a solution to many of these challenges. Aquatic microalgae biofilms constitute a hidden bank of chemical diversity and functional potential. Primarily phototrophic, biofilm-forming microalgae exhibit high metabolic flexibility with the capacity for complementary heterotrophy (mixotrophy) under low light conditions reported in several species. Biomass yields up to five-fold higher than those of suspension systems are also reported. ABC focusses on biofilm-forming species of diatoms, cyanobacteria, and chlorophytes species naturally rich in high-value organic and inorganic materials. In this review, authors report on ABC as a means for more consistent and cost-effective industrial biomass production. Principle physiochemical parameters affecting biofilm metabolism and metabolite biosynthesis are discussed, with special attention given to photobiology and the use of artificial LED light to enhance productivity in synthetic mixotrophic microalgae biofilms. Finally, inter-species microbe–microbe mutualism in natural biofilms is discussed and the challenges in harnessing positive interactions are reviewed.

Keywords

Attached biofilm cultivation · Mixotrophy · Diatoms · Cyanobacteria · Chlorophytes · Optogenetics · Synthetic biofilms · Bioproduction

18.1 Introduction

Under the "Green Deal" the European Union has set itself on a path to becoming the first "climate-neutral" continent in the world by 2050 (Fetting 2020). This ambitious task includes multilateral measures to catalyze a transition from fossil-dependent linear manufacturing processes toward a more bio-based circular economy. A fundamental prerequisite for this transition is massive yet environmentally sustainable increases in the supply of renewable bio-based feedstocks for conversion to food, feed, consumer products, and bioenergy. Exhibiting fast growth rates, an ability to grow on saltwater on waste land using waste nutrients, and generating substantial quantities of renewable biomaterials, microalgae hold great promise as such a feedstock (Udayan et al. 2021; Khavari et al. 2021; de Mendonça et al. 2021). The European microalgae biomass industry currently contributes approximately 1.25% to total global volume production per year (Araújo et al. 2021; Vazquez Calderon and Sanchez Lopez 2022; Verdelho-Vieira 2018) however, despite this small volume contribution, total product turnover in the European microalgae industry represents approximately 29% (750 million euros) of global market value (Verdelho-Vieira 2018). This reflects not only the high capacity of European producers to add value to the biomass produced but also the high premium that European consumers are willing to pay for sustainably produced renewable ingredients. This large market value share, combined with forecasts for continued growth in consumer demand (CAGR of 9.3% during the forecast period of 2023–2030, provides a compelling basis for the expansion of the European microalgae biomass industry. The Green

L. O'Higgins (✉)
Ellen Hutchins Building, Environmental Research Institute, University College Cork, Cork, Ireland
e-mail: linda.ohiggins@ucc.ie

I. Hamed
Biotechnlogy Research and Application Center, Cukurova University, Sarıçam/Adana, Turkey

A. de Jesús Martínez-Roldán (ed.), *Biotechnological Processes for Green Energy, and High Value Bioproducts by Microalgae, and Cyanobacteria Cultures*, Developments in Applied Phycology 13, https://doi.org/10.1007/978-3-031-43969-8_18

Deal, however, requires that this growth be inherently sustainable providing for carbon-neutral production with minimal loss of biodiversity or ecological health at industrial scales.

To meet these requirements, a significant increase in aerial biomass yields and attendant decrease in biomass production costs are required (Araújo et al. 2021; Benedetti et al. 2018; Khan et al. 2018). A persistent bottleneck in current commercial operations, however, is suboptimal biomass productivity and due largely to high-liquid volumes common to conventional suspension cultivation systems such as open ponds or enclosed photobioreactors (PBRs). Targeting production of whole dried biomass, single-cell proteins, and carotenoid extracts, most phototrophic suspension cultivation occurs under natural light. Contending with diel and seasonal variations in light quality, productivity is critically constrained by rapid attenuation of available light in dilute cultures (e.g., ~<1‰) and light-limitation of cell photosynthesis. According to the basic theory of photosynthesis, the upper limit of photosynthetic efficiency is 12.6% of full solar spectrum, corresponding to microalgal biomass productivity of 120–150 g m^2 day^{-1}. Averaging between 10 and 30 g m^2 day^{-1}, biomass yields from phototrophic open-ponds and enclosed PBRs, however, run up to 7.5 times lower (Brennan and Owende 2010; Mata et al. 2010). Similarly, phototrophic productivity in low-volume enclosed systems, such as flat-plate or tubular PBRs, is constrained by overexposure to incoming light and photodamage during the initial exponential growth phase and inhibition due to rapid light scattering at high cell densities during stationary phase. In both cases, low biomass productivities and the relatively high GHG emissions associated with energy-intensive dewatering of dilute biomass are key contributors to the high environmental costs reported for phototrophic suspension cultivation (Barr and Landis 2018; Smetana et al. 2017). Yielding up to 50 times more biomass than phototrophic culture, heterotrophic fermentation of microalgae-like aquatic thraustochytrids (e.g., *Schizochytrium* spp. and *Cryptothecodinium* spp.) is the cultivation strategy of choice for production of high-value products, most notably omega-3 fatty acids. Although having a lower environmental impact than phototrophic processes, heterotrophic fermentation is a CO_2-generating process requiring maintenance of elevated temperatures (≥30 °C), intensive O_2 aeration, and large and costly inputs of organic carbon. These processes are also constrained by the comparatively slow growth rate and low glucose-to-lipid conversion rates of thraustochytrid species (Ma et al. 2021). Moreover, using a suspension approach, heterotrophically produced biomass also requires energy-intensive dewatering and drying prior to processing—a factor accounting for up to 40% of total microalgae biomass production costs (Spruijt et al. 2014).

Exhibiting biomass productivities comparable to heterotrophic fermentation, Attached Biofilm Cultivation (ABC) of mixotrophic microalgae represents a promising approach for the sustainable expansion of the microalgae biomass industry. Involving the immobilization of cells on a fixed 2-D carrier, ABC provides for significantly reduced water use, enhanced exposure of cells to photosynthetically available radiation (PAR), improved nutrient distribution, and the production of a high-solids content biomass available for immediate processing without centrifugation (Dalirian et al. 2021). Recreating the close metabolic cooperation found between different components of natural multiphyletic assemblages, synthetic coculture of two or more biofilm species, for example algae–algae, algae–bacteria, algae–fungi, or algae–bacteria–fungi, is also shown to significantly improve biomass and lipid generation when compared to monospecific culture (Han et al. 2016; Martin et al. 2021; Mujtaba and Lee 2017; Rajendran and Hu 2016; Zhang et al. 2021). The suite of biotechnological tools to enable engineer functional synthetic cocultures has also advanced rapidly in recent years to include optogenetic techniques that use light to dynamically control biofilm formation, species composition and productivity. In the following sections, we review the ecophysiology of natural microalgae-dominated biofilms, focusing on-the eco-physiology and metabolic versatility of these understudied systems. Key physio-chemical aspects of the biofilm micro-environment are explored and the influence of phototactic biofilm-forming diatoms in modulating and enhancing overall biofilm productivity examined. Synergistic interactions between microalgae and other microbial biofilm components are also touched upon and emerging biotechnological tools to identify, characterize, and enhance beneficial species interactions are briefly reviewed. Finally, the potential application of mixotrophic ABC under artificial light to enhance biomass and metabolite productivity and improve the industrial utility of these versatile microbial systems is examined.

18.2 Natural Aquatic Biofilms and the Microphytobenthos

Consisting of eukaryotic microalgae (diatoms and green algae), prokaryotic bacteria (including cyanobacteria) and fungi, natural aquatic biofilms are complex syntrophic microbial consortia found attached to substrates in often turbid shallow-water environments. Being highly productive, and predominantly phototrophic, they represent an interesting alternative to unicellular suspension culture for more cost-effective microalgae biomass production. Small raphid diatoms such as those from the genera *Nitzschia*, *Navicula*, *Pleurosigma*, and *Amphora* represent the largest biomass component of aquatic biofilms (Admiraal 1984; Polizzi et al. 2017; Mantzorou and Ververidis 2019). These raphid diatoms are key biofilm architects, exuding massive quantities of adhesive extracellular polymeric substances (EPS) that mediate cell-to-cell cohesion and provide a protective hydrophilic matrix within which the multispecies biofilm community grows and develops (Admiraal et al. 1982; Hopkins 1964; Round 1971). Capable of doubling their biomass in as little as 3.5 h (average 7 h), biofilm-forming diatoms are shown to saturate photosynthetic light reactions at incident intensities more than four times lower than the minimum required by planktonic microalgae, i.e., ~30 μmol PAR m^{-2} s^{-1} versus ~150 μmol PAR m^{-2} s^{-1} (Admiraal 1977; Marra 2004; Ooms et al. 2016; Prelle et al. 2021). Highly productive, biofilm-forming diatoms are responsible for up to 50% of total primary production in turbid estuarine systems (Admiraal 1984; Smith and Underwood 2000; Lucas et al. 2001). Demonstrating active phototactic migrations these motile diatoms undertake daily vertical migrations optimizing their position within the biofilm matrix in response to surface irradiance intensity to enhance photon capture or reduce photodamage under light limited and supersaturated conditions respectively (Cartaxana et al. 2016; Consalvey et al. 2004; Jesus et al. 2006; Perkins et al. 2010). In addition to phototaxis, raphid biofilm diatoms have also developed intricate silica cell walls composed of symmetrical nanostructured arrays of pores that act as photonic waveguides for dynamic concentration and direction into the photosynthetic apparatus of available light (Fuhrmann et al. 2004). This combination of unique optical characteristics and phototactic motility enables benthic diatoms to maintain high net photosynthetic productivity within dense biofilm communities even at irradiance intensities considered too low tolight reactions in floating, planktonic species (Admiraal 1977).

Relative to the theoretical quantum limits of photosynthesis calculated by Emerson (1958) for planktonic microalgae, recent study also indicates an exceptional efficiency for photon capture and conversion among chlorophyte and cyanophyte species cultivated in attached biofilm state (Abu-Ghosh et al. 2016; Toninelli et al. 2016). For example, Martín-Girela et al. (2017) reported in excess of ten times higher rates of net photosynthetic activity within a chlorophyte-dominated synthetic biofilm grown under a low, near sub-saturating light intensity (51 μmol photons m^{-2} s^{-1}) compared to high light intensity (766 μmol photons m^{-2} s^{-1}). In this case, light was provided artificially using spectrally-precise light emitting diodes and, when applied as a pulsed, rather than continuous, beam, pulsed red light was found to elicit a remarkable 22.5% increase in biofilm quantum efficiency (measured as CO_2 assimilation per PAR photon absorbed) above Emersons' calculated theoretical maximum of 0.125 μmol CO_2 μmol $photons^{-1}$ (Emerson 1958; Falkowski and Raven 2007; Osborne and Geider 1987). In suspension culture, the quantum efficiency of planktonic microalage has been shown to fall to around 0.1 μmol CO_2 μmol $photons^{-1}$ under low intensity continuous irradiance, and lower still in the case of protein or lipid-rich biomass (0.034 μmol CO_2 μmol $photons^{-1}$) (Abu-Ghosh et al. 2016). Thus, while cultivation of microalgae under artificial light is shown to be costly, the cultivation of high photosynthetic efficiency synthetic microalgae biofilms under spectrally optimized and pulsed light regimes represents a potentially significant gains in terms of enhanced photosynthetic efficiency and biomass yields.

18.3 Mixotrophic Metabolism in Microalgae

While primarily photoautotrophic, the capacity for facultative heterotrophy at sub-saturating irradiances, either simultaneously or in sequence with photosynthesis, has been confirmed in numerous microalgae species (Kaplan et al. 1986; Jareonsin and Pumas 2021; Morales-Sánchez et al. 2015; Ukeles and Rose 1976; Chen and Chen 2006; Chojnacka and Marquez-Rocha 2004; Fabris et al. 2020; Godrijan et al. 2020; Kamalanathan et al. 2018; Liang et al. 2009; Litchman 2007; Patel et al. 2020; Smith et al. 2015; Ukeles and Rose 1976). Known as mixotrophy, this metabolic omnivory enables the catabolism of a wide variety of organic, as well as inorganic, carbon sources including glucose, fructose, glycerol, and others, greatly increasing the pool of carbon available for microalgae cell growth (Chojnacka and Marquez-Rocha 2004). Compared to monotrophic photo- or hetero-trophic cultivations, higher biomass and metabolite yields are frequently reported for mixotrophic cultures (Abiusi et al. 2020; Day and Tsavalos 1996; Sforza et al. 2012; Day and Tsavalos 1996; Li et al. 2014; Villanova et al. 2017; Yun et al. 2021). For example, comparing overall biomass and lipid yields under three trophic regimes, Kamalanathan et al. (2018) ranked the metabolic productivity of the chlorophyte Scenedesmus sp. as follows: mixotrophy > heterotrophy > photoautotrophy. These higher yields combined with significantly lower water use and vastly reduced harvesting costs in ABCs underscores the potential of mixotrophic cultivation strategies to improve the cost-competitiveness and sustainability of microalgae biomass production. The capacity for facultative heterotrophy in microalgae is thought to have first emerged through mutualistic relationships with heterotrophic eukaryotes and ancestral Cyanophytes, with transferred pathways subsequently retained in derivative Chlorophyte, or green algae, lineages (Oborník 2019). Among the diatoms (Bacillariophyta), heterotrophic metabolism is thought to have developed through a sequence of endosymbiotic interactions resulting in plastid recruitment from various heterotrophic prokaryotes, red (and potentially green) algae, and a variety of bacteria (Falciatore et al. 2020; Marella et al. 2021). In all cases however, three metabolic

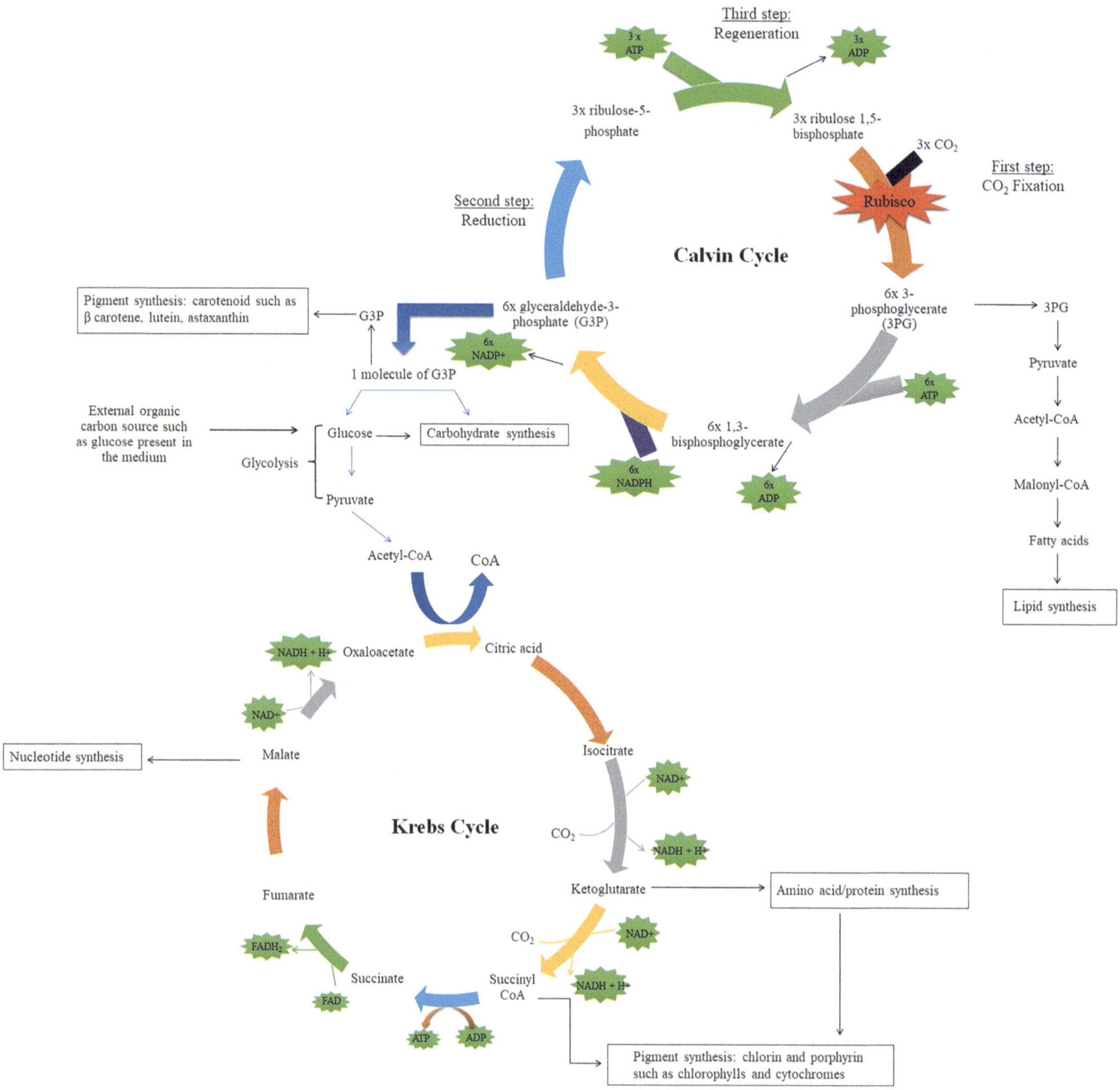

Fig. 18.1 Mixotrophic metabolism pathway in microalgae

mechanisms are associated with microalgal mixotrophy, namely the Calvin cycle (phototrophic), glycolysis, and the Krebs cycle (both heterotrophic) (Figure 18.1).

Photosynthesis occurs in the cell chloroplast and proceeds as a complex and tightly integrated pair of reactions termed the light and dark reactions. Known as the "Z-scheme for photosynthetic electron transport, photosynthetic light reactions occur between two coupled reaction centers (PSII to PSI) located on the thylakoid membrane of chloroplasts. Briefly, PSII to PSI use PAR energy captured by pigments (chlorophylls a and b) to oxidize water and to produce the molecular high-energy carriers ATP and NADPH2 (with O_2 as a by-product). A minimum of eight PAR photons are required for complete saturation of light reactions and fixation of 1 mol of CO_2. Light reactions occur on very short time intervals ranging from picoseconds to nanoseconds (charge separation) and microseconds to milliseconds (electron shuttling between PSII and PSI). Subsequent dark reactions occur in the chloroplast stroma and involve RuBisCo- catalyzed fixation, and conversion, of dissolved CO_2 into organic compounds in the Calvin–Benson cycle. Heterotrophic metabolism, on the other hand, occurs in the cell cytoplasm and involves the conversion of organic carbon, e.g., glucose ($C_6H_{12}O_6$,) into pyruvate ($CH_3COCOO-$ (pyruvic acid)) during glycolysis (Embden–Meyerhof–Parnas pathway) as well as NADPH and pentoses (5-carbon sugars) via the oxidative pentose phosphate (OPP) pathway. Products of both pathways are then metabolized in the Krebs cycle within the mitochondria, producing energy through oxidation of acetyl-CoA and releasing CO_2 as a by-product (Márquez-Rocha et al. 2020; Pang et al. 2019).

Most studies agree that photo- and hetero-trophic metabolisms function noncompetitively in microalgae, and while physically separate within the cell, they are thought to share electron transport systems (Cerón Garcí et al. 2000; Cheirsilp and Torpee 2012; Guminski et al. 1985; Vonshak et al. 2000) and thus, do not function independently of one another. As a consequence, mixotrophic growth is calculated as the sum of photo- and hetero-trophic metabolisms (Vonshak et al. 2000). Compared to monotrophic cultures, biomass yields reported for mixotrophic cultures are thus significantly higher. For example, compared to dark heterotrophic cultivation, early comparative study revealed biomass and lipid yields ×8.5 and ×5.8 higher in mixotrophic cultures of *Nannochloropsis salina* (Sforza et al. 2012) and *Tetraselmis* sp., respectively (Day and Tsavalos 1996). Investigating mixotrophic metabolism in the model diatom *Phaeodactylum tricornutum*, Villanova et al. (2017) reported two-fold higher biomass production in illuminated glycerol-fed mixotrophic cultures when compared to a phototrophic reference. Maintaining O_2-balance in chemostatic mixotrophic cultures of *Chlorella sorokiniana*, Abiusi et al. (2020) reported a doubling of biomass productivity and concentration compared to a phototrophic reference, with consumption of 94% of carbon substrate, making this a highly efficient close to carbon-neutral process. Similarly, investigating metabolic efficiency in *C. pyrenoidosa* under phototrophic, mixotrophic, and cyclic photoautotrophic/dark-heterotrophic (i.e. photoheterotrophic) conditions, Yang et al. (2000) reported highest biomass yield on supplied energy in the latter cyclic photoheterotrophic condition. Whether operating cyclically or synergistically, however, it is important to note that mixotrophy is a dual limited process whereby cell growth can be constrained by light intensity (phototrophic) and/or organic carbon substrate concentration (heterotrophic) (Cerón Garcí et al. 2000).

Thus, while microalgal metabolism responds to the interplay of many environmental factors, as primary energy sources for photosynthetic and heterotrophic growth respectively, balancing the availability of light availability with concentrations of O_2 is key to optimizing metabolic efficiency and productivity in mixotrophic cultures (Abiusi et al. 2020; Chojnacka and Marquez-Rocha 2004; Grama et al. 2016; Patel et al. 2020; Penhaul Smith et al. 2020; Yang et al. 2000). Maintaining this balance is central not only to optimizing biomass productivity but also to achieving maintaining a carbon neutral mixotrophic process. To summarize, light is paramount in determining the rate of phototrophic metabolism, which in turn provides oxygen required for heterotrophic respiration. In natural intertidal biofilms, for example, study by Walpensdorf et al. (2017) highlighted the importance of photosynthetic oxygen production to heterotrophic production. Here, investigators reported a 200% increase in biofilm volume specific heterotrophic O_2 consumption (at the depth interval 0.1 to 0.2 mm) under photosynthetically saturating daytime irradiances (~220 μmol photons m^{-2} s^{-1}). In addition, depth-integrated biofilm metabolism did not shift from net heterotrophic to net autotrophic until surface irradiance levels decreased to 45 μmol photons m^{-2} s^{-1}, (i.e. compensation irradiance at which respiration equals photosynthesis) underlining further the high photobiological efficiency of benthic microalgae. In contrast, studies by Abiusi et al. (2020) demonstrated that the balance between light-driven mixotrophic CO_2:O_2 consumption within suspension cultures was not reached until irradiance incident on the bioreactor surface exceeded 498 ± 17 μmol photons m^{-2} s^{-1}. These results indicate that balanced mixotrophic metabolism can be achieved at significantly lower irradiance in ABC compared to suspension cultivation, and highlight opportunities for the design of cost-effective low-intensity, spectrally-precise and pulsed artificial light regimes to improve biomass and product yields in mixotrophic ABC systems.

In addition to application of tailored regimes of artificial light and metabolic modalities, recent studies show further improvements in mixotrophic microalgae productivity in mixed species biofilms consist of multi-species consisting of multiple species from one or more microbial taxa. When compared to mono-cultures, microbial co-cultures are shown to improve the efficiency of complex chemical production processes by, for example, distributing enzymes across multiple strains and reducing metabolic burden, overcoming endogenous regulatory mechanisms or exploiting natural traits of different microbial species (Lalwani et al. 2021). Termed the interactome, growing knowledge of the biochemical mechanisms underpinning these symbiotic interrelationships is highlighting new opportunities for improved productivity in mixotrophic Attached Biofilm Cultivation (ABC) of symbiotic multi-species microbial biofilms. Table 18.1 provides details of studies using synthetic biofilm co-cultures comprised of algae–algae, algae–bacteria, algae–protozoa, or algae–bacteria–protozoa and which demonstrate benefits in terms of improved microalgae growth rate, metabolic efficiency, and overall resilience of the biofilm to pathogenic attack when compared to monocultures (de la Calle 2017; González-González and de-Bashan 2021; Kiran et al. 2018; Kim et al. 2020). The following sections review the current state of development in ABC systems technology and explores some current approaches to engineer synthetic co-cultures and harness

Table 18.1 A survey of synthetic microalgae coculture consortia, coculturing techniques used (including attached biofilm and suspension), target products, and applications

Cocultured microorganisms	Coculturing technique	Products and/or process of interest	Application	References
Biofilm coculture				
Algae–algae				
Microspora willeana, *Ulothrix zonata*, *Ulothrix aequalis*, and *Rhizoclonium hieroglyphicum*	Benthic Algae Growth Chambers (BAGCs)	Nitrogen removal	Bioremediation	Wilkie A, Mulbry W (2002) Recovery of dairy manure nutrients by benthic freshwater algae. Bioresour Technol 84(1):81–91
Algae–bacteria				
Unidentified natural consortium of phototrophic microalgae and *Pseudomonas aeruginosa*	CO_2 Sequestration Monitoring System (CSMS) biofilm reactor modules	Carbon dioxide and oxygen consumption	Bioremediation	
Navicula veneta and *Halomonas* sp.	Tanaka photobioreactor	Settlement of larval marine invertebrates	Aquaculture	RE Avendaño-Herrera, CE Riquelme. Production of a diatom-bacteria biofilm in a photobioreactor for aquaculture applications. Aquacultural Engineering 36 (2), 97–104
P. ticornutum, *Planothidium* sp., *Pseudomonas* sp. strain 8c, *Bacteroidetes* sp. strain 32, and *Proteobacterium* sp. strain 35	Novel flow lane incubation system	Surface colonization and growth dynamics	Ecology	Buhmann, M.; Kroth, P.G.; Schleheck, D. Photoautotrophic-heterotrophic biofilm communities: A laboratory incubator designed for growing axenic diatoms and bacteria in defined mixed-species biofilms. Environ. Microbiol. Rep. 2012, 4, 133–140
Monoraphidium sp., *Scenedesmus acutus*, and unclassified bacterial components	Gastight flow cell system	Nitrogen, phosphorous and acetate removal from wastewaters	Bioremediation	N.C. Boelee, H. Temmink, M. Janssen, C.J.N. Buisman, R.H. Wijffels, Balancing the organic load and light supply in symbiotic microalgal–bacterial biofilm reactors treating synthetic municipal wastewater, Ecological Engineering, Volume 64, 2014, Pages 213–221, ISSN 0925-8574
Algae–fungi				
Chlorella vulgaris and *Mucor indicus*	Fixed biofilm reactor	Linoleic acid	Bioremediation	Rodrigues Reis CE, Ogero D'Otaviano L, Rajendran A, Hu B. Co-Culture of Filamentous Feed-Grade Fungi and Microalgae as an Alternative to Increase Feeding Value of Ethanol Coproducts. *Fermentation*. 2018; 4(4):86
C. vulgaris and *Murcor circinelloides*	Novel mycoalgae biofilm reactor	Biofilm technology development	Biotechnology	Rajendran A, Fox T, Hu B. Nutrient recovery from ethanol co-products by a novel mycoalgae biofilm: attached cultures of symbiotic fungi and algae. J Chem Technol Biotechnol. 2017;92: 1766–76
Algae–bacteria–fungi				
Chlorella vulgaris, *Scenedesmus obliquus*, endophytic bacterial strain (S395-2), and fungi *Ganoderma lucidum (G. lucidum, 5.765)* and *Pleurotus ostreatus*	Enclosed photobioreactor	Nutrient purification and CO_2 removal	Biogas upgrading	Xuechang Dong, Jing Wei, Jing Huang, Chunzhi Zhao, Shiqing Sun, Yongjun Zhao, Juan Liu, Performance of different microalgae-fungi-bacteria co-culture technologies in photosynthetic and removal performance in response to various GR24 concentrations, Bioresource Technology, Volume 347, 2022, 126428, ISSN 0960-8524

(continued)

Table 18.1 (continued)

Cocultured microorganisms	Coculturing technique	Products and/or process of interest	Application	References
Suspension coculture				
Algae–algae				
Algae–algae: *Chlorella protothecoides* (heterotrophic and autotrophic)	Gas separation	CO_2 and O_2 exchange	Biofuels	Santos CA, Ferreira ME, Lopes Da Silva T, et al. A symbiotic gas exchange between bioreactors enhances microalgal biomass and lipid productivities: taking advantage of complementary nutritional modes. J Ind Microbiol Biotechnol. 2011;38(8): 909–917
Algae–algae: *Chlorella protothecoides* and *Tetraselmis suecica*	Pelletization and flocculation	Bio-flocculating compounds	Bioremediation	Muradov N, Taha M, Miranda AF, et al. Fungal-assisted algal flocculation: application in wastewater treatment and biofuel production. Biotechnol Biofuels. 2015;8:24
Algae–algae: *Chlorella vulgaris* and *Pseudokirchneriella subcapitata*	Direct mixing	Chlorellin	Ecology	DellaGreca M, Zarrelli A, Fergola P, et al. Fatty acids released by Chlorella vulgaris and their role in interference with Pseudokirchneriella subcapitata: experiments and modelling. J Chem Ecol. 2010;36(3): 339–349
Algae–bacteria				
Algae–bacteria: *Chlorella vulgaris*, *C. sorokiniana*, and *Azospirillum brasilense*	Alginate bead immobilized	Pigments and lipids	Biotechnology	L.E. de-Bashan, Y. Bashan, M. Moreno, V.K. Lebsky, J.J. Bustillos. Increased pigment and lipid content, lipid variety, and cell and population size of the microalgae Chlorella spp. when co-immobilized in alginate beads with the microalgae-growth-promoting bacterium Azospirillum brasilense. Can. J. Microbiol., 48 (6) (2002), pp. 514–521
Algae–bacteria: *C. vulgaris* and *Nitrobacter*	Erlenmeyer flask	Enhanced microalgal biomass and lipid production	Bioremediation	Sagar Grover, Shaily Rubina Tirkey, Veeramallegowda, Saroj Yadav and G. Sibi, 2020. Improved Biomass Through Mutualistic Co-culturing of *Chlorella vulgaris* with *Nitrobacter* in Sewage Water. *Biotechnology, 19: 1–9*
Algae–cyanobacteria: *Chlorella vulgaris* and *Microcystis aeruginosa*	Dialysis tube membrane	Linoleic acid and nitrous oxide	Ecology	Song, H., Lavoie, M., Fan, X. et al. Allelopathic interactions of linoleic acid and nitric oxide increase the competitive ability of *Microcystis aeruginosa. ISME J* 11, 1865–1876 (2017). https://doi.org/10.1038/ismej.2017.45
Algae–cyanobacteria: *Oocystis marsonii* and *Microcystis aeruginosa*	Membrane separation	Allelopathic metabolites not identified	Biotechnology	Dunker S, Althammer J, Pohnert G, Wilhelm C. A Fateful Meeting of Two Phytoplankton Species-Chemical vs. Cell-Cell-Interactions in Co-Cultures of the Green Algae Oocystis marsonii and the Cyanobacterium Microcystis aeruginosa. Microb Ecol. 2017 Jul;74 (1):22–32
Algae–cyanobacteria: *Desmodesmus subspicatus* and *Microcystis aeruginosa* (mycrocystins producing and nonproducing)	Dialysis tube membrane	Cell biomass, extracellular metabolites, and bioactives	Biotechnology	Azam Omidi, Maranda Esterhuizen-Londt, Stephan Pflugmacher, Interspecies interactions between Microcystis aeruginosa PCC 7806 and Desmodesmus subspicatus SAG 86.81 in a co-cultivation system at various

(continued)

Table 18.1 (continued)

Cocultured microorganisms	Coculturing technique	Products and/or process of interest	Application	References
				growth phases, Environment International, Volume 131, 2019, 105052, ISSN 0160-4120
Algae–other microorganism				
Algae–fungi: *Chlorella vulgaris* and *Mucor circinelloides*	Erlenmeyer flask	Biomass and lipids	Biotechnology	Zorn, S.M.F.E.; Reis, C.E.R.; Silva, M. B.; Hu, B.; De Castro, H.F. Consortium Growth of Filamentous Fungi and Microalgae: Evaluation of Different Cultivation Strategies to Optimize Cell Harvesting and Lipid Accumulation. *Energies* 2020, *13*, 3648. https://doi.org/10.3390/en13143648
Algae–yeast: *Chlorella vulgaris* and *Rhodotorula glutinis*	Erlenmeyer flask and membrane separation	Organic acids, propionic, pyruvic, and acetic acids	Biofuels	Zhiping Zhang, Hairui Ji, Guiping Gong, Xu Zhang, Tianwei Tan, Synergistic effects of oleaginous yeast Rhodotorula glutinis and microalga Chlorella vulgaris for enhancement of biomass and lipid yields, Bioresource Technology, Volume 164, 2014, Pages 93–99, ISSN 0960-8524

Adapted from Kapoore et al.

beneficial species interactions to improve microalgae biomass and product yields.

18.4 Attached Biofilm Cultivation Systems (ABCs)

A variety of Attached Biofilm Cultivation systems have been developed over the past 25 years. Originally developed for use in wastewater remediation, increasing industry understanding of the multiple uses of microalgae biomass has seen considerable development in ABCs technology, enabling production of higher quality biomass and metabolites suitable for direct human use (Cheng et al. 2017; Jairo Hernán et al. 2021; Ummalyma et al. 2023; Zhang et al. 2019). ABC involves the immobilization of cells on a fixed 2-D substrate usually immersed in small volumes of liquid media. Providing a large surface area for cell attachment, the distribution and uptake of both light and nutrients is vastly improved in ABC compared to suspension systems (Huang et al. 2016; Murphy and Berberoglu 2014). For example, ABCs provide more than three fold reduction in optical path length compared to the narrowest flat-plate suspension reactor, providing for more uniform exposure of cells to photosynthetically available radiation (PAR) and greatly enhancesd photosynthetic efficiency in conversion to useful biomass (mols.CO_2 reduced/O_2 evolved per unit light energy input). Photosynthetic biomass productivities up to seven times higher than well-operated suspension systems are reported for naturally-lit ABCs (Kandilian et al. 2017; Ooms et al. 2016; Podola et al. 2017; Wang et al. 2017). In terms of water use, while about 200 metric tons of water are required to produce 1 ton of microalgae in suspended cultivation, ABC system (ABCs) use only around 17 tons for nutrient media circulation and approximately 4 tons to maintain biofilm surface wetness (Katarzyna et al. 2015). These massive reductions in water use in turn greatly reduce costly biomass dewatering and drying costs (Singh and Patidar 2018), and, in many ABC scenarios, enables harvesting by simple scraping of a high-solids content biomass (75–85% water-content) suitable for immediate extraction without drying (Katarzyna et al. 2015; Spruijt et al. 2014; Wang et al. 2017).

ABCs designs are classified by the method of nutrient supply and reactor geometry (Fig. 18.2). These include perfusion systems (or porous substrate), constantly immersed systems, and intermittently immersed systems (alternating between gaseous and liquid phases). Using tiered attachment surfaces, horizontal or vertical stacking of ABCs also provides for more efficient use of space and higher areal productivity compared to suspension techniques (Mantzorou and Ververidis 2019; Roostaei et al. 2018). Reviewing performances among different ABCs, for example, Wang et al. (2017) reported marked improvements in biomass and lipid yields for perfusion (or porous-substrate) ABCs, and in

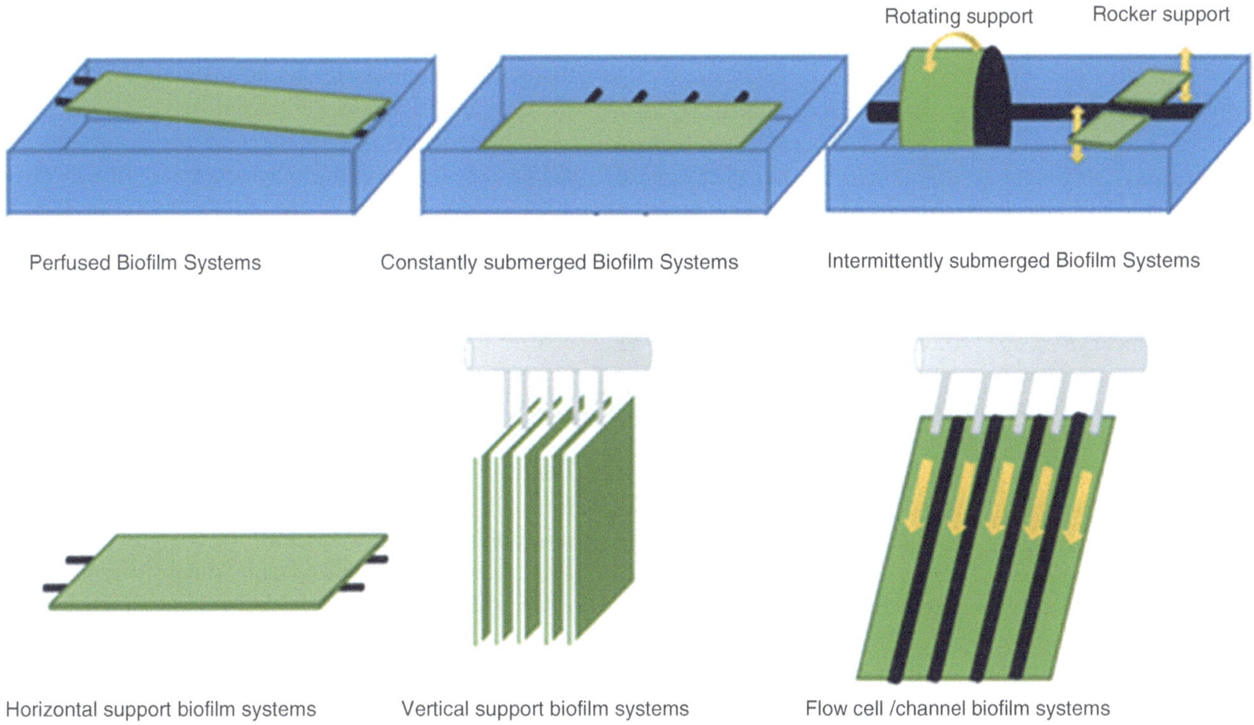

Fig. 18.2 Schematic of various Attached Biofilm Cultivation systems designs and arrangements

particular, twin-layer perfusion designs. Perfusion ABCs deliver nutrients and water though a microporous biofilm substrate sandwiched between a twin-layer biofilm support that exposes dual-biofilm surfaces directly to the surrounding atmosphere (Li et al. 2017; Naumann et al. 2013; Schultze et al. 2015; Wang et al. 2017). Exposing biofilm surfaces directly with the ambient gas phase in this way is shown to significantly improve gas exchange (CO_2 and O_2), reduce the vertical buildup of dissolved inorganic carbon (DIC) and help maintain depth-integrated neutral pH as the biofilm matures and thickens (Li et al. 2017). A significant advancement over both suspension PBRs and submerged ABCs, the absence of a continuous liquid barrier in perfusion ABCs eliminates shear stress common in gas-sparged suspension systems, eliminates the need for mechanical mixing, and minimizes/eliminates coupling of the solid retention time and the hydraulic retention time.

Cell immobilization can be achieved either through cell enclosure or through cell attachment (non-enclosed biofilms) (Moreno-Garrido 2008). While somewhat more time-consuming to establish, non-enclosed biofilms are more convenient since the end biomass is easily harvested by simple scraping, retaining an underlying seed stock for new colony formation in subsequent cultivations (Moghaddam et al. 2018). Non-enclosed cell immobilization can be passive (using cell's natural ability to attach to surfaces) or active (by means of gel encapsulations, flocculant agents, and chemical attachments). Active immobilization techniques include affinity immobilization, adsorption, confinement in liquid–liquid emulsion, covalent coupling, entrapment in polymers, and capture behind semipermeable membranes (de-Bashan and Bashan 2010). Biofilm attachment substrates can be natural such as wood, loofa sponge, natural polysaccharides (such as agar-agar, cellulose, and alginate), and lignocellulosic materials. Synthetic attachment substrates include stainless steel, polyacrylamide, polyurethane, polyvinyl chloride, polypropylene, polysulfone, and epoxy resin (Cheah and Chan et al. 2021; Chen et al. 2014; Zhang et al. 2017). Substrate materials selection however must not only take account of economic aspects including durability, cost, and availability but also the physiochemical properties of the material surface (Ozkan and Berberoglu 2013a,b; Venable and Podbielski et al. 2019). For example, studies show that cell adhesion and substrate colonization by biofilm-forming microalgae are strongly influenced by material texture (Huang et al. 2018), porosity and pore size, hydrophobicity (Gross et al. 2016), surface charge and free energy relative to that of the microalgae cell (Ozkan and Berberoglu 2013b; Zhang et al. 2020; Zheng et al. 2019). Surface hydrophobicity, charge and fee energy is in turn, influenced by the pH of the liquid medium (Ozkan and Berberoglu 2013b; Yuan et al. 2019; Zhang et al. 2020) while other environmental variables including light availability, temperature, hydrodynamic conditions, nutrient quality and quantity will also affect microalgae settlement and biofilm formation (Sabater et al. 2002).

18.5 Migratory Diatoms as a Tool to Optimize Mixotrophic Metabolism in ABC

In contrast to suspension systems which require energy intensive mechanical mixing to ensure adequate access to light and nutrients (Caldwell et al. 2021), mass transfer to cells inside ABCs relies largely on passive diffusion (Berner et al. 2015). As a biofilm matures and thickens (~>7 days old; ~100–200 μm), however, diffusion can become severely limited with tight packaging of cells often resulting in the development of steep, vertical gradients in light, DIC, O_2, and nutrients (N and P). These vertical gradients in turn create metabolically distinct and often growth-limited subpopulations, with light-limitation of photosynthetic surface populations caused by rapid absorption and scattering of incoming light and limitation of underlying heterotrophic metabolism due to low dissolved oxygen (DO) concentrations (Li et al. 2017). As described earlier, phototactic diatoms undertake vertical migrations to compensate for both supra- and sub-optimal irradiance intensity at the biofilm surface. Studies by Cook and Roy (2006), Cartaxana et al. (2016), Prins et al. (2020) and others have shown that these migrations however not only improve light capture in dense biofilms but also increase biofilm permeability, enhancing mass transfer, diffusion, and exchange of gaseous and dissolved constituents with depth and at the biofilm–cell interface. In the absence of mechanical mixing, these bioturbating migrations can provide advective relief of growth limiting conditions in mature biofilms, facilitating increased diffusion downward of O_2 and upward of DIC produced by photo- and heterotrophic populations, respectively (Fig. 18.3) (Cook and Roy 2006; Li et al. 2017). In addition, light signals are also shown to trigger massive changes in microalgal gene expression, eliciting metabolic shifts in response to different light wavelengths and photoperiods and directly regulating cellular enzyme production in some species (Depauw et al. 2012). Consequently, the use of artificial light to develop optogentetic tools and species-specific light recipes that can target enhanced production of industrially relevant metabolites is currently of great interest among microalgae biotechnologists (Hoffman et al. 2022; Lalwani et al. 2021; Pouzet et al. 2020).

Motility In addition to light-harvesting pigments, microalgae possess sophisticated arrays of spectrally precise photoreceptors that enable rapid responses to dynamic changes in light wavelength, intensity, and periodicity (Rockwell et al. 2014; Rockwell and Lagarias 2017). Typically composed of a protein photosensory module bound to a nonprotein chromophore, photoreceptors vastly extend the spectral light-sensing capabilities of algae from deep blue (434 nm) to near-infrared (734 nm) wavelengths. Due to the rapid attenuation of red light with depth in aquatic environments, blue-light receptors, i.e., aureochromes, are thought to predominate among microalgae (Schellenberger Costa et al. 2013; however, recent studies indicate that both cyanobacteria and eukaryotic algae also possess an array of red/far-red light-sensing phytochromes (Rockwell et al. 2014; Rockwell and Lagarias 2017). Delivering light stimuli downstream, these photoreceptors in turn, help to regulate photo-adaptive processes, cell movement, gene expression, metabolism, and cell division (Hegemann et al. 2001; Jaubert et al. 2017). Among the raphid diatoms, light quality-induced changes in physical motility have been reported under blue, green, and red wavelengths, indicating the presence of several band-specific photoreceptors (McLachlan et al. 2009; Cohn et al. 2015). Examining photo-physiological responses to light, for example, Prins et al. (2020) reported intense upward/downward migration of benthic raphid diatoms under low-intensity blue, and high-intensity red light, respectively. Similarly, early work by Nultsch (1971) showed strong upward migration in the raphid diatom *Nitzschia communis* under blue light, while recent study demonstrated measurable upward migration in the closely related species *Navicula perminuta* at all visible wavelengths up to, but not exceeding, 540 μm (McLachlan et al. 2009). Observing responses of natural intertidal biofilm diatoms, Wenderoth and Rhiel (2004) observed surface biomass accumulations 1.8 times higher in blue compared to white light. Collectively, these studies indicate that at low irradiance intensity, upward diatom migrations are strongest under monochromatic blue wavelengths, while at high irradiance intensity, red wavelengths are most efficient at stimulating downward migration.

Physiology Both growth and metabolism in microalgae are strongly light-dependent. For example, numerous studies show increased cell-specific growth rate, higher cell dry weight (DW), and marked changes in metabolite profile and specific yields in several species when cultivated under precisely controlled light regimes (Atta et al. 2013; Abu-Gosh et al. 2016; Baer et al. 2018; Maltsev et al. 2021; Sirisuk et al. 2018; Severes et al. 2017; Teo et al. 2014). For example, Prins et al. (2020) reported higher chlorophyll concentrations and PSII quantum efficiency in migrational diatom biofilms compared to non-migrational ones under low light intensity. Similarly, the use of spectrally precise *flashing* light delivered at flash-frequencies matching the timescales of electron shuttling in PSII has been shown to yield significant improvements in microalgae photon utilization and photosynthetic conversion efficiencies (Abu-Ghosh et al. 2016; Toninelli et al. 2016). Most notably, Martin-Girela et al. (2017) demonstrated a 22.5% increase ($P < 0.05$) in photosynthetic quantum yield in fixed biofilm cultures grown under low-intensity high flash-frequency red light

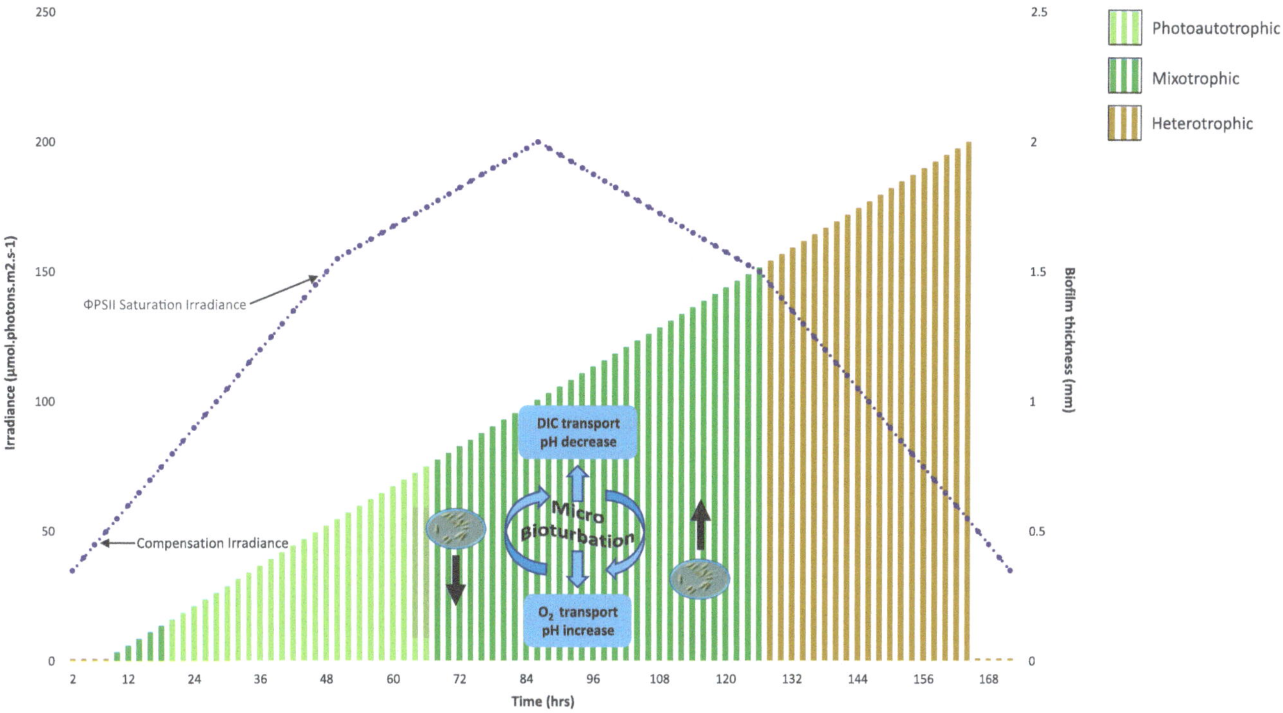

Fig. 18.3 Schematic representation of a precision cultivation strategy including coupled carbon-feeding and irradiance regime. The irradiance regime depicted is aimed at activating photokinetic migratory responses of motile diatom biofilms to optimize PAR capture in the surface euphotic zone, reduce vertical gradients in DIC and O_2 within the biofilm, and maintain depth-averaged pH to between 7.5 and 8.5 within a diatom dominated mixed synthetic biofilm

(51 µmol photons m^{-2} s^{-1}; 10 kHz; 630 nm). In contrast, recent studies indicate an increase in cell protein, PUFA, and pigment contents in two chlorophyte and one haptophyte species at low flash-frequency white light (Lima et al. 2021). Examining the physiological effects of light wavelength, the majority of studies report higher cell photon absorption under blue compared to red or green light, with Ruyters (1984) and others reporting a strong blue-light effect for enzyme activation, regulation of gene transcription, and energy derivation in microalgae. In particular, enhancement of lipid synthesis under monochromatic blue light is reported for species across multiple phyla including diatoms, haptophytes (Sirisuk et al. 2018), and chlorophytes (Teo et al. 2014; Zhang et al. 2022).

Upon review of the extensive literature investigating photobiological responses of biofilm-forming microalgae and, in particular, phototactic diatoms, therefore, authors have identified the following artificial irradiance protocol for to optimize the phototactic response and improve overall mixotrophic growth of a diatom-dominated microbial biofilm (Fig. 18.3):

1. Slight exceedance of the compensation irradiance as reported by Walpensdorf et al. (2017) for natural aquatic biofilms, i.e., ~45 µmol photons m^{-2} s^{-1}.
2. Gradual increase in irradiance intensity to photosynthetically saturating levels to stimulate upward cell migration, and intracellular phytopigment production. Upward phototaxis will also serve to increase biofilm permeability and inter alia, deliver CO_2 produced in heterotrophy to illuminated surface populations for use in photosynthesis.
3. Increase irradiance intensity above photosynthetically saturating levels, stimulating downward phototactic migration and improve delivery of photosynthetically-produced O_2 to underlying heterotrophic populations.
4. Maintenance of saturating intensity to maximize photosynthesis and O_2 production until light and/or O_2 limitation is reached.
5. Gradual reduction in irradiance intensity to compensation level to stimulate upward migration, switch depth-integrated biofilm net metabolic mode to heterotrophy and enhance overall biofilm biomass yields (Fig. 18.3).

18.6 Application of Artificial Light to Mixotrophic ABC

The refinement of artificial *light recipes* such as that suggested above to provide light precisely tailored to the specific (1) photosynthetic requirements, (2) phototactic, and (3) metabolic responses of industrially-relevant biofilm-forming microalgae represents a novel approach to optimizing mixotrophic biomass production. Within the terrestrial horticulture industry, for example, recent advances in LED technology have enabled development of new and sophisticated lighting systems capable of delivering highly precise light regimes tailored to maximize not only final biomass yields but also to control cropping, customize the metabolic profile and improve the attendant nutritional or medicinal value of commercially important plant species. Compared to traditional artificial light sources, state-of-the-art LED technologies exhibit significantly higher photoelectric conversion efficiency, and significantly longer lifespan (~100,000 h). LEDs are also far more adjustable in terms of the intensity, spectrum, and frequency of light delivered to enable, for example, close coupling of wavelength profiles and photon flux densities to exact photosynthetic requirements of photosynthesis and minimize energy wastage. However, despite extensive laboratory evidence of the potentially profound benefits in terms of improved biomass and bio-product yields, in the case of microalgae production, similar application of LED-enabled light recipes to improve the efficiency of production has been hindered by the rapid dissipation of light in conventional high-volume suspension-cultivation systems. In closed production systems for higher plants for example, artificial lighting costs represent about 30% of total OPEX (Olvera-Gonzalez et al. 2021). Using conventional high-volume enclosed suspension systems for microalgae production however, artificial lighting costs increase to about 85% of total OPEX, even when using low volume flat-panel PBRs (Spruijt et al. 2014). Thus, while enclosed systems production under artificial light currently provides significant benefits for the plant horticulture sector e.g. higher, more consistent and year-round harvests, artificial light use for microalgae cultivation is too costly under conventional suspension cultivation scenarios (Imran Khan et al. 2018). Given the higher photon capture rates and photosynthetic efficiencies reported for ABCs and the high photobiological efficiency of biofilm-forming microalgae, authors speculate that improvements in total biomass yield on light and the control of biomass metabolite profiles offered by mixotrophic cultivation in ABC modality under energy-efficient, and spectrally-precise LED lighting presents exciting new opportunities to improve the economic viability of micro-algae biomass and products production.

18.7 Conclusions

The utility of microalgae as a renewable bio-based feedstock has been the subject of great industrial interest since the 1970s; however, ensuring the economic and environmental sustainability of biomass production at large scale remains a challenge. Issues that need to be addressed relate primarily to the energy intensity of biomass production, the substantial CO_2 emissions related to operation processes, and the huge water volumes required during both cultivation and processing. Many researchers have been studying microalgae in terms of strain development, physiology, metabolic modeling, photobioreactor design and operation, biorefining, and renewable energy generation. To achieve a sustainable and circular bioeconomy with respect to ecological boundaries, a multidisciplinary approach should be employed for the efficient use of microalgae. In this review, immobilization of microalgae and biofilm cultivation using an efficient mixotrophic cultivation strategy under optimized artificial light was discussed. It was concluded that this approach can significantly improve productivity and lower costs, harnessing the high resource efficiency and productvities of mixotrophic biofilm-forming microalgae, greatly reducing water use and harvesting costs via use of perfusion ABCs, and employing state-of-the-art LED technology to provide for stable, and cost-effective production of high-value biomass. Given the environmental imperatives and growing market demand and environmental imperatives for sustainable, renewable and biobased alternatives to the myriad fossil-based materials consumers currently rely on, the opportunities for microalgae to support the green transition are only increasing.

References

Abiusi F, Wijffels RH, Janssen M (2020) Doubling of microalgae productivity by oxygen balanced mixotrophy. ACS Sustain Chem Eng 8(15):6065–6074

Abu-Ghosh S, Fixler D, Dubinsky Z, Iluz D (2016) Flashing light in microalgae biotechnology. Bioresour Technol 203:357–363. https://doi.org/10.1016/j.biortech.2015.12.057. Epub 2015 Dec 23

Admiraal W (1977) Influence of light and temperature on the growth rate of estuarine benthic diatoms in culture. Mar Biol 39:1–9

Admiraal W (1984) The ecology of estuarine sediment-inhabiting diatoms. In: Round FE, Chapman DJ (eds) Progress in phycological research biopress bristol, vol 3, pp 269–322

Admiraal W, Peletier H, Zomer H (1982) Observations and experiments on the population-dynamics of epipelic diatoms from an estuarine mudflat. Estuarine Coastal Shelf Sci 14:471–487

Araújo R, Vázquez Calderón F, Sánchez López J, Azevedo IC, Bruhn A, Fluch S, Garcia Tasende M, Ghaderiardakani F, Ilmjärv T, Laurans M, Mac Monagail M, Mangini S, Peteiro C, Rebours C, Stefansson T, Ullmann J (2021) Current status of the algae production industry in Europe: an emerging sector of the blue bioeconomy. Front Mar Sci 7:626389. https://doi.org/10.3389/fmars.2020.626389

Atta M et al (2013) Intensity of blue LED light: a potential stimulus for biomass and lipid content in freshwater microalgae *Chlorella vulgaris*. https://doi.org/10.1016/j.biortech.2013.08.162

Baer S et al (2018) Optimization of spectral light quality for growth and product formation in different microalgae using a continuous photobioreactor. https://doi.org/10.1016/j.algal.2016.01.011

Barr WJ, Landis AE (2018) Comparative life cycle assessment of a commercial algal multiproduct biorefinery and wild caught fishery for small pelagic fish. Int J Life Cycle Assess 23:1141–1150. https://doi.org/10.1007/s11367-017-1395-7

Benedetti M, Vecchi V, Barera S et al (2018) Biomass from microalgae: the potential of domestication towards sustainable biofactories. Microb Cell Factories 17:173

Berner F, Heimann K, Sheehan M (2015) Microalgal biofilms for biomass production. J Appl Phycol 27(5):1793–1804

Brennan L, Owende P (2010) Biofuels from microalgae—a review of technologies for production, processing, and extractions of biofuels and co-products. Renew Sust Energ Rev 14:557–577

Caldwell GS, In-na P, Hart R, Sharp E, Stefanova A, Pickersgill M, Walker M, Unthank M, Perry J, Lee JGM (2021) Immobilising microalgae and cyanobacteria as biocomposites: new opportunities to intensify algae biotechnology and bioprocessing. Energies 14 (9):2566. https://doi.org/10.3390/en14092566

Cartaxana P, Cruz S, Gameiro C, Kühl M (2016) Regulation of intertidal microphytobenthos photosynthesis over a diel emersion period is strongly affected by diatom migration patterns. Front Microbiol 7: 872

Cerón Garcí M, Fernández Sevilla J, Acién Fernández F et al (2000) Mixotrophic growth of *Phaeodactylum tricornutum* on glycerol: growth rate and fatty acid profile. J Appl Phycol 12:239–248. https://doi.org/10.1023/A:1008123000002

Cheah YT, Chan DJC (2021) Physiology of microalgal biofilm: a review on prediction of adhesion on substrates. Bioengineered 12(1):7577–7599. https://doi.org/10.1080/21655979.2021.1980671. PMCID: PMC8806711

Cheirsilp B, Torpee S (2012) Enhanced growth and lipid production of microalgae under mixotrophic culture condition: effect of light intensity, glucose concentration and fed-batch cultivation. Bioresour Technol 110:510–516

Chen G, Chen F (2006) Growing phototrophic cells without light. Biotechnol Lett 28:607–616

Chen X, Liu T, Wang Q (2014) The growth of Scenedesmus sp. attachment on different materials surface. Microb Cell Factories 13 (1):142. https://doi.org/10.1186/s12934-014-0142-z. PMCID: PMC4203884

Cheng P, Wang Y, Liu T, Liu D (2017) Biofilm attached cultivation of Chlorella pyrenoidosa is a developed system for swine wastewater treatment and lipid production. Front Plant Sci 8:1594

Chojnacka K, Marquez-Rocha F-J (2004) Kinetic and stoichiometric relationships of the energy and carbon metabolism in the culture of microalgae. Biotechnology 3:21–34

Cohn SA, Halpin D, Hawley N, Ismail A, Kaplan Z, Kordes T et al (2015) Comparative analysis of light-stimulated motility responses in three diatom species. Diatom Res 30:213–225. https://doi.org/10.1080/0269249X.2015.1058295

Consalvey M, Paterson DM, Underwood GJC (2004) The ups and down of life in a benthic biofilm: migration of benthic diatoms. Diatom Res 19:181–202

Cook PLM, Roy H (2006) Advective relief of CO2 limitation in microphytobenthos in highly productive sandy sediments. Limnol Oceanogr 51:1594–1601

Dalirian N, Najafabadi HA, Movahedirad S (2021) Surface attached cultivation and filtration of microalgal biofilm in a ceramic substrate photobioreactor. Algal Res 55:102239. ISSN 2211-9264. https://doi.org/10.1016/j.algal.2021.102239

Day JG, Tsavalos AJ (1996) An investigation of the heterotrophic culture of the green algaTetraselmis. J Appl Phycol 8:73–77. https://doi.org/10.1007/BF02186225

de la Calle F (2017) Marine microbiome as source of natural products. Microb Biotechnol 10(6):1293–1296

De-Bashan LE, Bashan Y (2010) Immobilized microalgae for removing pollutants: review of practical aspects. Bioresour Technol 101(6): 1611–1627

de Mendonça HV, Assemany P, Abreu M, Couto E, Maciel AM, Duarte RL, dos Santos MGB, Reis A (2021) Microalgae in a global world: new solutions for old problems? Renew Energy 165(Part 1):842–862. https://doi.org/10.1016/j.renene.2020.11.014. ISSN 0960-1481

Depauw FA, Rogato A, d'Alcalá MR, Falciatore A (2012) Exploring the molecular basis of responses to light in marine diatoms. J Exp Bot 63(4):1575–1591. https://doi.org/10.1093/jxb/ers005

Emerson R (1958) The quantum yield of photosynthesis. Annu Rev Plant Physiol 9:1–24

Fabris M, Abbriano RM, Pernice M et al (2020) Emerging technologies in algal biotechnology: toward the establishment of a sustainable, algae-based bioeconomy. Front Plant Sci 11:279. Published 2020 Mar 17

Falciatore A, Jaubert M, Bouly J-P, Bailleul B, Mock T (2020) Diatom molecular research comes of age: model species for studying phytoplankton biology and diversity. Plant Cell 32(3):547–572

Falkowski PG, Raven JA (2007) Aquatic photosynthesis, 2nd edn. STU - Student edition ed. Princeton University Press

Fetting C (2020) The European green deal. In: ESDN Report, December 2020. ESDN Office, Vienna

Fuhrmann T, Landwehr S, El Rharbi-Kucki M et al (2004) Diatoms as living photonic crystals. Appl Phys B Lasers Opt 78:257–260. https://doi.org/10.1007/s00340-004-1419-4

Godrijan J, Drapeau D, Balch WM (2020) Mixotrophic uptake of organic compounds by coccolithophores. Limnol Oceanogr 65: 1410–1421

González-González LM, de Bashan LE (2021) Toward the enhancement of microalgal metabolite production through microalgae-bacteria consortia. Biology (Basel) 10(4):282

Grama BS, Agathos SN, Jeffryes CS (2016) Balancing photosynthesis and respiration increases microalgal biomass productivity during Photoheterotrophy on glycerol. ACS Sustain Chem Eng 4:1611–1618

Gross M, Zhao X, Mascarenhas V et al (2016) Effects of the surface physico-chemical properties and the surface textures on the initial colonization and the attached growth in algal biofilm. Biotechnol Biofuels 9:38. https://doi.org/10.1186/s13068-016-0451-z

Guminski S, Jurajda K, Tatkowska E (1985) Increase in the biomass of some green algae species in nitrate and ammonium media depending on auto-, mixo- or heterotrophic conditions. Acta Soc Bot Pol 54:65–76

Han J, Zhang L, Wang S et al (2016) Co-culturing bacteria and microalgae in organic carbon containing medium. J Biol Res-Thessaloniki 23:8

Hegemann P, Fuhrmann M, Kateriya S (2001) Algal sensory photoreceptors. J Phycol 37:668–676

Hoffman SM, Tang AY, Avalos JL (2022) Optogenetics illuminates applications in microbial engineering. Annu Rev Chem Biomolecular Eng 13:1

Hopkins JT (1964) A study of the diatoms of the Ouse Estuary, Sussex-II. The ecology of the mud-flat diatom flora. J Mar Biol Assoc U K 44:333–341

Huang Y, Xiong W, Liao Q, Fu Q, Xia A, Zhu X, Sun Y (2016) Comparison of Chlorella vulgaris biomass productivity cultivated in biofilm and suspension from the aspect of light transmission and microalgae affinity to carbon dioxide. Bioresour Technol 222:367–373., ISSN 0960-8524. https://doi.org/10.1016/j.biortech.2016.09.099

Huang Y, Zheng Y, Li J, Liao Q, Fu Q, Xia A, Fu J, Sun Y (2018) Enhancing microalgae biofilm formation and growth by fabricating microgrooves onto the substrate surface. Bioresour Technol 261: 36–43

Jairo Hernán MO, Antonino P, Luigi F, Nicolaas Luc LP, Giovanni E (2021) A review of microalgal biofilm technologies: definition, applications, settings and analysis. Front Chem Eng 3:737710. https://doi.org/10.3389/fceng.2021.737710. ISSN=2673-2718

Jareonsin S, Pumas C (2021) Advantages of heterotrophic microalgae as a host for phytochemicals production. Front Bioeng Biotechnol 9:628597. https://doi.org/10.3389/fbioe.2021.628597

Jaubert M, Bouly J-P, Ribera d'Alcalà M, Falciatore A (2017) Light sensing and responses in marine microalgae. Curr Opin Plant Biol 37:70–77

Jesus B, Perkins RG, Consalvey MC, Brotas V, Paterson DM (2006) Effects of vertical migration by benthic microalgae on fluorescence measurements of photophysiology. Mar Ecol Prog Ser 315: 55–66

Kamalanathan M, Chaisutyakorn P, Gleadow R, Beardall J (2018) A comparison of photoautotrophic, heterotrophic, and mixotrophic growth for biomass production by the green alga Scenedesmus sp. (Chlorophyceae). Phycologia 57(3):309–317

Kandilian R, Jesus B, Legrand J, Pilon L, Pruvost J (2017) Light transfer in agar immobilized microalgae cell cultures. J Quant Spectrosc Radiat Transf 198:81–92. ISSN 0022-4073

Kaplan D, Richmond AE, Dubinsky Z, Aaronson S (1986) Algal nutrition. In: Richmond A (ed) Handbook for microalgal mass culture. CRC Press, Boca Raton, pp 147–198

Katarzyna L, Sai G, Singh OA (2015) Non-enclosure methods for non-suspended microalgae cultivation: literature review and research needs. Renew Sust Energ Rev 42:1418–1427

Khan MI, Shin JH, Kim JD (2018) The promising future of microalgae: current status, challenges, and optimization of a sustainable and renewable industry for biofuels, feed, and other products. Microb Cell Factories 17:36

Khavari F, Saidijam M, Taheri M, Nouri F (2021) Microalgae: therapeutic potentials and applications. Mol Biol Rep 48(5):4757–4765. https://doi.org/10.1007/s11033-021-06422-w. Epub 2021 May 24. PMCID: PMC8142882

Kim D, Hyun-Sik Y, Young-Saeng K, Jong-Guk K (2020) Effects of co-culture on improved productivity and bioresource for microalgal biomass using the floc-forming bacteria *Melaminivora jejuensis*. Front Bioeng Biotechnol 8. https://doi.org/10.3389/fbioe.2020.588210

Kiran GS, Ramasamy P, Sekar S, Ramu M, Hassan S, Ninawe AS, Selvin J (2018) Synthetic biology approaches: towards sustainable exploitation of marine bioactive molecules. Int J Biol Macromol 112:1278–1288

Lalwani MA, Kawabe H, Mays RL, Hoffman SM, Avalos JL (2021) Optogenetic control of microbial consortia populations for chemical production. ACS Synth Biol 10(8):2015–2029. https://doi.org/10.1021/acssynbio.1c00182

Li Y-R, Tsai W-T, Hsu Y-C, Xie M-Z, Chen J-J (2014) Comparison of autotrophic and mixotrophic cultivation of green microalgal for biodiesel production. Energy Procedia 52:371–376

Li T, Strous M, Melkonian M (2017) Biofilm-based photobioreactors: their design and improving productivity through efficient supply of dissolved inorganic carbon. FEMS Microbiol Lett 364(24):fnx218. https://doi.org/10.1093/femsle/fnx218

Liang Y, Sarkany N, Cui Y (2009) Biomass and lipid productivities of Chlorella vulgaris under autotrophic, heterotrophic and mixotrophic growth conditions. Biotechnol Lett 31:1043–1049

Lima S, Schulze PSC, Schüler LM, Rautenberger R, Morales-Sánchez D, Santos TF, Pereira H, Varela JCS, Scargiali F, Wijffels RH, Kiron V (2021) Flashing light emitting diodes (LEDs) induce proteins, polyunsaturated fatty acids, and pigments in three microalgae. J Biotechnol 325:15–24

Litchman E (2007) CHAPTER 16 - resource competition and the ecological success of phytoplankton. In: Falkowski PG, Knoll AH (eds) Evolution of primary producers in the sea. Academic Press, Burlington, pp 351–375

Lucas CH, Banham C, Holligan M (2001) Benthic-pelagic exchange of microalgae at a tidal flat. 2. Taxonomic analysis. Mar Ecol Prog Ser 212:39–52

Ma W, Wang YZ, Nong FT et al (2021) An emerging simple and effective approach to increase the productivity of thraustochytrids microbial lipids by regulating glycolysis process and triacylglycerols' decomposition. Biotechnol Biofuels 14:247

Maltsev Y, Maltseva K, Kulikovskiy M, Maltseva S (2021) Influence of light conditions on microalgae growth and content of lipids, carotenoids, and fatty acid composition. Biology 10(10):1060. https://doi.org/10.3390/biology10101060

Mantzorou A, Ververidis F (2019) Microalgal biofilms: a further step over current microalgal cultivation techniques. Sci Total Environ 651(Part 2):3187–3201

Marella TK, Bhattacharjya R, Tiwari A (2021) Impact of organic carbon acquisition on growth and functional biomolecule production in diatoms. Microb Cell Factories 20:135. https://doi.org/10.1186/s12934-021-01627-x

Márquez-Rocha JF, Palma-Ramírez D, García-Alamilla P, López-Hernández JF, Santiago-Morales IS, Flores-Vela AI (2020) Microalgae cultivation for secondary metabolite production. In: Microalgae - From physiology to application. IntechOpen. https://doi.org/10.5772/intechopen.88531

Marra J (2004) The compensation irradiance for phytoplankton in nature. Geophys Res Lett 31:L06305. https://doi.org/10.1029/2003GL018881

Martin N, Bernat T, Dinasquet J et al (2021) Synthetic algal-bacteria consortia for space-efficient microalgal growth in a simple hydrogel system. J Appl Phycol 33:2805–2815

Martín-Girela I, Curt MD, Fernández J (2017) Flashing light effects on CO2 absorption by microalgae grown on a biofilm photobioreactor. Algal Res 25:421–430

Mata TM, Martins AA, Caetano NS (2010) Microalgae for biodiesel production and other applications: a review. Renew Sust Energ Rev 14(1):217–232., ISSN 1364-0321. https://doi.org/10.1016/j.rser.2009.07.020

McLachlan DH, Brownlee C, Taylor AR, Geider RJ, Underwood GJ (2009) Light-induced motile responses of the estuarine benthic diatoms Navicula perminuta and Cylindortheca closterium (Bacillariophyceae) (1). J Phycol 45(3):592–599

Moghaddam E, Amirebrahim S, Harun R, Mokhtar MN, Zakaria R (2018) Potential of zeolite and algae in biomass immobilization. BioMed research international, 2018

Morales-Sánchez D, Martinez-Rodriguez OA, Kyndt J et al (2015) Heterotrophic growth of microalgae: metabolic aspects. World J Microbiol Biotechnol 31:1–9. https://doi.org/10.1007/s11274-014-1773-2

Moreno-Garrido I (2008) Microalgae immobilization: current techniques and uses. Bioresour Technol 99(10):3949–3964

Mujtaba G, Lee K (2017) Treatment of real wastewater using co-culture of immobilized Chlorella vulgaris and suspended activated sludge. Water Res 120:174–184

Murphy TE, Berberoglu H (2014) Flux balancing of light and nutrients in a biofilm photobioreactor for maximizing photosynthetic productivity. Biotechnol Prog 30(2):348–359

Naumann T, Çebi Z, Podola B et al (2013) Growing microalgae as aquaculture feeds on twin-layers: a novel solid-state photobioreactor. J Appl Phycol 25:1413–1420. https://doi.org/10.1007/s10811-012-9962-6

Nultsch W (1971) Phototactic and photokinetic action spectra of the diatom *Nitzschia communis*. Photochem Photobiol 14:705–712
Oborník M (2019) Endosymbiotic evolution of algae, secondary heterotrophy and parasitism. Biomolecules 9(7):266
Olvera-Gonzalez E, Escalante-Garcia N, Myers D, Ampim P, Obeng E, Alaniz-Lumbreras D, Castaño V (2021) Pulsed LED-lighting as an alternative energy savings technique for vertical farms and plant factories. Energies 14:1603. https://doi.org/10.3390/en14061603
Ooms MD, Dinh CT, Sargent EH, Sinton D (2016 Sep) Photon management for augmented photosynthesis. Nat Commun 1(7):12699. https://doi.org/10.1038/ncomms12699. PMCID: PMC5025804
Osborne BA, Geider RJ (1987) The minimum photon requirement for photosynthesis: an analysis of the data of Warburg & Burk (1950) and Yuan, Evans & Daniels (1955). New Phytol 106(4): 631–644
Ozkan A, Berberoglu H (2013a) Adhesion of algal cells to surfaces. Biofouling 29(4):469–482
Ozkan A, Berberoglu H (2013b) Cell to substratum and cell to cell interactions of microalgae. Colloids Surf B: Biointerfaces 112: 302–309
Pang N, Gu XY, Chen SL, Kirchhoff H, Lei HW, Roje S (2019) Exploiting mixotrophy for improving productivities of biomass and co-products of microalgae. Renew Sust Energ Rev 112:450–460. https://doi.org/10.1016/j.rser.2019.06.001
Patel AK, Joun J, Sim SJ (2020) A sustainable mixotrophic microalgae cultivation from dairy wastes for carbon credit, bioremediation and lucrative biofuels. Bioresour Technol 313:123681., ISSN 0960-8524. https://doi.org/10.1016/j.biortech.2020.123681
Penhaul Smith JK, Hughes AD, McEvoy L, Day JG (2020) Tailoring of the biochemical profiles of microalgae by employing mixotrophic cultivation. Bioresour Technol Rep 9:100321
Perkins RG, Lavaud J, Serôdio J, Mouget J-L, Cartaxana P et al (2010) Vertical cell movement is a primary response of intertidal benthic biofilms to increasing light dose. Mar Ecol Prog Ser 416:93–103. https://doi.org/10.3354/meps08787. ⟨hal-01095756⟩
Podola B, Tong L, Melkonian M (2017) Porous substrate bioreactors: a paradigm shift in microalgal biotechnology? Trends Biotechnol 35(2):121–132
Polizzi B, Bernard O, Ribot M (2017) A time-space model for the growth of microalgae biofilms for biofuel production. J Theor Biol 432:55–79
Pouzet S, Banderas A, Le Bec M, Lautier T, Truan G, Hersen P (2020) The promise of optogenetics for bioproduction: dynamic control strategies and scale-up instruments. Bioengineering (Basel) 7(4): 151. https://doi.org/10.3390/bioengineering7040151
Prelle LR, Albrecht M, Karsten U, Damer P, Giese T, Jähns J, Müller S, Schulz L, Viertel L, Glaser K (2021) Ecophysiological and cell biological traits of benthic diatoms from Coastal Wetlands of the Southern Baltic Sea. Front Microbiol 12:642811. https://doi.org/10.3389/fmicb.2021.642811. PMCID: PMC8072133
Prins A, Deleris P, Hubas C, Jesus B (2020) Effect of light intensity and light quality on diatom behavioral and physiological photoprotection. Front Mar Sci 7:203. https://doi.org/10.3389/fmars.2020.00203
Rajendran A, Hu B (2016) Mycoalgae biofilm: development of a novel platform technology using algae and fungal cultures. Biotechnol Biofuels 9:112
Rockwell NC, Lagarias JC (2017) Phytochrome diversification in cyanobacteria and eukaryotic algae. Curr Opin Plant Biol 37:87–93. https://doi.org/10.1016/j.pbi.2017.04.003. Epub 2017 Apr 23. PMCID: PMC5483197
Rockwell NC, Duanmu D, Martin SS, Bachy C, Price DC, Bhattacharya D, Worden AZ, Lagarias JC (2014) Eukaryotic algal phytochromes span the visible spectrum. Proc Natl Acad Sci U S A 111:3871–3876
Roostaei J, Zhang Y, Gopalakrishnan K, Ochocki AJ (2018) Mixotrophic microalgae biofilm: a novel algae cultivation strategy for improved productivity and cost-efficiency of biofuel feedstock production. Sci Rep 8(1):1–10
Round FE (1971) Benthic marine diatoms. Oceanogr Mar Biol Annu Rev 9:83–139
Ruyters G (1984) Effects of blue light on enzymes. In: Senger H (ed) Blue light effects in biological systems. Proceedings in life sciences. Springer, Berlin, Heidelberg. https://doi.org/10.1007/978-3-642-69767-8_32
Sabater S, Guasch H, Romaní A et al (2002) The effect of biological factors on the efficiency of river biofilms in improving water quality. Hydrobiologia 469:149–156. https://doi.org/10.1023/A:1015549404082
Schellenberger Costa B, Sachse M, Jungandreas A, Bartulos CR, Gruber A et al (2013) Aureochrome 1a is involved in the photoacclimation of the diatom *Phaeodactylum tricornutum*. PLoS One 8(9):e74451
Schultze LKP, Simon MV, Li T, Langenbach D, Podola B, Melkonian M (2015) High light and carbon dioxide optimize surface productivity in a Twin-Layer biofilm photobioreactor. Algal Res 8:37–44., ISSN 2211-9264. https://doi.org/10.1016/j.algal.2015.01.007
Severes et al (2017) Use of light emitting diodes (LEDs) for enhanced lipid production in micro-algae-based biofuels. https://doi.org/10.1016/j.jphotobiol.2017.04.023
Sforza E, Simionato D, Giacometti GM, Bertucco A, Morosinotto T (2012) Adjusted light and dark cycles can optimize photosynthetic efficiency in algae growing in photobioreactors. PLoS One 7(6): e38975. https://doi.org/10.1371/journal.pone.0038975. Epub 2012 Jun 20. PMCID: PMC3380057
Singh G, Patidar SK (2018) Microalgae harvesting techniques: a review. J Environ Manag 217:499–508
Sirisuk P, Sunwoo I, Kim SH, Awah CC, Ra CH, Kim J-M, Jeong G-T, Kim S-K (2018) Enhancement of biomass, lipids, and polyunsaturated fatty acid (PUFA) production in Nannochloropsis oceanica with a combination of single wavelength light emitting diodes (LEDs) and low temperature in a three-phase culture system. Bioresour Technol 270:504–511., ISSN 0960-8524. https://doi.org/10.1016/j.biortech.2018.09.025
Smetana S, Sandmann M, Rohn S, Pleissner D, Heinz V (2017) Autotrophic and heterotrophic microalgae and cyanobacteria cultivation for food and feed: life cycle assessment. Bioresour Technol 245(Pt A):162–170. https://doi.org/10.1016/j.biortech.2017.08.113. Epub 2017 Aug 20
Smith DJ, Underwood GJC (2000) The production of extracellular carbohydrates by estuarine benthic diatoms: the effects of growth phase and light and dark treatment. J Phycol 36:321–333. https://doi.org/10.1046/j.1529-8817.2000.99148.x
Smith RT, Bangert K, Wilkinson SJ, Gilmour DJ (2015) Synergistic carbon metabolism in a fast growing mixotrophic freshwater microalgal species Micractinium inermum. Biomass Bioenergy 82: 73–86
Spruijt J, Schipperus R, Kootstra M, Visser CLM de, Parker B (2014) AlgaEconomics: bioeconomic production models of micro-algae and downstream processing to produce bio energy carriers. Public Output report of the EnAlgae project, Swansea, June 2015, 67 pp
Teo CL, Atta M, Bukhari A, Taisir M, Yusuf AM, Idris A (2014) Enhancing growth and lipid production of marine microalgae for biodiesel production via the use of different LED wavelengths. Bioresour Technol 162:38–44. https://doi.org/10.1016/j.biortech.2014.03.113
Toninelli A, Wang J, Liu M et al (2016) Scenedesmus dimorphus biofilm: Photoefficiency and biomass production under intermittent lighting. Sci Rep 6:32305. https://doi.org/10.1038/srep32305
Udayan A, Pandey AK, Sharma P et al (2021) Emerging industrial applications of microalgae: challenges and future perspectives. Syst Microbiol Biomanuf 1:411–431. https://doi.org/10.1007/s43393-021-00038-8

Ukeles R, Rose WE (1976) Observations on organic carbon utilization by photosynthetic marine microalgae. Mar Biol 37:11–28

Ummalyma SB, Sirohi R, Udayan A et al (2023) Sustainable microalgal biomass production in food industry wastewater for low-cost biorefinery products: a review. Phytochem Rev 22:969–991. https://doi.org/10.1007/s11101-022-09814-3

Vazquez Calderon F, Sanchez Lopez J (2022) An overview of the algae industry in Europe – Producers, production systems, species, biomass uses, other steps in the value chain and socio-economic data. Publications Office of the European Union. https://data.europa.eu/doi/10.2760/813113

Venable ME, Podbielski MR (2019) Impact of substrate material on algal biofilm biomass growth. Environ Sci Pollut Res Int 26 (7):7256–7262. https://doi.org/10.1007/s11356-019-04148-8. Epub 2019 Jan 16

Verdelho-Vieira V (2018, April 25th – 26th) The volatility of the Algae Industry; Overview of the industry status in the world and in Europe. 8th European Algae Industry Summit. Vienna, Austria. https://doi.org/10.13140/RG.2.2.28277.91361

Villanova V, Fortunato AE, Singh D, Bo DD, Conte M, Obata T, Jouhet J, Fernie AR, Marechal E, Falciatore A, Pagliardini J, Le Monnier A, Poolman M, Curien G, Petroutsos D, Finazzi G (2017) Investigating mixotrophic metabolism in the model diatom Phaeodactylum tricornutum. Philos Trans R Soc Lond Ser B Biol Sci 372 (1728):20160404. https://doi.org/10.1098/rstb.2016.0404. PMCID: PMC5516113

Vonshak A, Cheung SM, Chen F (2000) Mixotrophic growth modifies the response of Spirulina (Arthrospira) platensis (cyanobacteria) cells to light. J Phycol 36(4):675–679

Walpersdorf E, Kühl M, Elberling B, Andersen T, Hansen B, Pejrup M et al (2017) In situ oxygen dynamics and carbon turnover in an intertidal sediment (Skallingen, Denmark). Mar Ecol Prog Ser 566: 49–65

Wang J, Liu W, Liu T (2017) Biofilm based attached cultivation technology for microalgal biorefineries—a review. Bioresour Technol 244(Part 2):1245–1253

Wenderoth K, Rhiel E (2004) Influence of light quality and gassing on the vertical migration of diatoms inhabiting the Wadden Sea. Helgoland Mar Res 58:211–215

Yang C, Hua Q, Shimizu K (2000) Energetics, and carbon metabolism during growth of microalgal cells under photoautotrophic, mixotrophic and cyclic light-autotrophic/dark-heterotrophic conditions. Biochem Eng J 6(2):87–102

Yuan H, Zhang X, Jiang Z, Wang X, Chen X, Cao L, Zhang X (2019) Analyzing the effect of pH on microalgae adhesion by identifying the dominant interaction between cell and surface. Colloids Surf B Biointerfaces 177:479–486., ISSN 0927-7765. https://doi.org/10.1016/j.colsurfb.2019.02.023

Yun H-S, Kim Y-S, Yoon H-S (2021) Effect of different cultivation modes (Photoautotrophic, mixotrophic, and heterotrophic) on the growth of Chlorella sp. and biocompositions. Front Bioeng Biotechnol 9:774143. https://doi.org/10.3389/fbioe.2021.774143

Zhang Q, Liu C, Li Y, Yu Z, Chen Z, Ye T, Wang X, Hu Z, Liu S, Xiao B, Jin S (2017) Cultivation of algal biofilm using different lignocellulosic materials as carriers. Biotechnol Biofuels 10(1):115

Zhang W, Ding W, Li YX et al (2019) Marine biofilms constitute a bank of hidden microbial diversity and functional potential. Nat Commun 10:517. https://doi.org/10.1038/s41467-019-08463-z

Zhang X, Yuan H, Wang Y, Guan L, Zeng Z, Jiang Z, Zhang X (2020) Cell surface energy affects the structure of microalgal biofilm. Langmuir 36(12):3057–3063

Zhang CS-H, Ho AL, Liang F, Zhou D (2021) Co-culture of Chlorella and Scenedesmus could enhance total lipid production under bacteria quorum sensing molecule stress. J Water Process Eng 39:101739

Zhang P, Xin Y, He Y et al (2022) Exploring a blue-light-sensing transcription factor to double the peak productivity of oil in Nannochloropsis oceanica. Nat Commun 13:1664. https://doi.org/10.1038/s41467-022-29337-x

Zheng Y, Huang Y, Xia A, Qian F, Wei C (2019) A rapid inoculation method for microalgae biofilm cultivation based on microalgae-microalgae co-flocculation and zeta-potential adjustment. Bioresour Technol 278:272–278. https://doi.org/10.1016/j.biortech.2019.01.083. Epub 2019 Jan 21

19 Interactions Between Organizations Regarding Management of River Water Basins at a Socially Vulnerable Context

Patricia L. Marconi and Laura I. de Cabo

Abstract

The hydrographic basin is the most accepted territorial unit for the integrated management of water resources. The Matanza-Riachuelo (MRB) and Reconquista (RB) basins are located in the most densely populated and industrialized region of Argentina. They are plain basins, with low-flow and low self-purification ability. They present the most serious and oldest urban environmental problems in Argentina in relation to the dumping of domestic and industrial effluents, pipelines, and a lack of land use planning. To improve the management capacities of both basins, two modalities were applied: institutional autarky in the MRB (Matanza-Riachuelo Basin Authority—ACUMAR) and adaptive concertation in the RB (Reconquista Basin Committee—COMIREC). In this context, an interaction was established between government agencies and research institutes that allowed the design and implementation of water remediation strategies applying *Chlorella vulgaris* immobilized in alginate. This experience demonstrated the potential of this simple and low-cost technology to remove urban-water contaminants, offering as an additional advantage of the possibility of an easy biomass recovery, which may become a source of alternative energy. The first approach was to push for the reduction of several chemical and microbial contaminants present in Cildáñez stream (MRB) using microalgae immobilized in alginate beads. However, these beads had a relatively short shelf life in the natural environment. To overcome this limitation, a second approach was designed: a 3D-printed PLA device to protect the microalgae in alginate matrix. The algae within the PLA devices were protected and able to bioremediate the highly contaminated Bancalari stream (RB).

P. L. Marconi (✉)
CONICET, CEBBAD-Univ. Maimónides, CABA, Argentina
e-mail: marconi.patricialaura@maimonides.edu

L. I. de Cabo
Museo Argentino de Ciencias Naturales "B. Rivadavia"-CONICET, CABA, Argentina

Keywords

Management of natural resources · Institutional organization · Government agencies · Remediation · Water contaminants

19.1 Introduction

In 2010, water and sanitation were explicitly recognized as universal human rights without distinction of any kind by the United Nations (UN) and was incorporated into the 2030 Agenda for Sustainable Development. In Latin America and the Caribbean, although in the last two decades there has been progress in access to safe water and sanitation services, there are still marked inequities between urban and rural populations, belonging to different ethnic groups, levels of education, and well-being, among others. Likewise, there is an underestimation of the degree of deprivation of services in certain population groups that are particularly susceptible to situations of vulnerability (Queiroz et al. 2020). People in vulnerable situations are frequently discriminated against in different ways (WWAP 2019a, b), which increases their chances of being deprived of water supply and sanitation services or facilities. In Argentina, it is estimated that 39.8 million people live in urban areas, of which 87% have access to water through the public network and only 58% to sewers. Although there are no reliable statistics regarding the level of treatment, it is estimated to be between 15 and 20% of the collected water (Lentini et al. 2017). Queiroz et al. (2020) estimated the inequality factors for access to water supply and sanitation services in Latin American and Caribbean countries from census data. For Argentina, Queiroz et al. (2020) concluded that there was hardly any difference in access to water supply and a greater inequality regarding sanitation services between the selected subgroups. It should

A. de Jesús Martínez-Roldán (ed.), *Biotechnological Processes for Green Energy, and High Value Bioproducts by Microalgae, and Cyanobacteria Cultures*, Developments in Applied Phycology 13, https://doi.org/10.1007/978-3-031-43969-8_19

be noted that Argentina has great regional heterogeneity. In addition, it is noteworthy that census data focus on access to facilities alone, not on the quality of water supply. Both facts can intensify the inequality at subnational level in the country.

The increase in the Argentine population (12% in the period 2001–2010) is associated with the consolidation and greater number of precarious settlements in urban areas, with the Metropolitan Area of Buenos Aires being one of those that presented the highest population growth. Consequently, it is the one with the greatest differences in the coverage of drinking water by network (70%) and in the sewer service (38%). Coverage gaps directly affect the most vulnerable social sectors. Likewise, in the peri-urban areas of large cities, there are risks of contamination in drinking water due to the poor state of maintenance of the networks and risks of groundwater contamination in areas without sewers (Lentini et al. 2017). In addition, the incorrect or nonexistent treatment of wastewater causes water pollution, which entails the loss of the ecosystem services in surface and groundwater environments.

According to the World Bank (2019) (https://www.worldbank.org/en/news/press-release/2019/08/20/worsening-water-quality-reducing-economic-growth-by-a-third-in-some-countries), there would be an inverse relationship between the income level of the countries of the world and the degree of organic contamination of their rivers, expressed through the level of dissolved oxygen in their waters: the lower the income, the lower the content of dissolved oxygen, that is, greater organic contamination. A combination of bacteria, sewage, chemicals, and plastics can suck oxygen from water supplies and transform water into poison for people and ecosystems. This relationship marks a deterioration that tends to be accentuated over time. Therefore, in areas with an unfavorable social context, water treatments that increase dissolved oxygen and decrease nutrients are a priority.

Historically, populations settled on the riverbanks. The humanity has found favorable conditions in these spaces for the development of agricultural, forestry, and urban activities, which condition their state of conservation. Consequently, riverbanks and floodplains are among the most altered landscapes in the world. Change in land use, in particular urbanization without proper planning, has led to the degradation of rivers and their corridors, even toward completely irreversible situations (Munné and Prat 2004). The human attitude toward riverbanks has been changing over time. Initially, these areas were considered unproductive and were linked to diseases of water origin (cholera, yellow fever, diarrhea, among others). They were used for open-air dumps and the riverbeds themselves near populated areas were used as routes for the transport and discharge of sewage effluents, solid waste, industrial effluents, and polluting substances that were transported to the sea or rivers with greater flow. Riparian areas have recently been considered as plausible recovery zones through remediation, mitigation, or restoration projects.

The "Sustainable Development Goal 6," adopted by the United Nations in 2015, ensures safe drinking water and sanitation for all, focusing on the sustainable management of water resources, wastewater, and ecosystems, and acknowledging the importance of an enabling environment. In this process, and in accordance with the recommendations of the different international organizations that are committed to promoting management models for hydrographic basins, various projects aimed at recovering water resources have been undertaken. The incorporation of community objectives in the decision-making process is crucial for the success of the recovery process of degraded environments. The inhabitants must identify with the project, and the greatest possible acceptance and participation by the population is required (Pfadenhauer 2001). Redmond (2004) argues that the inclusion of public participation in project planning and decision-making improves the quality of the results since the local public has specific information about the site. In addition, the intervention and support of local people in the design and execution of the project ensure continuity and success in future management. On the other hand, environmental management requires negotiations to resolve conflicts; therefore, cooperation between managers, decision-makers, the academic sphere, and the local population is necessary and beneficial when making decisions. For the recovery process to be sustainable, it is necessary that the recovered space be accepted, used, and valued by the local community.

The hydrographic basin is the most accepted territorial unit for the integrated management of water resources. Within the framework of the basin recovery process, actors from the scientific community have been called to develop and implement rehabilitation actions at the basin level. In this work, the interventions carried out in the Matanza-Riachuelo and Reconquista basins will be exposed from the interaction between autarchic organisms of the national public administration (ACUMAR), provincial (COMIREC), and decentralized organisms linked to the area of science and technique of the national state such as CONICET (National Council for Scientific and Technical Research) and universities (Buenos Aires University and Maimonides University).

19.2 Matanza-Riachuelo and Reconquista River Basins

The authorities and the inhabitants themselves dismissed the importance of conserving water resources and promoted the piping and channeling of many streams and rivers in urban areas of the metropolitan region of Buenos Aires (Argentina).

In rural areas, they were transformed and used for intensive agriculture or livestock farming. These interventions brought about the regulation of flows and, in general, have led to the reduction of lateral flows of water and sediments, modifying the processes of ecological succession (Ward 1998) and reducing the ecosystem services provided (de Cabo and Marconi 2021). The piping and channeling reduce the self-purification capacity of rivers and streams. The nonexistence of these lateral flows brings with it the loss of flow buffering capacity exerted by the riparian areas. In addition, the remedial capacity exerted by the vegetation and microorganisms of the riverbanks is noticeably diminished. On many occasions, the riverbanks have been deforested and/or reforested with exotic species (de Cabo et al. 2019), increasing the risk of erosion. Likewise, the piping of watercourses makes them more prone to clogging due to sedimentation and reduces the transport capacity. In addition, the invisibility of the course prevents adequate control of the discharge of effluents, increasing the health risk.

The discharges of liquid effluents and waste of rural, industrial, and domestic origin carried out for more than 200 years made the Matanza-Riachuelo basin one of the most polluted in the world (Mojica Duran and de Cabo 2021). In 2003, an environmental emergency was declared for the Matanza-Riachuelo basin. In 2004, a group of neighbors sued the Argentine state for the terrible state of the river and for the lack of interest in recovering it and, in 2006, the Supreme Court of Justice of the Nation ordered the State to present a sanitation plan. As a result of this, the Matanza-Riachuelo Basin Authority (ACUMAR) was created, which is an autonomous, autarchic, and interjurisdictional entity. Its objective is the promotion of infrastructure works for sanitation, cleaning and maintenance of public space, and control of environmental conditions and industrial activity. As of 2008, the Supreme Court of Justice of the Nation ordered ACUMAR to implement a sanitation plan whose objectives were aimed at improving the quality of life of the inhabitants of the Basin, recovering the environment in all its components (water, air, and ground) and prevent damage with a sufficient and reasonable degree of predictability (Mojica Duran and de Cabo 2021). Through three loans, the World Bank has financed US$ 1227 million for sanitation improvement, industrial control, territorial planning, and flood management of the basin (https://www.bancomundial.org/es/news/press-release/2021/06/01/el-banco-mundial-otorga-us-265-millones-para-finalizar-obras-clave-en-el-saneamiento-del-matanza-riachuelo).

The Reconquista River basin is located within a critical area with respect to the conservation status of the freshwater ecoregions of Latin America and the Caribbean (Olson et al. 1998). The main channel and its tributaries receive domestic and industrial effluent discharges, treatment plants, direct discharges from various industries (food, textile, rubber, chemical, tanneries), runoff from dumps and cultivated land, among others. Therefore, there is a significant discharge of organic matter and pollutants such as heavy metals as an example (de Cabo and Marconi 2021), which determines a high to very high degree of contamination in the middle and lower basin (Arreghini et al. 2007). In the 1980s, as a consequence of the results of the chemical and bacteriological analyses carried out, there were a series of demonstrations carried out by residents, athletes, and other sectors of the population who used the basin as a means of development of its activities. This information was presented to the rest of the community and to the leaders of the parties involved. The elaboration of a plan for the sanitation of the waters and regulation of the factories found in the municipalities was demanded. In 1994, an agreement was signed between the Nation and the province of Buenos Aires to execute the Sanitation and Flood Control Project for the Reconquista River Basin, subsidized by a loan from the Inter-American Development Bank (IDB). In 1997, the Reconquista River Basin Committee (COMIREC) was created. It is an autarchic entity whose purpose is to provide services and carry out actions leading to the comprehensive management and preservation of the water resources of the Reconquista River Basin. In 2014, the Province of Buenos Aires once again received a loan from the IDB to execute the Reconquista River Basin Environmental Sanitation Program. The purpose of this program is to develop works to increase the coverage of drinking water, sewerage and wastewater treatment, improve solid waste management, and control and reconversion of industries (https://www.gba.gob.ar/comirec/mision_y_vision).

From the biogeographic point of view, both basins are located in the Pampa Ondulada Ecosystem Complex of the Pampa Ecoregion (Matteucci 2012). According to the Köppen classification, the region's climate is temperate humid with annual rainfall of 1055 mm, with maximums in March (125 mm/month) and minimums in June (56 mm/month) (WMO 1996). The average temperature of the coldest month (July) is 10.4 °C and that of the warmest (January) 23.7 °C. Excess humidity is recorded from March to November (de Cabo and Marconi 2021). The ambient relative humidity is greater than 70% and the potential evapotranspiration is around 800 mm per year (Morello and Matteucci 1997). Frosts are rare. The reference landscape is characterized by riparian vegetation that corresponds to humid grasslands with isolated woody plants on the river terraces.

The physiographic characteristics of both basins reveal a shallow slope that ranges between 0.1 and 1%. Both rivers are short: 70 km (Matanza-Riachuelo) and 82 km (Reconquista). Its flows are controlled by rainfall and by the contribution of groundwater. In addition, its flows are affected by the tides of the Río de La Plata and by the regime of the southeast. The flow of the Matanza-Riachuelo river is 3–100 $m^3 s^{-1}$. The flow of the Reconquista River ranges approximately between 0.8 and 20 $m^3 s^{-1}$ (Salibián 2006).

In both basins, 22.5% of the country's total population lives: 6×10^6 inhabitants in the Matanza-Riachuelo basin and 4.2×10^6 inhabitants in the Reconquista basin. In the upper basins, livestock is the main economic activity and discharges of untreated sewage effluents are recorded. The middle and lower basins of both rivers are highly urbanized and present an important industrial activity and effluent discharges with a high load of organic pollutants and heavy metals. One of the most important sources of water and sediment pollution in terms of pollutant load are the liquid dumps from municipal effluent treatment plants and the direct dump of raw domestic wastewater in sectors that do not yet have sewers (Babay et al. 2014). In some municipalities belonging to the Matanza-Riachuelo and Reconquista basins, coverage is less than 18% for mains water and less than 10% for sewers (Lentini et al. 2017).

In this context, an interaction was established between government agencies and research institutes that allowed the design and implementation of water remediation strategies applying *Chlorella vulgaris* immobilized in alginate. Two experiences carried out by scientific and technical institutions convened by government entities with the purpose of seeking effective and efficient methods for the remediation of surface waters in both basins are described below.

19.2.1 Design and Implementation of Water Remediation Strategies

19.2.1.1 Cildáñez Stream (Matanza-Riachuelo River Basin) Case Study

The Cildáñez stream is part of the Matanza-Riachuelo basin. At the beginning of the twentieth century, many of the slaughterhouses of Buenos Aires were housed on its banks, which dumped the remains of the slaughter into its riverbed. The Cildáñez basin has an area of 3956 ha. It is a piped stream, except at its last segment, before pouring its water in the Matanza-Riachuelo (Fig. 19.1). It has a maximum flow of 257 $m^3 s^{-1}$, reaching 1 $m^3 s^{-1}$ in dry weather. It is located in the sector of the Autonomous City of Buenos Aires where the most vulnerable population is concentrated (Guitelman 2020). In 2017, a research work was started between CONICET (National Scientific and Technical Research Council of Argentine), Faculty of Pharmacy and Biochemistry (Buenos Aires University), Center for Biomedical, Basic, Applied and Development Studies (CEBBAD, U. Maimonides) and the Municipal Environmental Protection Agency (APRA).

Figure 19.1 shows the Cildáñez stream that it was chosen as a model for our studies due to the heavy contamination load, mainly composed of by-products of agriculture and waste materials derived from industries settled in the surroundings, which are often released to the environment without any treatment. The contamination present is originated in considerable amounts of nonbiodegradable inorganic and organic compounds. Moreover, industrial discharges include large amounts of nitrogen and phosphorus that can cause eutrophication of water and ecosystem imbalances (Le Moal et al. 2019; Nieto et al. 2019). According to the Cildáñez historical record, the concentration of heavy metals, with the exception of lead, remains within acceptable levels throughout the year. Instead of this, lead decreases in the fall and summer months (Fig. 19.2). Lead is found in the form of salt; however, it could dissolve, making stream water toxic to human health even at low exposure levels, because it tends to bioaccumulate in the body over time (EPA 2017).

Bibliographic reports show many articles describing the use of microalgae as effective bioremediation agents (de Cabo and Marconi 2021). In order to bioremediate the water of the Cildáñez stream, a native isolate of *Chlorella vulgaris* immobilized in an alginate matrix was evaluated (Trentini et al. 2017; Groppa et al. 2019; Marconi et al. 2020; Sánchez Novoa 2020; Orozco 2020).

First, we carried out the tests in bioreactors with *C. vulgaris* immobilized in alginate beads. The immobilization technology has many advantages, mainly an easy recovery of the biomass involved, avoiding eutrophication and adding the possibility of isolate by-products of economic value (Covarrubias et al. 2012; Kube et al. 2019). Likewise, alginate is a natural polysaccharide matrix for gastronomic use, and it is simple and economical for the manufacture of spherical capsules known as "beads." Immobilization prevent cells from suffering the toxic effects of various substances, extreme pH, high temperatures, allowing better survival and higher efficiency in biomass production (El-Sheekh et al. 2017).

The assays were carried out in bioreactors under mixotrophic conditions (PAR at 400 μmol photon $m^{-2} s^{-1}$), synthetic culture media supplemented with sucrose (3%, w/v) and indoleacetic acid (1 mg L^{-1}) as growth regulator (Murashige and Skoog 1962) and classic propellers (double blades Rushton turbines). In addition, marine style propeller blades were used as alternative mixing system. Figure 19.3 shows the growth kinetic of immobilized *C. vulgaris* cultures (batch system) with a typical S shape. The lag period lasted 2 days, followed by an exponential growth period, which were similar for both types of propellers until day 4. However, stirred cultures with the marine propeller grew until day 8, with a specific growth rate (μ) of 0.672 day^{-1} and a doubling time of 1.03 days. The cultures stirred with the Rushton propeller had a lower performance reflected in a lower growth rate ($\mu = 0.319$ day^{-1}) and a longer doubling time (2.17 days).

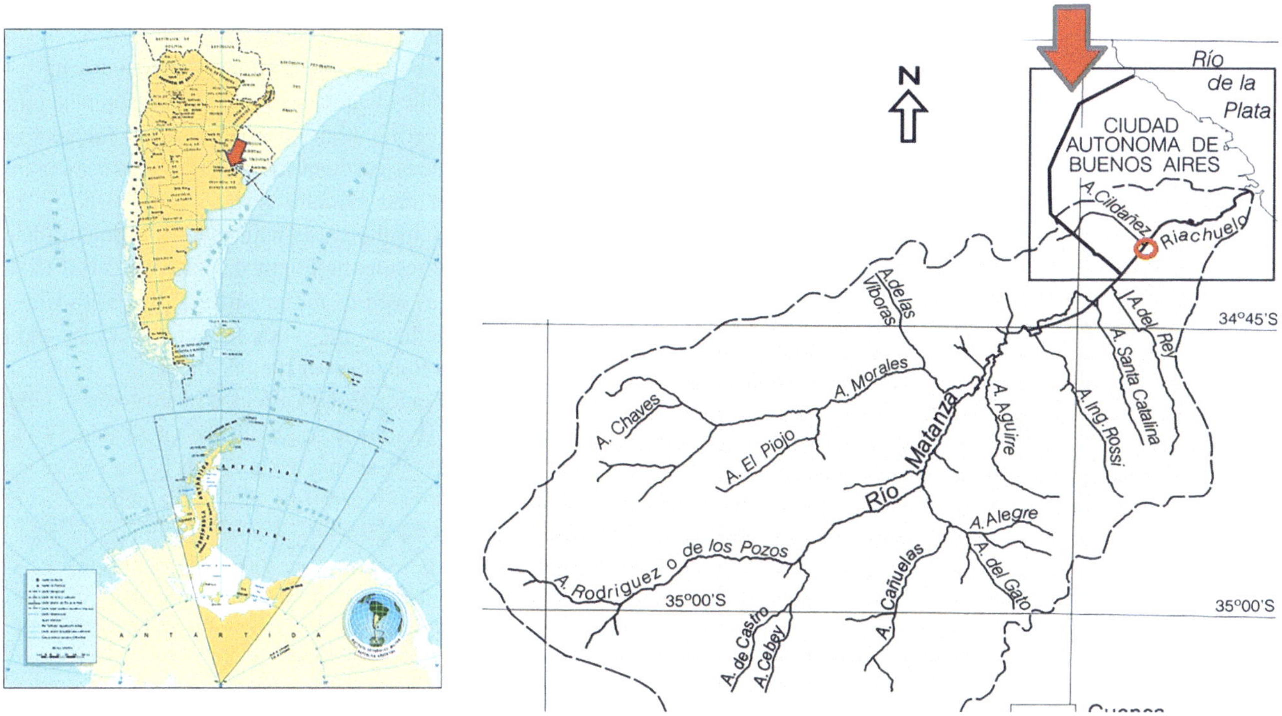

Fig. 19.1 Map of Cildáñez basin in the Matanza-Riachuelo

During the 8 days of culture, the biomass obtained was 11-fold higher in the bioprocess carried out with marine propellers compared to Rushton propellers. In summary, these results allowed to drive future bioprocesses using marine propellers and a batch system maintained for 6 days to bioremediate with immobilized Chlorella cultures. Under these conditions, the bioprocess allowed a high percentage of removal of ammonium (96%), nitrates (86%), nitrites (98%), total phosphorus (53%), and lead (95%), and the pH decreased from an average of 7.8–6.9 (Trentini et al. 2017). In this sense, the decrease in pH may be due to the semi-decomposition of carbonaceous sources as observed in similar experiments (Evans et al. 2017).

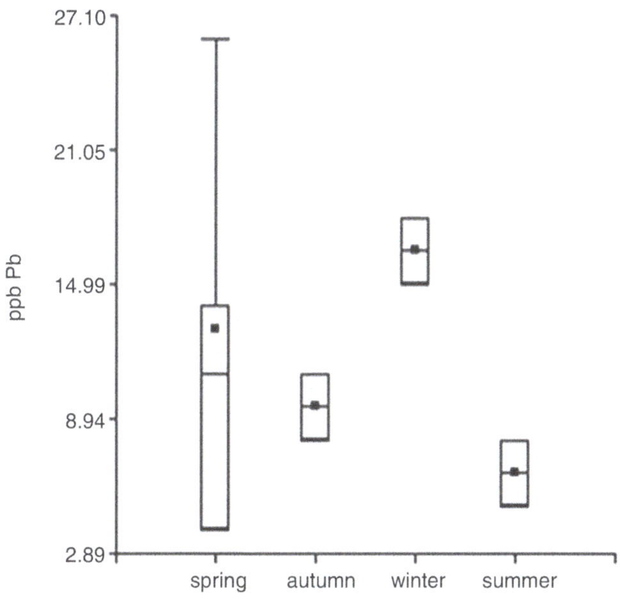

Fig. 19.2 Cildáñez historical record of Pb (ppb) by seasons

In the Cildáñez stream, anthropogenic activity originates considerable amounts of pathogenic microorganisms coming from sewer discharges. Figure 19.4 shows the colony-forming unit (CFU) for coliforms and *Escherichia coli* from the Cildáñez historical record. From these observations, it was decided to evaluate the efficiency in the removal of pathogenic bacteria using microalgae as described above. The bioprocesses were fed with water from Ao. Cildáñez sampled in the warmer months (spring or summer) where the amounts of contaminants detected frequently exceed the admissible limits proposed by the American Public Health Association (APHA 2005) (Fig. 19.4). The results showed that CFU for total coliforms and *E. coli* were significantly reduced after 6 days of treatment, with a remediation of about 99% in the presence of *C. vulgaris* immobilized in alginate matrix (Fig. 19.4). The reduction could be interpreted as a competition between microalgae and the microbiota of water samples that displace these populations of microorganisms (Sforza et al. 2014).

Lastly, a classic biological Allium test (using seeds of *Allium cepa*, onions) was used to corroborate the bioprocess efficiency. This test allows the simultaneous assessment of

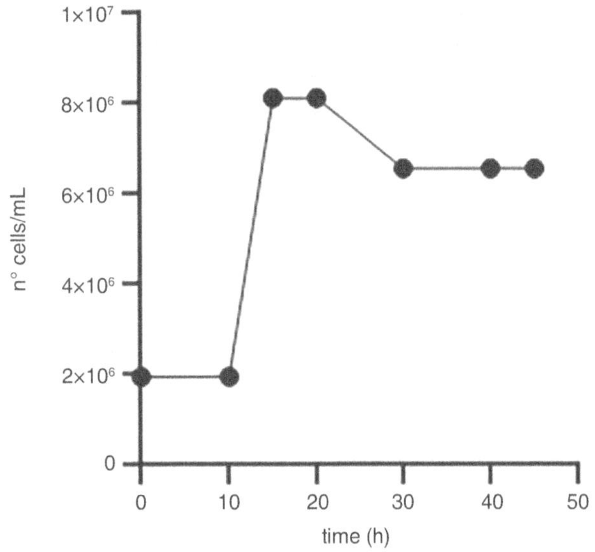

Fig. 19.3 Growth kinetic of immobilized *C. vulgaris* cultures in alginate beads (batch system) in Ao. Cildáñez water for 48 h in culture

cytotoxic, genotoxic, and mutagenic effects of environmental samples (Silveira et al. 2018). The onion's roots are large size and have reduced number of chromosomes ($2n = 16$), which is an advantage to a direct observation under microscope. In addition, the plant root is highly exposed to soluble contaminants, showing symptoms quickly (Leme and Marin-Morales 2009). The Allium test exposed the cytotoxicity of untreated Cildáñez stream water, showing reductions in the germination (55%) and mitotic rates at 48 h (Table 19.1). After the bioprocess with Chlorella cultures immobilized in alginate matrix, a significant increase in the growth index (GI) of onion roots irrigated with the treated water was observed (Table 19.1). The applied bioprocess allowed increasing the GI obtained with untreated water to 74%, thus reaching values similar to those observed with distilled water. After 7 days, almost all the seeds were already germinated when using the bioremediated water, while only about 70% of them were germinated when they were imbibed with untreated water. Additionally, this result indicates that the cytotoxicity of Cildáñez water does not lose over the time or attenuate with the time.

On the other hand, it was observed that while seeds exposed to distilled water had mycelial development (characteristic of the germination of environmental fungal spores), none of the seeds germinated in the presence of untreated Cildáñez water had mycelia, and seeds exposed to treated water had 40% of the seeds with mycelial growth (data not shown). These results highlight the toxic effect of Cildáñez stream water, as well as the treatment effect with Chlorella, which bioremediates the toxic effect of water and also prevents the germination of fungal spores.

The mitotic index (MI) is affected by the toxic of Cildáñez stream, showing similar results to the GI (Table 19.1). The untreated water has such a toxic effect, being the lowest MI obtained. This result could be due to the negative action of xenobiotics on the growth and development of the root apex (Leme and Marin-Morales 2009). In addition, it would explain the delay observed in the germination of onion seeds. Furthermore, the bioremediation effect of the bioprocess with immobilized *C. vulgaris* cells was confirmed.

19.2.2 Bancalari Stream (Reconquista River Basin)

Bancalari stream is located in the lower basin of the Reconquista River, an area with a high social vulnerability (https://www.gba.gob.ar/comirec). It is an urban contaminated stream that flows into the Reconquista River at 20 km north to the city of Buenos Aires. The zone of influence of the study area is mainly determined by categories of commercial and industrial land use. In order to evaluate the efficiency and dynamics of nutrient removal by *Chlorella vulgaris* immobilized in alginate within 3D structures, tests were developed under uncontrolled conditions. Prior to the trial, water samples were taken from the Bancalari stream for acclimatization and cultivation of *C. vulgaris* under controlled laboratory conditions. Cells were immobilized on

Fig. 19.4 *E. coli* historical record (CFU mL^{-1}) by seasons and after and before a bioprocess with immobilized *C. vulgaris*

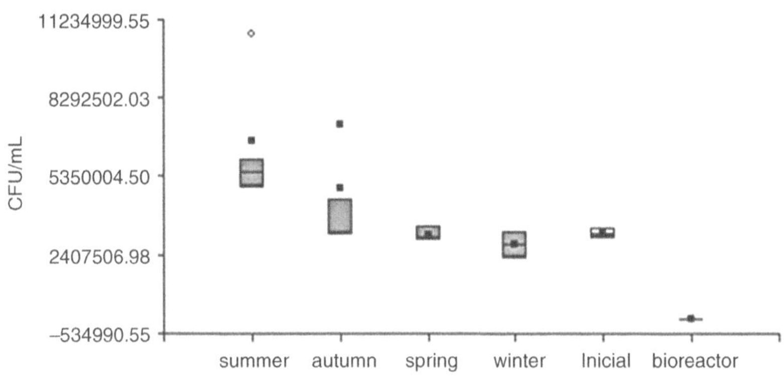

Table 19.1 Effect of bioremediation on the cytotoxicity of Ao. Cildáñez water

Treatment	GI (%) 48 h	GI (%) 7 days	MI (%) 48 h
Cildáñez	55 B	70 B	6.09 B
Cildáñez treated	74 A	93 A	11.01 A
Control water	86 A	96 A	9.88 A

Germination index (GI) and mitotic index (MI) in *A. cepa* seeds using water from Ao. Cildáñez untreated (Cildáñez), treated after bioprocess (Cildáñez treated), and distilled water as control. Different letters indicate significant differences ($p < 0.05$)

alginate beads in a stirred tank fermenter (Minifors, Infors HT®, Switzerland). Then, an assay was developed on the banks of the channel in geocells using water extracted from the Bancalari channel, with *C. vulgaris* immobilized in alginate within 3D structures. Inside each geocell, a 15 W submersible pump was placed to generate water currents that help the constant homogenization of the solution and aeration. This batch system was kept in the open air for 5 days. The bioprocesses were run in geocells on the side of Bancalari stream and feeding with their water (Sánchez Novoa 2020). Three treatments were applied:

- *G-Chl* geocells with a 3D-printed device containing *C. vulgaris* immobilized in the alginate matrix;
- *G-Al* geocells with a 3D-printed device containing the alginate matrix; and
- *G-Ctrl* geocells with a 3D-printed device.

In other tests, alginate beads with immobilized microalgae were placed in nets and submerged in contaminated surface water. However, we observed that alginate beads were degraded in a short time, limiting their applicability and efficiency. The beads were predated by birds, fish, and other local fauna. In order to protect the alginate matrix, we design a proof-of-concept based on the use of a 3D-printed device manufactured in polylactic acid (PLA), an eco-friendly polymer. This polymer allows gas diffusion and is stable at physiological temperatures and pH ranges, with the advantage of reduced weight and high mechanical strength resistance, among other interesting properties (Rosenzweig et al. 2015; Seol et al. 2018). In addition, once the treatment is finished, the algae immobilized in alginate and included in a 3D device can be easily removed from the environment, avoiding the dispersion of algae in natural environments, reducing the risk of eutrophication. In addition, it offers a possibility of an easy biomass recovery, which may become a source of alternative energy.

In the acclimatization test to the stream water, an increase in the algal biomass was observed. The biomass at initial time was 0.284 mg mL^{-1} and increased to 0.940 mg mL^{-1} after 2.83 days of duplication time. Therefore, the autochthonous strain of *C. vulgaris* showed tolerance to the water characteristics of the Bancalari stream and to the transition from the free state in suspension to the immobilization in alginate beads.

In the in situ test with geocells, the initial algal biomass was 0.104 g mL^{-1} on average per device. Five days later, an average of 0.763 g mL^{-1} algal biomass was obtained. The specific growth rate (μ) was 0.412 day^{-1} with 1.68 days of duplication time.

A high removal in ammonium, soluble reactive phosphorus (SRP), and dissolved inorganic nitrogen (DIN = nitrate + nitrites + ammonium) contents was achieved with the G-Chl (Table 19.2). It should be noted that the control treatment with only 3D devices (G-Ctrl), without alginate or algae, allows a significant reduction of ammonium (46.5%). This phenomenon could be due to the fact that, on the one hand, the nitrogenous species are adsorbed on the PLA and, in addition, nitrification and subsequent denitrification occur by the action of environmental bacteria (Wu et al. 2015; Wang and Wu 2016).

The nitrate removal in G-Chl (41%) and without algae, G-Al (47%), was similar, attributed to the microflora existing in the water (Wang and Wu 2016; Marconi et al. 2020). In all treatments, an increase in the final concentration of nitrites was observed, with respect to the concentration determined in the Bancalari stream (Table 19.2). The concentration of dissolved inorganic nitrogen (DIN) decreases so that the final removal of DIN for the treatments G-Al and G-Chl was 21.6% and 46.4%, respectively.

Most of the ammonium in wastewater can be eliminated by the action of bacteria associated with the intense oxygenation generated by the microalgae. This is due to the fact that the bacteria present in the aquatic environment use these nutrients as a substrate, thus they are usually decisive in the assimilation of nutrients and/or the formation of biomass in bioremediation systems (Lu et al. 2020). Ammonium, for example, is partly nitrified to nitrate that is available to microalgae. There are also denitrification processes by the action of bacteria in the suboxygenated microenvironments in the geocells (Zornoza-Zornoza 2017). On the other hand, the bacterial community developed better in the treatments that had some matrix that served as support (beads and 3D structures) such as those of the G-Al and G-Chl treatments. The immobilization of microalgae has been reported as an effective technique that increases the removal of nutrients compared to the same without immobilization (De-Bashan et al. 2002; Olguín 2003; De-Bashan and Bashan 2010; Ruiz-Marin et al. 2020).

Table 19.2 Ammonium, soluble reactive phosphorus (SRP), and dissolved inorganic nitrogen (DIN = nitrate + nitrites + ammonium) removal from the initial Bancalari stream (initial) and after 5 days of bioprocess with geocells containing a 3D-printed device (G-Ctrl) or geocells with a 3D-printed device containing *C. vulgaris* immobilized in the alginate matrix (G-Chl)

	Initial	G-Ctrl	G-Chl
Ammonium (mg L^{-1})	2.4	1.3	0.073
Soluble reactive phosphorus (SRP) (mg L^{-1})	0.78	0.91	0.014
Nitrates (mg L^{-1})	12.6	13.2	7.38
DIN (mg L^{-1})	15.1	15.1	8.06

The treatments G-Al and G-Chl showed removal of soluble reactive phosphorus of 3.57% and 98.2%, respectively. Then, the reduction in the concentration of SRP could be carried out mainly by *C. vulgaris*.

Chlorella vulgaris has the ability to grow under stress conditions, together with other microorganisms present in the water to be treated. The interaction of microalgae with bacteria and other microorganisms that uptake and/or transform nutrients can favor the growth of algal biomass and make the system more efficient. Bioremediation with *C. vulgaris* immobilized in alginate within 3D structures is an effective alternative to remove nutrients present in excess in contaminated aquatic environments.

19.3 Final Remarks

In Argentina, the series of governments at the national, provincial, and municipal levels generates changes in the formation of government agencies that bring about changes in the policies of each one of them after each poll. Often, the objectives for each area of government are frequently altered as a consequence of changes in the dominant political party. In particular, the changes in authorities of autarchic entities dedicated to the management of water basins such as ACUMAR and COMIREC imply modifications in the management goals for both urban basins. The main focus varies from efforts oriented purely to the sanitary aspect to others that are specifically dedicated to developing hydraulic works for flood control. In order to meet environmental objectives, a balance between both visions and the continuity of long- and medium-term policies is required. However, and despite these obstacles, this chapter shows two examples of interaction between scientific and technical institutions and government entities with good results. It is noteworthy that to achieve favorable results at the basin scale, greater interaction between various government areas is required. It is necessary the commitment of continuity of policies on the part of the autarchic entities. In addition, the interaction with provincial and municipal governments and local population is essential to ensure the achievement of long-term objectives.

References

APHA Standard methods for the examination of water and wastewater (2005) 21st ed. American Public Health Association, Washington, DC

Arreghini S, de Cabo L, Seoane R, Tomazin N, Serafini R, Fabrizio de Iorio A (2007) A methodological approach to water quality assessment in an ungauged basin, Buenos Aires, Argentina. GeoJournal 70:281–288

Babay PA, Itria RF, Romero Ale EE, Becquart ET, Gautier EA (2014) Ubiquity of endocrine disruptors nonylphenol and its mono-and di-ethoxylates in freshwater, sediments, and biosolids associated with high and low density populations of Buenos Aires, Argentina. Clean–Soil, Air, Water 42:731–737

Covarrubias SA, de Bashan LE, Moreno M, Bashan Y (2012) Alginate beads provide a beneficial physical barrier against native microorganisms in wastewater treated with immobilized bacteria and microalgae. Appl Microbiol Biotechnol 93:2669–2680

de Cabo L, Marconi PL (eds) (2021) Estrategias de remediacion para las cuencas de dos ríos urbanos de llanura: Matanza-Riachuelo y Reconquista. Ed. Fundación Azara. ISBN 978-987-3781-74-2. https://fundacionazara.org.ar/estrategias-de-remediacion-para-las-cuencas-de-dos-rios-urbanos-de-llanura-matanza-riachuelo-y-reconquista/

de Cabo L, Faggi A, Miguel S, Basilico G (2019) Rehabilitación de las riberas de un sitio de la cuenca baja del río Matanza-Riachuelo. Biología Acuática 33:1–14

De-Bashan LE, Bashan Y (2010) Immobilized microalgae for removing pollutants: review of practical aspects. Bioresour Technol 101:1611–1627

De-Bashan LE, Moreno M, Hernandez JP, Bashan Y (2002) Removal of ammonium and phosphorus ions from synthetic wastewater by the microalgae Chlorella vulgaris coimmobilized in alginate beads with the microalgae growth-promoting bacterium *Azospirillum brasilense*. Water Res 36:2941–2948

El-Sheekh MM, Metwally MA, Allam NG, Hendam HE (2017) Effect of algal cell immobilization technique on sequencing batch reactors for sewage wastewater treatment. Int J Environ Res 11:603

EPA WEB Archive (2017) United States environmental protection agency. Ground Water and Drinking Water, Basic Information about Lead in Drinking Water. https://www.epa.gov/

Evans L, Hennige SJ, Willoughby N, Adeloye AJ, Skroblin M, Gutierrez T (2017) Effect of organic carbon enrichment on the treatment efficiency of primary settled wastewater by *Chlorella vulgaris*. Algal Res 24:368–377

Groppa MD, Trentini A, Zawoznik M, Bigi R, Nadra C, Marconi PL (2019) Optimization of a bioremediation strategy for an urban stream of Matanza-Riachuelo basin. Int J Environ Ecol Eng 13:418–424

Guitelman A (2020) Arroyos de Buenos Aires: enterrados pero vivos / Adolfo Guitelman ; compilado por Gustavo Di Costa. - 1a ed compendiada. - Ciudad Autónoma de Buenos Aires: Consejo Profesional de Ingeniería Civil

Kube M, Mohseni A, Fan L, Roddick F (2019) Impact of alginate selection for wastewater treatment by immobilized *Chlorella vulgaris*. Chem Eng J 358:1601–1609

Le Moal M, Gascuel-Odoux C, Ménesguen A, Souchon Y, Étrillard C, Levain A, Pinay G (2019) Eutrophication: a new wine in an old bottle? Sci Total Environ 651:1–11

Leme DM, Marin-Morales A (2009) *Allium cepa* test in environmental monitoring: a review on its application. Mutat Res 682:71–81

Lentini EJ, García Larumbe JG, Zorrilla S, Ferro G, Palacios A, Biondini S, Ghiglione F, Luciani F (2017) Plan nacional de agua potable y saneamiento. Segunda versión. Ministerio del Interior, Obras Públicas y Vivienda, Presidencia de la Nación, República Argentina. https://www.argentina.gob.ar/sites/default/files/interior_agua_plan_agua_saneamiento.pdf (consulted in March 2022)

Lu W, Alam MA, Liu S, Xu J, Saldivar RP (2020) Critical processes and variables in microalgae biomass production coupled with bioremediation of nutrients and CO_2 from livestock farms: a review. Sci Total Environ 716:135247

Marconi PL, Trentini A, Zawoznik M, Nadra C, Mercadé JM, Sánchez Novoa JG, Groppa MD (2020) Development and testing of a 3D-printable polylactic acid device to optimize a water bioremediation process. AMB Express 10:1–9

Matteucci S (2012) Ecorregión Pampa. In: Morello J, Matteucci S, Rodríguez A, Silva M (eds) Ecorregiones y complejos ecosistémicos argentinos. Ed. Orientación Gráfia, Buenos Aires, pp 391–446

Mojica Duran M, de Cabo L (2021) Gestión del área ribereña urbana de la cuenca baja del ríoMatanza-Riachuelo. Brazilian J Anim Environ Res 4:3629–3641

Morello J, Matteucci SD (1997) El modelo agrícola del Núcleo Maicero como sistema complejo. En: J. Morello y O.T. Solbrig (comps.) La Pampa Ondulada: granero del mundo hasta cuando? Orientación Gráfica Editora, SRL, Buenos Aires, pp 201–231

Munné A, Prat N (2004) Defining river types in a Mediterranean area: a methodology for the implementation of the EU water framework directive. Environ Manag 34:711–729

Murashige T, Skoog F (1962) A revised medium for rapid growth and bioassays with tobacco tissue cultures. Physiol Plant 15:473–497

Nieto PG, García-Gonzalo E, Fernández JA, Muñiz CD (2019) Water eutrophication assessment relied on various machine learning techniques: a case study in the Englishmen Lake (Northern Spain). Ecol Model 404:91–102

Olguín EJ (2003) Phycoremediation: key issues for cost-effective nutrient removal processes. Biotechnol Adv 22:81–91

Olson DE, Dinerstein E, Canevari P, Davidson I, Castro G, Morisset V, Abell R, Toledo E (1998) Freshwater biodiversity of Latin America and the Caribbean: a conservation assessment. Biodiversity Support Program, Washington, DC

Orozco D (2020) Tesis de grado "Ingeniería metabólica en *Chlorella vulgaris* como posible estrategia para aumentar la eficiencia en la remediación de aguas residuales". Carrera de Biología, U. Maimonides

Pfadenhauer J (2001) Some remarks on the socio-cultural background of restoration ecology. Restoration Ecol 9:220–229

Queiroz VC, Carvalho RCD, Heller L (2020) New approaches to monitor inequalities in access to water and sanitation: the SDGs in Latin America and the Caribbean. Water 12:931

Redmond KM (2004) Bringing the river to life? Myths, motivations and practicalities of community involvement in urban river restoration. Unpublished MSc dissertation. University of London, 113 pp

Rosenzweig DH, Carelli E, Steffen T, Jarzem P, Haglund L (2015) 3D-printed ABS and PLA scaffolds for cartilage and nucleus pulposus tissue regeneration. Int J Mol Sci 16:15118–15135

Ruiz-Marin A, Canedo-López Y, Chávez-Fuentes P (2020) Biohydrogen production by Chlorella vulgaris and Scenedesmus obliquus immobilized cultivated in artificial wastewater under different light quality. AMB Express 10:1–7

Salibián A (2006) Ecotoxicological assessment of the highly polluted Reconquista River of Argentina. Rev Environ Contam Toxicol:35–65

Sánchez Novoa JG (2020) Bioprocesos de aguas contaminadas del Canal Bancalari, cuenca del Río Reconquista utilizando una cepa autóctona de *Chlorella vulgaris* inmovilizada en alginato dentro de estructuras 3D. Carrera Ciencias Ambientales Facultad de Historia, Geografía y Turismo Universidad del Salvador

Seol Kyoung SU, Zhao P, Shin B-C, Zhang S-U (2018) Infill print parameters for mechanical properties of 3D printed PLA parts. J Korean Soc Manufact Preciss Eng 17:9–16

Sforza E, Armandina Ramos-Tercero E, Gris B, Bettin F, Milani A, Bertucco (2014) An integration of *Chlorella protothecoides* production in wastewater treatment plant: from lab measurements to process design. Algal Res 6:223–233

Silveira MAD, Ribeiro DL, Vieira GM, Ribeiro Demarco N, Grégio d'Arce LP (2018) Direct and indirect anthropogenic contamination in water sources: evaluation of chromosomal stability and cytotoxicity using the *Allium cepa* test. Bull Environ Contam Toxicol 100: 2016–2220

Trentini A, Groppa MD, Zawoznik M, Bigi R, Perelman PE, Marconi PL (2017) Biorremediación del lago Lugano de la Ciudad Autónoma de Buenos Aires por algas unicelulares – estudios preliminares para su posterior utilización. Terra Mundus 4. http://dspace.uces.edu.ar:8180/xmlui/handle/123456789/4302

Wang J, Wu L (2016) Biological nitrate removal from water and wastewater by solid-phase denitrification process. Biotechnol Proc 34: 1103–1112

Ward J (1998) Riverine landscapes: biodiversity patterns, disturbance regimes, and aquatic conservation. Biol Conserv 83:269–278

WMO (1996) Normales climatológicas para el período 1961–1990. Reporte No. 847, Región 3: Sudamérica, Argentina, 257–270. World Meteorological Organization

Wu C, Tang D, Wang Q, Wang J, Liu J, Guo Y, Liu S (2015) Comparison of denitrification performances using PLA/starch with different mass ratios as carbon source. Water Sci Technol 71:1019–1025

WWAP (UNESCO World Water Assessment Programme) (2019a) World Water Assessment Programme: The United Nations World Water Development Report 4: Managing Water under Uncertainty and Risk

WWAP (UNESCO World Water Assessment Programme) (2019b) The United Nations World Water Development Report Leaving no one behind. UNESCO, Paris

Zornoza-Zornoza AM (2017) Estudio de la dinámica poblacional de protistas, metazoos y bacterias filamentosas y su interpretación ecológica en fangos activos (Doctoral dissertation).Universitat Politècnica de València Departamento de Ingeniería Hidráulica y Medio Ambiente Tesis Doctoral

SPRINGER NATURE

GPSR Compliance

The European Union's (EU) General Product Safety Regulation (GPSR) is a set of rules that requires consumer products to be safe and our obligations to ensure this.

If you have any concerns about our products, you can contact us on ProductSafety@springernature.com

In case Publisher is established outside the EU, the EU authorized representative is:

Springer Nature Customer Service Center GmbH
Europaplatz 3
69115 Heidelberg, Germany

The manufacturer's authorised representative in the EU is Springer Nature Customer Service Centre GmbH, Europaplatz 3, 69115 Heidelberg, Germany. If you have any concerns regarding our products, please contact ProductSafety@springernature.com

Printed and bound by CPI Group (UK) Ltd, Croydon, CR0 4YY
28/09/2026
02224095-0001